Another World

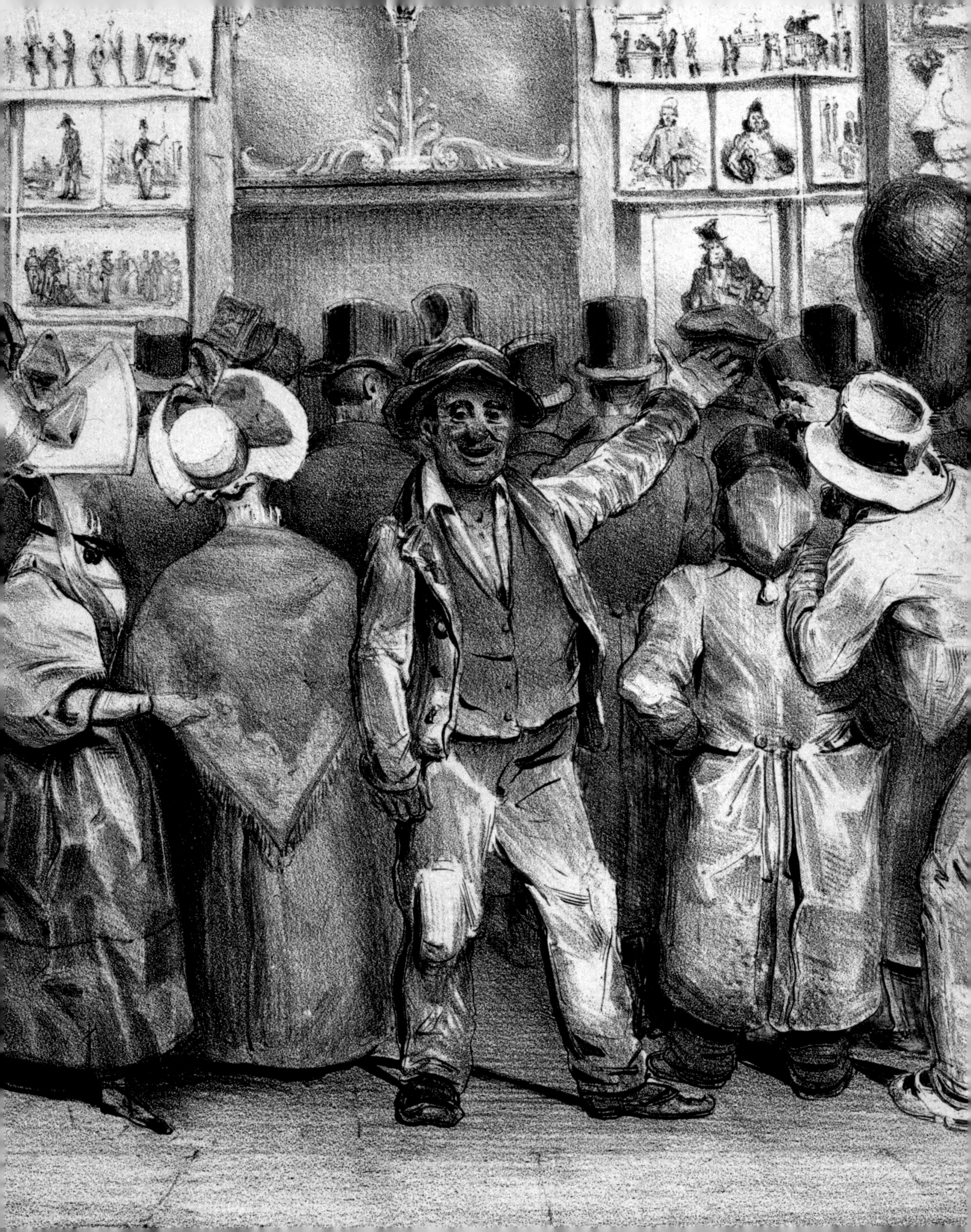

Another World

NINETEENTH-CENTURY ILLUSTRATED PRINT CULTURE

Patricia Mainardi

Yale University Press New Haven and London

yalebooks.com/art

Designed by Leslie Fitch
Set in Crimson and Source Sans Pro type by Leslie Fitch

Printed in China by Regent Publishing Services Limited

Library of Congress Control Number: 2016930356
ISBN 978-0-300-21906-7

A catalogue record for this book is available from the British Library.
This paper meets the requirements of ANSI/NISO Z 39.48-1992 (Permanence of Paper).
10 9 8 7 6 5 4 3 2 1

Jacket illustrations: (front) Grandville, *Cast Shadows* [*Les ombres portées*] (fig. 49); (back) Jules Gaildrau, *Lighted Kiosks. New Stalls for the Sale of Newspapers on the Boulevards* [*Kiosques lumineux. Nouveaux bureaux pour la vente des journaux sur les boulevards*] (fig. 83)

Frontispiece: Charles-Joseph Traviès, *You Have to Admit That the Head of State Looks Pretty Funny* (detail of fig. 48)

CONTENTS

ACKNOWLEDGMENTS

Another World began as a study of the early years of comics. I soon realized that they could not be considered in isolation because comics were part of a complex of interrelationships among artists who worked in several media at the same time, new technology, and new audiences. As I tried to understand the beginnings of this new art form, my work on comics gradually expanded to include other areas of illustrated print culture. This book is the result.

Along the way I have benefited enormously from the generosity of my peers. I am especially grateful to the Research Foundation of the City University of New York, to the Institut National d'Histoire de l'Art (INHA) in Paris, and to the Yale Center for British Art for their support, which enabled my research over several years. Colloquia and symposia gave me the opportunity to present the earliest drafts of my work and to continually revise it. A special thank you to Ségolène Le Men, whose writing has been an inspiration for my own, and who organized the 2006 colloquium at INHA, *Caricature: Bilan et recherche,* where I first presented aspects of this study. An appointment as Van Gogh Professor at the Van Gogh Museum and the University of Amsterdam in 2008 provided me with my earliest opportunity to survey the spectrum of nineteenth-century print culture and begin to formulate the outline of this book. Kevin Murphy's 2009 symposium *Ephemera: Impermanent Works in the Literary and Visual Culture of the Long Nineteenth Century* at the CUNY Graduate Center enabled me to discuss popular prints; Ben Katchor's 2010 symposium *The Artist as Author* at Parsons, The New School for Design, encouraged me to look at artist-driven texts; and Jennifer Tonkovich's 2011 Morgan Library symposium *Drawing in the Age of Revolution* gave me the opportunity to formulate what became my first chapter, "Drawing's Stepchild." Most recently Marie-Eve Thérenty's 2015 colloquium on *Presse et littérature du XIXe au XXIe siècle* at the NYU Institute of French Studies and Christiane Schwab's 2015 *Dissecting Society: Periodical Literature and Social Observation (1830–1850)* at NYU's Center for International Research in the Humanities and the Social Sciences (CIRHUS) permitted me to discuss my work with specialists in these areas. I am also thankful for the chance to offer lectures and presentations that contributed to this work at the Musée d'Orsay, the Metropolitan Museum of Art, the University of Pennsylvania Material Culture Workshop, NYU's Maison Française, and, several times, at the College Art Association annual conference.

Among the many friends and colleagues to whom I owe debts of gratitude, I am especially grateful to Herman Lebovics, whose incisive reading has greatly improved this text. Vanessa Schwartz's and Ruth Iskin's comments have proved invaluable. Colleagues in art history and especially in the world of prints have been generous with their time and advice: I thank warmly Philippe Bordes, Veronique Chagnon-Burke, Marie-Claude Chaudonneret, Elizabeth Childs, David Christie, Petra ten-Doesschate Chu, Therese Dolan, Stephen Edidin, Madeleine Fidell-Beaufort, Gillian Forrester, Andrea Immel, Ben Katchor, Diane Kelder, Lars Kokkonen, Marilyn Kushner, Heather Lemonedes, Karen Maguire, Constance McPhee, Nadine Orenstein, Caterina Pierre, Mary Frances Zawadzki. Friends, in particular Iska Alter, Faiza Bellounis, William Long, Pamela Mainardi, and Jessica Robinson, have helped in numerous ways; at Yale I thank Amy Canonico, Heidi Downey, Sarah Henry, and Leslie Fitch. In many ways they are all coauthors of this project, and whatever value it has I happily share with them.

❖ ❖ ❖

Where known, I have included the names of engravers (engr.) and lithographers (lith.), although they were not always identified. "Mr" as an abbreviation for *Monsieur* often appears in titles of early comics, so I have left it in my French citations wherever it occurs, while using "Mr." for translations. All translations are my own unless otherwise identified.

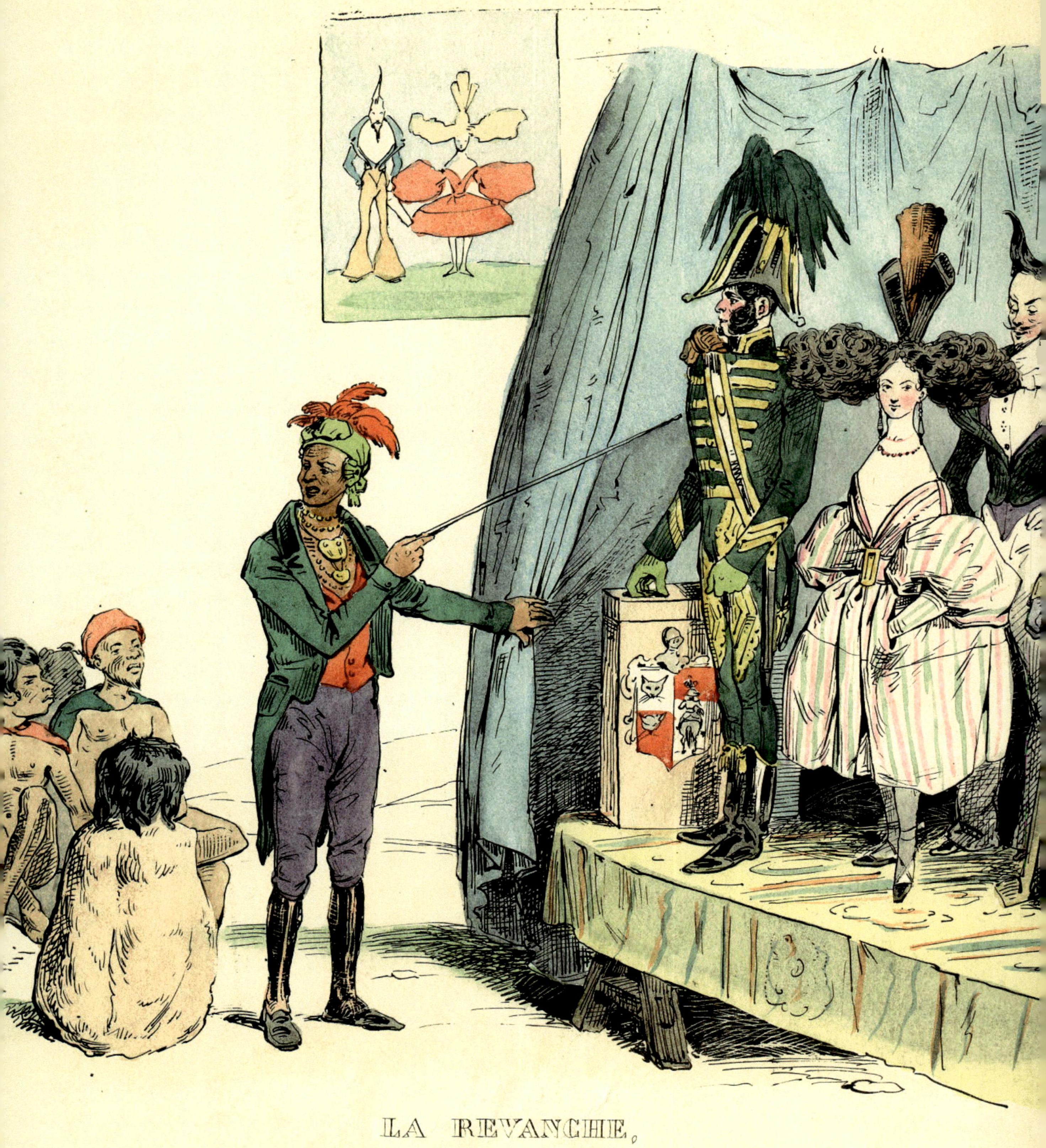

LA REVANCHE,

ou les français au Missouri.

Introduction

Another World

Between the late eighteenth century and the beginning of the twentieth, more prints, illustrated books, magazines, and newspapers were produced in the Western world than in previous centuries combined, an efflorescence comparable only to the period following Gutenberg's invention of movable type when printed material first circulated widely in Europe. This book is about the next wave of illustrated print culture, a tidal wave in comparison to the first, and one that ushered in the modern world. To borrow the title of the 1844 visionary novel by the artist known as Grandville, this period constitutes Another World, adjacent to our own, both familiar and vastly different at the same time.[1] Only gradually did that other world become our world, where we are bombarded with images on a daily basis. Not only was there a proliferation of imagery throughout the nineteenth century, there were new kinds of images, or old kinds of images in new media. Until recently only bits and pieces of this history had surfaced, often in relation to high-art production, when a particular image was discovered to have influenced a particular painting.

Much of this transformation took place in France, the epicenter of this media explosion and the main focus of this book. Nonetheless, art was becoming global in that period, as were steam shipping and the textile trade. International exchanges of technology and genres informed the development of illustrated print culture everywhere, and so this discussion must range across national borders. This is not a book about "illustration," however, if that word is understood to imply a hierarchy between word and image in which images are subordinate to text. Nor can this be an exhaustive study of illustrated print media, which would necessitate numerous volumes. It is, instead, a book about the development of the new graphic language enabled by advances in reproductive technology, a vernacular language of drawing that developed parallel to, and at times in opposition to, the language of high art. Rather than offering the reader yet another theoretical analysis of popular culture, although this literature has helped greatly in situating my theme, what I have written is more in the nature of an archaeology of the stratum of visual imagery lying just below the surface of high art.

DETAIL OF FIGURE 45

In any study of visual culture, the prime question is always how wide a net to cast.

As Vanessa Schwartz and Jeannene Przyblyski set forth in their essay "Visual Culture's History," the two polar positions are to include everything (visual culture), or to focus only on quality (art history).[2] In this book I chart a middle ground between the works of high art exhibited in the Salons and the ephemera of everyday life that scarcely merited a second look. I have chosen five key areas for discussion: lithographic imagery, particularly caricature; the illustrated press; comics; illustrated books; and popular prints. The images I discuss here were never considered high art, never destined for exhibitions, and yet they were designed for visual pleasure. Their projected audiences ranged widely. At one extreme were the illiterate or semiliterate classes who constituted the audience for popular prints in which the image could carry meaning exclusive of text. Caricature, which appealed to all classes, was always tightly regulated in France because the authorities feared its perceived accessibility to these same unlettered and potentially dangerous classes.[3] Artisans and the petite bourgeoisie could afford prints, some books, and the less expensive periodicals, and the bourgeoisie and aristocracy constituted the audience for richly illustrated and luxuriously bound publications of all genres. As a result, we cannot discuss illustrated print culture as though it had a clearly definable audience but must, instead, see it as a prism reflecting a multitude of facets.

Within the nineteenth century, technological innovation, motivated by the drive to reach new audiences either for ideology or for profit, had transfigured the world of the visual. Artists striving to reach wide audiences and to test the limits of new media transformed the visual world in lasting ways. The new graphic language that resulted, with its imaginative possibilities and visual strategies, is like a submerged continent that we are just beginning to discover, but which, when better known, will significantly remap our understanding of the role of the visual in that century.

Because I am interested in the symbiosis between technological innovation and advances in artistic expression, thence to the consumption of art, I have organized this study chronologically to highlight the invention of each new illustrated print medium. The fructification that forms the subject of this book was initiated and enriched by this technology, resulting in a new system of aesthetic production with specialists in each medium who catered to the tastes of a segment of the newly enfranchised public. This book traces out the network of alliances between new technologies, innovation, and new publics. By 1870, when the Republic was permanently established, new art forms for new classes of citizens were already in existence. The invention of lithography begot caricature, as well as an enormous amount of printed imagery of contemporary life and landscape; wood engraving made possible the illustrated press and encouraged the proliferation of illustrated books as well as their new formats. Both media made comics possible, at first lithographed, later wood engraved. In the course of their history, popular prints utilized all available print media, quickly abandoning each new process as it was replaced by something faster, cheaper, simpler. Although vastly different in content, all these media—caricatures, periodicals, comics, popular prints, and illustrated books and journals—utilized similar reproductive processes, all invented or developed in the course of the century.

I have focused on these major areas as the most fruitful for investigation because each arose at the inception of a particular technological advance. I highlight the ways in which technological developments informed and inspired the various media, their integration into contemporaneous visual culture, and the issues and arguments that greeted their early manifestations. In each case, the technological, the aesthetic, the political, and the cultural were intertwined, as each subsequent development interacted with what had come before. I focus exclusively on the fecund moment of invention of both new technology and new visualization because it is at this moment of inception and innovation that all things are possible; later, each medium will, inevitably, settle into its defined mode.

Were I to attempt to write a history of nineteenth-century illustration—even limited to France alone—it would run to dozens of volumes. That is not my intention here. I am interested in the moment of discovery or invention of these new illustrated print media, as each was greeted with criticism, enthusiasm, sometimes even euphoria, and as artists and publishers explored the limits of its potential. In each case the euphoria lasted, at most, a few decades before each new medium became assimilated and integrated into visual culture. It is my thesis that during these early years, when each medium was defining itself and finding its audience, many of the modes and tropes that subsequently became ubiquitous were invented. My intention is to explore the optimism and energy of these beginnings, as another world was rapidly becoming our world.

Several themes run throughout the entire book. The growth of the international print market is one such theme, tied to the globalization of visual media. Artists often worked in more than one country. Théodore Géricault traveled from his native France to England where his major print series, his *English Suite*, was conceived, drawn, and published. Richard Parkes Bonington went from his native England to work in France, where he contributed drawings to the greatest compilation of lithographic imagery of the century, *Picturesque and Romantic Travels through Old France*.[4] The earliest influence of the Swiss Rodolphe Töpffer's invention of the comic book took place in France. In all countries publishers became so aware of the benefits of an international audience that they often brought out translations of successful works or included legends on prints in two languages. The publishers of the illustrated press bought and sold plates internationally, and the growing awareness of the diversity of the world's peoples resulted in works in all media and in all countries describing and depicting different customs. As the nineteenth century was the period when many nations were establishing their modern borders and their national identity, this diversity was regional as often as it was global.

Another theme is the persistence of drawing, despite the new reproductive technology of photography. It is widely accepted that the invention of photography was the most important event in nineteenth-century visual culture, a reading that began with Walter Benjamin's classic 1936 essay "The Work of Art in the Age of Mechanical Reproduction," where he proposed a trajectory from lithography to photography to cinema. This reading—that everything tended toward the photograph—is, however, challenged by the history of high art, which did the opposite, with painting becoming ever less photographic by the decade. This alone should indicate that there is another story to be told. My focus here is on the persistence of the "hand-drawn" image despite the development of photography. What William Ivins referred to as "the pictorial statement without syntax"—photography—has never completely colonized the visual world, and has continued to cede space to the personal, the handmade, the particular.[5] A dialectic could here be proposed between the photographic image and caricature, with the exaggerations and distortions of caricature violating every norm that photography sought to establish. And yet here, in the twenty-first century, they continue to coexist.

Another theme, the primacy of genre over historical imagery, is apparent in high art painting from early in the century, first through genre vignettes inserted into history paintings, and later through the virtual collapse of history painting, replaced in public exhibitions and the public's affections by genre painting, images from everyday life. In that same way, the small pleasures and sorrows of daily life replaced the greater verities of religion and national epic in comics, caricature, and popular prints. Even book illustrations, which multiplied exponentially through the century, began to depict not only the great decisive narrative moments but numerous quotidian events along the way, attesting to the influence of broader markets and audiences.

The word-image relationship is a theme throughout, from the wordplay of caricatures and the brilliance of Grandville's punning images, to the allusive quality of the new kinds of book illustration and novella-like texts of late-century printed comics. Throughout the century, the balance of word and image was at stake, with illustrated novels and the earliest popular prints weighted toward text, while later, in comics and caricature, the narrative is carried mainly through imagery. Literary critics could lament this usurpation of the narrative function by the pictorial, but its speed and ease of consumption made it irresistible to newly expanded contemporaneous publics.

The book begins with "Drawing's Stepchild: Lithography and Caricature," which takes as its point of departure the invention of lithography in 1796 by the Bavarian Alöys Senefelder; this initiated the media explosion that has become characteristic of the modern age. Cheap and facile, lithography could produce thousands of images; this capacity for profusion in tandem with a rapidly changing society encouraged the production of topical caricature, a late transplant to France imported from England during the Revolutionary period. The allure of lithography was not limited to caricature, however, and it soon became the preferred medium for depictions of modern life, a position it kept throughout the nineteenth century until the new medium of photography was developed enough for the "snapshot." Taken up by the younger Romantic generation, lithography soon replaced engraving, which was time-consuming, laborious, and expensive. Prints and printmakers crossed national borders, contributing to a global market that expanded throughout the century, and print publishers established themselves internationally. This new availability of images was not universally accepted as an unalloyed benefit, however, and remained problematic in high art circles, identified with a most unwelcome democratization of what had formerly been elite culture. Chapter 1 discusses the technology and development of lithography, its themes, markets, and audiences, as well as the controversies at its inception, when it was praised by modernists as enhanced drawing but condemned by traditionalists as shoddy, industrial production.

Chapter 2, "Spreading the News: The Illustrated Press," focuses on the new concept of the illustrated universal survey periodical that appeared early in the 1830s, first in England, then in France. These replaced earlier publications that had few illustrations and were focused on limited subjects and even more limited audiences. The development of illustrated periodicals was enabled by technological advances such as the steam press, cheaper paper, and the use of wood engraving and stereotypes, as well as greater literacy among the citizenry. Although Walter Benjamin ignored wood engraving in the trajectory he charted from lithography to photography, it was the application of this critical invention that allowed text and images to be juxtaposed on a single page and thus established the format of the modern periodical press.

The earliest illustrated periodicals were published by social reformers in England and France who were attempting to raise the status of the working classes, but the medium soon attracted wealthier, more educated strata as well; within decades the illustrated press had spread throughout the world. This chapter considers the utopian promise of such publications at their inception, the material conditions of their publication, their audience, themes, and subjects, the technology that enabled them, and the debates that surrounded them in their early decades. In particular I examine their imagery, which included current events, scientific developments, scenes from literature, reproductions of high art, and cartoons. I conclude the chapter with the introduction of photography into the illustrated press, the one area where photography indisputably established hegemony over the hand-drawn image. And yet, despite this triumph, the intensely partisan, hand-drawn image has never disappeared from the periodical press, surviving in the form of editorial cartooning, what the French call a *dessin de presse,* a newspaper drawing. Even in the twenty-first century, these, and not photographs, remain the most controversial images that appear in journalism.

Caricature was well established internationally by the 1830s and was already appearing regularly in the new illustrated periodicals when it transmuted into the comic book, the subject of chapter 3, "The Invention of Comics: Stories in Pictures." Although comics in the form of graphic novels are a subject of scholarly interest today, their prehistory is less known and is often misunderstood. The first comics, by the Swiss Rodolphe Töpffer, were produced through lithography, which eliminated the labor-intensive engraving process used in earlier narrative series such as those of William Hogarth. Their success was manifest in their almost immediate translations and plagiaries in many languages. This third chapter traces comics from the works of Töpffer through the widespread adoption of the genre and its inclusion in periodicals. Perhaps here more than anywhere else, the invention of a new graphic language of signs is most apparent as artists sought ways to tell their stories with the greatest economy of means. By the mid-nineteenth century, comics were a familiar presence in the universe of illustrated print culture.

The same technological innovations that led to the illustrated press also made possible illustrated books, discussed in the fourth chapter, "Paths Forgotten, Calls Unheard: Pictures in Stories." These were far more serious and luxurious than the ephemeral comic book. They reached their zenith in France in the 1840s, later in England and America. Calling upon the talents of the finest artists of the period, these publications encouraged inventive varieties of word-image relationships rediscovered only in the twentieth century. The widespread use of wood engraving offered publishers of books as well as newspapers the possibility of layouts that interspersed images with text, producing rich tapestries of word and image impossible to achieve with lithography. Acrid debates arose from the literary establishment's negative reaction to this intrusion of the artist into the writer's domain because it resulted in a disruption of the traditional book layout where images had appeared infrequently, in isolation, and on separate pages, allowing the text to reign in pristine splendor. Artists, however, saw an opportunity to create new genres of illustrated novels where they—and not writers—took the lead. The resultant proliferation of images, which ranged into the hundreds for a single publication, forced a new kind of visual imagination, where the artist (and consequently the reader) envisioned minor moments of the narrative, namely genre subjects of the type that had already appeared as vignettes within history paintings. The result was a period of unparalleled creativity and innovation, a true golden age of illustrated books.

The final chapter, "The Curious History of Popular Imagery in France," surveys the development of what the French call *images d'Épinal* after the town in eastern France where most of these prints were produced. From their beginnings as crude woodcut playing cards and religious images, they were transformed during the century into an international trade in contemporaneous subjects, whimsical anecdotes of modern life indistinguishable from comics. As popular prints developed, they cross-fertilized with caricature, the illustrated press, comics, and illustrated books; they made forays into politics, education, and the new field of advertisement. The resulting amalgam of influences, modes, and styles is familiar to us today as characteristic of popular culture.

Some areas of visual culture that are worthy of study are not included here. The reproductive print is one such area. From being a completely neglected aspect of print culture under Modernism, considered merely a byway on the main route to photography, the study of reproductive prints has enjoyed something of a renaissance.[6] If the division in print studies has been between the reproductive print, which attempts to create a facsimile of a work of art, and the original print, a primary image designed for that medium, then we can look at illustrated print culture as a whole in the same way. The works I've chosen to study here are primary vehicles, meant to convey art value in themselves, not just in reference to another more "elevated" work of art. I propose that earlier scholars who saw photography as the be-all and end-all of visual culture were writing in the first flush of enthusiasm for that medium, when its veracity was emphasized over its subjectivity. Certainly in the twenty-first century we are much more convinced of photography's malleability, of all the ways in which it is not "true" but shares common ground with the hand-drawn image as an expressive medium. Nonetheless, it is probably this earlier central focus on photography as the ultimate destination of nineteenth-century print technology that has led to a plethora of studies of reproductive prints. Certainly Walter Benjamin's "The Work of Art in the Age of Mechanical Reproduction" is the cornerstone of this focus, but Ivins's *Prints and Visual Communication* also arrived at this same conclusion.

Color printing developed internationally from the eighteenth century onward, but until the 1860s, hand-coloring remained standard practice. Engelmann announced the process of color lithography in his *Album lithographique* of 1837, and Thomas Shotter Boys utilized it in his portfolio of twenty-five plates, *Picturesque Architecture in Paris, Ghent, Rouen, Antwerp, etc.*, published in London in 1839; nonetheless, color lithography did not come into widespread use for several decades. The first color lithograph produced specifically as a work of art is usually identified as Édouard Manet's *Polichinelle* (1874). Even then, as Michael Twyman shows in his definitive *History of Chromolithography*, the process was used primarily for reproductive imagery, and, until late in the century, earlier technologies such as stencil-colored lithographs or even the hybrid *gillotage*, metal-plate relief printing, persisted.[7] Because color existed in print media long before and concurrent with the development of chromolithography, usually in the form of hand-colored or stencil-colored imagery, I have considered such images in several chapters instead of according to color printing a separate chapter of its own. The first works of chromolithography with any pretensions to image value were produced as part of the poster movement in the late nineteenth century, but I have not included an extended discussion of posters because, from its beginnings, the poster movement aspired to high art status, with all the accoutrements thereof, including exhibitions, publications, and collectors.[8] This makes posters qualitatively different from the other media discussed in this book. Other areas, such as sheet music, menus, department store catalogues, and scientific illustration would be rewarding to study as visual culture, but I leave these subjects for investigation to other scholars. In this book I am interested in the handmade image that, like high art, was created to have image value; unlike high art, however, such images were destined not for art world exhibition and consumption but for this new public arena. And perhaps this is the most far-reaching consequence of this study, to point out that the desire for the pleasures of visual imagery, which had always existed in relation to the high arts of painting and sculpture, in the course of the nineteenth century expanded into new technology, new media, and new audiences.

Any history of such imagery must be interdisciplinary by nature, since its development was dependent on many factors. Technological advances introduced new processes and social reform new audiences. Social, political, and economic factors come into play, driving the technology and at the same time resulting from it. Urbanization and the rise of an increasingly literate working class with greater disposable income encouraged the spread of visual imagery to a wider audience than ever before. The burgeoning market for printed imagery opened new career paths as well: virtually all nineteenth-century artists worked in these media, supplying drawings to periodicals, print dealers, or publishers. As the same artists often worked in several different media, it is imperative to adopt a unified and integrated approach to understanding this history. The artist Grandville, for example, produced caricatures, drawings for the illustrated press, and illustrated books—and he was by no means unique. For many artists, the new illustrated print media encouraged experimentation in both form and content that expanded both the arena of visual imagery and the making of it, as well as the audiences for it.

Given the necessity of first establishing chronologies and cataloguing the material, art historical studies of the various genres of illustrated print culture have lagged behind scholarship in other areas of art history, not least because of the discipline's traditional preference for original versus reproductive works. While basic research forms a necessary foundation for further study, the consideration of artists or genres in isolation further reinforces their dislocation from the larger frame of art history. I hope to indicate some possibilities for the integration of the study of illustrated print media into the graphic arts and art history as a whole. I regard the illustrated print medium not as the poor relation of the more prestigious art of drawing—or "drawing's stepchild," as I call lithography in chapter 1—but as important in its own right, enabling artists to produce significant and memorable images. The imaginative possibilities and visual strategies developed by these nineteenth-century artists were often rediscovered by subsequent generations and assimilated into twentieth-century Modernism, so a study of their work will not only broaden our conception of nineteenth-century visual culture, but will also cast new light on more familiar art historical developments. Acknowledgment that these graphic media served as wellsprings of creativity for later artists whose work has been accorded higher status in art history is one intention of this study. A parallel can be drawn to the history of what was formerly called "primitive art," which until the later twentieth century was written from the point of view of Western high art and rarely as a history in itself. This book attempts to sketch out a history of its own for illustrated print media.

To give some idea of the novelty of this material, we should note that the word "illustrator" was not in common usage until well into the nineteenth century. Artists who produced drawings for books were simply artists, "designers" in nineteenth-century England, *dessinateurs* in France. We don't have an equivalent or adequate word for this in English since the root word *dessin*, like the Italian *disegno* means both drawing and design. Our word "draftsman" has a mechanical connotation, "designer" elides the actual production of the image, and "illustrator" implies the precedence of the literary over the visual. Until the nineteenth century, there was no separate vocation of illustrator; artists of all persuasions, usually painters, would provide drawings that would then be engraved by professionals. Only when illustrated print media became the lingua franca of the Western world was it possible for artists to become specialists, "illustrators" by trade. The development of this lingua franca of graphic media was greatly facilitated by the lack of copyright protection. Although William Hogarth lobbied successfully for such laws in England in the eighteenth century, and artists could be reasonably confident of protection within their home country, the first international copyright law was the Berne Convention of 1886; the United States did not ratify it until 1989. As a result, there is more than an accidental family resemblance among the graphic productions that propagated rapidly throughout the nineteenth-century world. To give one example, Rodolphe Töpffer's 1845 comic album *The Story of Mr. Cryptogame* was redrawn and reprinted in France in the periodical *L'Illustration* that same year; it was subsequently "borrowed" by writers and publishers in many different countries, resulting in *The Veritable History of Bachelor Butterfly* (1845) in England, *The Strange Adventures of Bachelor Butterfly* (1849) in the United States, and *Reizen en avonturen van mijnheer Prikkebeen* (1858) in Holland.[9] Graphic imagery traveled across national borders as swiftly as the railroad or steamship.

❖ ❖ ❖

While there is a great deal of theoretical scholarship on popular culture, there is too little scholarship in the visual aspects examined in this study. I acknowledge, gratefully, pioneering publications by a variety of scholars. Beatrice Farwell's magnum opus of twelve volumes of lithographic imagery is indispensible basic research, particularly since there is no French equivalent to the British Museum *Catalogue of Political and Personal Satires.*[10] Gordon Ray's catalogues of French and English book illustration are also valuable resources, as is Nicole Garnier's two-volume catalogue of French popular prints.[11] Rémi Blachon wrote the standard text on nineteenth-century wood engraving, and Michael Twyman has published extensively on lithography.[12] Philippe Kaenel's *Le métier d'illustrateur* was the first in-depth study of this profession and is still required reading on the subject. No one can work on comics without acknowledging a major debt to David Kunzle, whose *History of the Comic Strip* initiated study of the genre. Jean Adhémar's works on Romantic prints and books are essential, as is the work of Ségolène Le Men, whose numerous articles on albums and book illustration often constituted the first serious attention to her topics.[13] Other scholars have written definitive studies in specific areas: David Kerr on the publishing empire of Charles Philipon, Ruth Iskin on posters, Philip Dennis Cate on late nineteenth-century French lithographs.[14] Literary scholars such as Richard Sieburth and Margaret Cohen have contributed much to the theoretical study of nineteenth-century panoramic literature, and Maurice Samuels's *Spectacular Past* has provided a valuable reading of the shift to genre imagery in the depiction of history.[15] The work of Vanessa Schwartz on visual culture has set out the parameters of the field, encompassing the entire spectrum of the material world, and both she and Maurice Samuels have theorized the role of spectacle during this period.[16]

Of the media discussed in this book, comics have received a fair amount of attention in current scholarship, although that has focused mainly on contemporary graphic novels; caricature has received surprisingly little attention outside the work of Daumier, for whom there are numerous exhibitions and book-length studies. Gustave Doré, arguably the most important book illustrator of the nineteenth century, was accorded his first major museum exhibition only in 2014.[17] For an artist such as Grandville—who is at least as important as Daumier and Doré—there is not even a life and works study in English, and only one, from 1985, in French.[18] There is even less scholarship on other major figures such as Tony Johannot, Jean Gigoux, Achille Devéria, Cham, or Bertall. For the most part, studies of graphic media have focused on the *peintre-graveur* , the artist-printmaker, primarily a major painter such as Delacroix who also made prints—and this could explain why Grandville, who worked only in graphic media, has received so little attention. Within graphic media, political images have received the most attention; the political caricatures that Delacroix drew in his youth, for example, have been the subject of a book-length monograph, while his illustrations for literary texts have been treated only in exhibition catalogues.[19] Media that demand collaboration, such as wood engraving, have received the least attention of all. Other areas included in this study—book illustration, the illustrated press, and popular prints—have been discussed primarily by specialists in book arts, popular culture, or literature, rarely by art historians. My intention with this study is to provide a road map for the study of nineteenth-century illustrated print culture with its interrelationships of media, artists, and genres.

This investigation of the fabric of the historical and cultural milieu when illustrated print culture was becoming hegemonic, when Another World was becoming our world, is especially important now that the disciplines of history, art history, and literature are all being revised to include popular culture, which has itself become a recognized field of study. I hope that this book will contribute to that effort and will serve as a bridge between traditional art history and the new areas of visual studies. It is also my hope that this study will encourage art historians to accept all visual phenomena as their proper subject, that it will contextualize for historians the rise of illustrated print culture within the crucible of Modernism, provide a much-needed historical underpinning to the field of visual studies, and help to elucidate the inadequately understood ties that existed between word and image at the moment when high culture was first challenged by more popular movements. Finally—and not least—I hope that this study will be valuable to the general reader simply because it unearths a vast quantity of heretofore unfamiliar visual imagery that is rich and rewarding, and because it illuminates how our world of images, from tabloid newspapers to comics to the nightly news, came into being.

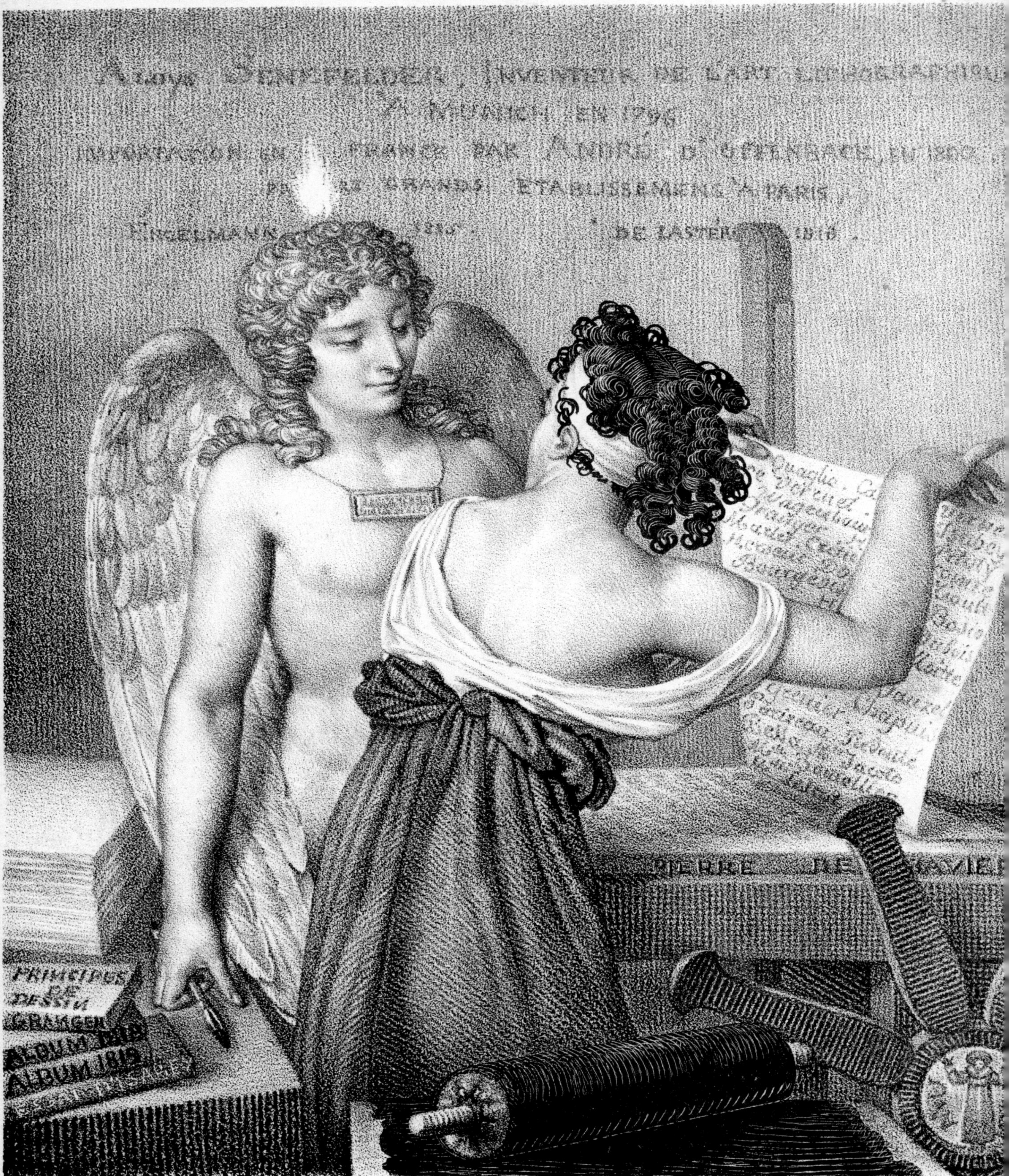

N. H. Iacob Dess.r de S.A.R. le Prince d'Eichstaedt

1 Drawing's Stepchild

Lithography and Caricature

When Alöys Senefelder (1771–1834) invented lithography in Bavaria in 1796, he intended it primarily for commercial usage, but artists soon discovered the advantages of this new medium.[1] It allowed them to draw directly on the stone plate, replacing the laborious process of creating a drawing on a metal plate by incising lines, as in engraving, or using repeated acid baths, as in etching. Lithography produced images that seemed miraculous in speed and fidelity, equally able to create line or shadow. In England, *Specimens of Polyautography* (the term first employed for the lithographic process) was published in 1803 with lithographic drawings by several artists, including Henry Fuseli (1741–1825). By 1819, Senefelder had commissioned a series of lithographic drawings to accompany his *Art of Lithography*; *The Genius of Drawing Encouraging the Art of Lithography* by Nicolas-Henri Jacob (1782–1871) effectively serves as its frontispiece (fig. 1).[2] Many contemporaneous critics felt that this sentiment should be reversed, that it was, in fact, the genius of lithography that encouraged the art of drawing. Lithography transformed drawing from a private into a public art, ensuring, as one critic wrote, that drawing would be "no longer shut away in portfolios but given to the public."[3] During these early decades, lithography did just that, finding the largest audience of any contemporaneous art form and initiating what has been termed "the nineteenth-century media explosion."[4]

In France before lithography, few artists made original prints, a term that defines a print that does not reproduce a work of art but is itself the work of art. French prints traditionally were the province of a highly trained cadre of engravers who first redrew an artist's work onto copperplate, then etched and engraved it into a facsimile of the painting or drawing.[5] These engravings demonstrate not only the high degree of skill that went into the reproductive print, but also show why such prints have been largely dismissed from the discussion of stylistic movements that has defined art history, a discipline traditionally focused on original production alone.

Engravers and traditionalists in general saw lithography as a shoddy substitute for the labor-intensive engravings that seemed to incarnate the eternal values of high art. An example of this is the 1819 editorial exchange between the journals *La Quotidienne* and *Le Constitutionnel*, wherein *Le Constitutionnel*, which supported both lithography and the 1789 Revolution, attacked *La Quotidienne* as the party of the aristocracy that

FIGURE 1
Nicolas-Henri Jacob, *The Genius of Drawing Encouraging the Art of Lithography*, Pl. IX from Alöys Senefelder, *The Art of Lithography. Collection of Several Examples in Drawing and Engraving to Supplement Practical Instruction in Lithography* [Le génie du dessin encourageant l'art de la lithographie / *L'art de la lithographie. Collection de plusieurs essais en dessins et gravures pour server de supplément à l'instruction pratique de la lithographie*]. Paris, 1819. Lith. Senefelder. The Getty Research Institute 2014.PR.8.

disdained lithography because it made art available to a wider audience; in return, *La Quotidienne* wrote: "Our opinion of this type of print-making has not changed; we continue to think it is catastrophic for the art of engraving and we persist in condemning the harm it has done in the last two years."[6] Nonetheless, lithography was soon identified as the medium par excellence for images of modern life because of the ease with which it could be executed, its responsiveness to individual stylistic temperaments, and its ability to depict the ephemeral and shifting scenes of a rapidly changing society. Quickly adopted by the younger generation of artists, it became identified with the Romantic movement in art and, in the course of the nineteenth century, lithographic images of all types became as widespread as photography in our own day.

THE HIGH ROAD: LITHOGRAPHY AND THE SALON

While Godefroy Engelmann (1788–1839) was not the first to publish lithography in France—that honor belongs to Peter-Friedrich André who was licensed in 1802 and whose brother Philipp published *Specimens of Polyautography* in London the following year—Engelmann was the most ambitious of the early French publishers.[7] In 1816, he moved his presses from his native Mulhouse to Paris and established the lithographic firm that he maintained there for several decades. The same year, he presented this new medium to the Institut de France with a portfolio of lithographic drawings. A committee of academicians was appointed to make an evaluation; their report stated: "In effect, when the painter draws on the stone, he alone creates and executes his drawing with the fire of genius or the love of perfection. It is his style, his personal manner, even the faults that one finds only in his work. We recognize in this work his sincere and spontaneous touch, the result, not of trial and error, but of the inspiration that guides the creative hand."[8] As a result of this official approbation, lithographs were shown in the Paris Salon, the official exhibition of art, beginning in 1817 as part of the section *Gravure*, and from then on the word *gravure* began to assume the generic meaning of "print" as opposed to its earlier limitation to the intaglio processes of etching and engraving.[9] Critics soon weighed in on this new medium. At the first Salon to include lithography, François Miel wrote: "The image on the page is the actual drawing. . . . The print is, so to speak, autograph; it is the hand of the master who drew it and it differs in no way from the original. It is itself an original."[10] The refrain that lithographs are not reproductions but original drawings multiplied was repeated constantly in the early decades of the medium's existence.

Most early lithographs exhibited at the Salon were portraits, views of landscape or cityscape, or reproductions of esteemed works of art; there was little genre imagery and no caricature at all although these categories were rapidly becoming predominant in lithographic production. Prints were listed under the name of the lithographer or publisher (often the same individual), while the artist who produced the lithograph was mentioned only secondarily, if at all. This is in the tradition of engraving, which always

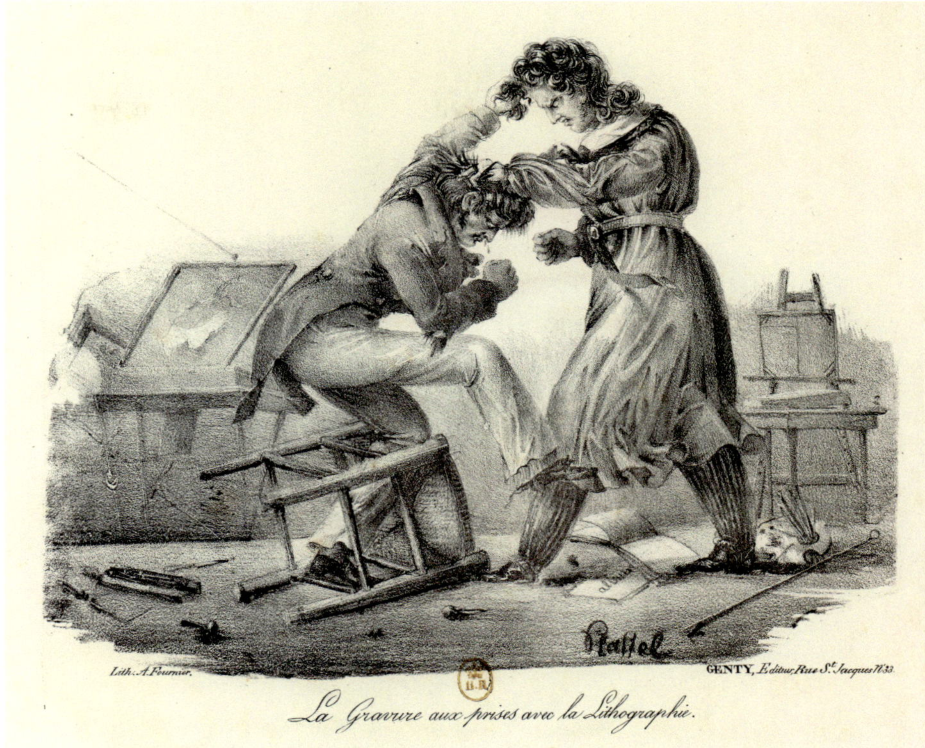

FIGURE 2
Henri Plattel, *Engraving Doing Battle with Lithography* [*La gravure aux prises avec la lithographie*], 1824. Lith. A. Fournier, Pub. Genty. Bibliothèque nationale de France.

appeared in the Salon as the work of the artist who produced the print, not the artist who produced the drawing or painting from which the print was made. The two careers of engraver and lithographer were not equivalent, however. The major determinant of their difference was that the engraver translated the artist's drawing or painting into the graphic medium by first redrawing it himself, then etching and engraving it with his own tools and in his own atelier. The resulting work was his as much as that of the originating artist, and, as a result, his equal status extended even to the French Academy, where engraving was recognized as a category for membership along with painting, sculpture, and architecture. Lithographers, however, had an altogether different status because the creative aspect of the medium—the drawing on the stone—was so clearly separate from the technical procedures of chemically fixing, inking, and printing it. This division alone, in the hierarchy of nineteenth-century aesthetics, would place lithography in the lower category of the mechanical arts. *Engraving Doing Battle with Lithography*, an 1824 caricature by Henri Plattel (1803–1859), demonstrates this by showing the engraver and lithographer, each with the tools of his trade (fig. 2). The engraver is dressed in an artist's smock and is depicted with his easel, palette, and maulstick, all signs of his status as artist, while the lithographer is shown as an artisan, clad in workman's trousers and surrounded by his tools.

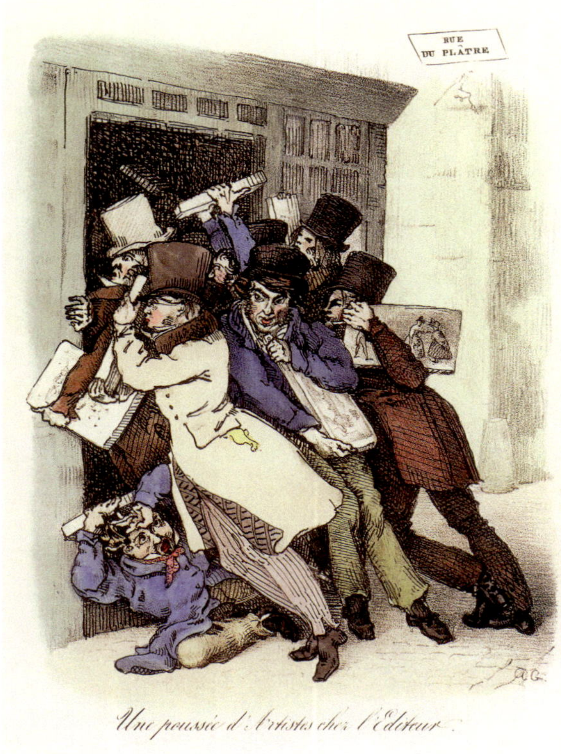

FIGURE 3
A. G., *A Crush of Artists at the Publisher's* [*Une poussée d'artistes chez l'éditeur*], ca. 1820s. Lith. Gobert, Pub. Genty.

Even the very definition of "lithographer" was ambiguous during these early years because it could denote various degrees of involvement with the print process. Thus, in these decades the status of lithographer fluctuated between artist and artisan before finally settling into that of a skilled artisan who executed drawings made by others.[11] In 1825 the critic E.-J. Delécluze (1781–1863) attempted some clarification by proposing a twofold division between what he called *dessinateurs lithographes*, artists who worked in lithography, and *imprimeurs lithographes*, lithographic printers who handled its technical aspects. Among lithographic artists, he established a further hierarchy between creative artists, those who drew original designs, and those who served merely as intermediaries, translating the drawings of others, as engravers had always done.[12] Such distinctions had been unnecessary when few artists made original prints or when engravers themselves carried out all the steps of the printmaking process. With the development of transfer paper in England in the 1820s, the lithographer's position was further diminished because artists could now draw directly on this specially prepared paper, give their drawings to the lithographer, and receive back the finished prints.[13] As a result of lithographers' further—some would say complete—separation from the creative aspects of the process, they never achieved the status that engravers enjoyed, and never were accepted as members of the Academy.

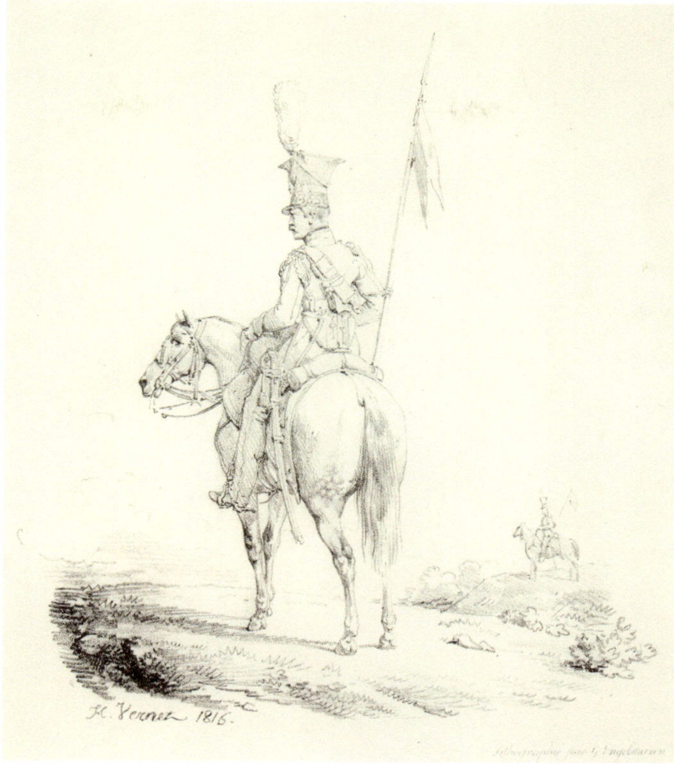

FIGURE 4
Horace Vernet, *The Lancer* [*Le lancier en vedette*], 1816. Lith. G. Engelmann. Yale University Art Gallery 1982, Everett V. Meeks, B.A. 1901, Fund.

Although the métier of composing drawings that would be seen only in reproduction as lithographic prints was scarcely more prestigious than the technical work of actually producing the prints, adventurous artists soon began producing lithographs in quantity. In *A Crush of Artists at the Publisher's*, the artists, each carrying a lithographic stone bearing his drawing, all try to crowd into the publisher's office at the same time (fig. 3). Among the pioneers were Carle Vernet (1758–1836) and his son Horace Vernet (1789–1863), Anne-Louis Girodet (1767–1824), Jean-Baptiste Isabey (1767–1855), Théodore Géricault (1791–1824), Nicolas-Toussaint Charlet (1792–1845), and Eugène Delacroix (1798–1863), all increasingly identified with the Romantic movement in art. In contrast, classical artists such as J.-A.-D. Ingres (1780–1867) had little interest in lithography or in making original prints of any kind, preferring the time-honored practice of having their works etched and engraved for reproduction by professionals.

The portfolio that Engelmann presented to the Institut de France in 1816 included Horace Vernet's *The Lancer*, which has been called the first commercially produced art lithograph in France (fig. 4).[14] His father Carle Vernet, perhaps the earliest French supporter of lithography, soon abandoned the intaglio processes that he had been using in favor of this new medium. He created the celebrated depiction of the print shop of François-Séraphin Delpech (1778–1825), one of lithography's first dealers and publishers,

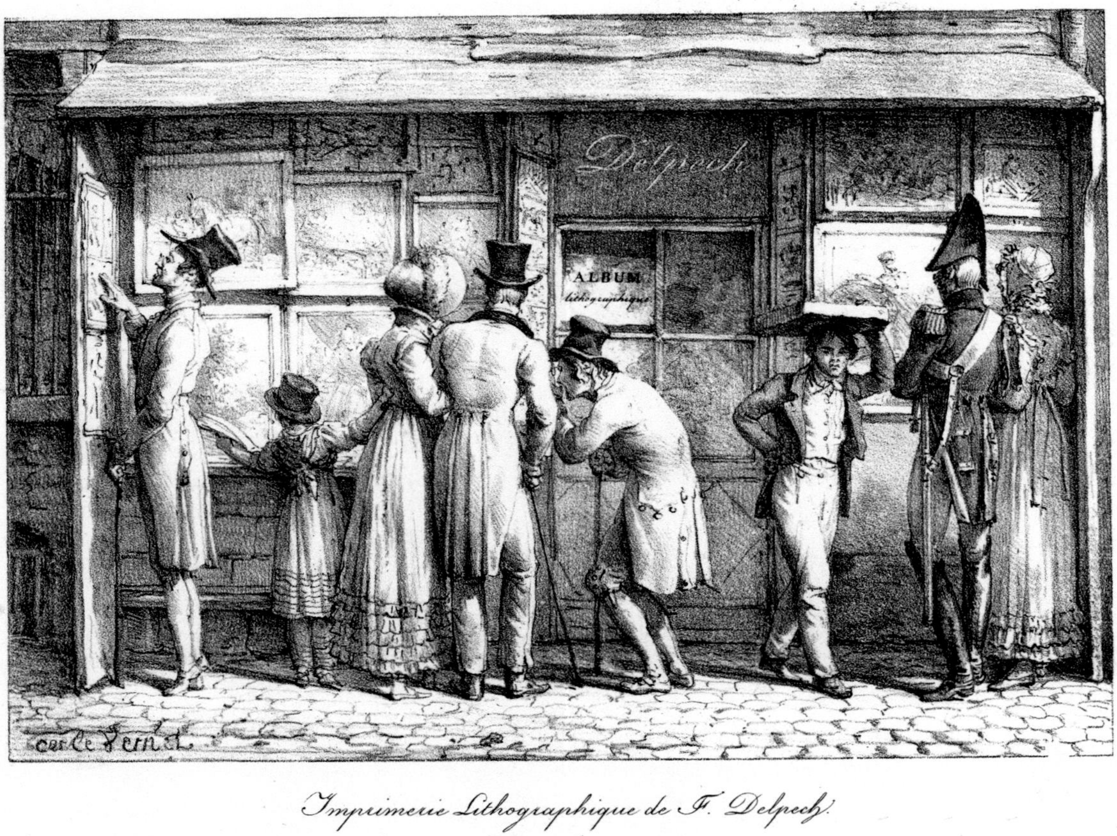

FIGURE 5
Carle Vernet, *F. Delpech's Lithographic Print Shop* [*Imprimerie lithographique de F. Delpech*], ca. 1818. Lith. & Pub. Delpech.

showing the customary crowd of Parisians admiring Delpech's latest editions (fig. 5). Both Vernets were such ardent supporters of lithography that Senefelder had their names listed among lithography's early champions on the print that the young woman is showing to the Genius of Drawing in figure 1.

Not many academicians were interested in this new process, but Pierre-Narcisse Guérin (1774–1833), who had served on the Institut committee that reported on Engelmann's portfolio, made a few lithographs, as did Antoine-Jean Gros (1771–1835) whose studio produced many prominent lithographers in subsequent decades.[15] In the context of academician lithographers, Anne-Louis Girodet is noteworthy as he contributed a lithograph to Engelmann's 1816 portfolio, a portrait of Coupin de Couperie, writing across it, *Si pingi melius, non potuit melior*—"No matter how well depicted, it cannot be better than this."[16] For Girodet, lithography revised the entire concept of printmaking as a reproductive medium. His friend Firmin-Didot wrote: "Haven't I seen Girodet applying himself to drawing a great number of his compositions on lithographic stones, and following the progress of this new technique with great interest? Lithography fascinated him, he said, just because it allowed him to reproduce his own work without the help of an interpreter and with his own hands."[17] Girodet not only worked in lithography himself, he also trained his students in the medium. The most renowned of them was Hyacinthe-Louis Aubry-Lecomte (1787–1858), one of the greatest French lithographers of the nineteenth century, with whom Girodet worked to create a portfolio of sixteen prints based on his painting of Ossian.[18] Shown at the 1822 Salon under the name of its publisher, Engelmann, the portfolio was titled *A Collection of Study Heads, after the 1801 painting by Mr. Girodet-Trioson, Member of the Institute, Representing the Shades of French Heroes Received in the Heavenly Palaces of Ossian. Lithographed under His Direction by Aubry-Lecomte, His Student).*[19] The sixth in that series, *Starno* (fig. 6), depicts the fierce king of Lochlin who, in James Macpherson's epic *Ossian*, killed his own daughter. Starno is almost insignificant in Girodet's painting, one of many figures swarming the periphery of the picture, but here we are given a haunting image of the brooding king, his gnarled fingers clutching his weapon while other fingers grope for his throat and a sword pierces his arm. Images such as this, based on fragments of a preceding work and made in collaboration, go far beyond the tradition of reproductive printmaking to create new and unique images in themselves.

Among the few recognized artists who exhibited lithographs at the Salon, Delacroix stands out as the heir to Girodet. Like Girodet, Delacroix produced prints

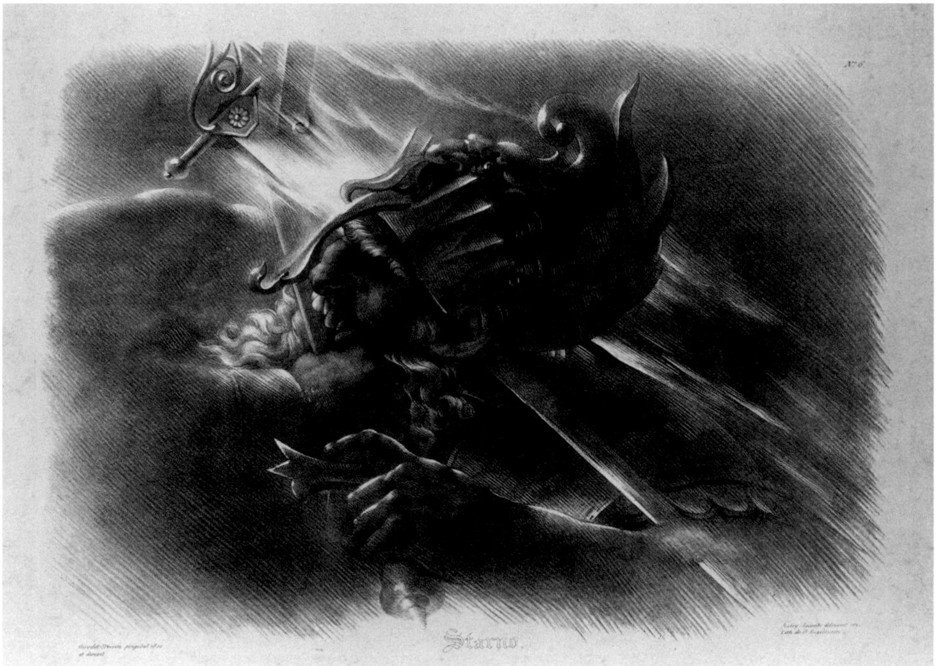

FIGURE 6
Hyacinthe-Louis Aubry-Lecomte, *Starno*, no. 6 in *A Collection of Study Heads, after the 1801 Painting by Mr. Girodet-Trioson, Member of the Institute, Representing the Shades of French Heroes Received in the Heavenly Palaces of Ossian. Lithographed under His Direction by Aubry-Lecomte, His Student* [*Collection de têtes d'étude, d'après le tableau peint en 1801, par Mr Girodet-Trioson, Membre de l'Institut & représentant les ombres des héros français reçues dans les palais aëriens d'Ossian. Lithographiées sous sa direction par Aubry-Lecomte, son elève*], 1821. Lith. Aubry-Lecomte and G. Engelmann. Harvard Art Museums/Fogg Museum. Gift of Charles H. Taylor M5119.

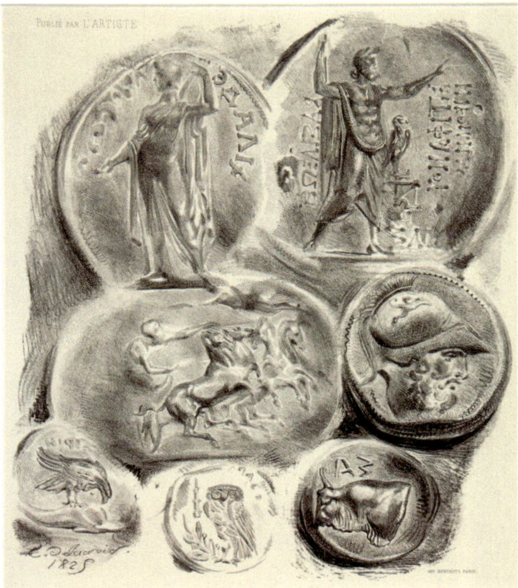

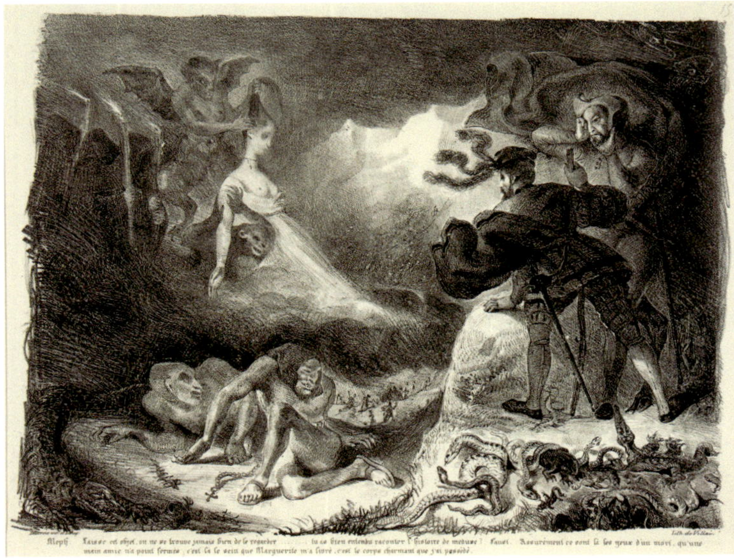

FIGURE 7
Eugène Delacroix, *Sheet of Seven Antique Medals* [*Feuille de sept médailles antiques*], 1825. Lith. G. Engelmann. Yale University Art Gallery, Gift of Ralph Kirkpatrick, Hon. M.A. 1965.

FIGURE 8
Eugène Delacroix, *Marguerite's Ghost Appears to Faust* [*L'ombre de Marguerite apparaissant à Faust*], 1828, from Goethe, *Faust*, 1826–1827. Lith. Villain, Pub. Ch. Motte. Yale University Art Gallery, Anonymous gift.

based on literary texts throughout his career; both artists found the conjunction of word and image felicitous. Unlike Girodet, however, Delacroix did not see the process as inherently reproductive; for him it was an especially fruitful kind of drawing. Lithography allowed Delacroix to give free rein to his imagination, far beyond the scope of his painting. He produced over 130 prints, mostly lithographs, but also including youthful caricatures in hand-colored etching that, like those of Jacques-Louis David, owed much to English precedents.[20] He did sixteen lithographs for Shakespeare's *Hamlet*, individual prints after other plays of Shakespeare as well as works of Byron and Sir Walter Scott, seven lithographs after Goethe's *Götz von Berlichingen* and a portfolio of seventeen after Goethe's *Faust.* In addition he did numerous lithographs of animals—lions, tigers, horses—and several still lifes that are unlike anything in the history of art. *Sheet of Seven Antique Medals* uses the soft tones of lithography to depict the classical body that was, at the time, more commonly drawn in the academic linear style (fig. 7). The coins themselves are deformed, and the image as a whole seems to be a commentary on classicism, relegating it to an antiquarian past, nostalgic to be sure, but definitely diminished in stature. We should not be surprised by Delacroix's statement to the critic Philippe Burty (1830–1890): "The *Medals* (whose proofs I retrieved for you) were shown by the print dealers but no one wanted them."[21] Who, indeed, would want these images that existed on some conflicted plane between Classicism and Romanticism, at a time when battle lines between these contemporaneous aesthetic movements were so clearly drawn?

And yet, Delacroix actually seemed to think in terms of lithography, as though it could give him something that mere drawing could not. In 1824, after a visit to his friend Soulier, he wrote: "He inspired me to do some lithographs of animals—a tiger on its prey, for example, some vultures, and so on."[22] When Burty asked him about his *Faust* series, Delacroix wrote:

> I recall that, around 1821, I saw some compositions by Retzsch that really impressed me, but it was especially the performance of a dramatic opera based on *Faust* that I saw in London in 1825 that excited me to do something like that.... You know that Motte was the publisher; he had the unfortunate idea of publishing the lithographs with a text that killed them, not to even mention the weirdness of my plates which provoked several caricatures and set me up more and more as one of the leaders of the *School of Ugliness*.... I don't remember how much I made from that: something like a hundred francs and an engraving by Lawrence, the *Portrait of Pius VII.* All my speculations have been like this. *Hamlet* even more so: I had the series printed at my own expense and published it myself. It cost me five or six hundred francs, and I didn't get back even half of my expenses.[23]

The "weirdness" of his imagery, apparent even to Delacroix himself, was far more extreme than anything he ever did in his paintings. In *Marguerite's Ghost Appears to Faust*, the image is roughly drawn, bleeding out of its margins, its highlights scratched out from the blackness, with fantastic creatures not only surrounding Marguerite but dominating the lower third of the plate (fig. 8). Even Goethe, who acquired most of Delacroix's *Faust* series, wrote: "I must confess that Mr. Delacroix has in some scenes surpassed my own notions."[24] Delacroix clearly used lithography as a direct conduit to the wellsprings of his imagination, whether or not he followed through in painting.

THE LOW ROAD: LITHOGRAPHY IN THE STREETS

The portfolio of lithographs that Engelmann presented to the Institut in 1816 included a variety of subjects in order to demonstrate the breadth of the new medium's scope and to render it respectable. Other early French lithographers, such as Charles-Philibert de Lasteyrie (1759–1849) and even Senefelder himself, also availed themselves of the practice of publishing lithographs in portfolios in order to demonstrate the possibilities of this new medium.[25] A plate from Senefelder's *New Invention of Lithography* of 1823 shows the traditional advertising image for the printing trade, a trompe l'oeil collage of lithographic products: business forms, playing cards, a map, sheet music, botanical illustration, an architectural illustration, landscape and cityscape views, a portrait, and genre images (fig. 9).[26] This image illustrates the complex nature of the lithographic trade, which included commerce, art, and popular imagery. Of these categories, art prints made up the smallest segment of the business of lithography, barely represented here. The works of Delacroix, for example, were neither profitable nor widely disseminated; despite their unquestionable value, crowds did not gather in the streets in front of print shop windows

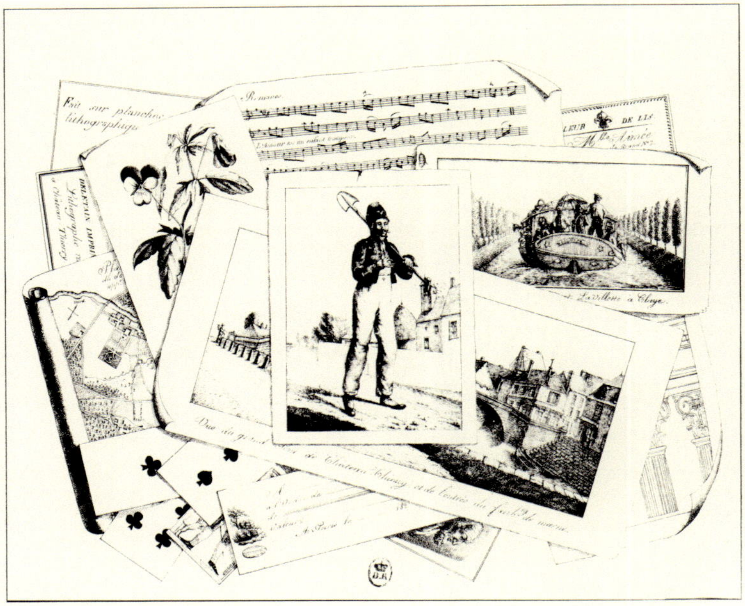

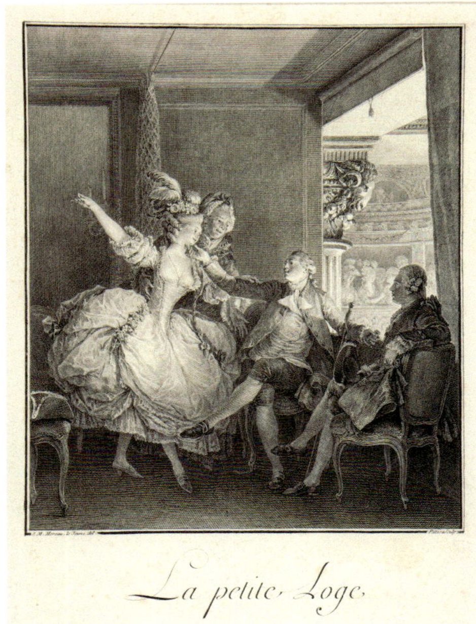

FIGURE 9
Alöys Senefelder, Plate from *New Invention of Lithography. Lithographic Portfolio, Collection of Subjects of Various Kinds, Drawn and Printed from Lithographic Plates* [*Nouvelle invention lithographique. Porte-feuille lithographique, recueil de sujets de divers genres, dessinés et imprimés sur planches lithographiques*], 1823. Lith. Senefelder.

FIGURE 10
J. M. Moreau le Jeune, *The Small Loge*. Engr. Jean-Baptiste Patas from *Third Suite of Prints Showing the History of French Manners and Costume in the Eighteenth Century* [*La petite loge / Troisième suite d'estampes pour servir à l'histoire des moeurs et du costume des français dans le dix-huitième siècle*], 1783. Etching and engraving. Pub. Prault. Yale University Art Gallery, purchased by the University.

such as Delpech's (see fig. 5) to see such images. They came instead to enjoy the vision of contemporary life, popular images of manners and mores that provided unexpected pleasure to a public that was unaccustomed to seeing itself depicted. When images had been expensive and laborious to produce, the pleasures of such recognition were out of reach for most people, and, indeed, earlier series such as the *Suite of Prints Showing the History of French Manners and Customs in the Eighteenth Century* (1775–1783) were limited to depictions of the upper classes.[27] *The Small Loge*, for example, shows four elegantly dressed patrons of the opera, clearly wealthy enough to have secured a private box (fig. 10). Lithography, on the other hand, provided images as up-to-date as newspapers, and far cheaper—in an era that had not yet witnessed the beginnings of the illustrated press. The importance of this was confirmed by the art critic and statesman Adolphe Thiers (1797–1877), who wrote in 1824:

> The discovery of lithography has caused a true revolution. The facility that artists have to seize the crayon themselves, to spontaneously jot down on stone the ideas that strike them and to deliver these rapid sketches to an even more rapid press, this has led them to let nothing escape that strikes them, to set down all the images that appeal to them, to seize all the fleeting aspects that nature offers them. An attitude, a gesture, an unusual figure, a scene of manners, a popular farce, or a military action, they are all reproduced with a marvelous spontaneity. Each of their impressions, rendered with the courage, directness, and truth of immediate inspiration, comes to us without the intermediary of a cold and unfeeling engraver. This is the very drawing of the master

that, flung down onto the stone and reproduced an infinite number of times by the lithographic press, has been transmitted to us with all the genius of the original, with his touch, his idiosyncrasies, his expression, his verve.[28]

The types of works that so impressed Thiers, quickly drawn images for a rapidly changing society, rarely appeared in the Salon, but were, instead, on view in the streets, in the windows of print shops such as Delpech's. At this point the history of lithography becomes prismatic, with the high road to the Salon taken by Girodet and Delacroix increasingly isolated from the busy thoroughfares where lithography thrived. To comprehend the scope of the revolution initiated by lithography, we must look at these popular prints, published and sold by the thousands, and exhibited, it seems, every place except the Salon. The "media explosion" that has often been dated to the 1830s began, in fact, more than two decades earlier, with the spread of lithography.

It is in this context that we must see the work of Géricault for, although his lithographs are considered the first real masterpieces of the medium, predating those of Delacroix, he never exhibited them at the Salon. Instead, he hoped to exploit the increasing popularity of the medium, as the Vernets and his friend Charlet were already doing. The role of Charlet is important here: while not enjoying the esteem today of major Romantic figures like Delacroix, he was influential and celebrated in his own time. A student of Gros and a licensed lithographic printer himself, he worked in the medium all his life and introduced many of his fellow artists to the process, including Edmé-Jean Pigal (1798–1872), Henry Monnier (1799–1897), Eugène Lami (1800–1890), Hippolyte Bellangé (1800–1866), Charles Philipon (1800–1861), and Denis-Auguste Raffet (1804–1860)—a roll call of early lithographic artists.[29] Charlet was known principally for his depictions of Napoleonic military subjects. In his sardonic *Hah! What Fun... to Be a Soldier!* we see two ragtag veterans of combat, one wounded with his arm in a sling, the other trudging along virtually shoeless (fig. 11). Charlet had, however, a broad range in both style and subject. His political caricatures were both biting and influential. *My Dear Children, I Hold You All in My Heart* of 1823, shows the unmistakable contours of King Louis XVIII (1755-1824) with his "heart" clearly inscribed on his derrière (fig. 12). The image retained its relevance and was updated for *Le Charivari* ten years later merely by changing the "portrait" to that of the then reigning monarch, King Louis-Philippe (1773–1850).[30] Delacroix greatly admired Charlet, wrote an essay praising him and even criticized Baudelaire for disparaging him.[31]

Géricault made his first lithograph in Paris in 1817, one year after the Institut report; he was then twenty-six years old. In the six years remaining in his brief life, he produced almost a hundred lithographs.[32] In 1820, he went to London with Charlet, where they both drew scenes from urban life. Charlet's *The Nymph of the Thames* and Géricault's *The Piper* show their close collaboration (figs. 13, 14). This trip resulted in Géricault's masterpiece of lithography, a portfolio of twelve drawings printed by Charles Joseph Hullmandel (1789–1850), the major London lithographer. Entitled *Various Subjects Drawn from Life*

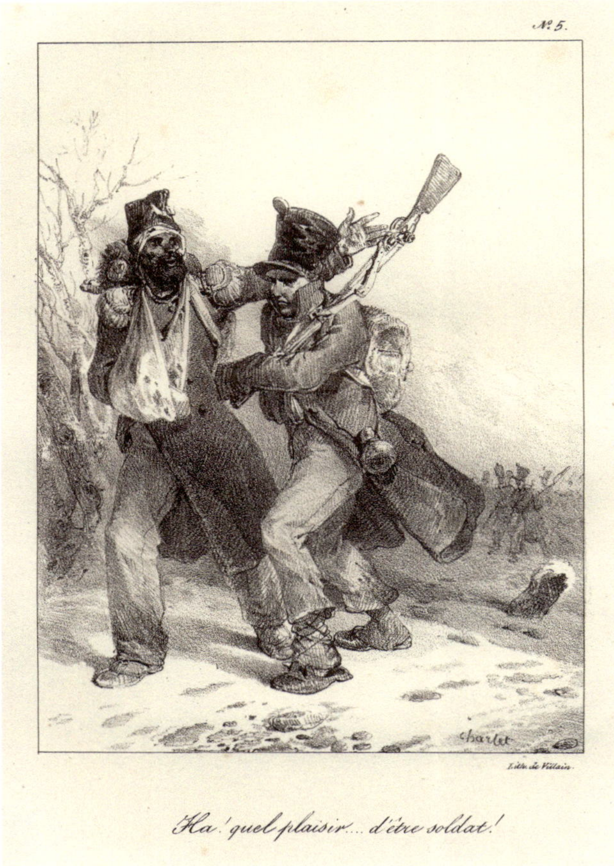

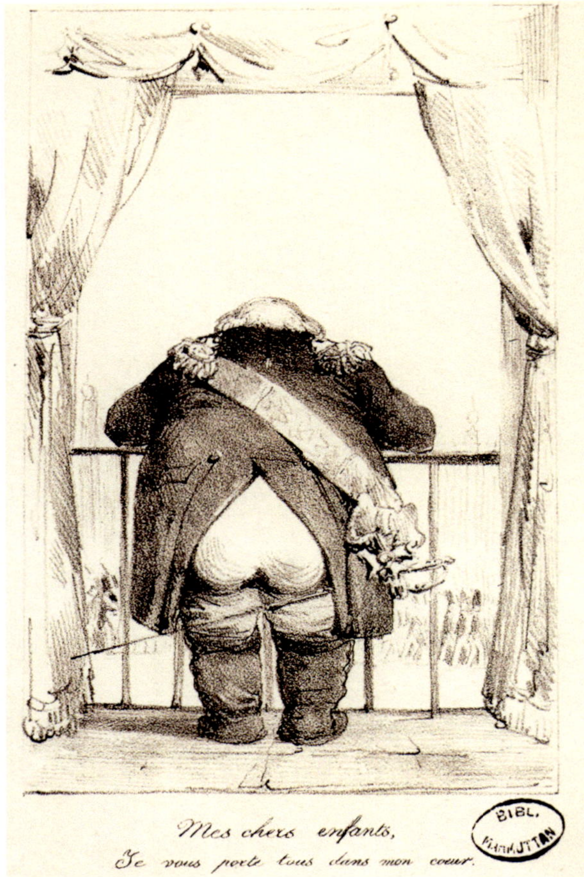

FIGURE 11
Nicolas-Toussaint Charlet, *Hah! What Fun . . . to Be a Soldier!* [*Hah! Quel plaisir . . . d'être soldat!*], 1828. Lith. Villain, Pub. Gihaut frères. Brown University Digital Repository, Anne S. K. Brown Military Collection.

FIGURE 12
Nicolas-Toussaint Charlet, *My Dear Children, I Hold You All in my Heart* [*Mes chers enfants, je vous porte tous dans mon coeur*], 1823. Lith. Villain.

and on Stone, and including, besides *The Piper*, such well-known images as *Adelphi Wharf* and *Pity the Sorrows of a Poor Old Man*, the portfolio has always been called, simply, *The English Suite*.[33] It is an excellent example of the international collaborations that would become characteristic of the nineteenth-century trade in printed images. From London, Géricault wrote to a friend:

> I am more reasonable than you, since at least I'm working hard and making lithographs. I am dedicated for the time being to this genre that, being brand new in London, has a popularity here that is inconceivable. With a bit more tenacity than I have, I am certain that one could make a considerable fortune. I flatter myself that, for me, this will be merely a form of advertisement, and that soon the taste of the true art lovers who will get to know me this way will employ me in work that is more worthy of me. You call this ambition; but my God, it is nothing more than striking while the iron is hot; and since I've already been encouraged, I send to the devil all those Sacred Hearts of Jesus; it's a fool's game to die of hunger. I'm abandoning the tragedienne's

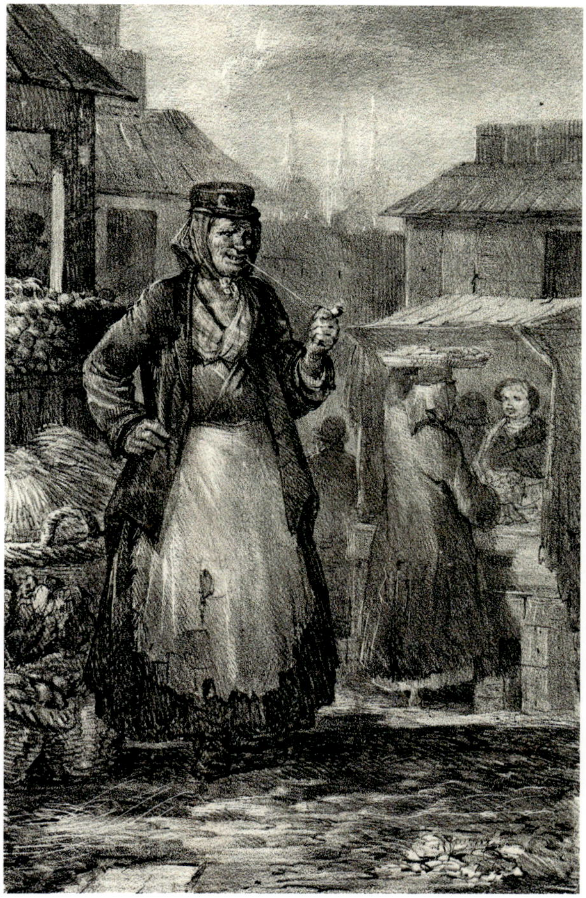

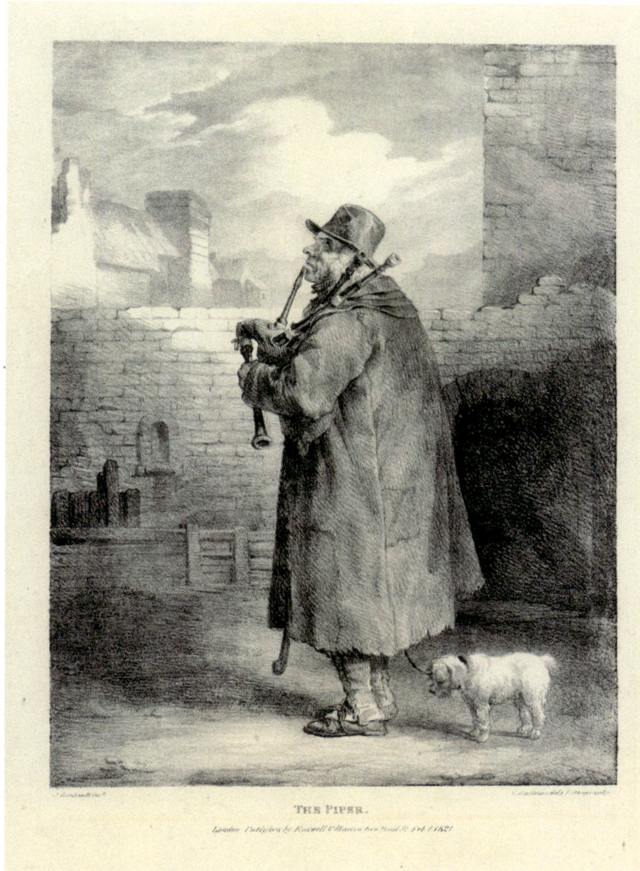

FIGURE 13
Nicolas-Toussaint Charlet, *The Nymph of the Thames* [*La nymphe de la Tamise*], 1820. Lith. C. Hullmandel. British Museum.

FIGURE 14
Théodore Géricault, *The Piper*, 1821, from *Various Subjects Drawn from Life and on Stone by J. Géricault*, 1821. Lith. C. Hullmandel, Pub. Rodwell and Martin. Yale University Art Gallery, Gift of Charles Y. Lazarus, B.A. 1936.

stage and the Holy Scriptures to shut myself up in the stables, from which I'll emerge covered with gold.[34]

While this might seem to show Géricault's disdain for his prints, we should also note his even greater disdain for the traditional subjects of high art, religious and history painting. Géricault saw quite clearly that "the stables," as he put it, would furnish the subject matter for a reinvigorated art of the future.[35]

Géricault's images are unique in their hybrid stance between high art and popular imagery. His lithograph *The Boxers*, for example, depicts a famous English prizefight (fig. 15). It was clearly inspired by popular English prints, most notably an illustration by George Cruikshank (1792–1878) of *The Second Contest between Crib and Molineux, September 28, 1811*, which appeared in Pierce Egan's 1812 *Boxiana: Sketches of Ancient and Modern Pugilism* (fig. 16).[36] The visual language that Géricault developed in *The Boxers* as well as in his other lithographs occupies a position somewhere between caricature and naturalistic drawing. Today we make a clear distinction between these two modes

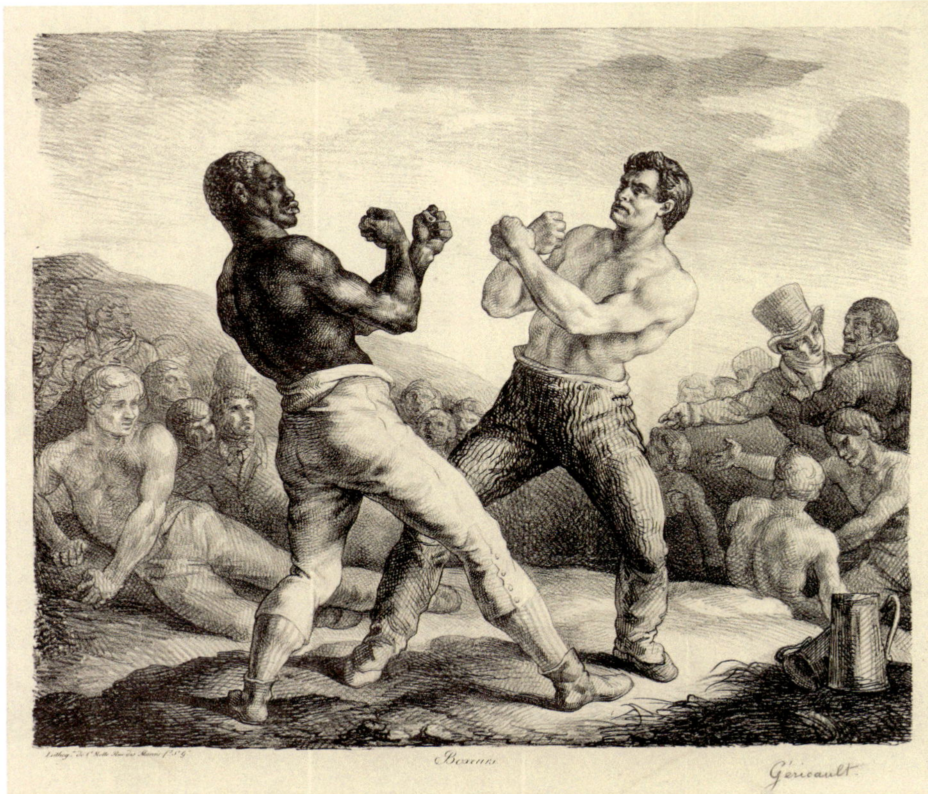

FIGURE 15

Théodore Géricault, *The Boxers* [*Les boxeurs*], 1818. Lith. C. Motte. Yale University Art Gallery, Everett V. Meeks, B.A. 1901, Fund.

of representation, but in early nineteenth-century France they were considered part of the same phenomenon—imagery of modern life—no doubt because both maintained their distance from academic drawing in form and content. We would also, today, draw sharp lines of demarcation between illustration, which is naturalistic reportorial drawing, visual satire, which is naturalistic in style but presents humorous situations, and caricature, which always involves distortion and exaggeration as well as ridicule.[37] We can better understand the nineteenth-century conflation of caricature and naturalism if we think of the contemporaneous range of drawing possibilities as a continuum that could be distinguished clearly only at its poles: academic drawing based on idealized antique models and poses at one extreme, vicious antigovernment caricature with exaggerated, even deformed, physiognomies at the other. Most lithographic drawings fell somewhere between these extremes; as one early critic wrote, "some are pleasing, some moral, others bawdy."[38] As a result, in nineteenth-century France, all were regarded as caricature.

To comprehend this phenomenon, we must digress for a moment to review the history of what we would call caricature today, but was then the *non-dit* of lithography. Although caricature was profitable and widely disseminated, published even by Engelmann, neither he nor anyone else ever sent it to the Salons. Its history begins

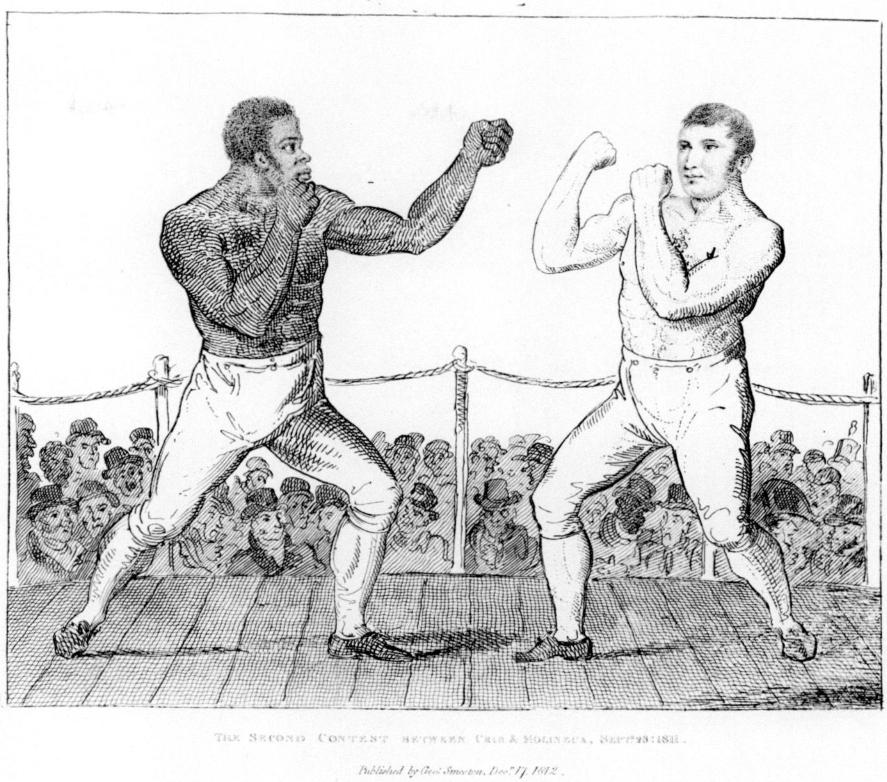

FIGURE 16

George Cruikshank, *The Second Contest between Crib and Molineux, September 28, 1811*, from Pierce Egan, *Boxiana; or, Sketches of Ancient and Modern Pugilism. From the Days of the Renowned Broughton and Slack to the Heroes of the Present Milling Era*. London: G. Smeeton, 1812, vol. 1, 412–13. Etching. Yale University, Beinecke Library.

in Renaissance Italy and continues through eighteenth-century England with William Hogarth (1697–1764) and his successors: Thomas Rowlandson (1756–1827), James Gillray (1757–1815), and George Cruikshank.[39] The battle lines between "serious" drawing and caricature are the subject of Hogarth's *Characters and Caricaturas* (fig. 17), where he contrasts the depiction of character shown in examples from Raphael, on the lower left, with the exaggerations and distortions characteristic of caricature in the examples on the lower right, adapted from his Italian predecessors Leonardo da Vinci (1452–1519), Agostino Carracci (1560–1609), and Pier Leone Ghezzi (1674–1756).[40] France had Jacques Callot (1592–1635) and Spain had Francisco Goya (1746–1828), but before the nineteenth century no country had the rich comic traditions of English caricature. The upheavals of the Revolutionary decades in France changed this, however, as quantities of vicious anti-French images began circulating in Europe. James Gillray's *French Liberty, British Slavery* is a typical image, contrasting the ragged, emaciated, but "liberated" Frenchman, with nothing to eat but onions, to the "enslaved" but plump, well-dressed Englishman feasting on his roast beef (fig. 18). To combat broadsides such as this, the Revolutionary Committee of Public Safety commissioned the august classical painter Jacques-Louis David (1748–1825) to produce three counterattacks in 1793.[41] *English Government* is by far the most extreme

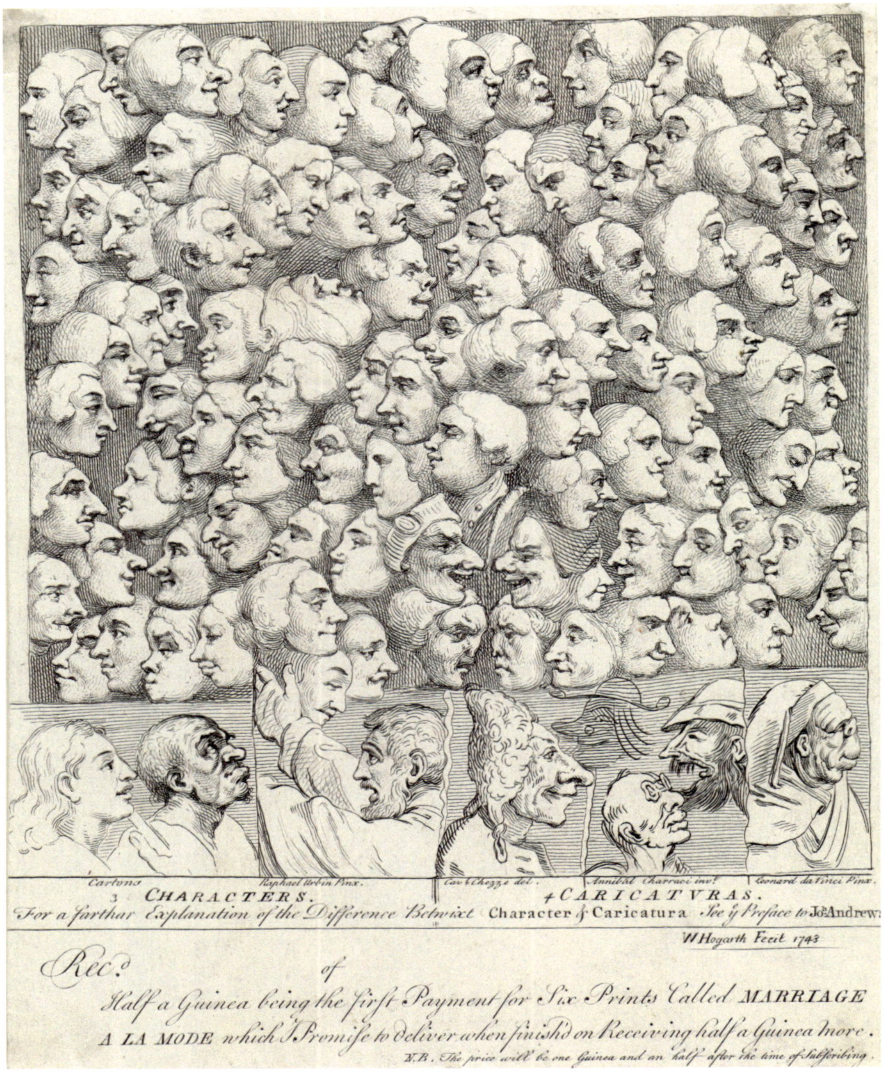

FIGURE 17

William Hogarth, *Characters and Caricaturas*, 1743. Etching. Yale University, Lewis Walpole Library.

of these, the image accompanied by a legend explaining, in case there was any doubt, that English government is here depicted as a flayed devil wearing royal decorations, with the portrait of the king, George III, inscribed on his derrière, from which a multitude of taxes spew forth to overwhelm the citizenry (fig. 19).[42] And yet, despite images such as these, the English style of grotesque caricature had little resonance in France.[43] In what might well be the earliest history of caricature in France, Étienne de Jouy (1764–1846) in 1814 attempted to explain, in a backhanded way, why his country lagged so far behind its traditional rival, England, in producing such imagery: "The Revolution inundated France with a deluge of caricatures, in which, daily, every event, every session of the assembly, every circumstance in the life of the major deputies was, in turn, exposed to public ridicule.

FIGURE 18
James Gillray, *French Liberty, British Slavery*, 1792. Hand-colored etching. Pub. H. Humphrey. Yale University, Lewis Walpole Library.

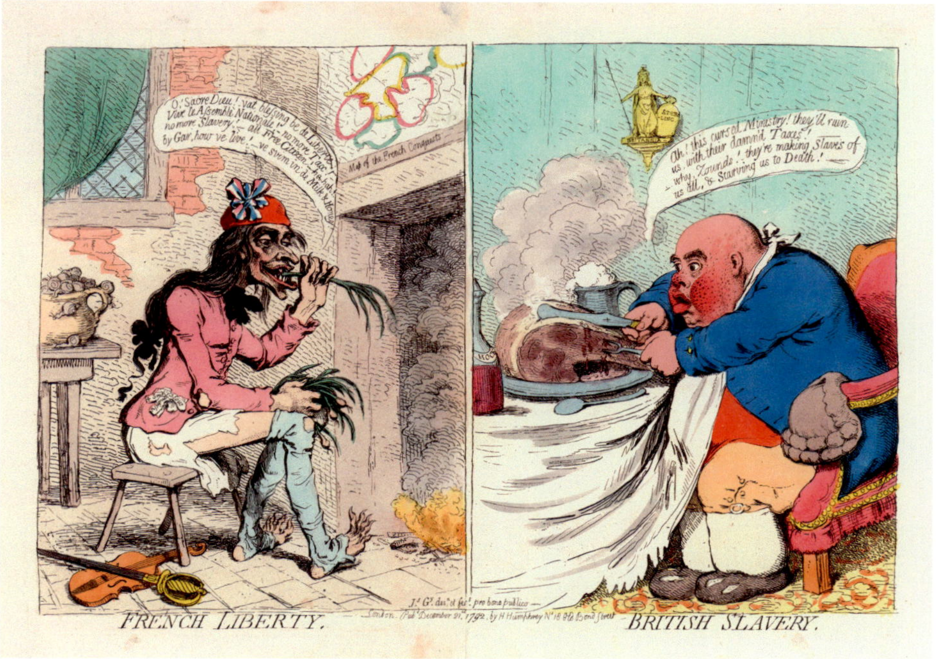

FIGURE 19
Jacques-Louis David, *English Government* [*Le gouvernement anglois*], 1793. Hand-colored etching. Pub. Bance. British Museum.

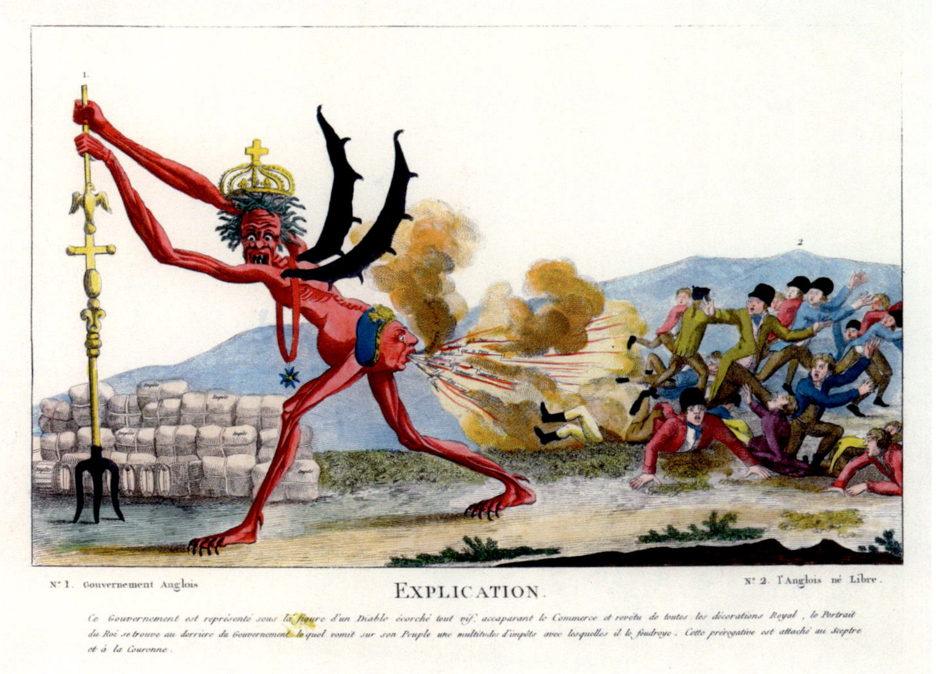

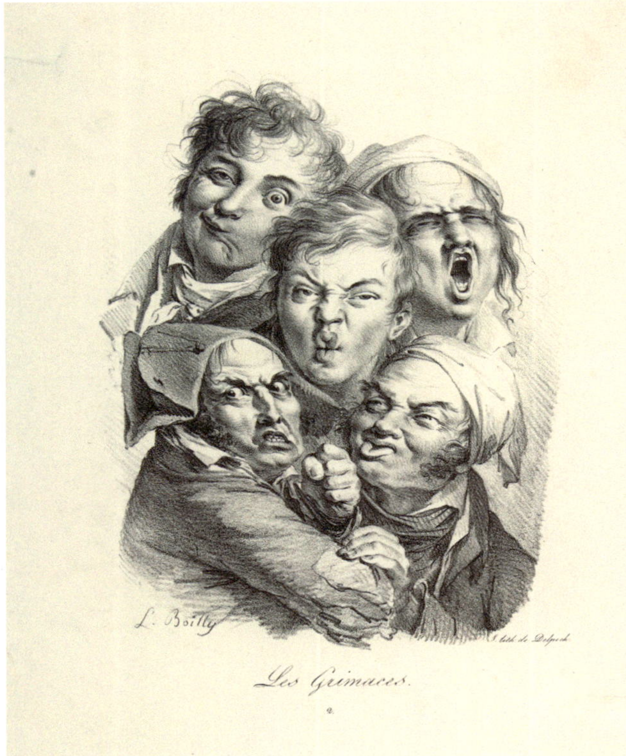

FIGURE 20
Louis-Léopold Boilly, *The Grimaces*, from *Collection of Grimaces* [*Les grimaces / Recueil de grimaces*, 1823. Lith. & Pub. Delpech. Metropolitan Museum of Art, The Elisha Whittelsey Collection, The Elisha Whittesey Fund, 1959.

This made us realize our inferiority in the kind of political caricature at which the English excelled. Their lack of taste, which caused them to fall so far behind in the arts, far from being a shortcoming, is for them a path to success. With no hesitation for fear of breaking rules, offending propriety, or insulting good judgment, they allow their wild imagination full flight and produce humorous monstrosities with unequaled fecundity."[44] Michel Melot, in his study of caricature of the revolutionary period, attributes its late start in France to censorship, to the absence of the bourgeois audience that supported it in England and—like Jouy—to the engrained French preference for classicism with the consequent distaste for the exaggerations and distortions of caricature.[45]

For the French, the determining characteristic of English caricature was this taste for the grotesque. But if "humorous monstrosities" such as David's *English Government* were not really to French taste, they were soon replaced with a gentler style of drawing that was, nonetheless, often coupled with acerbic texts. An explanation of sorts was given in 1828 by Charles-François Farcy (1792–1867), editor of *Journal des artistes et des amateurs*: "Certainly, if we are going to praise this verve, this comic force, this quality that the English call humor and that they know how to put into these kinds of sketches, we should not praise the exaggerations of form or gesture that overshoot the mark and turn a drawing that ought to be witty into a drawing that is only extravagant. The French are better draftsmen and so almost always avoid this excessiveness, and since, thank God, they are never lacking in wit, they let wit dominate the form as well as content of their caricatures."[46]

Farcy went on to explain that there were three levels of caricature. While all three employed ridicule as a weapon, the lowest form (at which, he felt, the British excelled) was based on the grotesque and utilized physical deformation. A slightly more elevated type of caricature did not employ the grotesque, but nonetheless ridiculed a humorous situation or comic event. Most elevated of all—and truly French, in his opinion—was a drawn moral or philosophical critique. Notwithstanding his attempt to impose this hierarchy on visual humor, the grotesque deformations borrowed from the English were by no means absent in early French caricature: the most popular series by Louis-Léopold Boilly (1761–1845) was his *Collection of Grimaces* (1823–1828), where he seems to take great pleasure in challenging the serenity of the classical visage by depicting its antipodes (fig. 20).[47] Aaron Martinet (1762–1841) published *The Grotesque Museum* from 1817 to 1830, a series of sixty-four hand-colored etchings by Godissart de Cari (active 1803–1829), and, by 1830, the hunchback Mayeux had already become a stock figure.[48] Despite this, for Farcy the highest level of caricature, the moral critique, was attained only by the artist known as Grandville (Jean-Ignace-Isidore Gérard, 1803–1847).[49] And indeed, Grandville's series of seventy-one hand-colored lithographs produced in 1828 and 1829, *The Metamorphoses of the Day*, is today acknowledged as the major work of caricature of the 1820s, even being—the ultimate accolade—plagiarized in England.[50] In *Misery, Hypocrisy, Avarice*, none of the physiognomies is distorted or grotesque (fig. 21). Grandville has merely portrayed each figure in the guise of the animal appropriate to its behavior: Misery is the poor dying mouse, lying under its patched bed quilt and surrounded by broken-down furniture; Hypocrisy is the clergy and municipal officials, ravens who feed on the dead and dying; Avarice is the landlord, the well-dressed "fat cat" shedding phony tears.

By the late 1820s, with Rowlandson and Gillray dead and Cruikshank shifting his interests to book illustration, England's golden age of caricature was virtually over, but France's was just beginning. France would soon lead the world in lithographic production of all types.[51]

Lithography's ability to reproduce the exact quality of an artist's line, depicting up-to-the-minute subjects in thousands of prints with no loss of quality, resulted in a widespread taste for the rapid sketch, or *croquis*. *Croquis* meant authentic, spontaneous, done from life, not labored, not artificial, and certainly not done from museum models; the word appeared as a badge of honor in the title of numerous portfolios of lithographs. This flood of lithographic imagery, cheap, plentiful, responsive to the artist's every whim, gradually developed into the visual language of contemporary life so familiar to us through later nineteenth-century painting, particularly Impressionism.

Since political caricature was severely controlled through censorship, other types of imagery predominated.[52] Images of costumes and customs were the most prevalent; in an

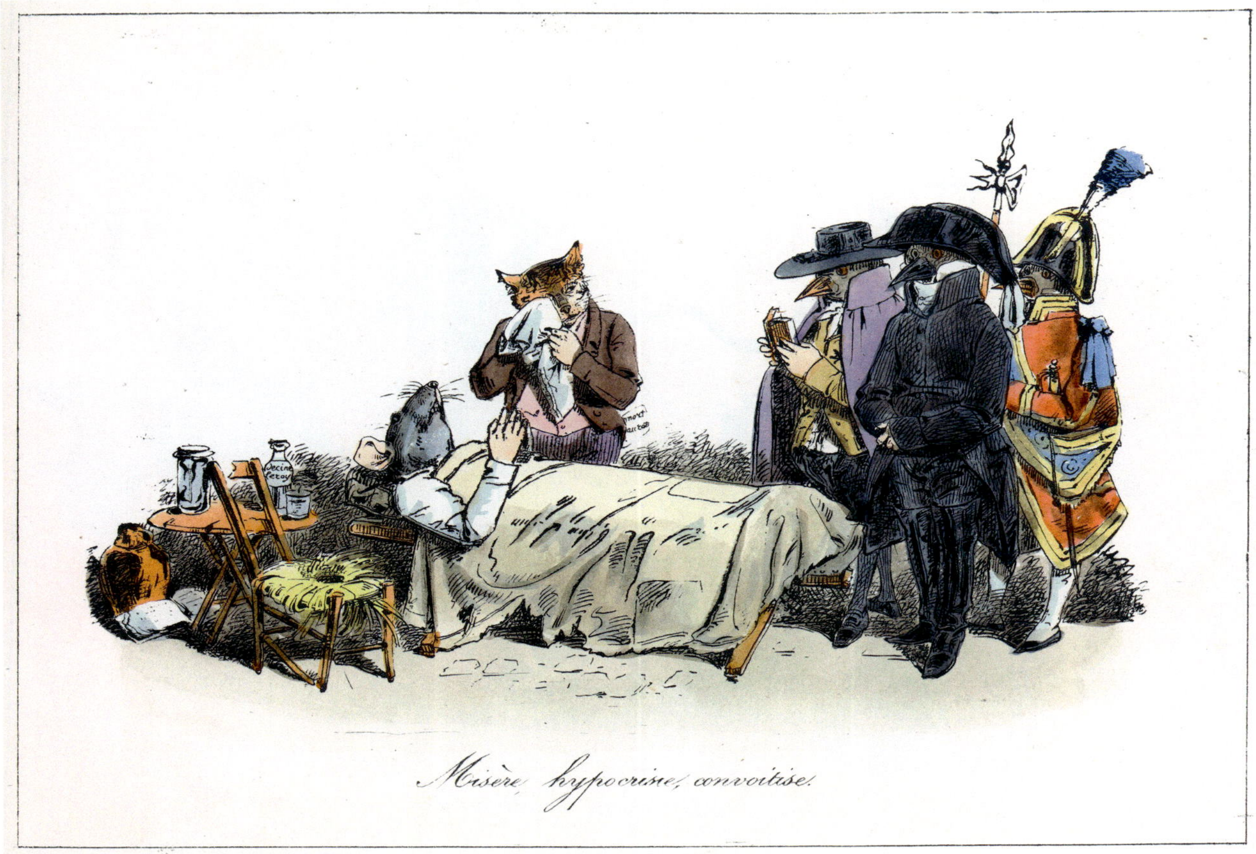

FIGURE 21

Grandville, *Misery, Hypocrisy, Avarice*, from *The Metamorphoses of the Day* [*Misère, hypocrisie, convoitise / Les métamorphoses du jour*], 1828–1829. Hand-colored lithograph. Lith. & Pub. Aubert. Metropolitan Museum of Art, Gift of Lincoln Kirstein, 1970.

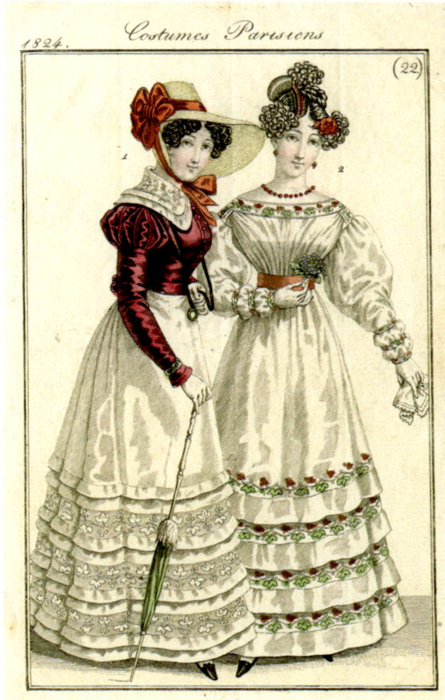

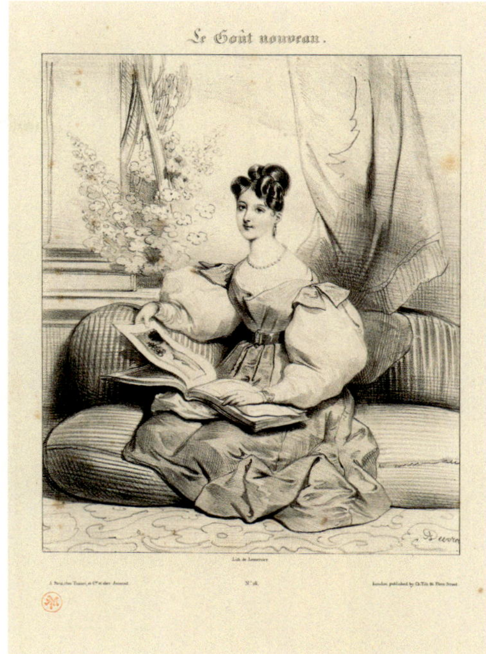

FIGURE 22

Parisian Fashions [*Costumes parisiens*], *Journal des dames et des modes*, April 20, 1824, no. 22. Hand-colored engraving.

FIGURE 23

Achille Devéria, *The New Taste* [*Le goût nouveau*], no. 18, 1831. Lith. Lemercier, Pub. Tessari and Aumont, Paris; Ch. Tilt, London. Santa Barbara Museum of Art, Gift of Albert and Dana Broccoli.

age of rigid social stratification and sumptuary laws, the two concepts of dress (costume) and behavior (custom) were closely linked; they shared the same etymological roots, and only gradually assumed differentiated meanings.[53] Fashion plates (costume) existed before the nineteenth century, of course, in the form of hand-colored etchings or engravings, but because they were expensive, they were viewed principally by the upper classes and their dressmakers.[54] As sumptuary codes and rigid class distinctions diminished, an avid interest in costume was the result, with lithographs by the thousands representing various styles of dress—civilian, theatrical, and military.[55] Early fashion plates focused almost solely on the clothing, as can be seen in the 1824 *Parisian Fashions* from *Journal des dames et des modes*, where the two young women are shown posing, modeling their outfits, but in no identifiable setting (fig. 22). Within a decade, there had been a major transformation, apparent in the 1831 album *The New Taste* by Achille Devéria (1800–1857), where elegantly clad young women are depicted in genre settings. A model lounges on a low chaise before a bouquet of flowers and a drapery-framed window; her fashionable clothing, the furnishings, even the patterned rug, all conspire to create an attractive scene of bourgeois luxury (fig. 23). Devéria has resituated the fashion plate in the world of art as well as commerce by showing his model leafing through an album very like those that were fashionable at the time, indeed very like the one in which this image appeared. Lithography encouraged artists to work across what had previously been distinct modes of illustrated print culture. The result was cross-fertilization: the fashion plate became a genre theme.

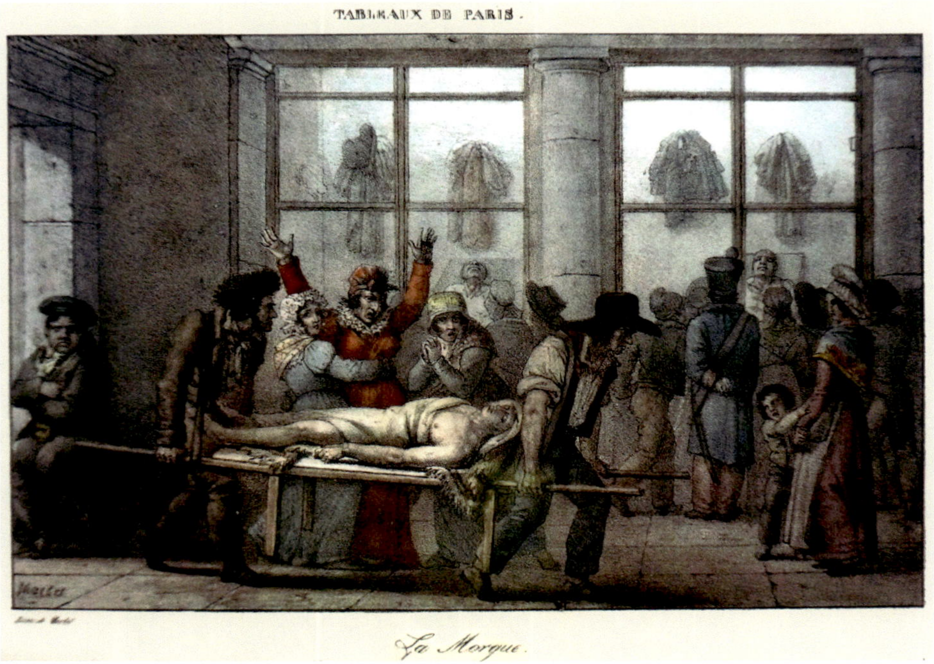

FIGURE 24
Jean-Henri Marlet, *The Morgue* [*La morgue*], from *Tableaux de Paris*, 1821–1824. Lith. Marlet.

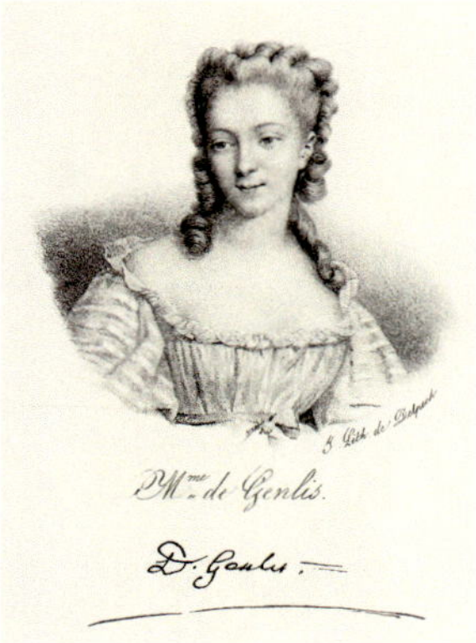

FIGURE 25
Mme de Genlis, from *Iconography of Contemporaries* [*Iconographie des contemporaines*]. Lith. Delpech. Paris: Delpech, 1823–1832, vol. 1, n. p.

Costume was by no means the only area of print culture to undergo a major transformation. The ambitious project of Jean-Henri Marlet (1771–1847), *Tableaux de Paris* (1821–1824), comprised seventy-two hand-colored lithographs representing the life of the capital across the social spectrum, and might well be seen as a more vernacular response to the renowned suite of prints showing French manners and customs in the eighteenth century (see fig. 10). Unlike Moreau le Jeune's depiction of upper-class life, Marlet's *The Morgue* depicts the arrival of a corpse in the Paris morgue, with spectators exhibiting a variety of reactions, from curiosity to horror (fig. 24).[56] Lithographed portraits became exceptionally popular: it is easy to forget that, before the illustrated press and before photography, portraits of celebrated individuals were rare, their physiognomies virtually unknown to the general public. Lithography made such portraiture cheap, widely available, and immensely popular: one of the most successful early series was Delpech's *Iconography of Contemporaries*, which carried a long subtitle announcing as its subject *Portraits of Individuals whose Names Are Linked Most Particularly Either by Their Actions or by Their Writings to the Various Events that Have Taken Place in France from 1789 to 1829*.[57] Beginning publication in 1823, by 1832 two hundred lithographed portraits of political and cultural celebrities had been issued, including the popular author *Mme de Genlis* (1746–1830) (fig. 25).

Images of cityscape and landscape, both near and far, were also staples of the print trade. Although topographical prints had been engraved earlier in small editions, they proved too costly for most individuals. The avid taste for seeing the various regions of one's country that drove the multiplication of such images in the nineteenth century can be attributed to a newfound sense of national identity, the growing ease of travel, and

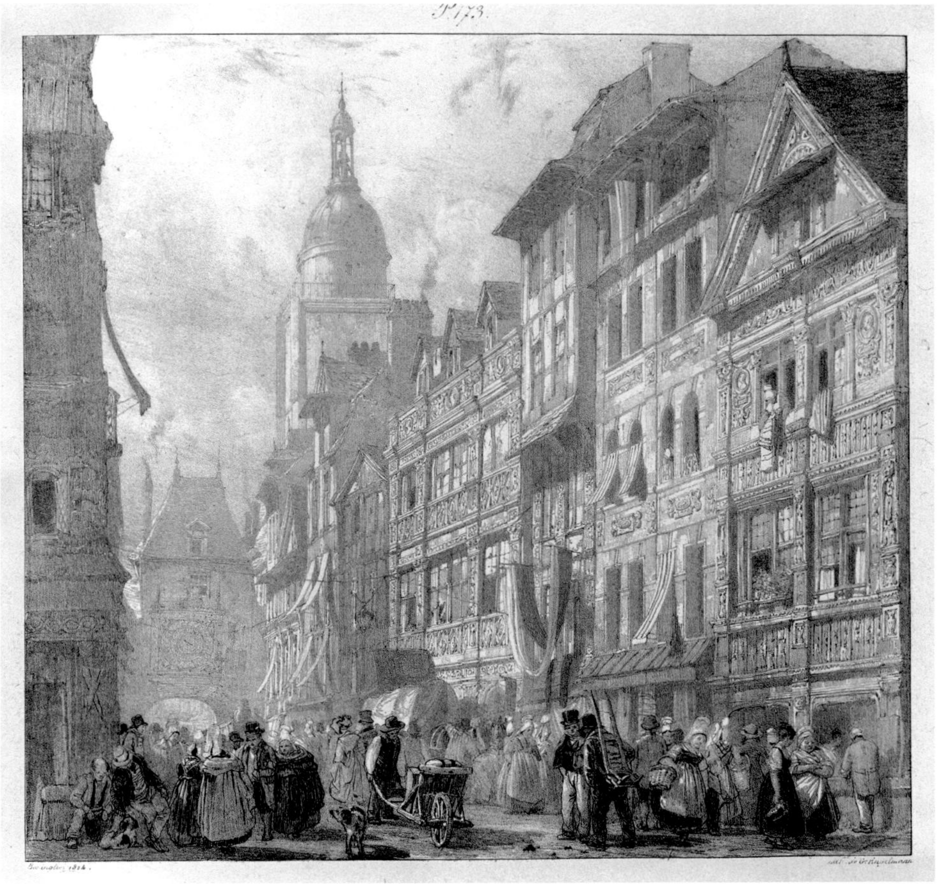

FIGURE 26
Richard Parkes Bonington, *Rue du Gros Horloge, Rouen*, from Nodier, Taylor, and Cailleux, *Picturesque and Romantic Travels through Old France, Normandy* [*Voyages pittoresques et romantiques dans l'ancienne France, Normandie*], vol. 2, 1825. Lith. Engelmann, Pub. Gide fils. Harvard Art Museums/Fogg Museum, Gray Collection of Engravings Fund G8298.

the advantages of lithography. The major achievement here is the series produced by the writer Charles Nodier (1780 –1844) in collaboration with Justin Taylor (1789–1889) and Alphonse de Cailleux (1788–1876), *Picturesque and Romantic Travels through Old France*; issued in twenty-three grand folio volumes from 1820 to 1878, each volume or set of volumes was devoted to a specific region.[58] With explanatory texts alongside the work of leading graphic artists of the period, including Richard Parkes Bonington (1803–1828), Jean-Baptiste Isabey, Théodore Géricault, Thomas Shotter Boys (1803–1874), and A.-E. Fragonard (1780–1850), the series both catalyzed and crystallized newfound feelings of nationalism, encouraging citizens to think of themselves as French, not merely Parisian or Breton or Norman. Richard Parkes Bonington's *Rue du Gros Horloge, Rouen* appeared in the first set of volumes, devoted to Normandy, and is typical in its atmospheric rendering of a locale replete with historic significance (fig. 26).[59] Similar but less ambitious compilations treated every continent of the world during the nineteenth century, although it was most often Europeans who created and consumed these images. They should be understood both as indications of colonialism and, at the same time, as the beginnings of global consciousness.[60]

FIGURE 27
James Gillray, *Very Slippy Weather*, 1808. Hand-colored etching. Pub. H. Humphrey. Yale University, Lewis Walpole Library.

FIGURE 28
Pierre Nolasque Bergeret, *The Idlers of the Rue du Coq* [*Les musards de la rue du coq*], 1805. Hand-colored lithograph. Pub. Imprimerie Lithographique, rue St. Sébastien no. 24. Metropolitan Museum of Art Mary Martin Fund, 1984.

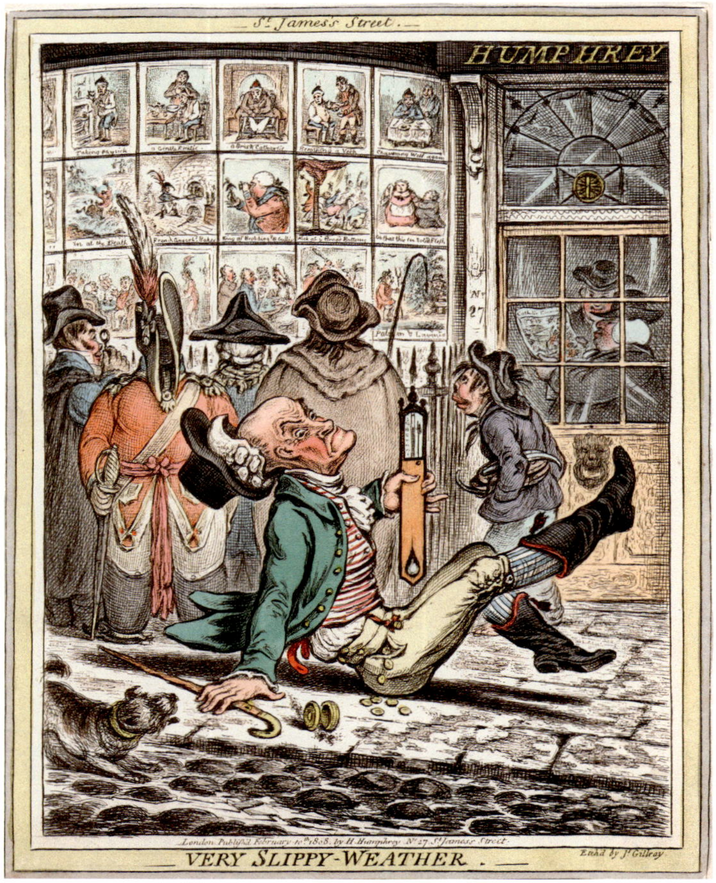

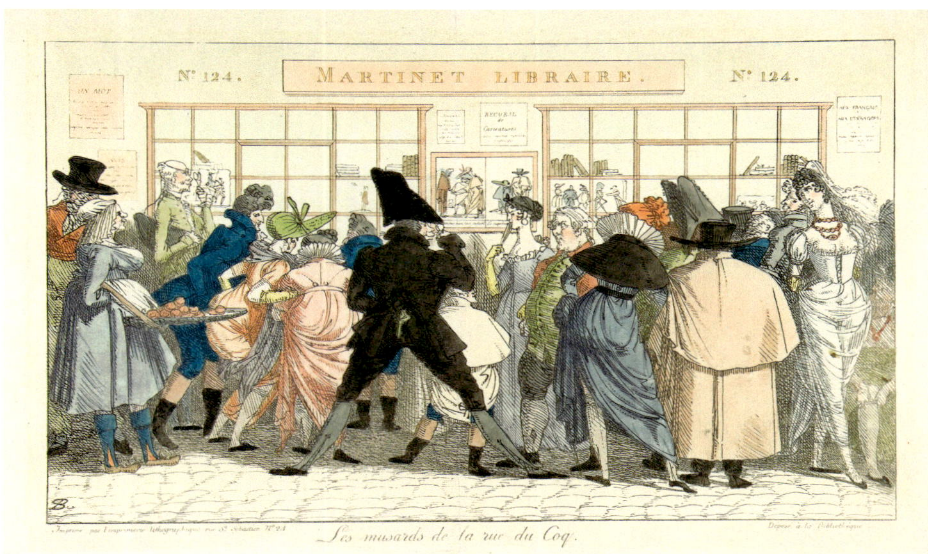

These categories—costume, custom, portraiture, and topography—encompassed the vast majority of lithographic prints, but because such subjects seem inconsequential from the vantage point of politics, they have received much less attention than the category of political caricature, which was always the smallest grouping quantitatively. We might say that political caricature, with its distortions, its harsh and cruel emphasis on public life, is coded male, while social imagery, whose observations of social customs and private lives were conveyed in a more naturalistic style, has been coded female. In terms of scholarly interest, the result has been that political caricature has been considered a major genre, while social imagery has been considered minor and thus has received relatively little attention.[61] This difference in evaluation was not shared by contemporaries, however, although they were well aware of the political necessity behind the focus on social rather than political subjects. In 1814 Étienne de Jouy noted: "The burin is no freer than the pen; the Argus of censure, a magnifying glass in each of its eyes, watches over prints with as much attention as books. Our draftsmen are limited to sketching our manners: in this category, the collections of *The Incredible Ones, Marvellous Women, The Elite of Fashionable Society* are sought after as monuments to our customs, so much the more precious since renowned artists have not disdained to imprint them with the seal of their talent."[62] And so, before dismissing the images of social life as less important than political caricature, let us remember *The Human Comedy* by Honoré de Balzac (1799–1850), which he began writing in the 1820s, and which shared with contemporaneous lithographic imagery an interest in the full spectrum of social mores of his period. He wrote in its preface: "By compiling an inventory of vices and virtues, by assembling the principal elements of the passions, by depicting characters, by choosing the main events of social life, by creating types through a distillation of similar characteristics, I might perhaps succeed in writing that history forgotten by so many historians, that of manners and morals."[63] I propose that if we consider the lithographic imagery of these decades in this same light–Charlet's *Nymph of the Thames* (fig. 13) is a good example here—we will have a more accurate idea of its importance, for it preceded even literature in its depiction of all aspects of the rapidly changing nineteenth century world.

THE BUSINESS OF LITHOGRAPHY

In eighteenth-century England, print shops maintained salesrooms fronting onto the street with windows where their latest wares were displayed. James Gillray's *Very Slippy Weather* shows one of the most successful of these, Hannah Humphrey's London establishment (fig. 27). Aaron Martinet, who published the collections Jouy praised so highly, is usually credited with being the first to institute this custom in Paris. His shop is depicted in the famous 1805 lithograph by Pierre Nolasque Bergeret (1782–1863), *The Idlers of the Rue du Coq* (fig. 28).[64] Martinet came from a family of engravers and, typical of the time, was a printer, publisher, and book dealer who also maintained a *cabinet de lecture*, a public

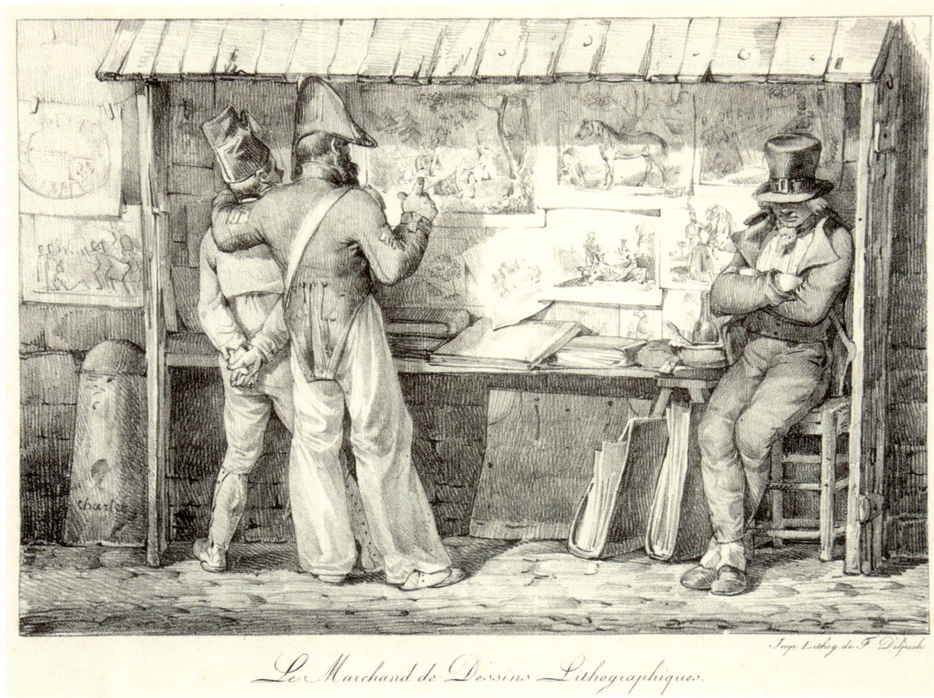

FIGURE 29
Nicolas-Toussaint Charlet, *The Seller of Lithographic Drawings* [*Le marchand de dessins lithographiques*], 1818. Lith. & Pub. Delpech. Yale University Art Gallery, Stephen Carlton Clark, B.A. 1903, Fund.

reading room where, for a small fee, the public could read or borrow books and journals. From 1797 he published etchings and engravings but soon shifted to lithography.[65] Étienne de Jouy described Martinet's shop thus: "But all this luxury of modern shops scarcely merits a second glance from the public, while it crowds in around the modest shop window of the bookstore on rue du Coq. This boutique has its regulars who have never set foot inside. They are happy to scrutinize through the windows all the lovely things offered for their amusement, to survey the new caricatures, theatrical representations, portraits of actors and musicians, uniforms of French and foreign soldiers, the latest fashions and furnishings, and we could name someone of excellent taste who, in his opinion passes an hour more agreeably in front of Martinet's boutique than at a performance of one of Molière's masterpieces."[66] Jouy's reference to Molière is especially telling, since French theater was also polarized at the time between the high road of classical theater epitomized by Racine and the brilliant but often vulgar comedies of Molière, which, although they might be enjoyable, were nonetheless ranked by literati as a lower form of literature, as popular entertainment.

As lithography developed, increasing numbers of printers opened shops in the area around the Louvre. In 1814, *Journal des arts, des sciences et de la littérature* noted, "Martinet isn't the only one attracting idlers. Everywhere in the streets large groups gather around *exhibitions of caricatures*."[67] The following year Engelmann and Charles de Lasteyrie set up lithographic printing houses in Paris and commenced to train the entire next generation of French lithographers. François-Séraphin Delpech, whose shop is seen in figure 5, was a student of Lasteyrie. He became a direct rival to Martinet in the production of these popular images, and was so successful that upon his death his widow Marie-Marguerite-Brigitte Naudet Delpech (1781–1851) continued the firm, becoming one of the few licensed female lithographers in France.[68] Only gradually did these three professions of printer, publisher, and print seller become separate enterprises. Even the Maison Aubert, which, established in 1829, would become the major publisher of caricature in nineteenth-century France, combined in its headquarters in the Galerie Véro-Dodat all three functions of printing shop, publishing house, and salesroom.[69]

In the production of popular lithographic imagery during this period, single-sheet prints were cheapest and most prevalent, sold in outdoor stalls as well as in more upscale print shops. The type of open-air market depicted by Charlet in figure 29, *The Seller of Lithographic Drawings*, persists to this day in Paris where along the Seine, as in Charlet's day, prints are pinned up, piled up, or stored in portfolios. In Charlet's day, the same print could be purchased either in black and white or in a more expensive hand-colored version—true color printing would have to wait until the 1860s before it would become commercially viable.[70] In drawing style these prints could be extremely caricatured, meaning that some features were cruelly exaggerated, even grotesque, or the drawing could be simply illustrational, naturalistic, or only slightly overstated. As Jouy's description of Martinet's window attests, these images represented contemporary life in modes that were either reportorial or satiric in intent. While these prints, which represent the majority of lithographic images produced during the period, were perhaps less cruel than British productions in their physiognomic attack on individuals, they were often more cruel in their depictions of social customs and behavior. With greater or lesser editorial comment, greater or lesser emphasis on naturalistic reportage, all, nonetheless, fell under the contemporaneous rubric of caricature.

Besides single-sheet prints, there were prints in series on related themes; these were published either individually over a long period of time or all at once and sold as a portfolio. Production in series was a traditional marketing technique that, in France, went back centuries to works such as Jacques Callot's *Miseries of War* (1633). In lithography, such portfolios were commissioned by the publisher, often from several different artists under a general title with their subjects only tangentially related.[71] The dominant personality was that of the publisher, not the artist, to the extent that early series often lack completely any mention of the creator of the drawing. One particularly long-lived series, Martinet's *Little Theater Gallery*, began in 1796 as hand-colored etchings of theatrical characters in costume, and then continued in lithography until 1880, by which time almost three thousand had been published.[72] Collectors could choose among the prints of a series to make up their own portfolio and could even have their prints bound into an album should they so desire. A clever innovation was that lithographic series were usually numbered directly on the plate; this decision created an added incentive to purchase all the works in the series,

FIGURE 30
Departure of Mr. Calicot for Combat in the Mountains [Le départ de Calicot pour le combat des montagnes], 1817. Lith. Cte de Lasteyrie.

which otherwise would seem incomplete. Publishers issued such prints a few at a time, often producing a frontispiece to the series only if and when it sold well. Publication of the series *Infatuations*, for example, was announced on October 27, 1827 in the *Bibliographie de la France*, the official government record of publication; after eighteen lithographs were issued, the frontispiece was announced on December 5, with the title *Eighteen Scenes of Infatuation*. It must have been extremely popular because a new series of eighteen was announced on February 16, 1828, after which the series continued even longer under the title *Memoirs of Infatuation*.[73]

A particularly fruitful innovation facilitated by lithography—and one essential for the development of the comic book, which will be discussed in a subsequent chapter—was the series based on a recurrent character, a readily identifiable individual representing a social type who rapidly became a recognized part of social discourse. In England, Thomas Rowlandson had already invented the character of Doctor Syntax, who made his first appearance in 1809 in "The Schoolmaster's Tour," published in Rudolph Ackermann's *Poetical Magazine*. The work was republished in book form three years later as *The Tour of Doctor Syntax, in Search of the Picturesque*. Since Doctor Syntax appeared in a coherent narrative sequence of drawings, he will be included later, in my discussion of the comic book. Here I am concerned with the invention of characters whose escapades never coalesced into a coherent narrative, but instead were presented in single-sheet format, continuing over months and years. The first such figure in French popular imagery was Mr. Calicot, who made his debut in a theatrical production of 1817.[74] Calicot (Calico) was a salesclerk in a fabric shop and represented a satire on the pretensions of all those who, once danger had passed, adopted the personae of veterans of Napoleon's military campaigns. In one image we see him sallying forth, decked out in a military uniform made of pin cushions for epaulets, a feather plume for a sword, and a draper's hook for a lance (fig. 30). The title has him departing for the *combat des montagnes*, literally combat in the mountains, but also the name of an amusement park ride much like the modern roller coaster, a pun impossible to translate into English. By attacking those who merely pretended to be Bonapartist veterans, caricaturists managed at the same time to support their exiled hero and to criticize the current social order. Over sixty Calicot prints have been catalogued, the earlier ones etched, the later ones lithographed, drawn by a variety of artists, and all attesting to the birth—and success—of this new type of theme in France.

The hunchback Mayeux (or Mahieu) was invented in the 1820s, but he was made famous by the artist Charles-Joseph Traviès de Villers (1804–1859), known simply as Traviès, in dozens of humorous drawings. The concept of finding humor in the juxtaposition of the unfortunate Mayeux with situations that underscore his physical limitations is unacceptable today, but at that time it also functioned on a symbolic level: standing for those whose self-satisfaction blinds them to the reality of their situation, as well as those who are able to rise above their limitations, Mayeux represented Everyman. In one example, Mayeux (here Mahieu) does guard duty, carrying a bayonet almost twice

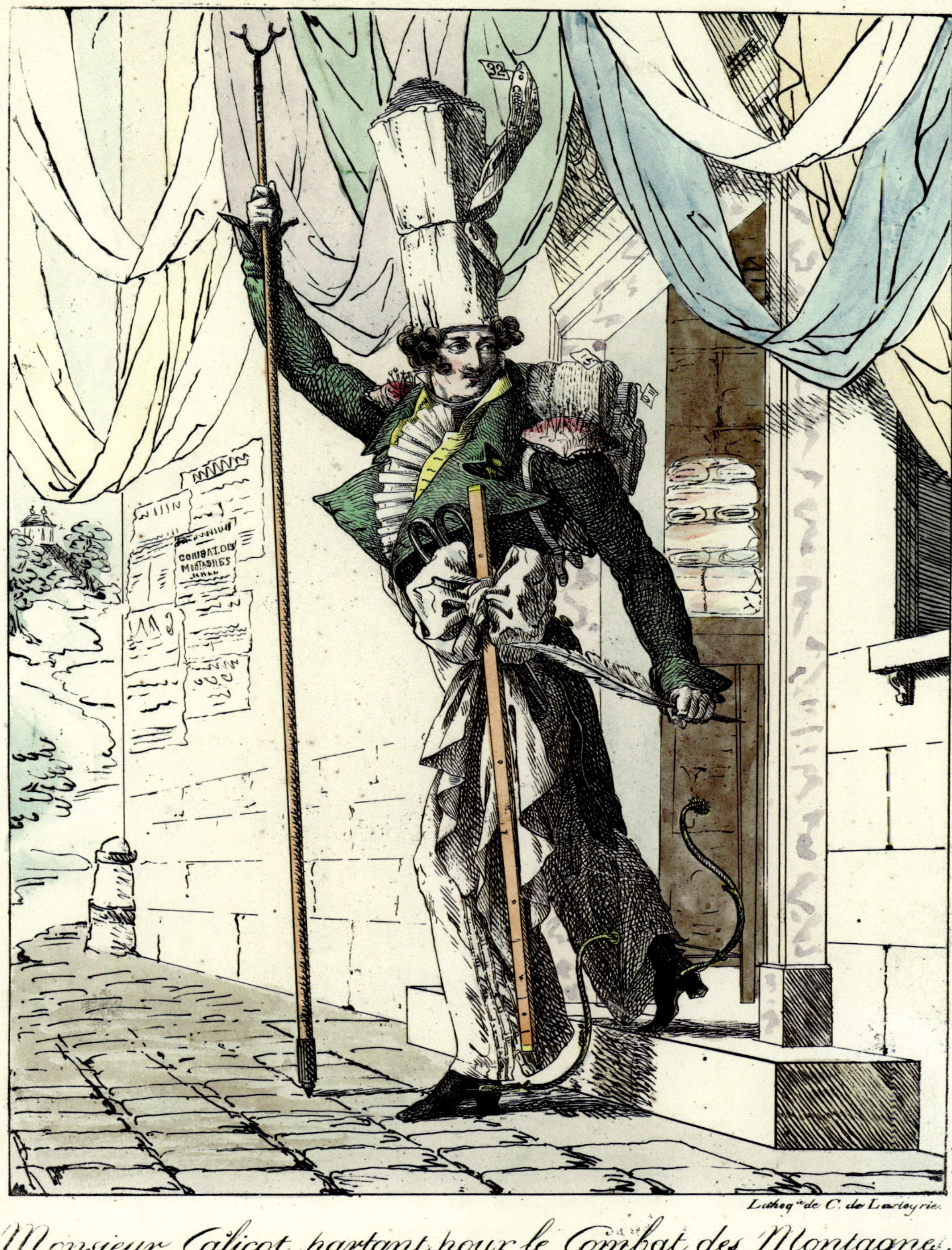

Lithog.ie de C. de Lasteyrie.

Monsieur Calicot partant pour le Combat des Montagnes.

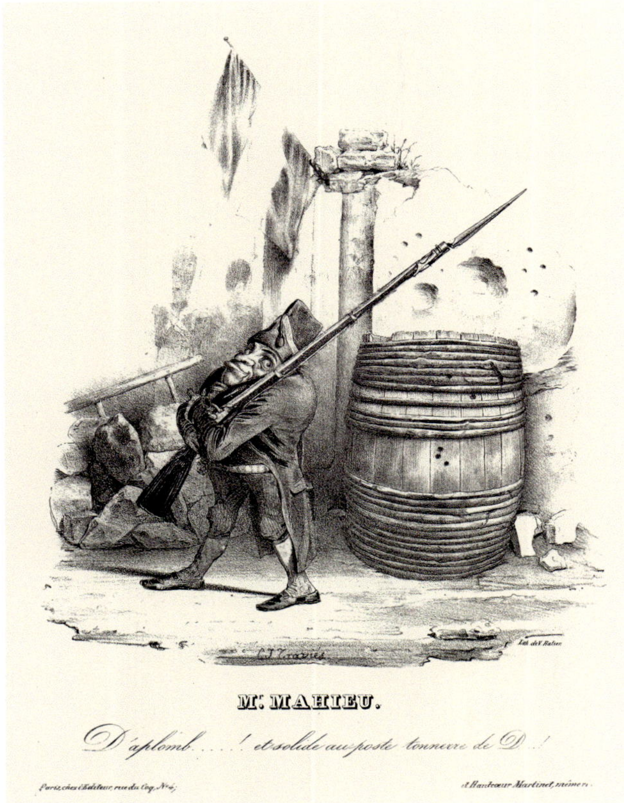

FIGURE 31
Charles-Joseph Traviès, *Mr. Mahieu. Steady . . . ! and Faithful at His Post, by God . . . !* [*M. Mahieu. D'aplomb . . . ! et solide au poste tonnerre de D . . . !*], 1830. Lith. Ratier, Pub. Hautecoeur-Martinet.

OPPOSITE **FIGURE 32**
Honoré Daumier, *Gentlemen and Ladies!* [*Messieurs et dames!*], from *Caricaturana*, no. 7. Hand-colored lithograph. Lith. & Pub. Aubert. *Le Charivari*, July 10, 1836. Metropolitan Museum of Art, Gift of Edwin De Turck Bechtel, 1953.

his height and announcing that he is *Steady . . . ! and Faithful at His Post, by God . . . !* (fig. 31). The drawing was published during the summer of 1830, after the July revolution that toppled the legitimist monarchy. It proposes, visually, that the citizen's army that overthrew the king was truly a popular movement, composed of individuals from all ages, classes, and conditions.[75] From the 1830s on, such characters became familiar. The hustler Robert Macaire was popularized by Honoré Daumier (1803–1879), who created him in collaboration with Charles Philipon who wrote the captions; there were eventually 101 drawings in the series.[76] *Gentlemen and Ladies!* shows Macaire waving a fistful of no-doubt bogus stock certificates in front of what he hopes will be a gullible audience (fig. 32). While his accomplice beats a drum, Macaire intones *Silver mines, gold mines, diamond mines are like watery soup, nothing in comparison . . .*[77] Calicot and Robert Macaire were already familiar through literature and legend when they were first adopted by artists, but there were also inventions that had no earlier precedents. Henry Monnier, a writer as well as artist, invented Mr. Prudhomme, the ultimate bourgeois, featuring him in plays, stories, and drawings. He even assumed the persona, presenting himself in that role on stage, in art, and in real life.[78] In Monnier's *Self-Portrait as Mr. Prudhomme*, his stance exaggerates his big belly, the consummate mark of the self-satisfied bourgeois (fig. 33).

Caricaturana N°7.

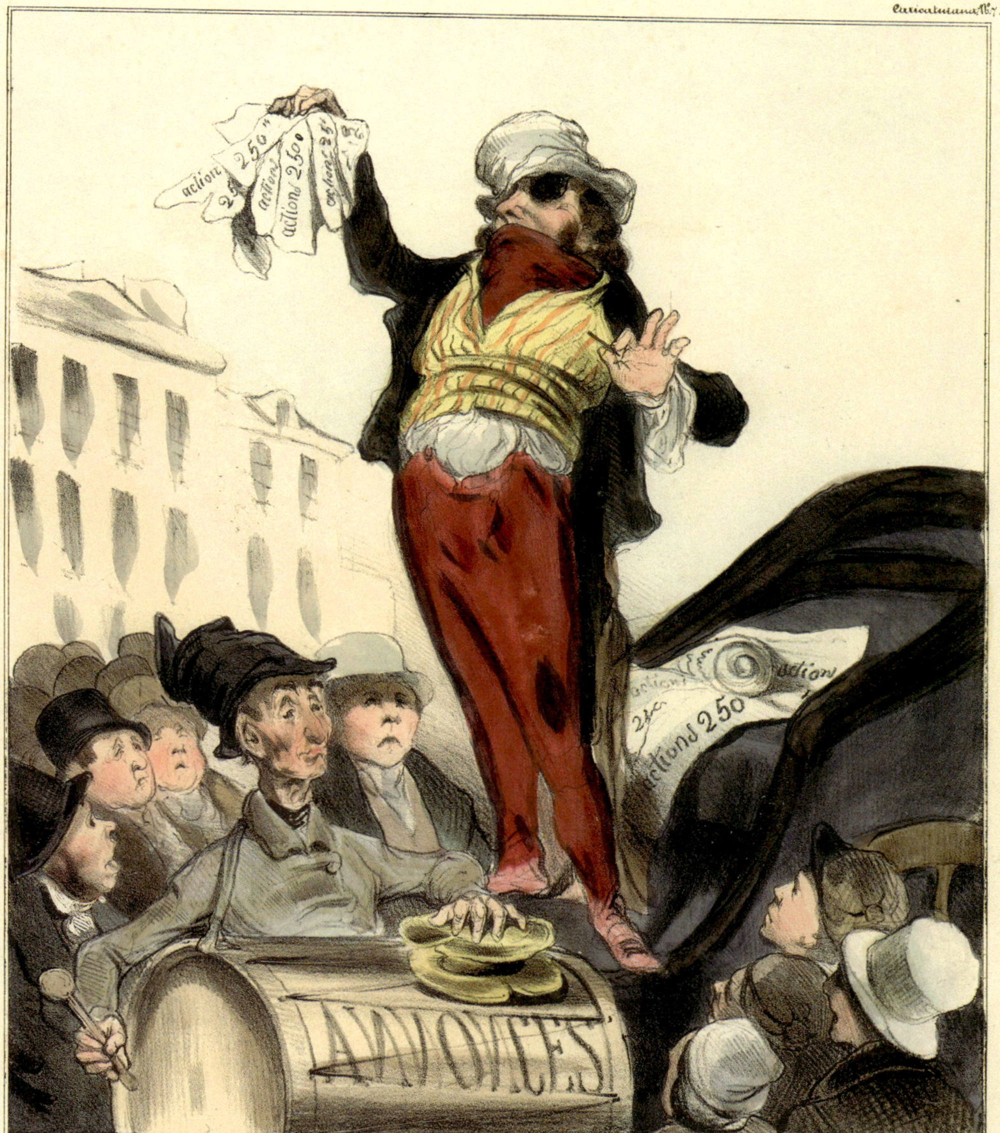

Ch. Ph. inv. H. D. lith. Chez Aubert, gal. véro-dodat. Imp. d'Aubert et de Junca, gal. Colbert.

Messieurs et Dames!

Les mines d'argent, les mines d'or, les mines de diamant ne sont que de la roupie Bouille de la ratatouille en comparaison de ma bouille. Mais (que vous m'allez dire) tu vends alors les actions un million? Mes actions, Messieurs, je ne les vends pas je les donne pour 200 miserables francs j'en donne deux pour une, je donne une aiguille, un cure-oreille un passe lacet et je vous donne encore ma benediction par dessus le marché. En avant la grosse caisse!

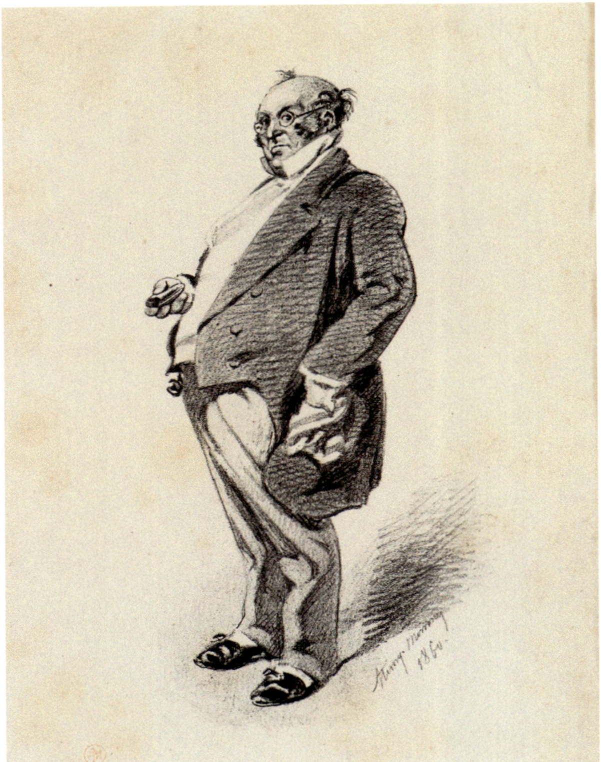

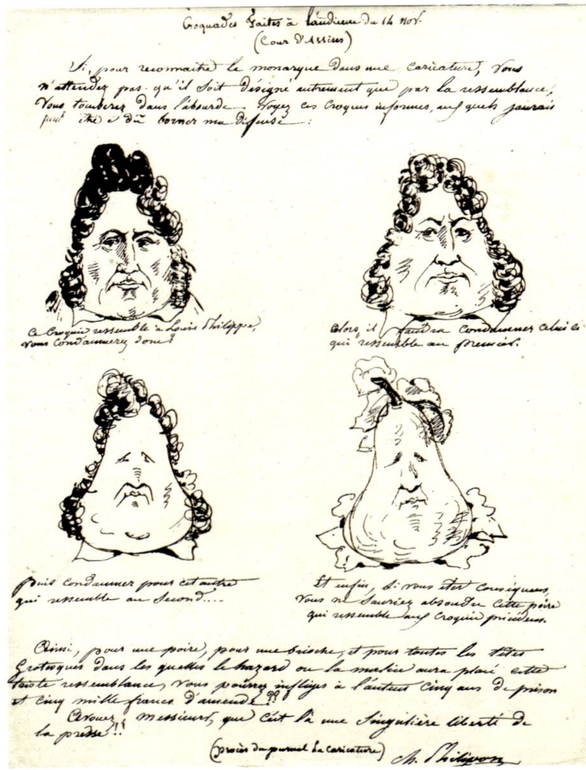

FIGURE 33
Henry Monnier, *Self-Portrait as Mr. Prudhomme*, 1860. Lithograph. Santa Barbara Museum of Art, Gift of Albert and Dana Broccoli.

FIGURE 34
Charles Philipon, *Sketches Made during the Session of November 14 (Criminal Court)* [*Croquades faites à l'audience du 14 novembre (Cour d'assises)*]. Lith. & Pub. Aubert. *La Caricature* 1, no. 56 supplement, November 24, 1831.

OPPOSITE **FIGURE 35**
Charles-Joseph Traviès, *The Pear Has Become Popular!* [*La poire est devenue populaire!*]. Hand-colored lithograph. Lith. Bénard, Pub. Aubert. *Le Charivari*, April 28, 1833.

The best-known recurrent character of the century, however, was not a character at all but a fruit—Charles Philipon's pear, a "portrait" of King Louis-Philippe (1773–1850). Philipon conceived of it during his trial in November 1831 for having published the caricature *Dupinade* that showed the king as a plasterer covering up the slogans and promises that had brought him to power the previous year.[79] Philipon's defense was that his cartoon depicted merely a plasterer and that it wasn't his fault if critics could find the king's likeness in anything. To prove his point to the court, he drew the king's visage, and with a few subtractions, transformed him into a pear. While he was, nonetheless, condemned to six months in prison and a fine, in revenge he published his drawing in *La Caricature* (fig. 34).[80] The pear soon began to appear everywhere as a surrogate for the forbidden image of the king. All the major cartoonists drew pears in order to comment on current political issues with impunity. In *The Pear Has Become Popular!* Traviès shows the result: by 1833 even street urchins were spreading the message that the king is a pear, slang for a gullible fool (fig. 35). And in case anyone missed the point, Traviès has drawn a second pear with the comment *voleur* (thief) inscribed beneath it, as well as the word *cochon* (pig) written on the wall next to the boy's work in progress.[81] By utilizing a full panoply of tools in word and image alike, these artists conveyed subtle (and not so subtle) messages to a growing audience that—as Traviès's drawing demonstrates—was not limited to those who

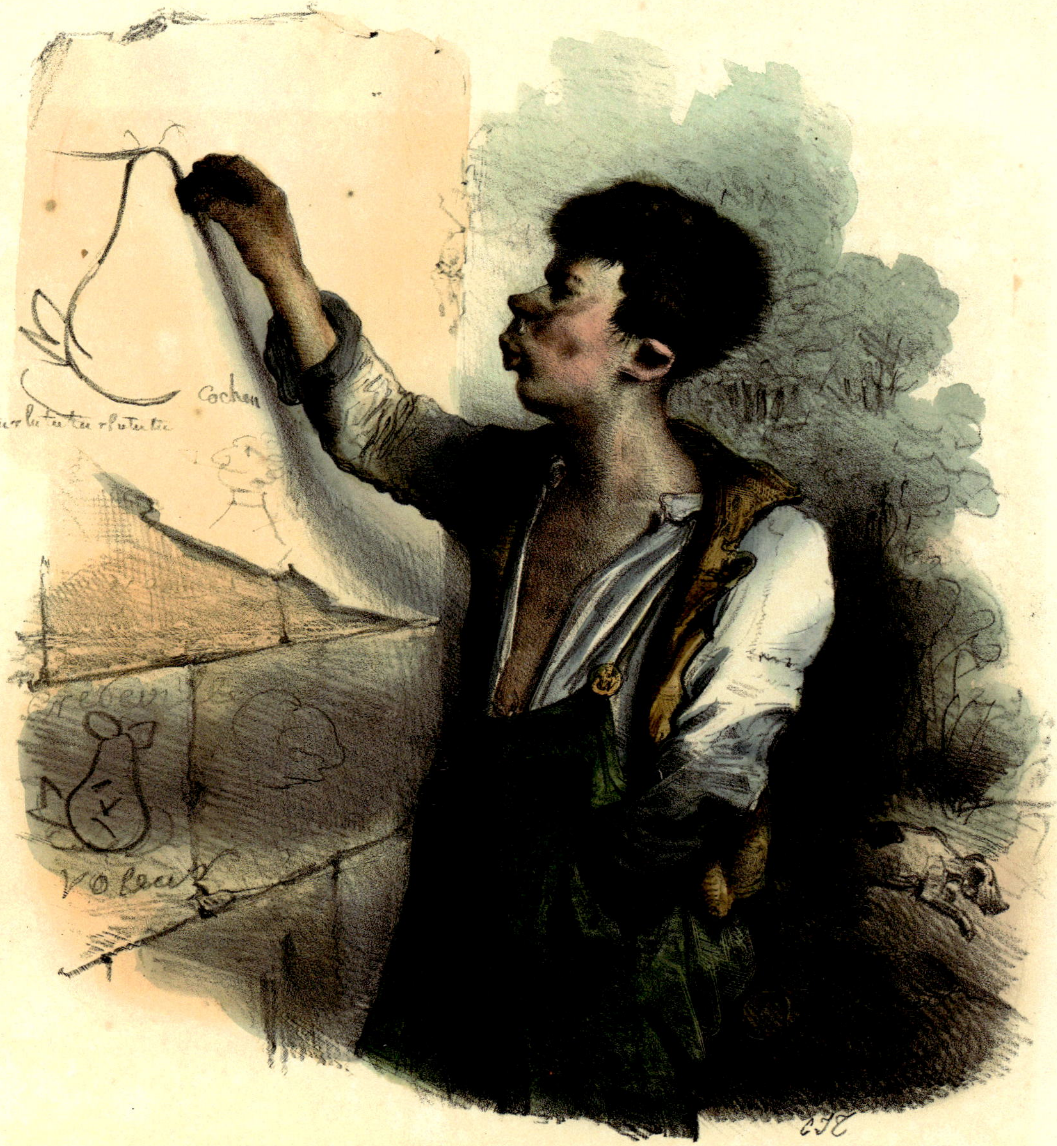

LA POIRE EST DEVENUE POPULAIRE!

Le voyoux employé aux trognons de pommes dans les théâtres des boulevards, la croque sur les murailles pendant ses nombreux loisirs, c'est ainsi que Paris s'embellit tous les jours.

L. de Benard, rue de l'Abbaye N.º 4.

chez Aubert, galerie véro dodat.

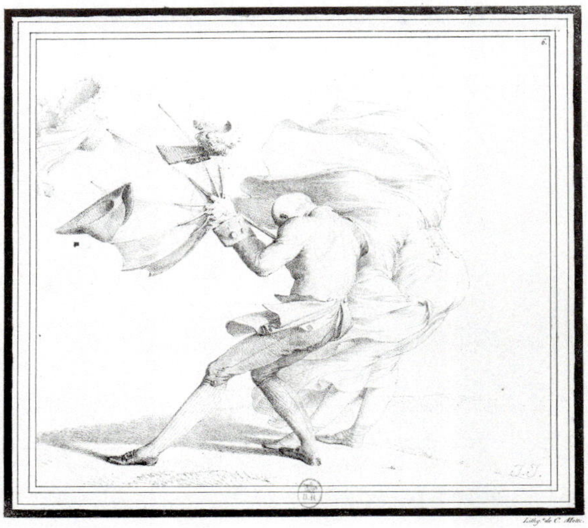

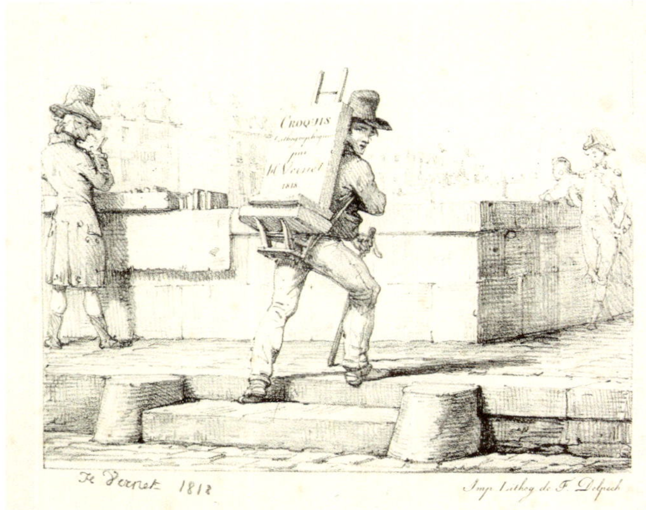

FIGURE 36
Jean-Baptiste Isabey, *The Gust of Wind* [*Le coup de vent*], from *Caricatures de J. J. à Paris*, no. 6, 1818. Lith. C. Motte, Pub. Alphonse Giroux. Bibliothèque nationale de France.

FIGURE 37
Horace Vernet, frontispiece to *Lithographic Sketches of H. Vernet* [*Croquis lithographiques de H. Vernet*], 1818. Lith. & Pub. Delpech. Princeton University Art Museum, Bequest of Henry K. Dick, Class of 1909.

could purchase a print, or even to those who could read, but increasingly encompassed all social classes.

While the earliest portfolios of lithographs were in the nature of sample books or collections of heterogeneous prints by various artists assembled by publishers to show off the possibilities of the medium, soon individual artists began producing entire portfolios themselves. One of the earliest was Jean-Baptiste Isabey, whose *Caricatures of J. J. in Paris*, published in 1818, was a collection of untitled vignettes of society types. *The Gust of Wind*, for example, shows an elegant couple struggling against a strong wind that has blown their parasol inside-out (fig. 36). The collection proved so popular that it was reissued in London in a hand-colored edition of stipple engravings in 1819, this one retitled *A Turn Up*.[82] Isabey, a respected painter though not an academician, clearly did not want to be publicly identified with caricature. The same year he published a portfolio of "serious" lithographic drawings that displayed his full name; it is clearly visible among the albums on the table in Jacob's *Genius of Drawing* (see fig. 1).[83] A parallel could be drawn to the literary history of the novel, which, despite its success with the public, was not considered a respectable art form; early novelists, too, often concealed their identity. The Vernet family was more forthcoming. Carle Vernet openly produced a twelve-part portfolio of ninety-six equestrian images in 1817–1818, and Horace Vernet published numerous portfolios beginning with his *Lithographic Sketches* of 1818, for which he provided a handsome frontispiece on which his name is prominently displayed (fig. 37).[84] Both exhibited lithographs in the first Salons to accept them, although their works were shown under the name of Engelmann, their lithographer and publisher. Many of their prints exploited the major market for the Napoleonic imagery that, masquerading as nostalgia, actually provided a forum for political dissent during the repressive years of the Restoration. Vernet's 1823 *Rotten Weather* could refer either to meteorological or to political conditions (fig. 38).

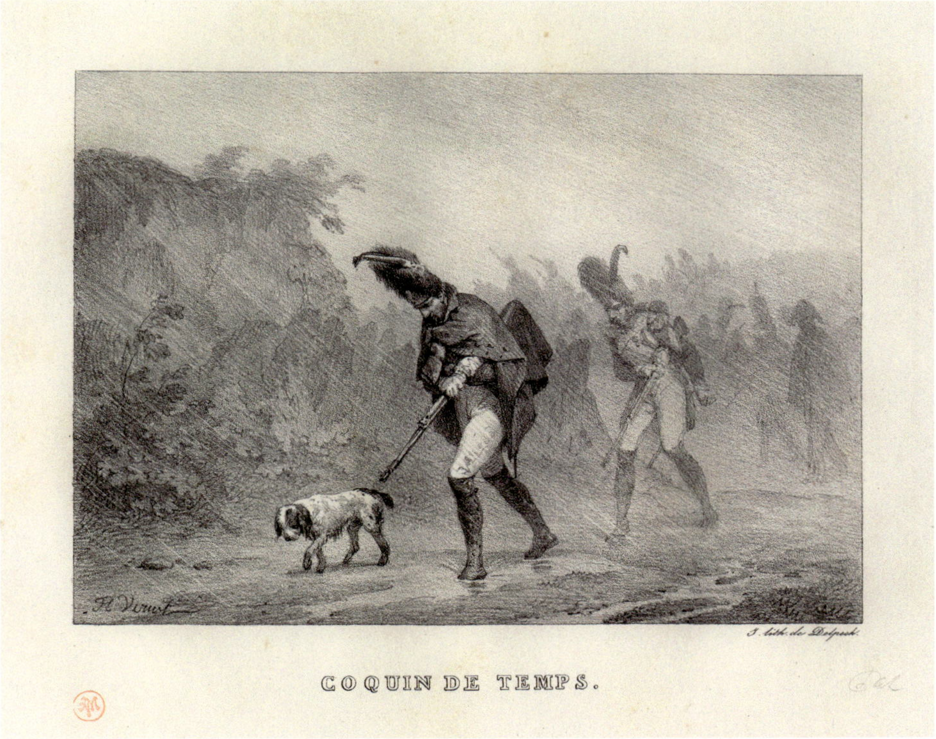

FIGURE 38

Horace Vernet, *Rotten Weather* [*Coquin de temps*], 1823. Lith. & Pub. Delpech. Brown University Digital Repository, Anne S. K. Brown Military Collection.

Along with Charlet, and, later, Raffet, the Vernets became identified with this political stance.[85] Géricault, after his *English Suite*, went on to produce additional portfolios and individual prints, always featuring his name prominently displayed.[86]

One of the earliest artists to establish a personal market for imagery of modern urban life was Henry Monnier.[87] His first collection, published in 1824, consisted of forty lithographs with cover sheet and frontispiece, its captions printed in both French and English, with the lengthy title *Parisian Encounters. Picturesque Medley. Sketched from Life, among the Pleasures, Fashions, Activities, Labors, Leisure, Faults, Vices, Miseries, Luxuries and Extravagances of the Inhabitants of the Capital among All Stations of Life and All Classes of Society*.[88] The title itself demonstrates why Balzac was so taken by Monnier that he wanted to adopt his character of Mr. Prudhomme for his own work.[89] Monnier published at least one series of humorous drawings each year: *Contrasts* (1824), *Fashions and Absurdities* (1825), *Recreations* (1826), *Parisian Sketches* (1827), *Administrative Manners* (1828), to cite just a few.[90] His masterpiece was the 1828 *Administrative Manners* which, by chronicling the daily routine of a Paris office, introduced a new modern subject into French art. Each of the eighteen lithographs in the series, available either in color or in black and white, shows typical office characters and activity; at two o'clock, for example, the office remains deserted because no one has yet returned from lunch. In *Messrs. the Director, Chiefs, Assistant-chiefs, Clerks, Supernumeraries, etc., etc., etc., Going to Compliment*

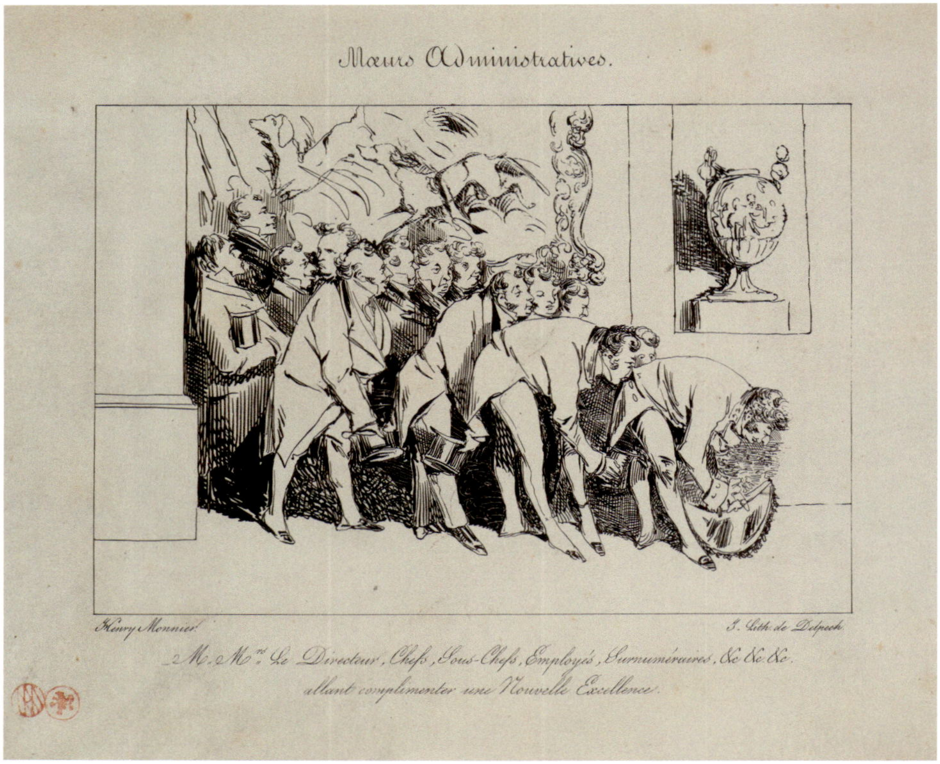

FIGURE 39

Henry Monnier, *Messrs the Director, Chiefs, Assistant-chiefs, Clerks, Supernumeraries, etc., etc., etc., Going to Compliment a New Excellency*, from *Administrative Manners* [*M. M. le directeur, chefs, sous-chefs, employés, surnuméraires, etc., etc., etc., allant complimenter une nouvelle excellence / Moeurs administratives*], 1828. Lith. & Pub. Delpech. Santa Barbara Museum of Art, Gift of Albert and Dana Broccoli.

a New Excellency, office personnel are virtually falling over each other to demonstrate their loyalty to a new boss (fig. 39). Whereas the themes of Isabey's drawings were virtually timeless, Monnier focused specifically on institutions characteristic of modern life.

England had seen the earliest industrial revolution and by the nineteenth century seemed to be the home of all modern technology; in addition, the absence of a powerful central arts administration made it the Shangri-La that beckoned young French artists whose ambitions diverged from the strictures of academic art. Monnier fully exploited the transchannel trade in prints, making numerous voyages to England and publishing prints with captions in both languages. In 1829 he even produced, with Eugène Lami (1800–1890), the handsome hand-colored album *Travels in England* in collaboration with the major English publishers Colnaghi and Charles Tilt.[91] Other French artists—Géricault, Charlet, Philipon, Isabey, even Delacroix—all sought inspiration and opportunities across the Channel, and, of course, English caricature, no matter how strongly criticized as vulgar and grotesque, long remained the gold standard of the genre. As one periodical admitted in 1833, "The genre of grotesques, transmitted to us by England, along with constitutional government and roast beef and potatoes, is strongly admired by our artists."[92]

Portfolios of related prints became albums bound with covers as well as frontispieces. Although albums had existed earlier as blank-sheeted books to be used for notes and sketches, the lithographic process allowed them to be produced as facsimiles of artists' sketchbooks. They usually included eight to twelve drawings of unrelated subjects with minimal captions, the whole bound by a single thread. Albums are advertised on the door of Delpech's print shop in Vernet's 1818 depiction (see fig. 5), and can be seen piled up next to the press in Jacob's 1819 image for Senefelder (see fig. 1). The first identified album was published by Delpech in 1817; titled simply *Lithographic Album,* it included twenty prints by, among others, Carle and Horace Vernet, Antoine-Jean Gros, and Hortense Lescot.[93] From 1822 the firm of Antoine-François Gihaut (active 1815–1839) popularized these collections, issuing numerous albums as inexpensive and amusing New Year's gifts.[94] Many albums maintained the format of earlier portfolios, which comprised drawings contributed by several different artists, either as miscellanies or as themed collections. These soon became the most popular form of lithographic collection, imitating the informality of an artist's sketchbook and attesting both to the success of lithography as a drawing medium and to the public's desire to enter into the artist's private world. Albums did not completely replace either single-sheet prints or portfolios, however, since their cost was many times greater than that of individual prints; their appeal was mostly to a wealthier audience. Albums included both full-page images and *macédoines,* the medley prints that incorporated several small vignettes on a single page. The macédoine had its debut in the eighteenth century in literary anthologies that included snippets from different sources, an analogy to the fruit salad after which it was named. By the 1820s, the principle of the macédoine had been translated from literature into the new medium of lithography.[95] These lithographs quickly became a staple of the print trade; when published as single-sheet images, they were often cut apart and pasted into scrapbook albums or used as decals for decoration. When included in lithographed albums, they could be used by drawing teachers or just enjoyed for their varied imagery.

Album covers were usually plain wrappers whose ornate typeface simply announced title and artist, but their frontispieces became increasingly elaborate and witty, often self-referentially commenting upon the popularity of the album phenomenon. Charlet's 1824 *It's the End of the World!* shows a popular entertainer, a *saltimbanque,* overwhelmed by a deluge of flying albums, his livelihood now under siege by this new form of entertainment, his parasol and drum in tatters (fig. 40). That same year, Bellangé created a more developed carnivalesque motif, with the sideshow barker shouting: "Enter Gentlemen and Ladies, right in there you will see the family of the famous Lithographantoccini, brought here from Senegal by the famous Captain Crayonizinkhotzp! These little creatures have arrived despite deprivation of sleep and especially of nourishment, in order to give you the most beautiful collection of Albums, Collections of Sketches, Landscapes, Civil and Military Subjects, Caricatures, Scenes of the People as well as of Society, Drawing Manuals, Portraits in Tone, in Cross-hatching, in Stipple, etc. etc. Enter, Gentlemen and Ladies, it's the moment of their daily exercise" (fig. 41).[96] On the wall behind him is an assortment of the types of prints he is advertising—landscape, genre, military—and on the floor at his feet is a display of his wares, albums whose covers are labeled *recueil lithographique* (lithographic collection) and *croquis* (sketches). Through the open tent entrance

PAGE 50 **FIGURE 40**
Nicolas-Toussaint Charlet, *It's the End of the World!*, frontispiece to *Lithographic Sketches by Charlet* [*C'est la fin du monde! / Croquis lithographiques par Charlet*], 1824. Lith. Villain, Pub. Gihaut. The Cleveland Museum of Art, Anonymous gift in honor of John Bonebrake 1997.228.

PAGE 51 **FIGURE 41**
Hippolyte Bellangé, *Enter, Gentlemen and Ladies…*, frontispiece to *Lithographic Sketches by H. Bellangé* [*Entrez, messieurs et dames… / Croquis lithographiques par H. Bellangé*], 1824. Lith. Villain, Pub. Gihaut. British Museum.

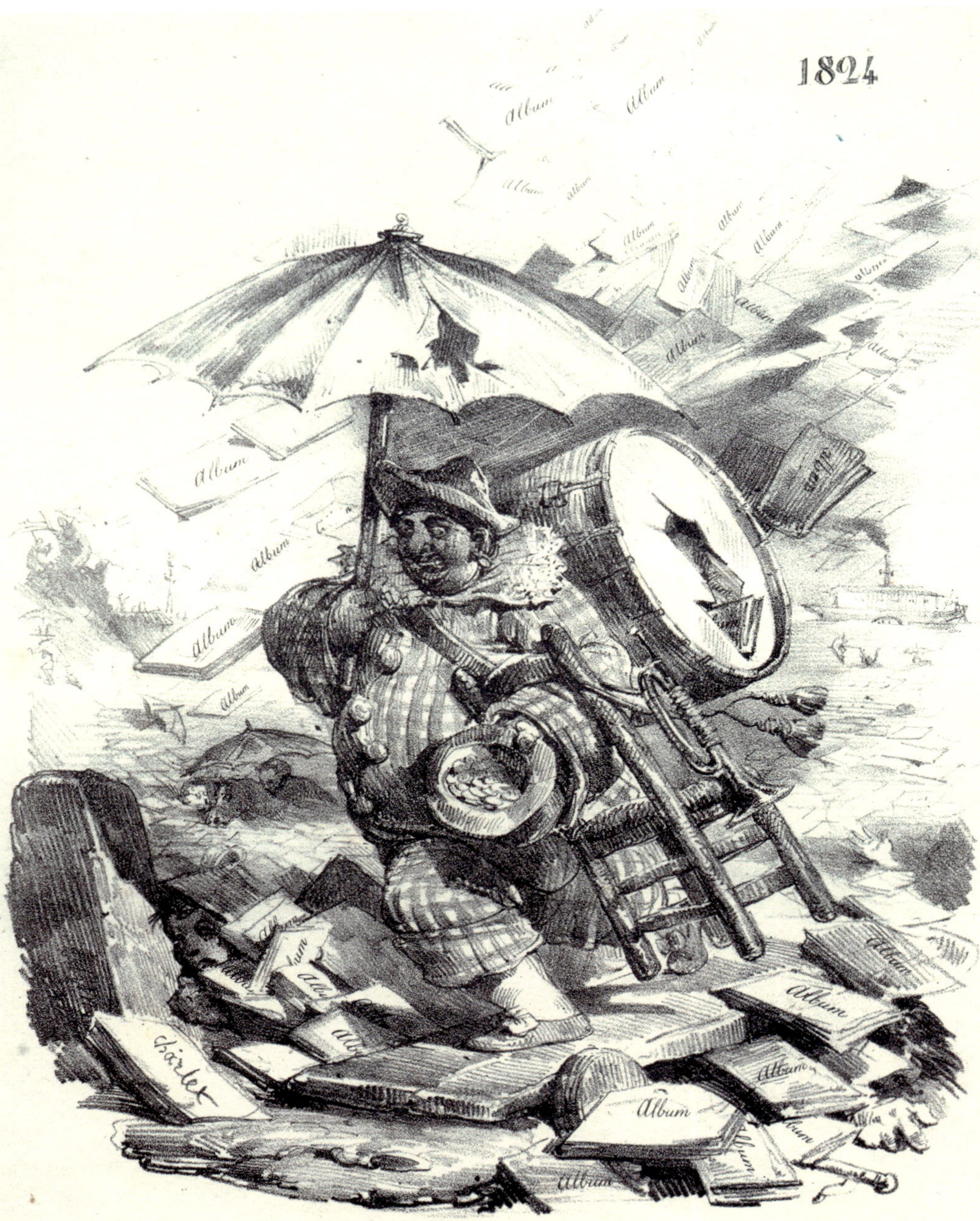

C'est la fin du monde!

Lith. de Villain

chez Gihaut, éditeur, Bd des Italiens, N.° 5

FIGURE 40

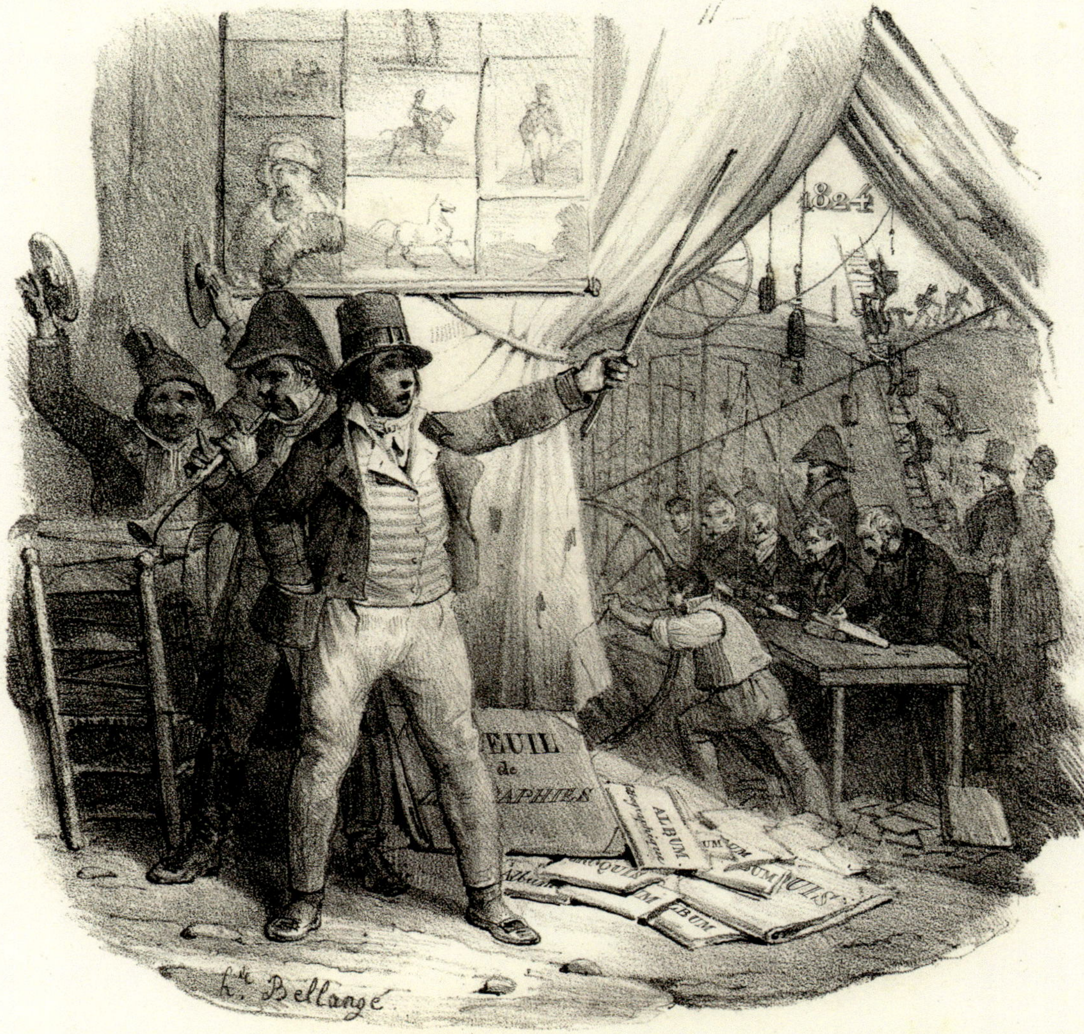

Entrez Messieurs et Dames, C'est ici là dedans que se fait voir la famille des fameux **Lithographantoccini** *apportée du Sénégal par le célèbre Capitaine* **Crayonizinkbotzy!** *Ces petits animaux sont parvenus à force de privations, de sommeil et surtout de nourriture, à former une très belle collection d'Album, Receuil de Croquis, Paysages, Sujets civils et militaires, Carricatures, Scènes populaires, Idem de Société, Principes de dessin, Portrait au grainé doux, à l'hachure, au pointillé. &c. &c. Entrez Messieurs et Dames C'est le moment de leurs exercices.*

FIGURE 41

we can see artists drawing at their worktable, while a worker turns the lithography wheel for printing and overhead the chains and pulleys of factory production are apparent; all this is accompanied by the horn tooting and cymbal clashing of the carnival musicians. This lithograph, through word and image, aptly conveys the excitement that resulted from this new availability of quantities of pictures of contemporary life while, at the same time, it presents a sardonic commentary on the industrialized production of imagery that lithography encouraged.

Beginning in the 1820s, numerous artists produced albums annually. Charlet was the most enterprising, publishing at least one a year beginning in 1822, usually under the title of *Croquis lithographiques*, lithographic sketches. His albums featured a variety of genre scenes whose subtle exaggerations of features and gesture teeter between genre imagery, visual satire, and caricature, apparent in *Hah! What Fun… to Be a Soldier!* (see fig. 11). Their mixture of styles and subjects is understandable only when we realize that caricature had as yet no clear definition apart from the grotesques identified with the English school. As an example of this, in 1825 Goya's etchings were republished as an album by the pioneer lithographer Charles Motte (1785–1836) under the title *Spanish Caricatures, No More, No Less, by Goya.*[97] The album included images from several of his series, *Los Caprichos* interspersed with *The Disasters of War*—which no one today would identify as caricature. The distinctions between caricature and other types of drawing were clearly more fluid at this time than ever before or since.

Tracing the history of albums is virtually impossible, however, since so few have survived intact. There was mandatory registration of printed matter in France and this *dépot légal* was announced in the official weekly publication *Bibliographie de la France*. At least in principle, one copy of each publication, whether book or print, was deposited with the state and sent to the Bibliothèque nationale de France.[98] Albums and portfolios, however, were not legally books unless they were bound; since publishers often sold their prints in various formats, they preferred to fulfill their legal obligation as cheaply as possible by giving the state single-sheet prints instead of bound albums. In the *Bibliographie de la France*, books and prints were treated differently. Books were always listed by author, but prints could be listed collectively by publisher or printer (e.g., "ten lithographs, Delpech"), or individually by title and artist. The publisher, who was legally responsible, was always identified, the work's title sometimes, the artist less often. The consequence of this uncertain status is that, in the 1820s, when the production of albums was at its height, there are virtually no listings for them; as a result, often we do not know their titles and cannot estimate their numbers. In addition, the legal deposit requirement seems to have been fulfilled rather casually with regard to prints; it is rarely possible to trace complete series even when the individual prints were numbered on the plate. Publishers seem to have felt that the legal requirement would be fulfilled if they deposited at least some of the prints in a series, and excellent hand-colored specimens were rarely squandered on the state. Nor is there any way to estimate the size of editions since, until late in the nineteenth century, prints were sold in unnumbered editions and reprinted according

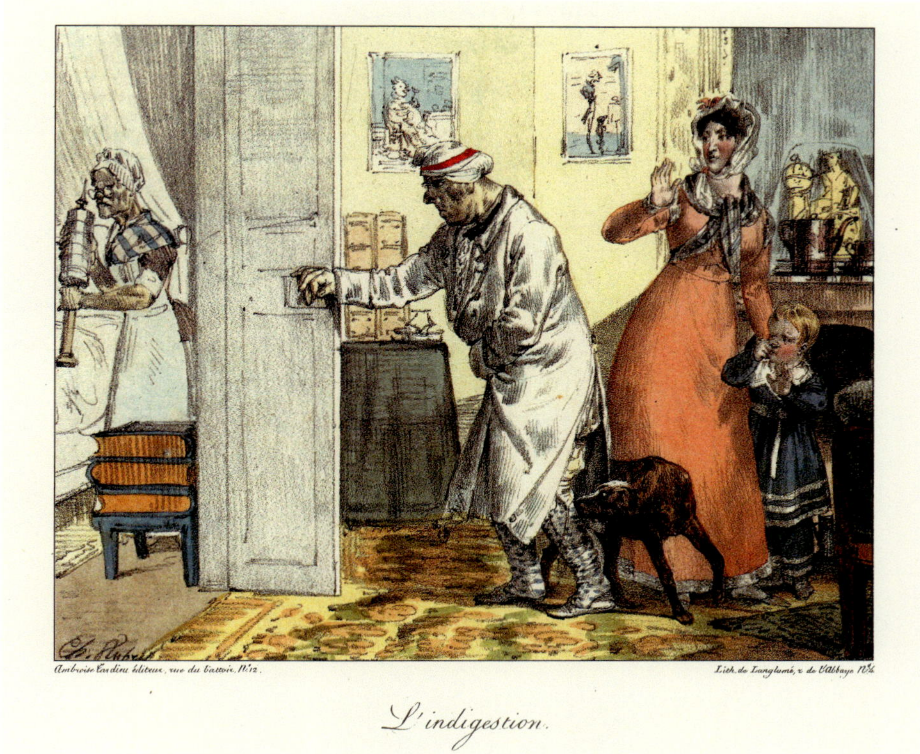

FIGURE 42

Charles Aubry, *Indigestion*, no. 12 in *The Comic Album of Picturesque Pathology*, [*L'indigestion / Album comique de pathologie pittoresque*], 1823. Lith. Langlumé, Pub. Ambroise Tardieu.

to demand; there was no incentive for a publisher to announce a new edition of a popular print and then have to duplicate all the legal requirements. Jean Adhémar, longtime curator of prints at the Bibliothèque nationale de France, noted that albums were often disassembled either by the individual collector who wanted to frame the prints, or later, by libraries and museums that wanted to classify them by subject or by artist.[99] As a result, the Bibliothèque nationale de France, which in principle received the official dépôt légal and should have everything published in France, has few complete albums: title sheets and individual prints are often separated, most of the title sheets are lost, and individual prints have been catalogued in various ways—by artist, by date, by subject. Since these publications were always regarded as ephemeral, they have fared no better elsewhere, and it is ironic that what is perhaps the best extant collection is at the Morgan Library in New York City, to which the late Gordon N. Ray bequeathed his superb collection of French and British illustrated books.[100]

One of the only albums classified as a book—and therefore remaining intact—is the 1823 *Comic Album of Picturesque Pathology*.[101] It contained witty depictions of different ailments and their sufferers by, among others, Charles Aubry (1803–1883), Hippolyte Bellangé, and Edmé-Jean Pigal, accompanied by short texts "explaining" each ailment in a witty pseudoscientific style. Aubry's *Indigestion* shows the pathetic sufferer, clutching his gut, on his way to the "cure" offered by the domestic brandishing a clyster, an early enema (fig. 42). Undoubtedly the texts on facing pages, anonymous and brief as they are, allowed this publication to be preserved as a book instead of being cut up for individual prints. The text accompanying *Indigestion* offers a counterpoint to the sympathy evoked by the image by pointing out that this is a malady of the privileged classes: "Ambition usually has eyes bigger than its belly."[102] This strategy of word-image counterpoint can operate in either direction, with credence given to the word, as here, or to the image, as in Charlet's *Hah! What Fun... to Be a Soldier!* (see fig. 11). In each case the meaning can be deciphered only through the complex interplay of the visual and the literary.

The increasing numbers of print series that insisted on being read as a narrative sequence had the best likelihood of being preserved in entirety; a prime early example is the 1824 *A Year in the Life of a Young Man. A True Story in Seventeen Chapters, Written by Himself* by Victor Adam (1801–1867). Adam's narrative parallels Hogarth's *A Rake's Progress* (1735), whose eight engravings recount the rise and fall of a young man who inherits a fortune, squanders it on gambling, drinking, and dissolute behavior, and ends up in the insane asylum of Bedlam. There are important differences, however, in that, while Adam's young man's life follows the same trajectory, at the end he saves himself by marrying a rich but plain heiress; in the last drawing she gestures toward a table laden with banknotes, coins, and jewelry, and he responds *This Has to End! I'll Marry Her* (fig. 43). The same year, Émile Wattier (1800–1868) published a similarly cynical graphic tale, *A Year in the Life of a Young Girl*.[103] Wattier also borrowed from Hogarth, in this case *A Harlot's Progress*, his series of six engravings published in 1732. But whereas Hogarth's young woman meets her end diseased and alone, Wattier's, though similarly "ruined" by a cad, survives and even prospers: in the last drawing she is elegantly dressed and married. The French always admired Hogarth's prints and their narratives, but, as we can see here, definitely not his moral messages.[104] Compared to Hogarth's fully worked-out engravings, laden with detail and symbolism, these French versions were rapidly drawn and cheaply printed in lithography, ephemeral, quickly read, and just as quickly discarded. Nonetheless, even when lacking bindings and covers, such series, sequential and always numbered on the plate, have had better survival odds than single-sheet images.

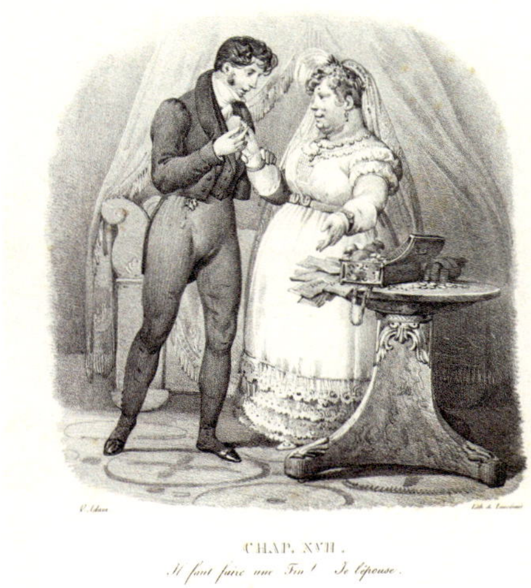

FIGURE 43
Victor Adam, *Chapter 17, This Has to End! I'll Marry Her*, from *A Year in the Life of a Young Man. A True Story in 17 Chapters, Written by Himself* [*Chapitre 17. Il faut faire une fin! Je l'épouse / Un an de la vie d'un jeune homme. Histoire véritable en 17 chapitres, écrit par lui-même*], 1824. Lith. Langlumé, Pub. Sazerac & Duval. Metropolitan Museum of Art, The Elisha Whittelsey Collection, The Elisha Whittelsey Fund, 1962.

By the late 1820s, albums were no longer a novelty, and the fad for albums as miscellaneous collections of sketches had run its course. They continued to be published throughout the century, however, becoming longer and more substantial. They developed into two quite different ratios of word to image: richly illustrated books, often novels, where an extensive text was accompanied by far fewer images, and comic books that recounted a story in numerous sequential images but with minimal text. These two developments will be discussed in subsequent chapters, but the most immediate result of the tidal wave of printed imagery that had begun with the invention of lithography was the creation of illustrated periodicals. Since these early periodicals have more in common with print portfolios and albums than with the illustrated newspapers and magazines that will be discussed in the next chapter, they will be seen first in this context. These publications disseminated to even wider audiences the images of modern life that had become endemic in France.

LITHOGRAPHIC PERIODICALS

Although there were earlier periodicals that included images, the first that can truly be described as the illustrated press are the journals *La Silhouette* (1829–1831) and *La Caricature* (1830–1835). *Le Nain jaune* (1814–1815) included an etched caricature each month. *L'Album. Journal des arts, des modes, et des théâtres* (1821–1823) featured a fashion plate or illustration. Several others, such as *La Pandore. Journal des spectacles, des lettres, des arts, des moeurs et des modes* (1823–1830) included individual lithographs as a subscriber bonus.[105] What distinguished *La Silhouette* and *La Caricature*, however, was the size, quantity, quality, and importance of their images, and the integrated relationship of text to image that they established.[106] In all these early periodicals, however, images were printed on separate pages because the letterpress technique used for the text could not be combined with the illustrational processes of either intaglio or lithography, both of which required separate plates and thus separate pages. The advantage of this was that in the case of large-format periodicals such as *La Silhouette* and *La Caricature*, the size of their images rivaled those available from print dealers and thus added to the periodicals' attraction as *de facto* lithographic albums.

La Silhouette was modeled on the English humor magazines that Monnier—who was instrumental in its founding—had, no doubt, seen on his trips to London.[107] *Caricature Magazine, or Hudibrastic Mirror* (1807–1812), published by George Moutard Woodward (1760–1809) in London consisted of a weekly portfolio of caricatures, but without accompanying text, while *The Scourge, or Monthly Expositor of Imposture and Folly* (1811–1816) was mostly text but did include a topical hand-colored etched caricature by George Cruikshank in every issue, much like *Le Nain jaune* in France. *La Silhouette*, however, was a new kind of periodical, one that has since become familiar. Larger in size than earlier journals, its articles were a mélange of culture and politics, and its images, always identified as *dessins*, drawings, were related to and discussed in its texts, not just in a brief note accompanying the image. Each weekly eight-page issue of *La Silhouette* offered two full-page lithographs, one of which was hand-colored, by well-known artists such as Daumier, Grandville, Devéria, and Monnier; its back cover announced, "Each issue, composed of a sheet of satin vellum paper in four pages, printed in two columns, is accompanied by two lithographs. Thirteen issues every three months will form an album."[108] Its genealogy from the albums of the previous decade is apparent not just from this notice, but also from its relatively large size and from the subjects of its images, which ran the gamut from political caricature to genre, landscape, and theater scenes. Most revelatory,

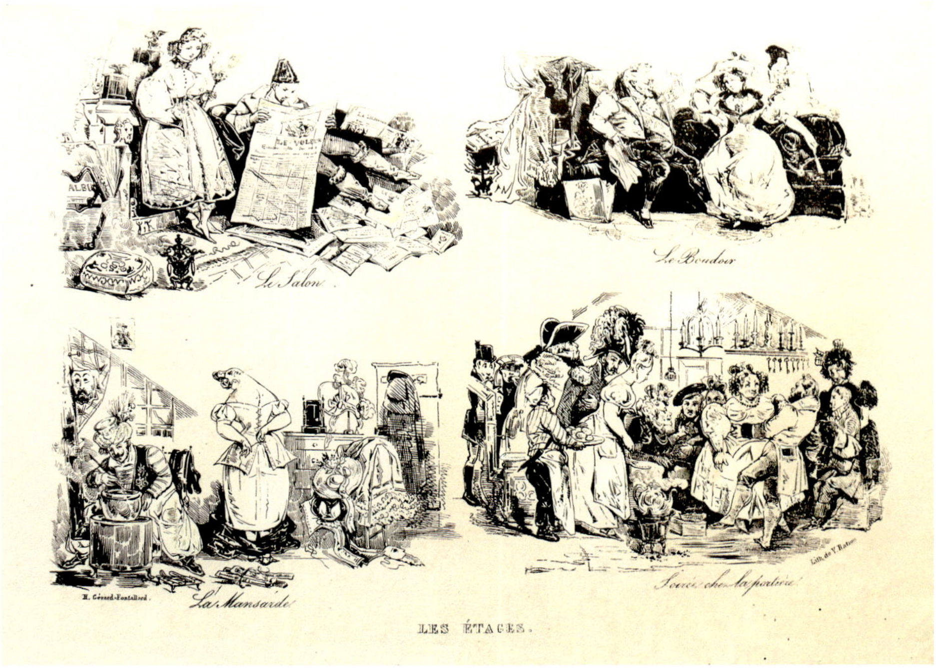

FIGURE 44
Gérard-Fontallard, *Elevations. The Salon. The Boudoir. The Garret. A Party at the Porter's* [*Les étages. Le salon. Le boudoir. La mansarde. Soirée chez la portière*]. Lith. V. Ratier. *La Silhouette* 1 (1830), no. 9.

though, was its subtitle: *Album lithographique.*[109] Founded by the caricaturist Charles Philipon along with the lithographic printer Victor Ratier (1807–1899) and the publisher Benjamin-Louis Bellet (1805–1882), *La Silhouette* opened a new market and audience for printed imagery.[110] It sold for fifty-two francs annually, more than other journals, and even offered a luxury edition that, printed on higher quality paper, cost a hundred francs. Balzac published drafts of his *Scenes of Private Life* there, and, besides political caricature, *La Silhouette* offered a rich assortment of imagery and commentary on contemporary life.[111] An image by Gérard-Fontallard (Henri-Gérard Fontallard, b. 1798), *Elevations. The Salon. The Boudoir. The Garret. A Party at the Porter's* showed the activities of the different classes inhabiting the levels of a typical Parisian residence; the theme soon became a staple of French graphic art (fig. 44).[112] An article on the Missouri tribe of Native Americans was accompanied by one of Grandville's most perceptive images, *Revenge, or the French in Missouri,* where colonialist roles are reversed and the French become the sideshow, the Missouri the spectators (fig. 45).[113] Although the French "natives" have extravagant costumes and coiffures, Grandville's image shows little caricature in the English sense of anatomical exaggerations and distortions; it amply fulfills Farcy's dictum that the highest form of caricature should be philosophical, not physical. Caricatures of extravagant costumes and coiffures were familiar themes, but here they are pressed into service in order to underscore the absurdity of the European sense of superiority over other cultures.

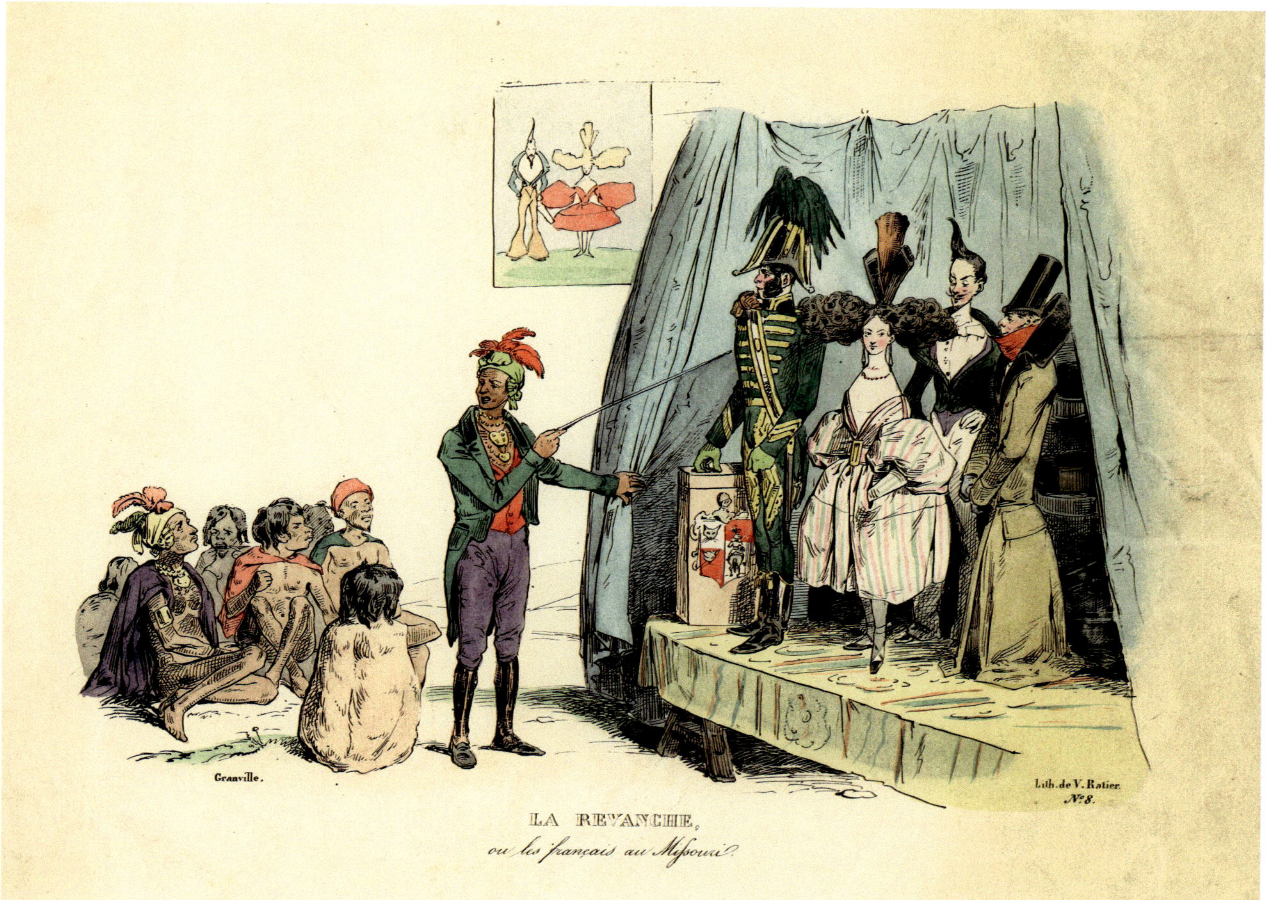

FIGURE 45

Grandville, *Revenge, or the French in Missouri* [*La revanche, ou les Français au Missouri*]. Lith. V. Ratier. *La Silhouette* 1 (1830), no. 4. British Museum.

La Silhouette is known today principally as the precursor of the better-known periodicals Charles Philipon later founded, *La Caricature* and *Le Charivari*, as well as for Philipon's caricature of King Charles X, *A Jesuit* (fig. 46). Philipon published *A Jesuit* in *La Silhouette* well before the July Revolution of 1830, and incurred a fine and a prison sentence as a result; undeterred, he republished it twice again.[114] *La Silhouette* certainly deserves credit for its pivotal role in the creation of the illustrated press, but, in addition, it provided a venue for many young artists whose work would soon become celebrated; the young Honoré Daumier, for example, published some of his earliest and most memorable drawings there. *Blessed Are Those Who Hunger and Thirst, for They Shall Be Satisfied* takes as its text one of the Christian Beatitudes that offer blessings for the poor, the meek, and the persecuted (fig. 47). Here Daumier juxtaposes this familiar prayer with a "snapshot" of contemporary reality: a fat priest walks past a starving man lying in the street; passing before the "Pension Bourgeoise," the priest offers a blessing—and nothing else. The harshness of the contrast between smug hypocrisy and abject poverty is made almost unbearable by the loveliness of the pastel coloration of the image. A stark black-and-white drawing would have been, in this context, less moving.

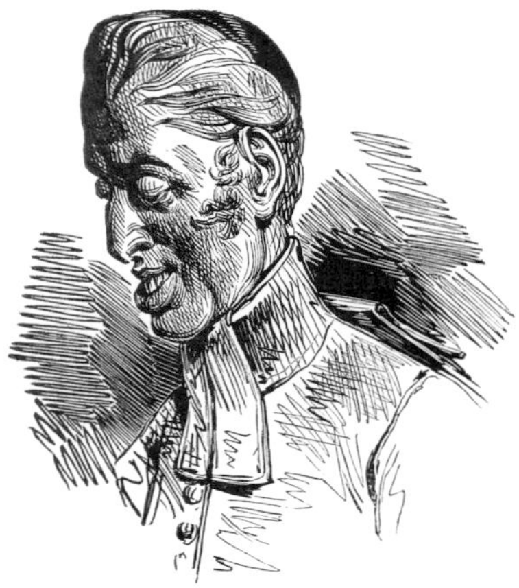

UN JÉSUITE.

FIGURE 46
Charles Philipon, *A Jesuit*. Lith. V. Ratier. *La Silhouette* 2 (1830), no. 2.

In any discussion of French caricature and popular lithography, Charles Philipon is the pivotal figure. A student of Gros, he, like so many others, had been introduced to lithography by his fellow student Charlet.[115] His first prints were humorous genre scenes and, while his career as an artist lasted only a few years, within this brief time he created two of the most incendiary images in nineteenth-century art, Charles X as a Jesuit (see fig. 46) and Louis-Philippe as a pear (see fig. 34). With talents more literary than visual, he dictated the contents of many of his artists' caricatures, even writing captions for Daumier and making certain that, along with the credit lines for artist, lithographer, and publisher, there would also be an entry *Philipon inv*, i.e., *From an idea by Philipon.*[116] Nor was Philipon's involvement with lithography limited to the creation of journals and prints: two weeks before the establishment of *La Silhouette*, he opened a print shop with his half-sister Marie-Françoise-Madeleine Philipon (1794–1868) and brother-in-law Gabriel Aubert (1800–1847), in the fashionable Passage Véro-Dodat near the Louvre.[117] The Maison Aubert, as it was called (though Philipon was always its guiding spirit), would develop into the largest lithographic printing house in Paris, with a virtual monopoly over the publication of caricature in all its forms. Traviès's *You Have to Admit that the Head of State Looks Pretty Funny* shows the façade of this establishment in what had by 1831 become a standard trope—crowds of spectators representing all classes, pressing up against the windows of a print shop to see the latest images—but with an added fillip: the raggedy gent in the foreground is gesturing toward a prominently displayed image of The Pear (fig. 48).

When censorship was (temporarily) abolished in the wake of the 1830 revolution, Philipon lost no time in abandoning *La Silhouette* and registering his intention of

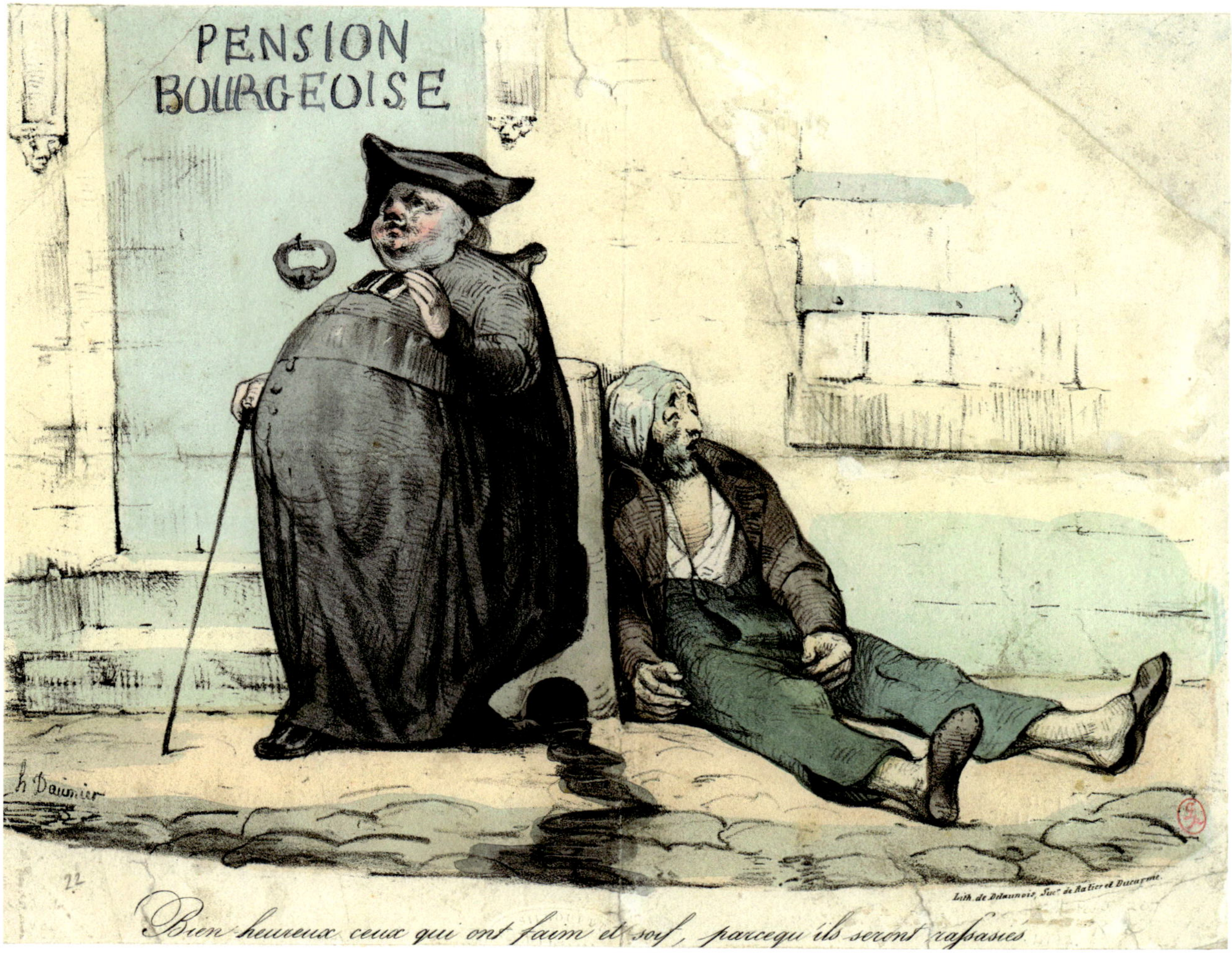

FIGURE 47

Honoré Daumier, *Blessed Are Those Who Hunger and Thirst, for They Shall Be Satisfied* [*Bien heureux ceux qui ont faim et soif, parce qu'ils seront rassasiés*]. Lith. Delaunois. *La Silhouette* 4 (1830), no. 10. Hand-colored lithograph. British Museum.

publishing a new periodical, this time without partners. He named it, appropriately enough, *La Caricature*; its subtitle, *morale, politique et littéraire*, signaled the breadth of topics it would address. A weekly of four pages with two lithographs, one hand-colored, it first appeared on November 4, 1830. Its prospectus stated: "Since 1789, caricature has been a necessity for our country. It is extremely popular, and if, until now, it has not been published as regularly as opinion and humor, it is because the cost of engraving has prohibited this enterprise. It is not that the prints lacked an audience, but the audience lacked the prints. Today the process of lithography has rendered practically commonplace this exquisite pleasure that Parisians alone used to enjoy daily in the streets or on the boulevards."[118] Philipon's focus on the imagery is evident in the announcement that appeared in every issue: "La Caricature includes ONE HUNDRED AND FOUR lithographs per year executed by the most renowned artists. . . . The lithographs that appear in the journal are

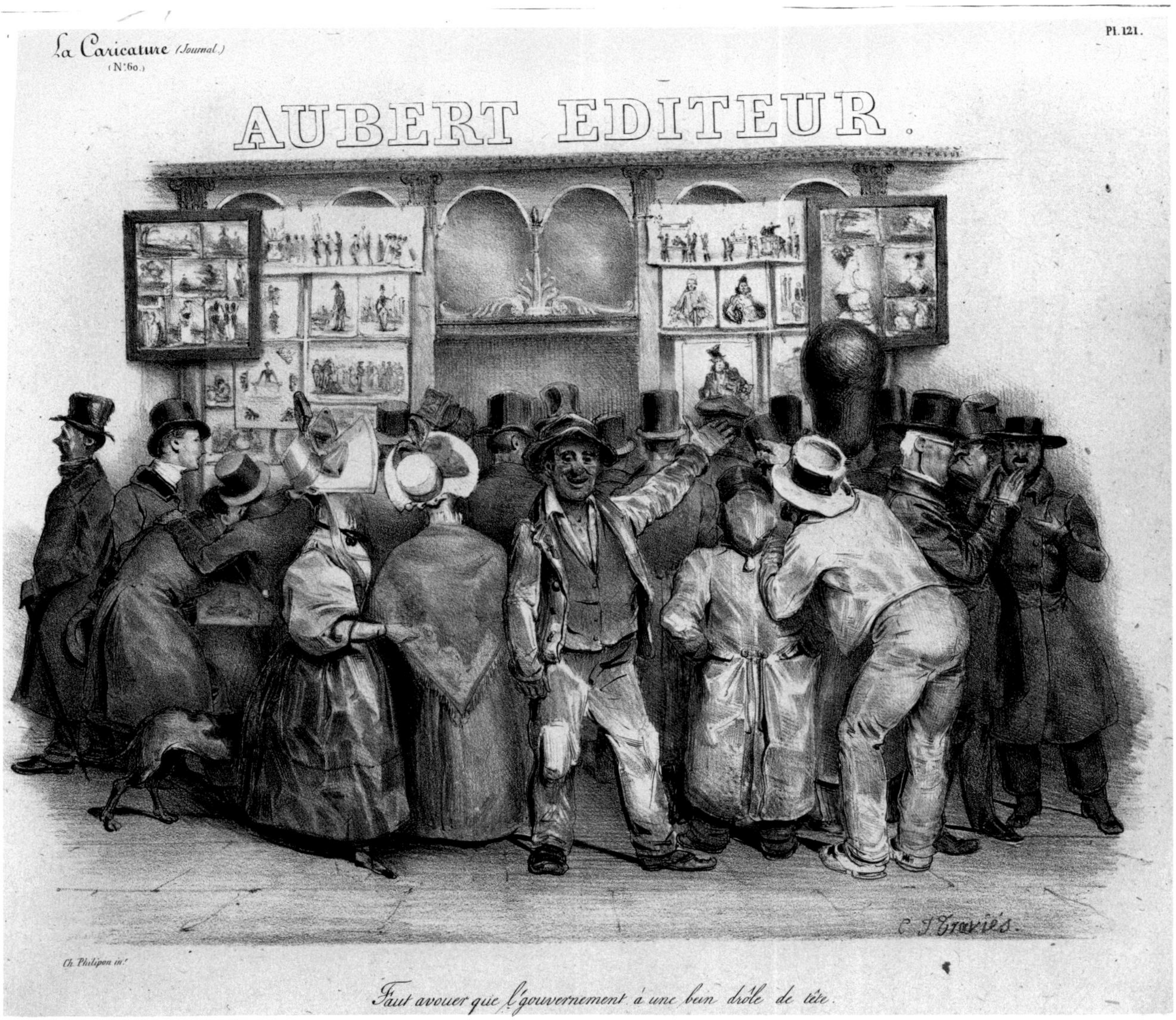

FIGURE 48

Charles-Joseph Traviès, *You Have to Admit That the Head of State Looks Pretty Funny* [*Faut avouer que l'gouvernement à une bien drôle de tête*], 1831. Inv. Ch. Philipon, Lith. Delaporte, Pub. Aubert. *La Caricature* 3, no. 60 (December 22, 1831), pl. 121. Metropolitan Museum of Art, The Elisha Whittelsey Collection, The Elisha Whittelsey Fund, 1966.

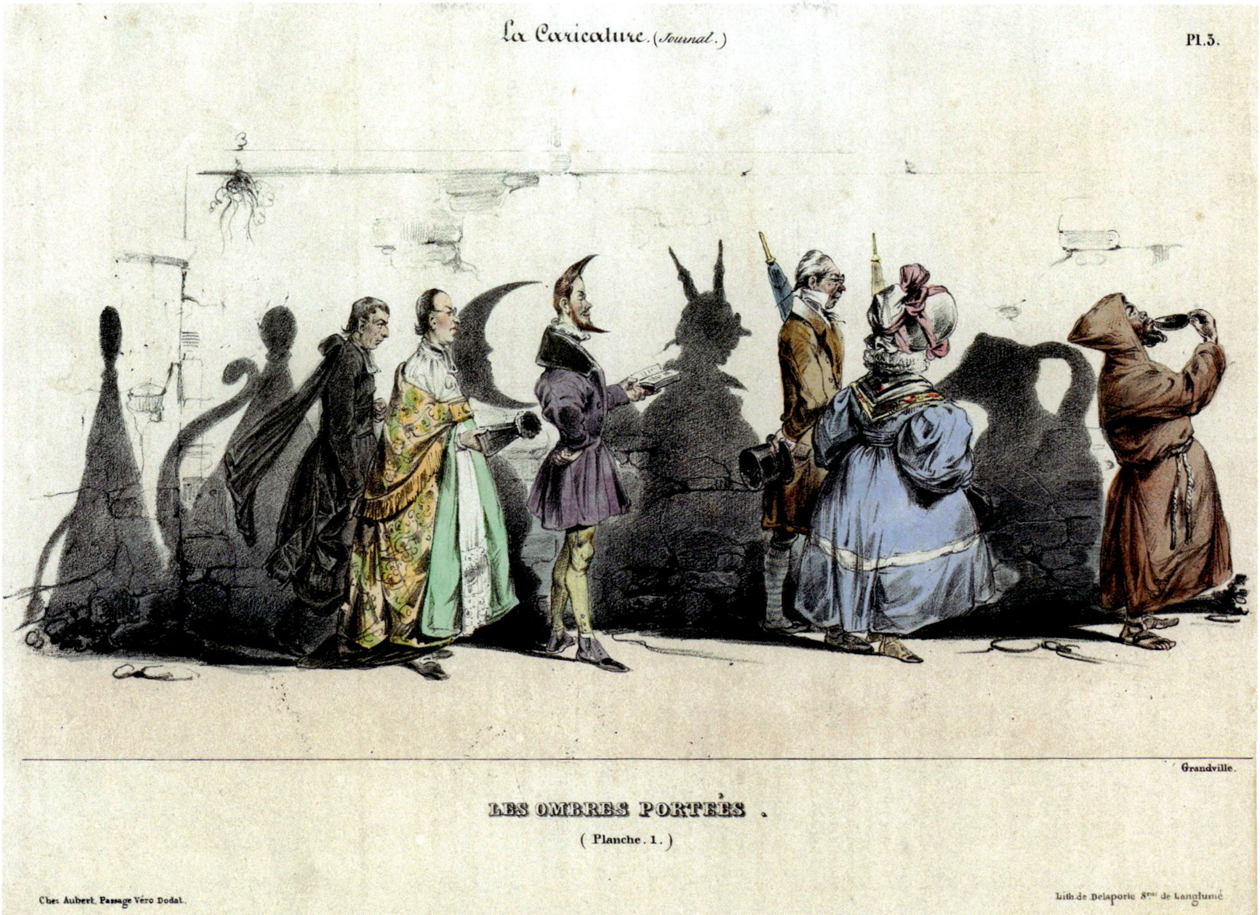

FIGURE 49
Grandville, *Cast Shadows* [*Les ombres portées*], pl. 1. Lith. Delaporte, Pub. Aubert. *La Caricature* 1, no. 2 (November 11, 1830), pl. 3. Hand-colored lithograph. Heidelberg University Library.

not for sale. Dealers can obtain them only by subscribing to the journal."[119] The images are undeniably among the finest lithographs produced in the century. Grandville's *Cast Shadows* is emblematic of the entire enterprise of caricature, showing the reality beneath appearances (fig. 49). Here each individual's shadow reveals his inner character: the friar is a *cruche* (jug), in slang a dunce; the husband is a cuckold, his "horns" formed by his and his wife's parasols; the young fop is a *lune* (moon), an ass. Printed on heavy paper, with carefully applied hand-coloring, *La Caricature* was indeed a luxury production, although the price was the same as *La Silhouette*, fifty-two francs annually.

Balzac abandoned *La Silhouette*, which soon ceased publication, to become the editor of *La Caricature*, where he continued to publish articles and reviews as well as excerpts and drafts from his novels.[120] Early images in *La Caricature* were varied; like *La Silhouette*, Philipon's new journal included genre images, depictions of manners, costume, and theater, in addition to the political images for which it is better known. We can see Balzac's influence in the inclusion of images such as Pigal's *Popular Entertainment*, which shows

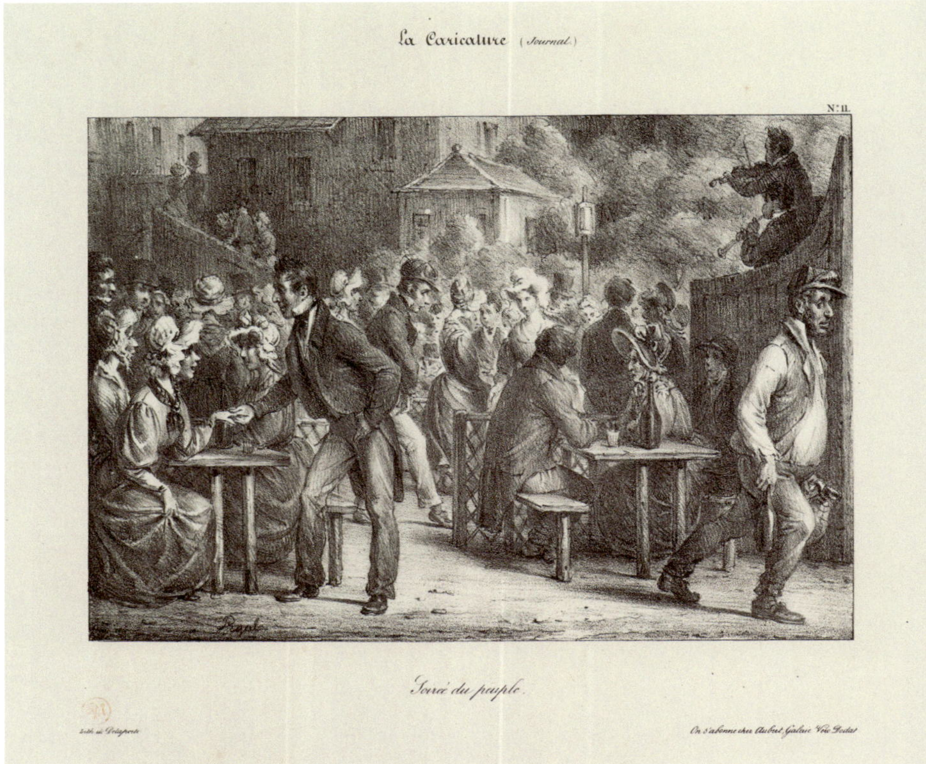

FIGURE 50
Edmé-Jean Pigal, *Popular Entertainment* [*Soirée du peuple*]. Lith. Delaporte, Pub. Aubert. *La Caricature* 1, no. 6 (November 4, 1830), pl. 11. Santa Barbara Museum of Art, Gift of Mr. and Mrs. Michael G. Wilson.

representatives of all classes at an outdoor dance (fig. 50). Exiting right is a paunchy workman; in the left foreground a foppish young man is chatting up a young woman whom he clearly hopes might be his next conquest; young people dance, musicians play, an elderly couple watches. Balzac would have been delighted by this "human comedy." In early 1831, Balzac resigned, however, when *La Caricature*'s orientation shifted away from the social observation that had always formed the basis of his own writing to an exclusively political focus.[121] With him went its team of well-known artists, most of whom had come from Gros's studio: Charlet, Decamps, Devéria, Victor Adam, Bellangé, and Raffet. Others—Grandville, Traviès, and Daumier—continued with *La Caricature*, publishing memorable images until 1835 when it ceased publication after draconian press laws made political caricature impossible.[122]

La Caricature never turned a profit and was always subsidized by Aubert's print trade, so even while he was still publishing it Philipon conceived of another periodical of wider circulation that would be less political and more profitable, the daily *Le Charivari* (1832–1937). Its genealogy was manifest by its prospectus, which featured a jester surrounded by a heap of publications, each labeled "Album." Lithographic albums inspired all three of Philipon's journals. This is underscored by the image used regularly in advertisements for *Le Charivari* and on the cover of its annual volume (fig. 51). It depicted fashion plates, genre scenes, and caricatures, Philipon's version of the traditional print publisher's advertising medley seen in figure 9.

In announcing his new publication, Philipon outlined the differences between the two periodicals.

> *La Caricature*, which appears only once a week, is and always must be a journal of the highest quality; its drawings, made carefully, prepared attentively and printed on vellum paper, make it a journal for collectors and libraries. Those of our friends who have been most honored and encouraged by the public will, from now on, be commissioned *exclusively* to execute drawings for this publication, which we promise to maintain above all others that have been and will be produced in this genre. For public places, for those art lovers who prefer variety to political satire, we are creating a daily journal that, each day, will publish a new sketch; printed more quickly than *La Caricature* and featuring less fastidiously completed drawings, it will be sold at the standard price of literary journals.[123]

Le Charivari made its debut on December 1, 1832 and had four pages, one of which displayed the daily drawing. Its innovation was that its image page (always page three) was no longer detached but was run through the press twice, once to print the image, a second time for the letterpress text, recto and verso. Like Philipon's two earlier periodicals, *Le Charivari*'s main selling point was its art—365 drawings annually. Within three weeks it had set the masthead that always announced its prime feature, "Journal publishing a new drawing each day" (fig. 52)—until 1835, when, in the wake of the new repressive press laws,

FIGURE 51
Advertisement for *Le Charivari*, April 1, 1840.

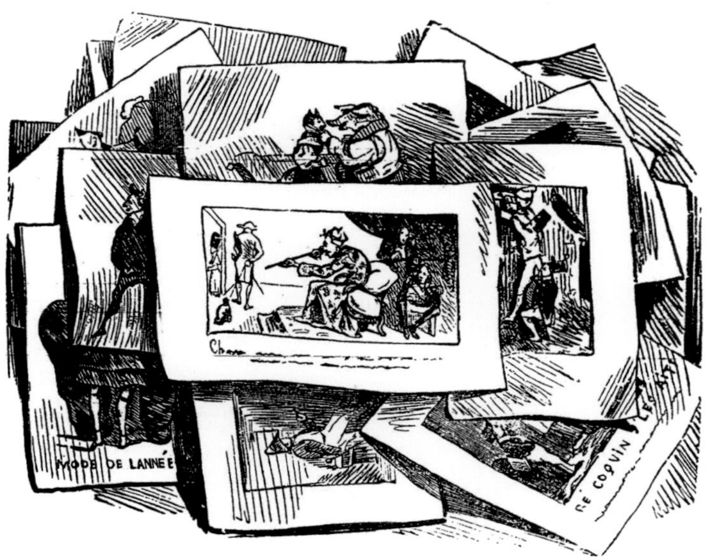

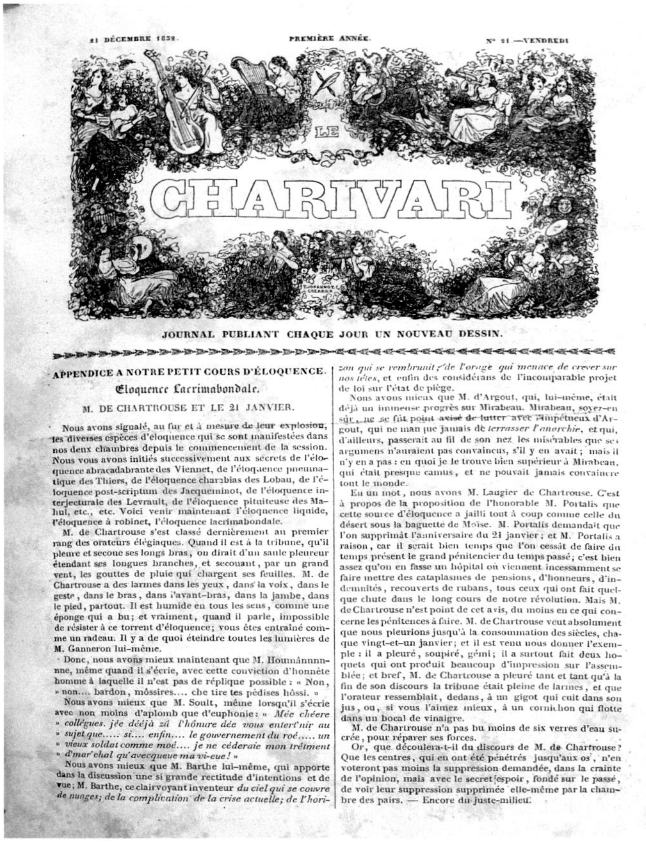

21 DÉCEMBRE 1832. PREMIÈRE ANNÉE. N° 21 —VENDREDI

LE CHARIVARI

JOURNAL PUBLIANT CHAQUE JOUR UN NOUVEAU DESSIN.

APPENDICE A NOTRE PETIT COURS D'ÉLOQUENCE.

Éloquence Lacrimabondale.

M. DE CHARTROUSE ET LE 21 JANVIER.

Nous avons signalé, au fur et à mesure de leur explosion, les diverses espèces d'éloquence qui se sont manifestées dans nos deux chambres depuis le commencement de la session. Nous vous avons initiés successivement aux secrets de l'éloquence abracadabrante des Viennet, de l'éloquence pneumatique des Thiers, de l'éloquence charabias des Lobau, de l'éloquence post-scriptum des Jacqueminot, de l'éloquence interjecturale des Levrault, de l'éloquence pituiteuse des Mahul, etc., etc. Voici venir maintenant l'éloquence liquide, l'éloquence à robinet, l'éloquence lacrimabondale.

M. de Chartrouse s'est classé dernièrement au premier rang des orateurs élégiaques. Quand il est à la tribune, qu'il pleure et secoue ses longs bras, on dirait d'un saule pleureur étendant ses longues branches, et secouant, par un grand vent, les gouttes de pluie qui chargent ses feuilles. M. de Chartrouse a des larmes dans les yeux, dans la voix, dans le geste, dans le bras, dans l'avant-bras, dans la jambe, dans le pied, partout. Il est humide en tous les sens, comme une éponge qui a bu; et vraiment, quand il parle, impossible de résister à ce torrent d'éloquence; vous êtes entraîné comme un radeau. Il y a de quoi éteindre toutes les lumières de M. Ganneron lui-même.

Donc, nous avons mieux maintenant que M. Hounânnnnnne, même quand il s'écrie, avec cette conviction d'honnête homme à laquelle il n'est pas de réplique possible : « Non, » non.... bardon, môssires.... che tire tes pédises hôssi. »

Nous avons mieux que M. Soult, même lorsqu'il s'écrie avec non moins d'aplomb que d'euphonie : « *Mée chéere » collègues. jée dééjà zà l'hônure dée vous entert'nir au » sujet que..... si.... enfin.... le gouvernement du roé..... un » vieux soldat comme moé.... je ne céderaie mon trétment » d'mar'chal qu'avecqueue ma vi-eue!* »

Nous avons mieux que M. Barthe lui-même, qui apporte dans la discussion une si grande rectitude d'intentions et de vue; M. Barthe, ce clairvoyant inventeur *du ciel qui se couvre de nuages; de la complication de la crise actuelle; de l'horizon qui se rembrunit; de l'orage qui menace de crever sur nos têtes*, et enfin des considérans de l'incomparable projet de loi sur l'état de piège.

Nous avons mieux que M. d'Argout, qui, lui-même, était déjà un immense progrès sur Mirabeau. Mirabeau, soyez-en sûr, ne se fût point avisé de lutter avec l'impétueux d'Argout, qui ne manque jamais de *terrasser l'anarchie*, et qui, d'ailleurs, passerait au fil de son nez les misérables que ses argumens n'auraient pas convaincus, s'il y en avait; mais il n'y en a pas : en quoi je le trouve bien supérieur à Mirabeau, qui était presque camus, et ne pouvait jamais convaincre tout le monde.

En un mot, nous avons M. Laugier de Chartrouse. C'est à propos de la proposition de l'honorable M. Portalis que cette source d'éloquence a jailli tout à coup comme celle du désert sous la baguette de Moïse. M. Portalis demandait que l'on supprimât l'anniversaire du 21 janvier; et M. Portalis a raison, car il serait bien temps que l'on cessât de faire du temps présent le grand pénitencier du temps passé; c'est bien assez qu'on en fasse un hôpital où viennent incessamment se faire mettre des cataplasmes de pensions, d'honneurs, d'indemnités, recouverts de rubans, tous ceux qui ont fait quelque chute dans le long cours de notre révolution. Mais M. de Chartrouse n'est point de cet avis, du moins en ce qui concerne les pénitences à faire. M. de Chartrouse veut absolument que nous pleurions jusqu'à la consommation des siècles, chaque vingt-et-un janvier; et il est venu nous donner l'exemple : il a pleuré, soupiré, gémi; il a surtout fait deux hoquets qui ont produit beaucoup d'impression sur l'assemblée; et bref, M. de Chartrouse a pleuré tant et tant qu'à la fin de son discours la tribune était pleine de larmes, et que l'orateur ressemblait, dedans, à un gigot qui cuit dans son jus, ou, si vous l'aimez mieux, à un cornichon qui flotte dans un bocal de vinaigre.

M. de Chartrouse n'a pas bu moins de six verres d'eau sucrée, pour réparer ses forces.

Or, que découlera-t-il du discours de M. de Chartrouse? Que les centres, qui en ont été pénétrés jusqu'aux os, n'en voteront pas moins la suppression demandée, dans la crainte de l'opinion, mais avec le secret espoir, fondé sur le passé, de voir leur suppression supprimée elle-même par la chambre des pairs. — Encore du juste-milieu.

FIGURE 52
Le Charivari, December 21, 1832.

this statement was revised to "Journal publishing a drawing whenever the censors permit."[124] Although the journal, like its two predecessors, *La Silhouette* and *La Caricature*, has since become known primarily for its political caricatures—the very word *charivari* signifies a noisy public demonstration of disapproval—from its beginnings Philipon insisted on striking a balance. *Le Charivari*'s images, Philipon announced, would include:

> the latest and most elegant fashions; the most picturesque costumes and striking scenes from the most successful plays; places, monuments, and landscapes on which some event of the day may focus attention; of scenes of daily life; sketches of major pictures in public museums or privately owned, and others attracting notice in various exhibitions; views of the most interesting sessions of the two Chambers of Parliament with the features, postures, mannerisms, gestures, and, in a word, all the parliamentary ways of the different speakers; and finally portraits of actors, actresses, artists of renown, scientists, men of letters, politicians, ministers, diplomats, peers, deputies, princes, kings, and any personality who rightly or wrongly, in France or abroad, may for a moment arouse public curiosity.[125]

Philipon continually stressed his intention of making this journal appeal to all tastes. In its prospectus, he wrote: "Political, literary, dramatic, and moral satire will not entirely invade our efforts. After a charivari, we will often present a serenade. Jeers for some, harmony for others.... Thus *Le Charivari* will offer a complete orchestra of virtuosos of all kinds."[126] It would be, he wrote, "a complete panorama where we continuously reproduce, with crayon and pen, every aspect of this kaleidoscopic world we live in.[127] Philipon's program echoed the high hopes expressed a decade earlier by Thiers, who prophesized that lithography would reproduce all aspects of contemporary life. Philipon's determination to include the social as well as the political can be seen in the numerous drawings by Victor Adam commissioned for the new journal. Adam was prolific in many genres (see fig. 43), but his scenes of contemporary life were especially admired. *Paris. Royal Swimming School* is a good example of Philipon's program although, in view of the less than ideal physiognomies on display around the swimming pool, it is uncertain whether this image was intended to represent jeers or harmony (fig. 53). In any case, genre subjects such as this would soon inspire new generations of French artists to abandon the antique in favor of contemporary life.

Although outlined in *Le Charivari*, Philipon's program would not come to fruition until the following decades, with the establishment of the great wood-engraved illustrated weeklies. Meanwhile, he had found a way to capitalize on the images that *Le Charivari* published daily. With *La Caricature*, Philipon had attempted to establish a luxury market for lithographs, announcing: "We will sand down the lithographic stones three months after their first publication and so will establish a high price for this collection, which will thus become rare."[128] Within two years, however, he realized that the mass market offered much better opportunities than the luxury trade. The prospectus for *Le Charivari* listed Aubert's latest editions for sale, and every issue thereafter carried a notice stating "All the lithographs that appear in *Le Charivari* are available at the Aubert print shop."[129] This brief notice was soon expanded to lengthy advertisements listing the different titles and formats: reprinted on better paper, hand-colored or black and white, sold individually, in portfolios, or in bound albums. Many of the best-known graphic artists contributed entire series of drawings to *Le Charivari* that were then reprinted for sale either as albums or as individual prints. Edmé-Jean Pigal, a regular contributor, published numerous series in *Le Charivari*: his *Familiar Scenes* proved so popular that after the first six lithographs appeared in 1833, *Le Charivari* announced: "In order to complete the series of *Familiar Scenes* by Pigal, we felt we had to offer our subscribers the drawing that will serve as the cover for this charming collection."[130] The publisher Ostervald aîné had done much the same with Philipon's series *Infatuations* in 1827, issuing the drawings individually until sales justified the printing of a frontispiece. Aubert's marketing innovation went even further: he first published Pigal's drawings in *Le Charivari*; then, after the series was complete, began to advertise its availability as a more luxurious portfolio: "Familiar Scenes by Pigal, 24 compositions of manners, 75 centimes each in color."[131] For decades,

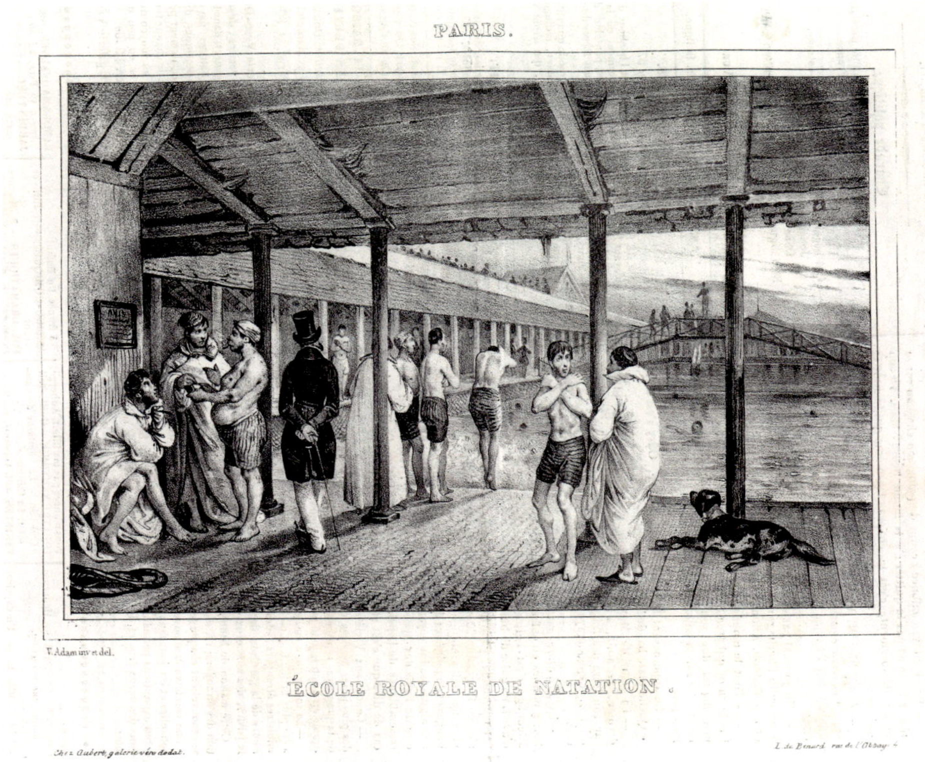

FIGURE 53
Victor Adam, *Paris. Royal Swimming School* [*Paris. École royale de natation*]. Lith. Bénard, Pub. Aubert. *Le Charivari*, August 3, 1834. Metropolitan Museum of Art, Gift of Arthur Sachs, 1923.

Daumier published three lithographs weekly in *Le Charivari*, with virtually all his series reissued by Aubert in various formats. The hundred lithographs of his *Caricaturana*, for example (see fig. 32), were published in *Le Charivari* from 1836 to 1838, then reissued in black and white as *Album Caricaturana (The Robert Macaires)*, and, a year later, in a hand-colored two-volume edition as *A Hundred and One Robert Macaires.*[132] Since the lithographs printed in *Le Charivari* had text on their verso with the ink often bleeding through, prior publication did not interfere with the market for the complete portfolio, which was always printed on higher quality paper, available hand-colored, and sold either in bound albums or by the sheet. If anything, the lower quality of the prints in *Le Charivari* served as advertisements for Aubert's premium wares.

Within a few years, *La Silhouette, La Caricature,* and *Le Charivari* had caused a seismic shift in French cultural life and set the stage for the development of the illustrated press as we know it. No longer would a single image per issue suffice; now periodicals needed to offer their readers numerous drawings and of a higher quality than ever before.

In 1831, *L'Artiste* was founded, a nonpolitical cultural weekly that included lithographs with each issue *hors texte,* meaning that they were printed separately and could be removed to be placed in portfolios or framed.[133] Other periodicals rapidly followed suit.

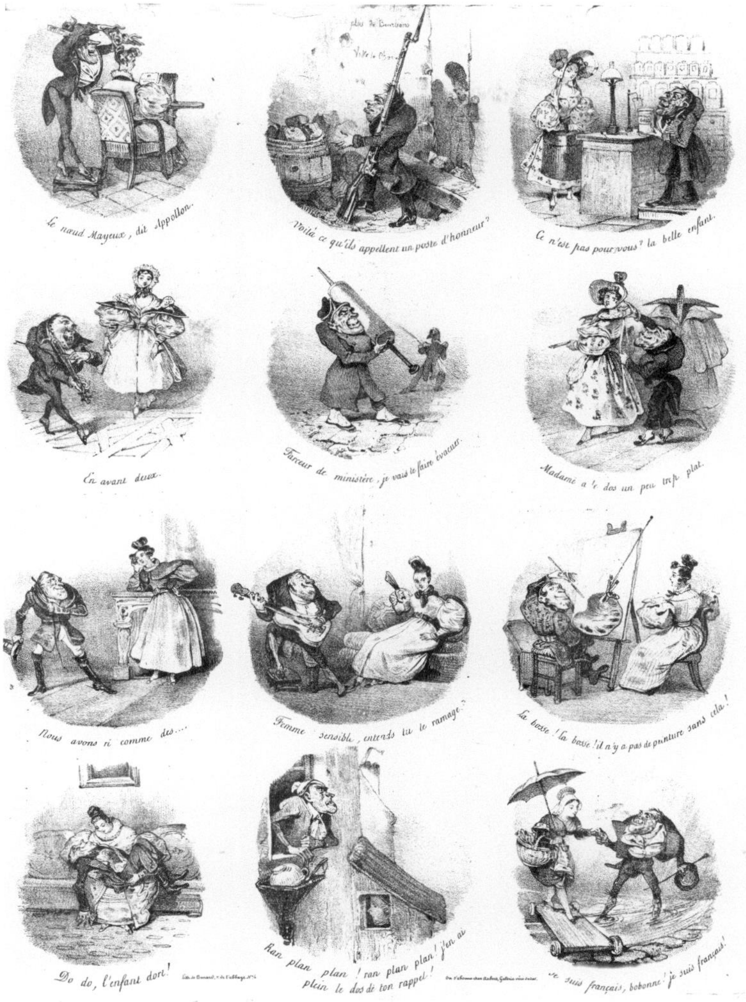

FIGURE 54

Charles-Joseph Traviès, *Mayeux*. Lith. Bénard, Pub. Aubert. *Le Charivari*, December 10, 1832. Metropolitan Museum of Art. Gift of Arthur Sachs, 1923.

All of these periodicals fostered artistic innovation and encouraged the drawing of images of contemporary life. *La Silhouette* opened the way, and disseminated the earliest work of Daumier and Grandville; *La Caricature* published quality hand-colored lithographs, bringing the medium to new heights; *Le Charivari* was the earliest French journal to include full pages of vignettes on a single theme, a format that had developed out of the macédoine, the medley print that had figured in lithographed albums for at least a decade and that would develop into the sequential narratives of comics. Its very first issue included a page of twelve Mayeux vignettes by Traviès, interrelated thematically but not sequentially (fig. 54). *Le Charivari* was also the first periodical to publish the "Salon in Caricature" that soon would develop into a regular accompaniment to

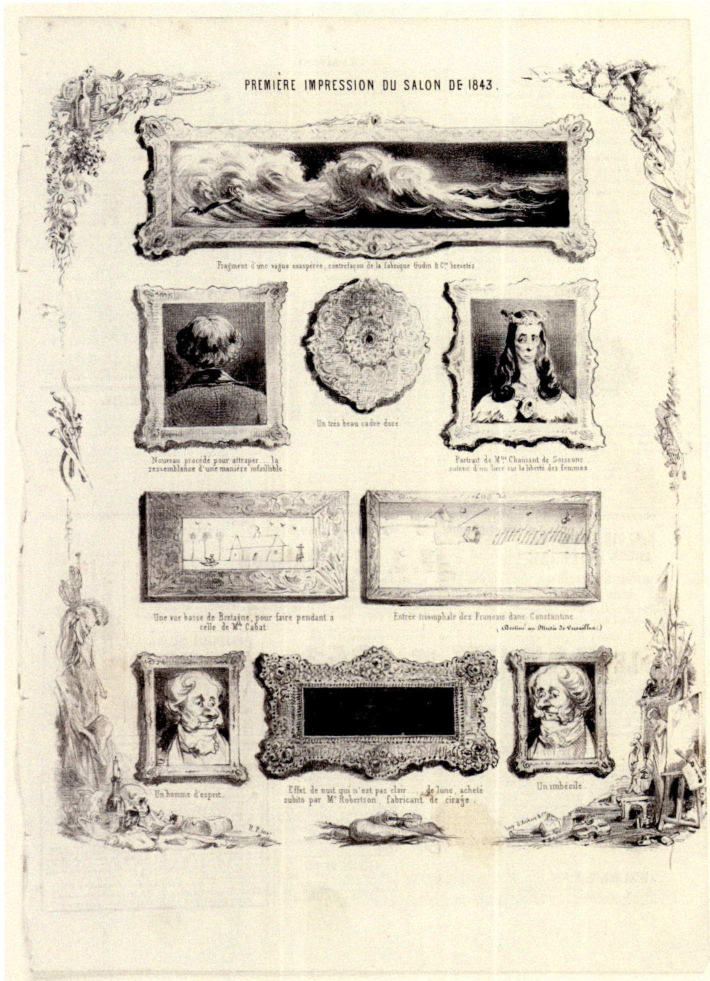

FIGURE 55

R. P. [Raimond Pelez], *First Impression of the Salon of 1843* [*Première impression du salon de 1843*]. Pub. Aubert. *Le Charivari*, March 19, 1843. Harvard Art Museums/Fogg Museum, Gift of Charles H. Taylor, M5119.

every art exhibition throughout the century. The first, by Raimond Pelez in 1843, set the tone by caricaturing a variety of images in both word and image (fig. 55). Here we see *A Beautiful Gilt Frame,* an enormous extravaganza that encloses a microscopic painting, and an all-black canvas titled *Effect of Night, Not Moonlit, Quickly Bought by Mr. Robertson, Manufacturer of Shoe Polish.*[134] On a more serious note, *L'Artiste* popularized the work of Romantic artists, disseminating a taste for the vernacular French landscape through its weekly lithographs. The Barbizon school of landscapists was especially prominent, with Jules Dupré contributing numerous lithographs suitable for framing, such as *The Mill in the Sologne* (fig. 56).

Increasingly, the lithographed drawings of modern life that had proved so exciting in the early decades of the nineteenth century became an accepted aspect of everyday life. The images that appeared initially only in print shop windows were now

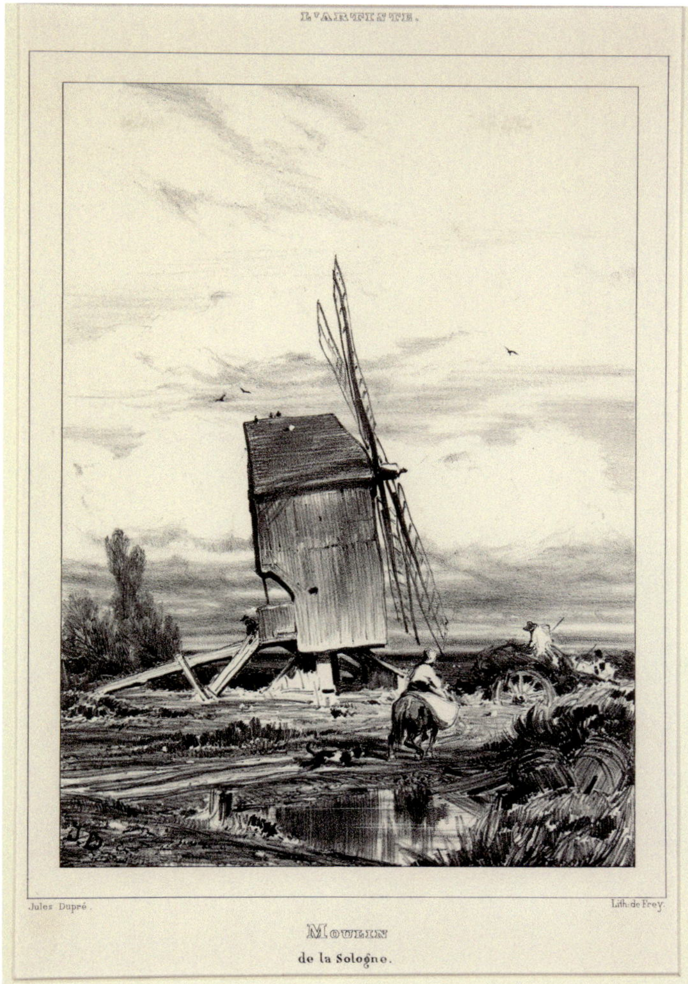

FIGURE 56
Jules Dupré, *The Mill in the Sologne* [*Le moulin de la Sologne*]. Lith. Frey. *L'Artiste* 10 (1835), no. 17. Yale University Art Gallery, Gift of G. Allen Smith for the G. Allen Smith Collection.

available everywhere, even in the daily or weekly press. The euphoria that had greeted the new invention of lithography early in the century and the successive innovations brought by prints, portfolios, albums, and periodicals gradually became a normal part of daily life. In 1834 *L'Artiste* would complain: "It's to French artists that lithography owes its most remarkable progress; it has become popular here with a marvelous rapidity. It has invaded the entire art of drawing, it has exploited all tastes, all whims, every interest of the public: art, literature, science, industry, fashion. It has lent the assistance and the facility of its crayon to everything. Lithography has popularized itself to such an extent by prostituting itself in this way, that it has ended by being disparaged and by provoking the scorn and disdain of serious artists who want to preserve the dignity of art."[135]

As a result of the success of *Le Charivari* (and despite Philipon's protestations to the contrary), lithographed periodicals increasingly became identified with caricature and

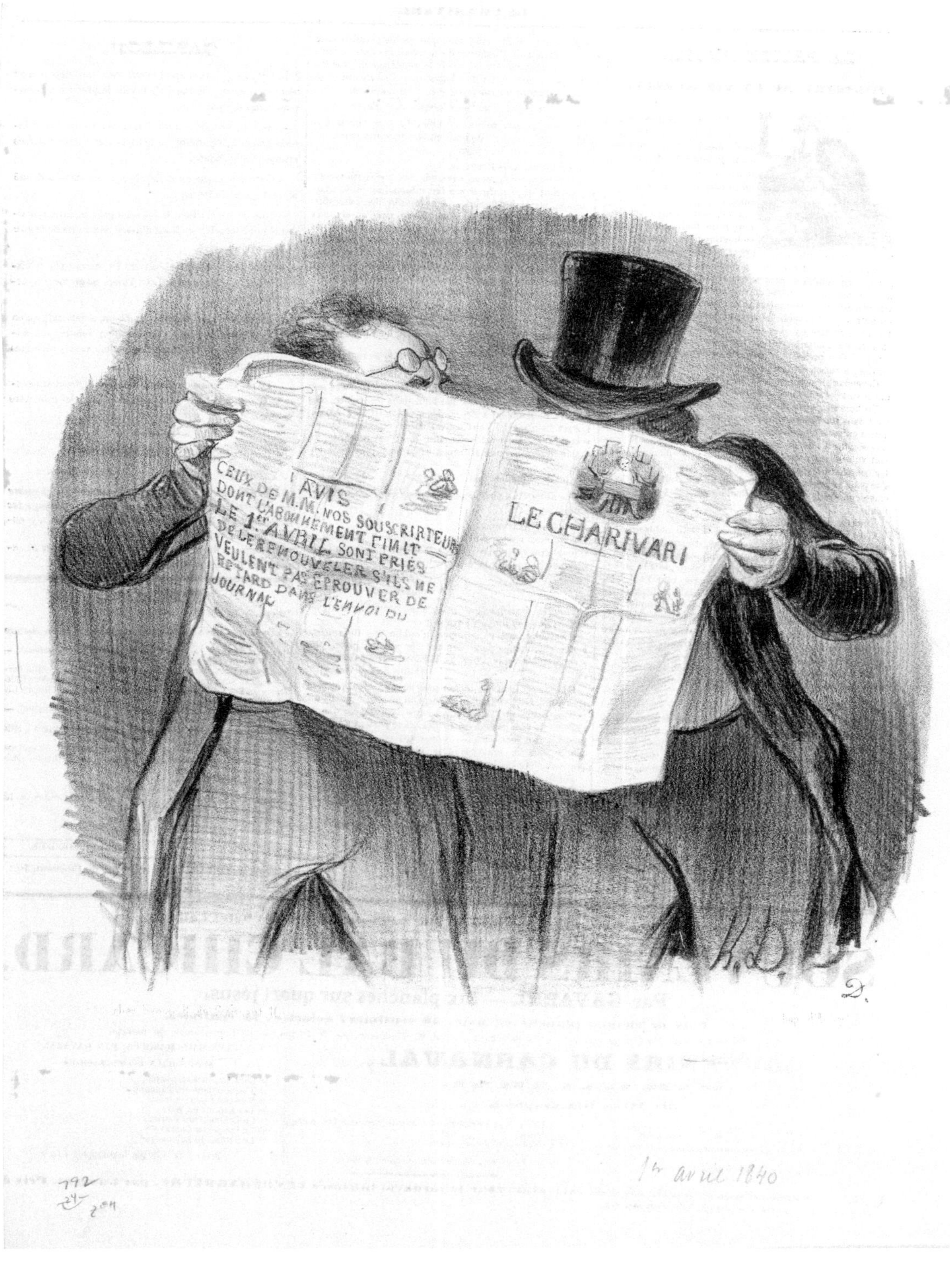
AVIS
CEUX DE M.M. NOS SOUSCRIRTEUR
DONT L'ABONNEMENT FINIT
LE 1er AVRIL SONT PRIÉS
DE LE RENOUVELER S'ILS NE
VEULENT PAS ÉPROUVER DE
RETARD DANS L'ENVOI DU
JOURNAL
LE CHARIVARI
1er avril 1840

FIGURE 57
Honoré Daumier, *Notice. Those gentlemen among our readership whose subscription expires on April 1st are requested to renew it if they wish to avoid a delay in receiving the journal* [*Avis. Ceux de M.M. nos souscripteurs dont l'abonnement finit le 1er avril sont priés de le renouveler s'ils ne veulent pas éprouver de retard dans l'envoi du journal*]. Lith. Bénard, Pub. Aubert. *Le Charivari*, April 1, 1840. Metropolitan Museum of Art, Bequest of Grace M. Pugh, 1985.

humorous images in general. The great age of illustrated humor magazines had begun. Soon there was *La Caricature provisoire* (1838), *Le Musée Philipon* (1840), *Paris Comique* (1845), and *Le Journal pour rire* (1847), to name just a few. Although the French had been slow to initiate the observation of daily life in caricatures and genre prints, they quickly surpassed the English, their old rivals, to the point where *Punch*, established in London in 1841, chose as its subtitle *The London Charivari*.[136]

The process of lithography had inspired the first great age of printed imagery in France, making prints cheaper and more widely disseminated than ever before, ultimately leading to the illustrated press. Even while these early lithographed periodicals were in their first flower, however, in the early 1830s the innovative adaptation of wood engraving would revolutionize printing by enabling text and image to be combined on the same page. Wood engraving proved irresistible, even to *Le Charivari*, whose masthead had utilized this medium from its inception, even while its full-page drawings were always lithographed. In 1838, its subtitle was changed to read, "Publishing every day a new drawing, either lithographed or engraved, along with wood-engraved vignettes."[137] As can be seen in the page layout visible in the 1840 issue of *Le Charivari* (fig. 57), it had joined the ranks of the new illustrated press that it had itself inaugurated—but that is the subject for the next chapter.

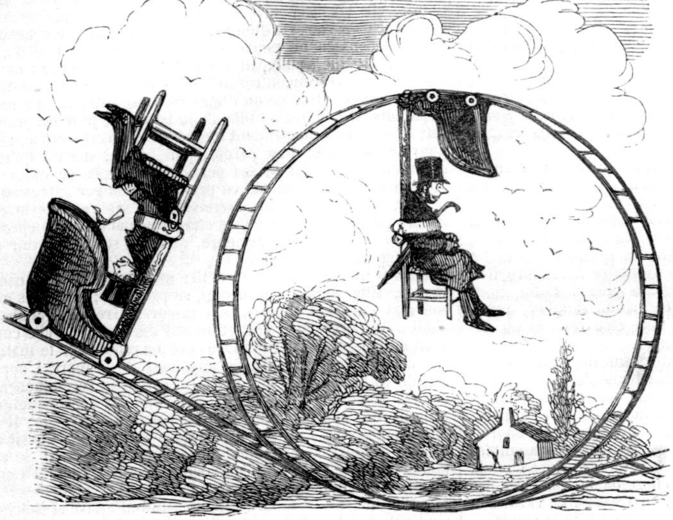

(Comment on devrait placer les voyageurs afin qu'ils n'aient pas la tête en bas pendant le voyage.)

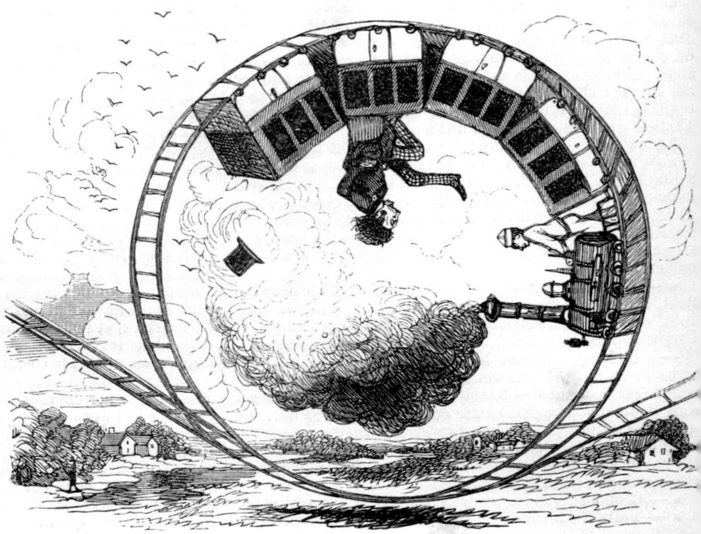

(Sapristi ! moi qui croyais avoir pris ma place sur l'impériale de la voiture !)

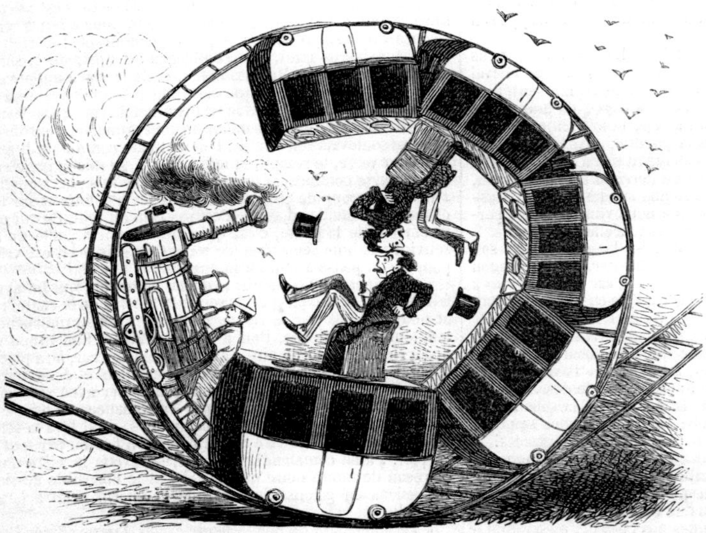

(Inconvénient de voyager dans un cercle d'un trop petit diamètre.)

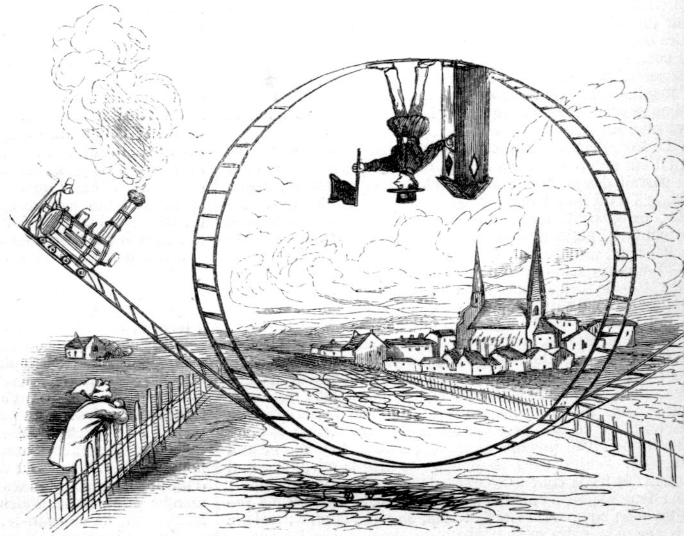

(Un cantonnier dans l'exercice de ses fonctions sur le chemin de fer centrifuge.)

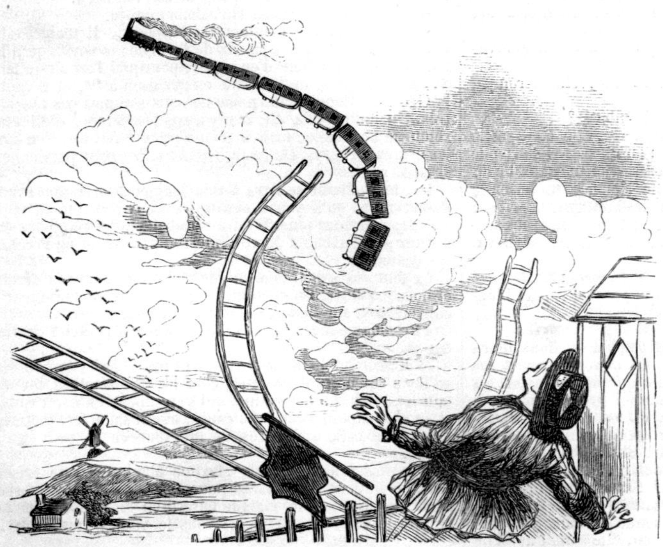

(Un convoi jeté hors de la voie par la rupture du cercle.)

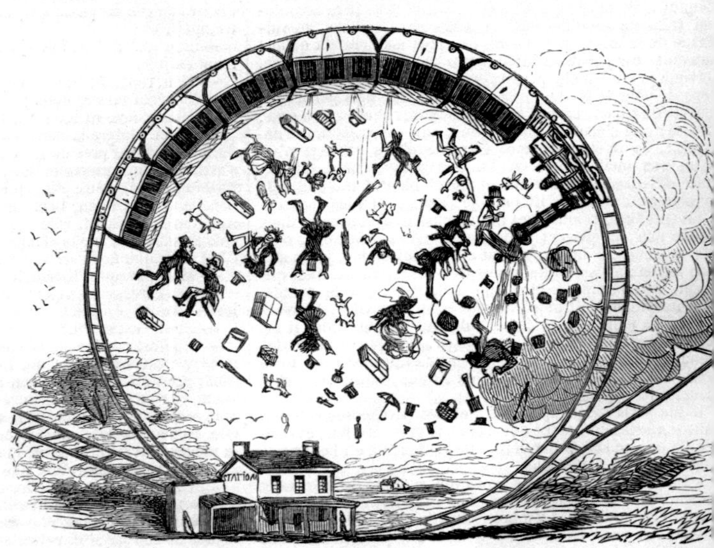

(De quelle façon, le convoi venant à ralentir sa marche, les voyageurs arrivent à la station avant le train.)

2 Spreading the News

The Illustrated Press

The early illustrated periodicals introduced in the previous chapter incorporated one or more full-page lithographs in each issue, an innovation that resulted in the widespread dissemination of images of modern life to a public broader than ever before. Nonetheless, it was only the development of wood engraving into a process allowing text and images to be combined on a single page that resulted in the transformation of the illustrated press from the lithographic albums of the 1820s into modern newspapers. Within a few decades, hundreds of illustrated periodicals were founded throughout the world devoted to virtually every subject; as the century progressed, they were ever more lavishly illustrated.[1] Alongside journals focused on fashion, politics, or the arts, a new type of periodical began to appear—the illustrated general survey magazine, what the British call a miscellany. Unlike earlier publications, this new format aspired to encyclopedic content and featured both illustrations and text on virtually every page. While such publications were made possible by technological innovations such as the industrial production of paper, the use of stereotypes from wood engravings, and fast steam-driven presses, it was the existence of an ever-growing literate public that encouraged these innovations and made them profitable.

A discussion of the illustrated press might well begin with the broadsheets called *canards* in France, sold by what was sardonically called a *marchand des crimes*, a hawker of crime (fig. 58). These combined an explanatory text with a crude woodcut image—often the same image recycled, on the principle that one murder, flood, or fire looked very like another. Annual almanacs published images along with calendars and other important information. It is generally acknowledged, however, that it was the fashion press that produced the first regularly published periodicals featuring illustrations: *Mercure galant* (1678) was the earliest such publication.[2] In its wake almost 150 such periodicals were established in France alone before 1850.[3] Some, to be sure, were short-lived, but others, such as *Journal des dames et des modes* (1797–1837) (see fig. 22) or *Petit courrier des dames* (1822–1868), continued for decades, often including articles on topics other than fashion. Illustrations in the fashion press, however, as in all publications until the 1830s, were *hors texte*, with the etchings, engravings, or later, lithographs, printed on separate sheets of paper, the caption incised or drawn in reverse on the plate; letterpress, composed of movable type, presented a less demanding alternative but required the plate to be run

DETAIL OF FIGURE 87

through the press twice, once for the image and a second time for the text. Until the establishment of the lithographed journals discussed in the previous chapter, one plate per issue had been the norm for periodicals. Since dailies or weeklies were customarily either four or eight pages (one or two sheets, folded and printed on both sides), binding wasn't necessary, although subscribers often had their issues bound annually into individual volumes. When printed on fine paper and hand-colored, the result was a luxury limited-edition publication. Rarely did such periodicals have a subscriber list greater than a thousand, nor could they, given the limitations of print production.[4]

When wood engraving came into general use, the look of the illustrated press changed dramatically to the format familiar to us today, with columns of type interspersed with images of varying sizes. We have seen how the page format of *Le Charivari* changed from 1832 to 1840 (see figs. 52, 57). Since the advantages conferred by the wood engraving process are at the heart of the nineteenth-century expansion and transformation of the illustrated press, wood-engraved images will form the focus of this chapter.

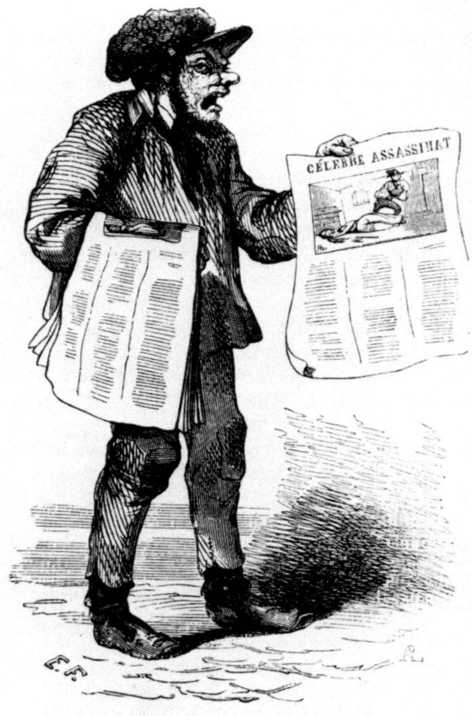

(Le Marchand de Crimes.)

FIGURE 58
E. F. [Eugène Forest], *The Hawker of Crime* [*Le marchand des crimes*], *L'Illustration*, March 23, 1844. Wood engraving.

WOOD ENGRAVING

In wood engraving, the image is incised on the dense end grain of hardwood, usually boxwood, instead of softer side-planks that are inclined to splinter along the grain. This shift allowed images of finer detail, and the plate could be clamped into a frame with letterpress type to print a page of combined text and image. The British again took the lead here, as they had earlier in the production of caricatures.[5] Although wood engraving was known earlier, it was Thomas Bewick (1753–1828) who brought the medium into common usage and trained the engravers who would revolutionize illustrated print culture. Bewick's own publications, such as his *General History of Quadrupeds* (1790) and his *History of British Birds* (1797–1804), stand as monuments to the art of wood engraving; *The Chillingham Bull* is regarded as one of his finest works (fig. 59). When Bewick's student Charles Thompson (1789–1843) arrived in France in 1816, he brought his métier with him and mentored the young engravers who would become the most sought-after practitioners of the art: Louis-Henri Brevière (1797–1869), Henri-Désiré Porret (b. 1800), and the team of John Andrew (1817–1870), Jean Best (1808–1879), and Isidore Leloir (b. ca. 1802/3), whose monogram ABL became synonymous with fine wood engraving.[6] By the early 1830s, wood engraving was common enough in practice that it was feasible to conceive of its use in an illustrated periodical.

The process began with a drawing that was first transferred to and then engraved on the woodblock. This plate was so laborious to produce that engraving a large one could take a week of full-time labor to complete. Since even end-grain hardwood is relatively fragile, the resultant plate was rarely used for printing. Instead it served as a matrix for the stereotype, a metal plate that produced an exact copy—and gave us our pejorative term for unimaginative repetitive concepts. Stereotypes were produced through a mechanical process by which a plaster mold was made from the original woodblock, and from that plaster the metal plate was produced for the actual printing. This stereotype could withstand the rigors of mass-production printing and had the advantage that it could be replaced easily if damaged or broken, whereas original woodblocks would have to be recut. In addition, since many identical stereotypes could be cast from the original wood plate, larger and faster editions were possible with numerous presses churning out multiple, identical, pages.[7]

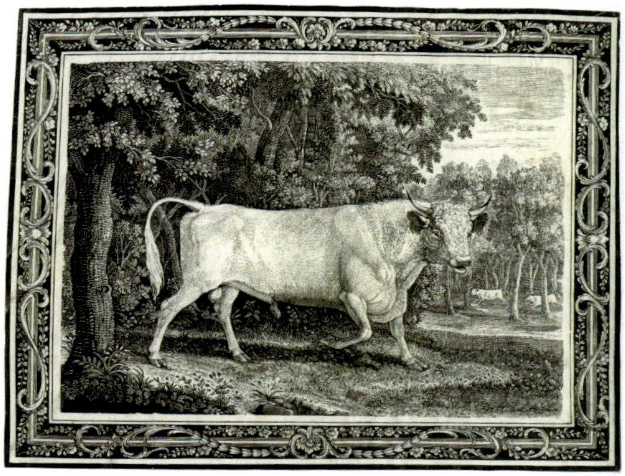

FIGURE 59
Thomas Bewick, *The Chillingham Bull*, 1789. Wood engraving. Southwestern University Special Collections.

At the same time, presses themselves were undergoing a transformation. For centuries, presses had been operated by one or two printers who laboriously turned the huge wheel forward and back to produce each sheet. Production was static, and so, by the end of the eighteenth century, this procedure could produce around 150 sheets per hour, no faster than in preceding centuries.[8] With the development of the mechanized steam-driven press with continuous cylindrical rotation, a thousand sheets per hour could be printed, and with the use of stereotypes, production could be multiplied further. *The Times* of London installed these new presses in 1814, and steam-driven presses were brought to France nine years later. These *mécaniques*, as they were called, rapidly transformed the periodical press and enabled ever-larger and cheaper editions, although not without the opposition of printers who saw their livelihood threatened.[9]

THE ILLUSTRATED *PENNY PRESS*

The first illustrated general survey periodical was the *Penny Magazine*, published in London beginning March 31, 1832 (fig. 60). *Le Charivari*, by contrast, didn't begin publication until December 1, 1832 and was always more polemical than newspapers or general survey periodicals. There is, however, a French contender for the honor of primacy, cited often in French histories: *Journal des connaissances utiles*, a monthly whose first issue was published in October 1831 by the National Society for Intellectual Emancipation. Émile de Girardin (1806–1881) was behind this effort; he would found numerous journals in his career, the best known of which was the first mass-circulation daily newspaper in France, *La Presse* (1836–1952).[10] Girardin's strategy—to sell cheaply in order to sell in quantity, to sell in quantity in order to sell cheaply, and to sell advertising in each issue to ensure a profit—would completely transform the French publishing industry. This "Journal of Useful Knowledge" was his first effort in this direction; it offered thirty-two pages monthly for only four francs per year. In comparison, the elite monthly *Revue des deux mondes* charged twelve times that amount and *Le Charivari* cost eighteen times that amount. The claim that *Journal des connaissances utiles* was the first mass-circulation illustrated periodical is based on its inclusion of wood-engraved illustrations within its text. These, however, were small technical diagrams. The first—images of a sundial and of siphons—appeared in the April 1832 issue, which made this journal, at best, coeval with the first issue of the *Penny Magazine*.[11] Only beginning in August 1832, six months after the establishment of the *Penny Magazine*, did *Journal des connaissances utiles* regularly include illustrations; nonetheless, by December it could proudly announce that each issue featured ten illustrations, the work of a Mr. Leblanc, who was identified only as the draftsman of the Paris Conservatoire des arts et métiers.[12] Despite the small size and low quality of its illustrations, however, the success of *Journal des connaissances utiles* was immediate: by the end of 1832 it had 130,000 subscribers, an unprecedented number for a periodical but one attained through the expedient of not actually collecting annual subscription fees.[13] Nonetheless, it did reach a truly impressive number of readers at a time when the average circulation of most French daily newspapers was under fifteen thousand.[14] Its range of articles was indicated by the complete title printed on its cover: *Journal of useful political, agricultural, and commercial knowledge, explaining to all men who can read: their duties, as citizen, juror, national guardsman, mayor or adjunct mayor, field or forest guard; their rights, as taxpayer, municipal elector, municipal councilor, voter, candidate; their interests, as head of household, landlord, farmer, manufacturer, merchant, worker*.[15] The journal provided information and advice of all kinds, from the latest international news to precautions against cholera, from how to preserve eggs to lessons in rudimentary mathematics. Articles featured brief biographies of major historical figures, explained the workings of the latest inventions, and published and discussed all legislation important to a newly enfranchised citizenry. In a signed article promoting the importance of an educated public, Girardin wrote, "Pour education over the heads of the people; you owe them this baptism!"[16] The astonishing success of *Journal des connaissances utiles* was noted by the newspaper *Le Temps*, which explained: "Almanacs, because of their modest price, have been until now the only reading possible for two million citizens, who are thus deprived of the most useful ideas. This lacuna between newspapers and almanacs has now been partially filled by a monthly publication, precise and organized, costing *four francs per year*."[17]

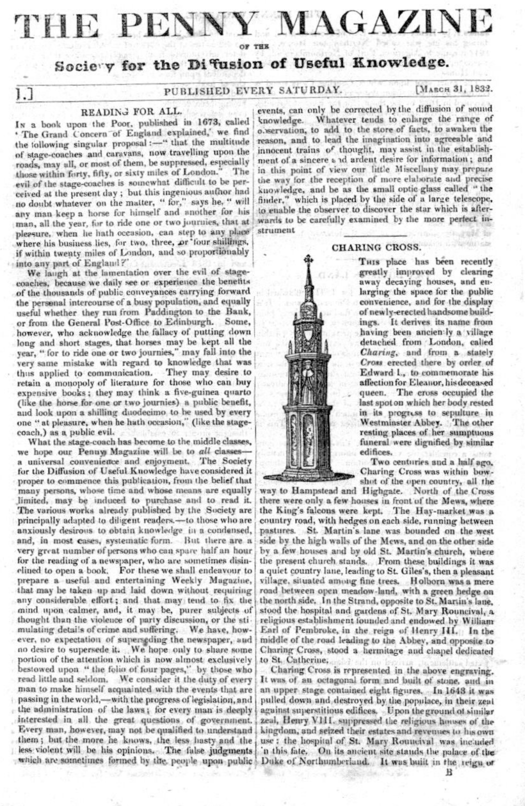
THE PENNY MAGAZINE
OF THE
Society for the Diffusion of Useful Knowledge.

1.] PUBLISHED EVERY SATURDAY. [March 31, 1832.

READING FOR ALL.

In a book upon the Poor, published in 1673, called 'The Grand Concern of England explained,' we find the following singular proposal:—"that the multitude of stage-coaches and caravans, now travelling upon the roads, may all, or most of them, be suppressed, especially those within forty, fifty, or sixty miles of London." The evil of the stage-coaches is somewhat difficult to be perceived at the present day; but this ingenious author had no doubt whatever on the matter, "for," says he, "will any man keep a horse for himself and another for his man, all the year, for to ride one or two journies, that at pleasure, when he hath occasion, can step to any place where his business lies, for two, three, or four shillings, if within twenty miles of London, and so proportionably into any part of England?"

We laugh at the lamentation over the evil of stage-coaches, because we daily see or experience the benefits of the thousands of public conveyances carrying forward the personal intercourse of a busy population, and equally useful whether they run from Paddington to the Bank, or from the General Post-Office to Edinburgh. Some, however, who acknowledge the fallacy of putting down long and short stages, that horses may be kept all the year, "for to ride one or two journies," may fall into the very same mistake with regard to knowledge that was thus applied to communication. They may desire to retain a monopoly of literature for those who can buy expensive books; they may think a five-guinea quarto (like the horse for one or two journies) a public benefit, and look upon a shilling duodecimo to be used by every one "at pleasure, when he hath occasion," (like the stage-coach,) as a public evil.

What the stage-coach has become to the middle classes, we hope our Penny Magazine will be to *all* classes—a universal convenience and enjoyment. The Society for the Diffusion of Useful Knowledge have considered it proper to commence this publication, from the belief that many persons, whose time and whose means are equally limited, may be induced to purchase and to read it. The various works already published by the Society are principally adapted to diligent readers,—to those who are anxiously desirous to obtain knowledge in a condensed, and, in most cases, systematic form. But there are a very great number of persons who can spare half an hour for the reading of a newspaper, who are sometimes disinclined to open a book. For these we shall endeavour to prepare a useful and entertaining Weekly Magazine, that may be taken up and laid down without requiring any considerable effort; and that may tend to fix the mind upon calmer, and, it may be, purer subjects of thought than the violence of party discussion, or the stimulating details of crime and suffering. We have, however, no expectation of superseding the newspaper, and no desire to supersede it. We hope only to share some portion of the attention which is now almost exclusively bestowed upon "the folio of four pages," by those who read little and seldom. We consider it the duty of every man to make himself acquainted with the events that are passing in the world,—with the progress of legislation, and the administration of the laws; for every man is deeply interested in all the great questions of government. Every man, however, may not be qualified to understand them; but the more he knows, the less hasty and the less violent will be his opinions. The false judgments which are sometimes formed by the people upon public events, can only be corrected by the diffusion of sound knowledge. Whatever tends to enlarge the range of observation, to add to the store of facts, to awaken the reason, and to lead the imagination into agreeable and innocent trains of thought, may assist in the establishment of a sincere and ardent desire for information; and in this point of view our little Miscellany may prepare the way for the reception of more elaborate and precise knowledge, and be as the small optic glass called "the finder," which is placed by the side of a large telescope, to enable the observer to discover the star which is afterwards to be carefully examined by the more perfect instrument.

CHARING CROSS.

This place has been recently greatly improved by clearing away decaying houses, and enlarging the space for the public convenience, and for the display of newly-erected handsome buildings. It derives its name from having been anciently a village detached from London, called *Charing*, and from a stately *Cross* erected there by order of Edward I., to commemorate his affection for Eleanor, his deceased queen. The cross occupied the last spot on which her body rested in its progress to sepulture in Westminster Abbey. The other resting places of her sumptuous funeral were dignified by similar edifices.

Two centuries and a half ago, Charing Cross was within bow-shot of the open country, all the way to Hampstead and Highgate. North of the Cross there were only a few houses in front of the Mews, where the King's falcons were kept. The Hay-market was a country road, with hedges on each side, running between pastures. St. Martin's lane was bounded on the west side by the high walls of the Mews, and on the other side by a few houses and by old St. Martin's church, where the present church stands. From these buildings it was a quiet country lane, leading to St. Giles's, then a pleasant village, situated among fine trees. Holborn was a mere road between open meadow-land, with a green hedge on the north side. In the Strand, opposite to St. Martin's lane, stood the hospital and gardens of St. Mary Rouncival, a religious establishment founded and endowed by William Earl of Pembroke, in the reign of Henry III. In the middle of the road leading to the Abbey, and opposite to Charing Cross, stood a hermitage and chapel dedicated to St. Catherine.

Charing Cross is represented in the above engraving. It was of an octagonal form and built of stone, and in an upper stage contained eight figures. In 1643 it was pulled down and destroyed by the populace, in their zeal against superstitious edifices. Upon the ground of similar zeal, Henry VIII. suppressed the religious houses of the kingdom, and seized their estates and revenues to his own use: the hospital of St. Mary Rouncival was included in this fate. On its ancient site stands the palace of the Duke of Northumberland. It was built in the reign of

B

FIGURE 60
The first issue of *The Penny Magazine*, March 31, 1832.

The similarities between *Journal des connaissances utiles* and the later *Penny Magazine* were not accidental, as the National Society for Intellectual Emancipation, the publisher of *Journal des connaissances utiles*, acknowledged that it was modeled after the English Society for the Diffusion of Useful Knowledge, the publisher of the *Penny Magazine*.[18] Philanthropic societies promoting social reform through education were operative in both countries during these years because the political turmoil, riots, and revolutions of previous decades had instilled in less radical reformers everywhere the same imperative: to educate "the people," and to disseminate healthy—as opposed to revolutionary—ideas that would transform the populace into moral, law-abiding citizens. For Girardin, education was the key to political stability; he wrote: "What the Society for Elementary Instruction has done for the primary education of children, another association, the National Society for Intellectual Emancipation, has just undertaken for the great number of men upon whom our new institutions have conferred political rights and municipal duties before they are capable of carrying them out intelligently."[19] Despite the paucity of illustrations at its debut, *Journal des connaissances utiles* did take precedence over the *Penny Magazine* in one respect: it was the first illustrated mass-circulation periodical specifically aimed at the working classes.[20] Before the establishment of the *Penny Magazine*, the publication program of the British Society for the Diffusion of Useful Knowledge had been limited to almanacs, books, and pamphlets; as a result, Émile de Girardin's *Journal des connaissances utiles*, established six months earlier, must be credited, if not as the first illustrated periodical, then at least as the first to include a new class of readers in the audience for the periodical press.

THE *PENNY MAGAZINE*

The English reformer Charles Knight (1791–1873) shared Girardin's populist sentiments, as well as his mission of social reform.[21] As superintendent of publications for the Society for the Diffusion of Useful Knowledge, Knight was effectively both publisher and editor of the *Penny Magazine*. His audience, as he wrote in its first issue, was "a very great number of persons who can spare half an hour for the reading of a newspaper, who are sometimes disinclined to open a book. For these we shall endeavor to prepare a useful and entertaining Weekly Magazine, that may be taken up and laid down without requiring any considerable effort; and that may tend to fix the mind upon calmer, and, it may be, purer subjects of thought than the violence of party discussion, or the stimulating details of crime and suffering."[22]

The major difference between its articles and those of *Journal des connaissances utiles* was the omission of practical advice and current events; the *Penny Magazine* aimed, instead, to raise the general level of education of its readership and to combat the influence of the radical press among the working classes. Knight's own account of its inception emphasizes this: while walking with his neighbor, Matthew Davenport Hill, Member of Parliament for Hull, he and Hill were talking of "cheap and offensive publications." "'Let us,' he exclaimed, 'see what something cheap and good can accomplish! Let

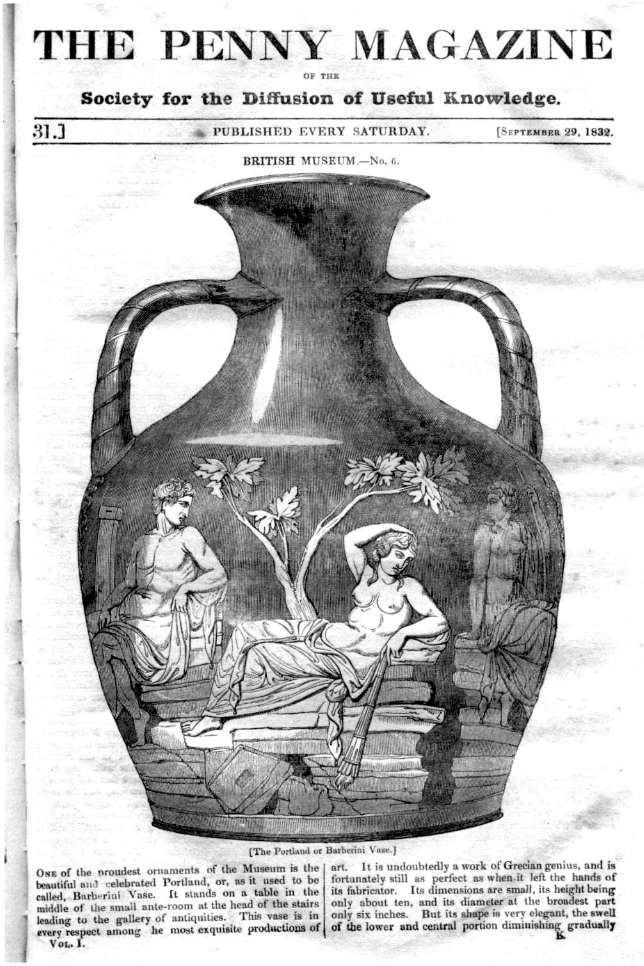

THE PENNY MAGAZINE

OF THE

Society for the Diffusion of Useful Knowledge.

31.] PUBLISHED EVERY SATURDAY. [September 29, 1832.

BRITISH MUSEUM.—No. 6.

[The Portland or Barberini Vase.]

One of the proudest ornaments of the Museum is the beautiful and celebrated Portland, or, as it used to be called, Barberini Vase. It stands on a table in the middle of the small ante-room at the head of the stairs leading to the gallery of antiquities. This vase is in every respect among he most exquisite productions of art. It is undoubtedly a work of Grecian genius, and is fortunately still as perfect as when it left the hands of its fabricator. Its dimensions are small, its height being only about ten, and its diameter at the broadest part only six inches. But its shape is very elegant, the swell of the lower and central portion diminishing gradually

Vol. I. K

FIGURE 61

The Portland or Barberini Vase, *The Penny Magazine*, September 29, 1832. Wood engraving.

us have a Penny Magazine!' 'And what shall be its title?' said I. 'THE PENNY MAGAZINE.'"[23] Knight took Hill's idea to the Society for the Diffusion of Useful Knowledge, which agreed to act as publisher, with Knight as editor.

By the end of its first year of publication, Knight could state:

> The subjects which have uniformly been treated have been of the broadest and simplest character. Striking points of Natural History—Accounts of the great Works of Art in Sculpture and Painting—Description of such Antiquities as possess historical interest—Personal Narratives of Travellers—Biographies of Men who have had a permanent influence on the condition of the world—Elementary Principles of Language and Numbers—established facts in Statistics and Political Economy—these have supplied the materials for exciting the curiosity of a million of readers. This consideration furnishes the most convincing answer to the few (if any now remain) who assert that General Education is an evil.[24]

Sales figures of the *Penny Magazine* showed that he had correctly gauged his readership. In his preface to the first volume, he wrote: "It was considered by Edmund Burke about forty years ago, that there were eighty thousand *readers* in this country. In the present year it has been shown, by the sale of the 'Penny Magazine,' that there are two hundred thousand *purchasers* of one periodical work. It may be fairly calculated that the number of readers of that single work amounts to a million."[25] Lively, attractive, and well illustrated right from its beginnings, the *Penny Magazine* immediately enjoyed international fame, inspiring emulation everywhere. But it was not only its articles that proved attractive to readers. Even Charles Knight had to admit: "It must not, however, be forgotten that some of the unexampled success of this little work is to be ascribed to the liberal employment of illustrations by means of Wood-cuts."[26] This was the feature that set the *Penny Magazine* apart from its rivals, and, most especially, from *Journal des connaissances utiles.* With the *Penny Magazine,* the era of the illustrated press had truly begun.

Knight himself seems to have been somewhat unprepared for the attraction exercised by his "liberal employment" of wood-engraved images. He wrote: "At the commencement of the publication, before the large sale it has reached could at all have been contemplated, the cuts were few in number, and partly selected from another work of the Society—the 'Library of Entertaining Knowledge.' But as the public encouragement enabled the conductors to make greater exertions, to give permanency to the success which the 'Penny Magazine' had attained, it became necessary to engage artists of eminence, both as draughtsmen and as wood engravers, to gratify a proper curiosity, and cultivate an increasing taste, by giving representation of the finest Works of Art, of Monuments of Antiquity, and of subjects of Natural History, in a style that had been previously considered to belong only to expensive books."[27] Indeed, in an age where images were few, expensive, and, with the exception of popular prints, reserved for the affluent classes, it is no wonder that Knight's *Penny Magazine* proved so appealing.

Its first issue contained only three small engravings, all borrowed from previous publications. An image of Charing Cross appeared beneath the masthead on the first page, accompanying an article on the monument (see fig. 60). Inside, images of the horns of the wapiti and of a bear mauling a horse bore little relationship to the accompanying article on the zoological gardens of Regent's Park, London. The images steadily improved in size, number, and quality, however, as did the page layout and print quality. By the issue of April 14, 1832, Knight had realized that a two-column illustration was more attractive than the small ones of the first issue, and from then on each issue contained four to six illustrations, one or more of which would span the page. Within a few months, the *Penny Magazine* had invented the cover format that virtually all illustrated periodicals would adopt henceforth—a dramatic full-page illustration under the masthead. The first of these, the cover of September 29, 1832, bore a wood engraving, *The Portland or Barberini Vase,* and was indeed impressive (fig. 61).

In his memoirs, Knight boasted, and with reason: "I availed myself—perhaps more than most of the publishing of that period—of the revived process of wood engraving,

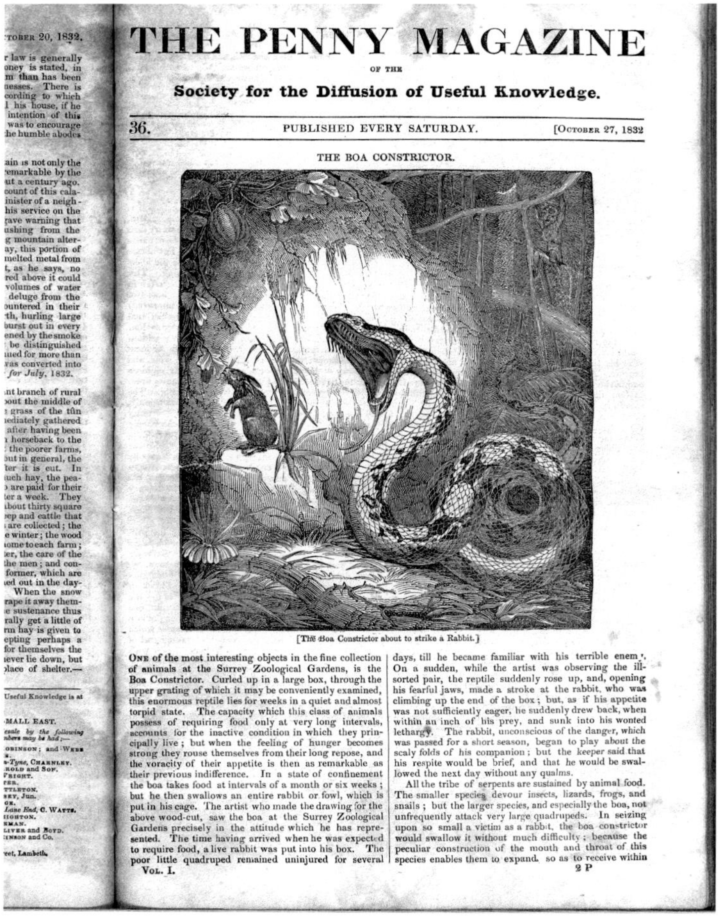

THE PENNY MAGAZINE

OF THE

Society for the Diffusion of Useful Knowledge.

36. PUBLISHED EVERY SATURDAY. [OCTOBER 27, 1832

THE BOA CONSTRICTOR.

[The Boa Constrictor about to strike a Rabbit.]

ONE of the most interesting objects in the fine collection of animals at the Surrey Zoological Gardens, is the Boa Constrictor. Curled up in a large box, through the upper grating of which it may be conveniently examined, this enormous reptile lies for weeks in a quiet and almost torpid state. The capacity which this class of animals possess of requiring food only at very long intervals, accounts for the inactive condition in which they principally live; but when the feeling of hunger becomes strong they rouse themselves from their long repose, and the voracity of their appetite is then as remarkable as their previous indifference. In a state of confinement the boa takes food at intervals of a month or six weeks; but he then swallows an entire rabbit or fowl, which is put in his cage. The artist who made the drawing for the above wood-cut, saw the boa at the Surrey Zoological Gardens precisely in the attitude which he has represented. The time having arrived when he was expected to require food, a live rabbit was put into his box. The poor little quadruped remained uninjured for several days, till he became familiar with his terrible enemy. On a sudden, while the artist was observing the ill-sorted pair, the reptile suddenly rose up, and, opening his fearful jaws, made a stroke at the rabbit, who was climbing up the end of the box; but, as if his appetite was not sufficiently eager, he suddenly drew back, when within an inch of his prey, and sunk into his wonted lethargy. The rabbit, unconscious of the danger, which was passed for a short season, began to play about the scaly folds of his companion; but the keeper said that his respite would be brief, and that he would be swallowed the next day without any qualms.

All the tribe of serpents are sustained by animal food. The smaller species devour insects, lizards, frogs, and snails; but the larger species, and especially the boa, not unfrequently attack very large quadrupeds. In seizing upon so small a victim as a rabbit, the boa constrictor would swallow it without much difficulty; because the peculiar construction of the mouth and throat of this species enables them to expand, so as to receive within

VOL. I. 2 P

FIGURE 62

The Boa Constrictor about to Strike a Rabbit, *The Penny Magazine*, October 27, 1832. Wood engraving.

to diffuse popular Art as well as popular Literature. In this species of enterprise 'The Penny Magazine' led the way."[28] Elementary art appreciation, "to familiarize the people with great works of art," was one of his major goals, and soon the *Penny Magazine* was publishing what he called "engravings of a costly character."[29] He more than made good on his promise to provide quality illustrations of great works of art: within its first year, there were full-page engravings of the Colosseum, the Parthenon, the Laocoön, Raphael's tapestry cartoons, and the Apollo Belvedere.[30]

Knight had sworn that in his journal there would be "no excitements for the lovers of the marvellous," a regular feature of broadsheets, but he more than made up for this omission by including images that would certainly have seemed marvelous to his readers: a zebra, a hippopotamus, a kangaroo, a crocodile, a giraffe (twice!), an elephant, and, in a tour de force of wood engraving, a full-page cover illustration of a boa constrictor (fig 62).[31] Faraway peoples, exotic customs, and natural history were described at length and illustrated, as in *Wild Bushman, Singular Dexterity of a Goat,* and *Catching Turtles on the Coast of Cuba* (fig. 63).[32]

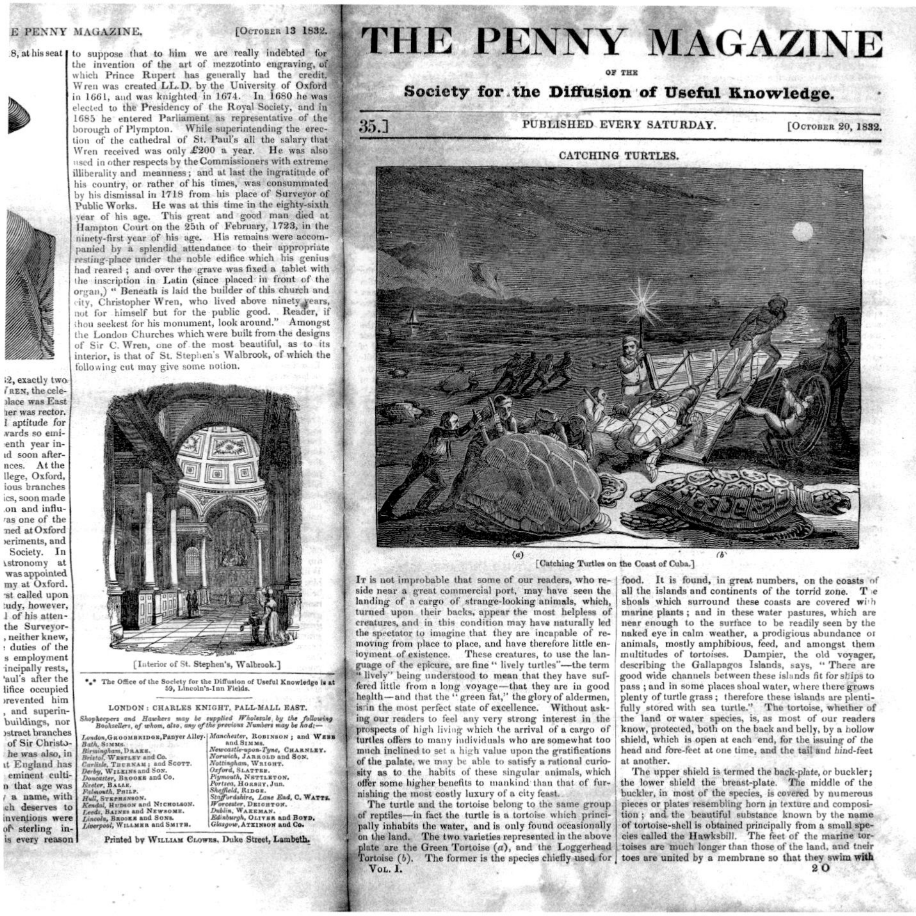

E PENNY MAGAZINE. [October 13 1832.

to suppose that to him we are really indebted for the invention of the art of mezzotinto engraving, of which Prince Rupert has generally had the credit. Wren was created LL.D. by the University of Oxford in 1661, and was knighted in 1674. In 1680 he was elected to the Presidency of the Royal Society, and in 1685 he entered Parliament as representative of the borough of Plympton. While superintending the erection of the cathedral of St. Paul's all the salary that Wren received was only £200 a year. He was also used in other respects by the Commissioners with extreme illiberality and meanness; and at last the ingratitude of his country, or rather of his times, was consummated by his dismissal in 1718 from his place of Surveyor of Public Works. He was at this time in the eighty-sixth year of his age. This great and good man died at Hampton Court on the 25th of February, 1723, in the ninety-first year of his age. His remains were accompanied by a splendid attendance to their appropriate resting-place under the noble edifice which his genius had reared; and over the grave was fixed a tablet with the inscription in Latin (since placed in front of the organ,) "Beneath is laid the builder of this church and city, Christopher Wren, who lived above ninety years, not for himself but for the public good. Reader, if thou seekest for his monument, look around." Amongst the London Churches which were built from the designs of Sir C. Wren, one of the most beautiful, as to its interior, is that of St. Stephen's Walbrook, of which the following cut may give some notion.

[Interior of St. Stephen's, Walbrook.]

*** The Office of the Society for the Diffusion of Useful Knowledge is at 59, Lincoln's-Inn Fields.

LONDON: CHARLES KNIGHT, PALL-MALL EAST.

Shopkeepers and Hawkers may be supplied Wholesale by the following Booksellers, of whom, also, any of the previous Numbers may be had:—

London, Groombridge, Panyer Alley.	*Manchester*, Robinson; and Webb and Simms.
Bath, Simms.	*Newcastle-upon-Tyne*, Charnley.
Birmingham, Drake.	*Norwich*, Jarrold and Son.
Bristol, Westley and Co.	*Nottingham*, Wright.
Carlisle, Thurnam; and Scott.	*Oxford*, Slatter.
Derby, Wilkins and Son.	*Plymouth*, Nettleton.
Doncaster, Brooke and Co.	*Portsea*, Horsey, Jun.
Exeter, Balle.	*Sheffield*, Ridge.
Falmouth, Philp.	*Staffordshire, Lane End*, C. Watts.
Hull, Stephenson.	*Worcester*, Deighton.
Kendal, Hudson and Nicholson.	*Dublin*, Wakeman.
Leeds, Baines and Newsome.	*Edinburgh*, Oliver and Boyd.
Lincoln, Brooke and Sons.	*Glasgow*, Atkinson and Co.
Liverpool, Willmer and Smith.	

Printed by William Clowes, Duke Street, Lambeth.

THE PENNY MAGAZINE

OF THE

Society for the Diffusion of Useful Knowledge.

35.] PUBLISHED EVERY SATURDAY. [October 20, 1832.

CATCHING TURTLES.

(a) (b)

[Catching Turtles on the Coast of Cuba.]

It is not improbable that some of our readers, who reside near a great commercial port, may have seen the landing of a cargo of strange-looking animals, which, turned upon their backs, appear the most helpless of creatures, and in this condition may have naturally led the spectator to imagine that they are incapable of removing from place to place, and have therefore little enjoyment of existence. These creatures, to use the language of the epicure, are fine "lively turtles"—the term "lively" being understood to mean that they have suffered little from a long voyage—that they are in good health—and that the "green fat," the glory of aldermen, is in the most perfect state of excellence. Without asking our readers to feel any very strong interest in the prospects of high living which the arrival of a cargo of turtles offers to many individuals who are somewhat too much inclined to set a high value upon the gratifications of the palate, we may be able to satisfy a rational curiosity as to the habits of these singular animals, which offer some higher benefits to mankind than that of furnishing the most costly luxury of a city feast.

The turtle and the tortoise belong to the same group of reptiles—in fact the turtle is a tortoise which principally inhabits the water, and is only found occasionally on the land. The two varieties represented in the above plate are the Green Tortoise (*a*), and the Loggerhead Tortoise (*b*). The former is the species chiefly used for food. It is found, in great numbers, on the coasts of all the islands and continents of the torrid zone. The shoals which surround these coasts are covered with marine plants; and in these water pastures, which are near enough to the surface to be readily seen by the naked eye in calm weather, a prodigious abundance of animals, mostly amphibious, feed, and amongst them multitudes of tortoises. Dampier, the old voyager, describing the Gallapagos Islands, says, "There are good wide channels between these islands fit for ships to pass; and in some places shoal water, where there grows plenty of turtle grass; therefore these islands are plentifully stored with sea turtle." The tortoise, whether of the land or water species, is, as most of our readers know, protected, both on the back and belly, by a hollow shield, which is open at each end, for the issuing of the head and fore-feet at one time, and the tail and hind-feet at another.

The upper shield is termed the back-plate, or buckler; the lower shield the breast-plate. The middle of the buckler, in most of the species, is covered by numerous pieces or plates resembling horn in texture and composition; and the beautiful substance known by the name of tortoise-shell is obtained principally from a small species called the Hawksbill. The feet of the marine tortoises are much longer than those of the land, and their toes are united by a membrane so that they swim with

Vol. I. 2 O

FIGURE 63

Catching Turtles on the Coast of Cuba, *The Penny Magazine*, October 20, 1832. Wood engraving.

Numerous articles and illustrations, such as that of Charing Cross in the first issue, attempted to familiarize the British citizenry with its history and monuments; there were illustrated features on St. Paul's Cathedral and Westminster Bridge, as well as Scotland's Falls of Clyde and Holyrood Chapel.[33] Engraved portraits of celebrated individuals worthy of emulation invariably accompanied accounts of their accomplishments: the artists Rubens, Flaxman, and Poussin; writers and poets Petrarch, Shakespeare, Sir Walter Scott, and Milton; and men who had, in the opinion of Knight, improved the world, such as Adam Smith, William Penn, Erasmus, and Isaac Newton.[34] In addition, numerous illustrated features described material culture and recent technology: the history of tea, tobacco, and the silk trade; the diving bell and suspension bridges (fig. 64).[35] There were even lessons on mathematics, such as the lengthy introduction to decimal fractions that extended over three pages in the issue of November 17, 1832. In effect, the *Penny Magazine* served as a substitute for the primary schooling that was just becoming a normal and required stage of life. Never before had articles and illustrations of this quality been available to the working classes, and for merely a penny.

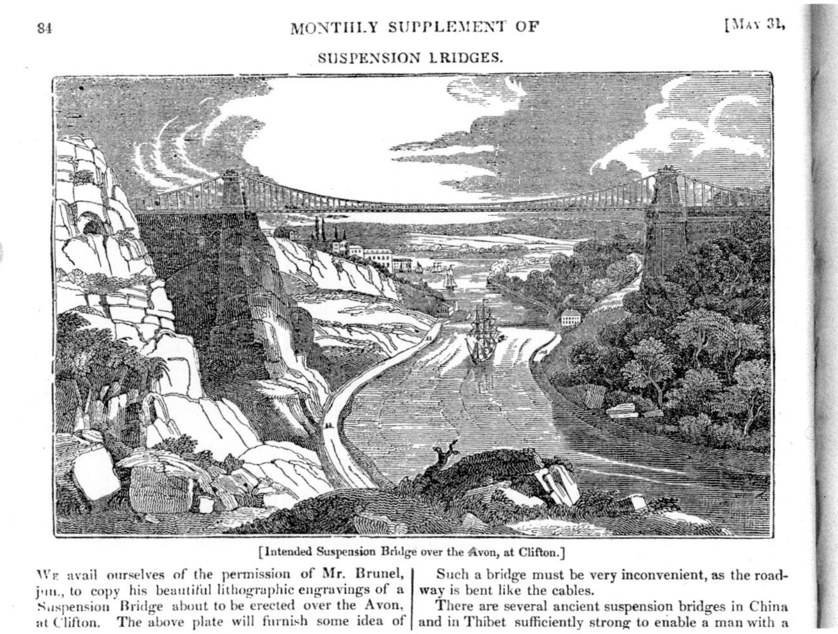
84 MONTHLY SUPPLEMENT OF [MAY 31,

SUSPENSION BRIDGES.

[Intended Suspension Bridge over the Avon, at Clifton.]

We avail ourselves of the permission of Mr. Brunel, jun., to copy his beautiful lithographic engravings of a Suspension Bridge about to be erected over the Avon, at Clifton. The above plate will furnish some idea of

Such a bridge must be very inconvenient, as the roadway is bent like the cables.

There are several ancient suspension bridges in China and in Thibet sufficiently strong to enable a man with a

FIGURE 64

Intended Suspension Bridge over the Avon at Clifton, after the lithograph by Isambard Kingdom Brunel, The Penny Magazine, monthly supplement, April 30 to May 31, 1832. Wood engraving.

Knight was tireless in carrying out his program, and his ambitions did not stop at national borders. In his autobiography, *Passages of a Working Life*, he proudly announced: "At this period, 1836, the 'Penny Magazine' was producing a revolution in popular Art throughout the world. Stereotype casts of its best cuts were supplied by me for the illustration of publications of a similar character, which appeared in eleven different languages and countries. . . . The entire work was also reprinted in the United States from plates sent from this country."[36] His list of countries that received his plates included, besides the United States: "Germany—France—Holland—Livonia (in Russian and German)—Bohemia (Sclavonic)–Italy—Ionian Islands (modern Greek)—Sweden—Norway—Spanish America—the Brazils."[37] Among these nations, however, it was in France that his ideas bore fruit almost immediately: on February 9, 1833, less than a year after the *Penny Magazine* began, *Le Magasin pittoresque* published its first issue, continuing until 1938 (fig. 65).

LE MAGASIN PITTORESQUE

While the ideas of the Comte de Saint-Simon (1760–1825) on the social utility of art have received serious scholarly attention, his influence on the illustrated press has scarcely been noted.[38] And yet, within a few months of the foundation of Knight's *Penny Magazine*, this publication attracted the attention of a group of Saint-Simonians who worked on its journal, *Le Globe*, but who had broken with the free-love philosophy of Prosper Enfantin in the "great schism" of 1831. The initial proposal for a French penny magazine, a *journal à deux sous*, came from Alexandre Lachevardière (1795–1855), the printer for both *Le Globe* and *Journal des connaissances utiles*, who became acquainted with the *Penny Magazine* on his 1832 trip to London.[39] Lachevardière had been one of the founders of *Le Globe*, and was the first printer in France to import the steam-driven Cowper presses that had made the *Penny Magazine* possible. Printers were crucial to the establishment of the periodical press for, although publishers could contract with printers for a single book, periodicals demanded an ongoing commitment for daily, weekly, or monthly production; only a printer could guarantee this continuity, and so Lachevardière became the publisher of *Le Magasin pittoresque*.[40] Édouard Charton (1807–1890) became its editor and remained the intellectual force behind the magazine for over fifty years. He later recalled the circumstances of its founding:

> November 1832: My friends Jean Reynaud, Pierre Leroux, Sainte-Beuve, advise me to accept the direction of a weekly illustrated digest, at ten centimes, that Mr. Lachevardière, printer, proposes to publish in imitation of the *Penny Magazine*, recently founded in London by the historian Charles Knight. I hesitate, I'm a lawyer. My friends point out that this would be a service to the cause of instruction and education and that I am qualified to do this by my previous work as editor of the *Bulletin de la Société pour l'instruction de l'éducation élémentaire* and the *Journal de la Société de la morale chrétienne*. I accept after receiving assurance that I would have the collaboration of my friends, most of whom graduated from the major institutions of France. One of them, Euryale Cazeaux, who has since become Inspector General of Agriculture, temporarily acts as my coeditor.[41]

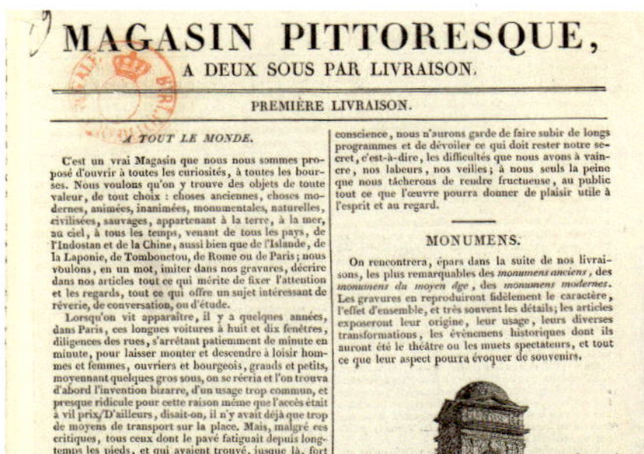

MAGASIN PITTORESQUE,
A DEUX SOUS PAR LIVRAISON.

PREMIÈRE LIVRAISON.

A TOUT LE MONDE.

C'est un vrai Magasin que nous nous sommes proposé d'ouvrir à toutes les curiosités, à toutes les bourses. Nous voulons qu'on y trouve des objets de toute valeur, de tout choix : choses anciennes, choses modernes, animées, inanimées, monumentales, naturelles, civilisées, sauvages, appartenant à la terre, à la mer, au ciel, à tous les temps, venant de tous les pays, de l'Indostan et de la Chine, aussi bien que de l'Islande, de la Laponie, de Tombouctou, de Rome ou de Paris; nous voulons, en un mot, imiter dans nos gravures, décrire dans nos articles tout ce qui mérite de fixer l'attention et les regards, tout ce qui offre un sujet intéressant de rêverie, de conversation, ou d'étude.

Lorsqu'on vit apparaître, il y a quelques années, dans Paris, ces longues voitures à huit et dix fenêtres, diligences des rues, s'arrêtant patiemment de minute en minute, pour laisser monter et descendre à loisir hommes et femmes, ouvriers et bourgeois, grands et petits, moyennant quelques gros sous, on se récria et l'on trouva d'abord l'invention bizarre, d'un usage trop commun, et presque ridicule pour cette raison même que l'accès était à vil prix. D'ailleurs, disait-on, il n'y avait déjà que trop de moyens de transport sur la place. Mais, malgré ces critiques, tous ceux dont le pavé fatiguait depuis longtemps les pieds, et qui avaient trouvé, jusque là, fort coûteux de se faire rouler en carrosse, se sont montrés moins scrupuleux. Le vil prix ne les a pas effrayés; les moyens de transport ne leur ont point paru trop nombreux. Ils ont estimé que l'invention était agréable et utile, et l'invention a réussi. De plus riches qu'eux ont à la fin partagé leur opinion. Maintenant on ne s'étonne plus de voir toutes ces machines à trois et quatre roues traverser la ville en tous sens, et s'avancer de loin, de conserve et en bonne intelligence, au milieu des tilburys, des landaux, des fiacres, et des cabriolets.

De même, notre *Magasin* à deux sous, dans un ordre d'entreprise bien différent, se recommande *à tout le monde;* mais il est plus particulièrement destiné à tous ceux qui ne peuvent consacrer qu'une humble somme à leurs menus plaisirs.

Notre grande ambition sera d'intéresser, de distraire : nous laisserons l'instruction venir à la suite sans la violenter, et nous ne craignons pas que jamais elle reste bien loin en arrière; elle évitera seulement de revêtir les formes arrêtées, sévères, de l'enseignement spécial et méthodique, et son influence s'exercera à la manière de cette éducation générale que les classes de la société riches en loisirs doivent à des relations habituelles avec les hommes distingués, à des lectures variées, choisies, et aux souvenirs des voyages.

Ces relations, ces lectures, ces voyages, interdits au grand nombre, notre recueil aura pour but constant de chercher à en tenir lieu. Nous aurons bien du malheur si, devant ce tableau toujours changeant du monde entier, que nous déroulerons continuellement sous les yeux de nos lecteurs, ils ont des pensées, des désirs que nous ne puissions satisfaire. A toute question nous espérons avoir une réponse prête, en nous tenant attentivement à la hauteur des connaissances, des découvertes, des productions des beaux arts, en appelant tour à tour nos artistes, nos écrivains, à représenter, à dire ce qui est vrai, ce qui est beau, ce qui est utile, sans mélange d'exagération ou d'imaginations mensongères. Ces promesses faites, résolus à les tenir avec conscience, nous n'aurons garde de faire subir de longs programmes et de dévoiler ce qui doit rester notre secret, c'est-à-dire, les difficultés que nous avons à vaincre, nos labeurs, nos veilles; à nous seuls la peine que nous tâcherons de rendre fructueuse, au public tout ce que l'œuvre pourra donner de plaisir utile à l'esprit et au regard.

IMPRIMERIE DE LACHEVARDIÈRE. — TOME I.

MONUMENS.

On rencontrera, épars dans la suite de nos livraisons, les plus remarquables des *monumens anciens*, des *monumens du moyen âge*, des *monumens modernes*. Les gravures en reproduiront fidèlement le caractère, l'effet d'ensemble, et très souvent les détails; les articles exposeront leur origine, leur usage, leurs diverses transformations, les évènemens historiques dont ils auront été le théâtre ou les muets spectateurs, et tout ce que leur aspect pourra évoquer de souvenirs.

FONTAINE DES INNOCENS.

RECONSTRUCTION DE LA FONTAINE DES INNOCENS. — LE CIMETIÈRE. — LE PASSAGE DES CHARNIERS. — L'ÉGLISE ET LES RECLUSES. — TRANSLATION D'OSSEMENS AUX CATACOMBES.

Cette fontaine n'avait autrefois que trois côtés, et était adossée à l'angle de la rue Saint-Denis et de la rue aux Fers. Sa première construction datait du treizième siecle; en 1550 elle fut réparée : les travaux d'architecture en furent alors confiés à Pierre Lescot, abbé de Clagny, et les sculptures à notre célèbre Jean Goujon, tué le jour de la Saint-Barthélemy.

Dans le mois de mars 1788, sur la proposition d'un ingénieur nommé Six, toutes les parties de ce monument dignes d'être conservées furent transportées au milieu du marché des Innocens, qu'on commençait à établir, et la fontaine fut réédifiée d'après un plan nouveau. Comme il fallait l'agrandir et lui donner quatre faces pour pouvoir l'isoler au centre d'une place carrée, on fut obligé de compléter l'architecture. Jean Goujon avait sculpté cinq naiades, on eut soin de les laisser entre les pilastres des arcades, où les artistes admirent encore ces figures d'un caractère si naïf et si gracieux :

1

FIGURE 65
The first issue of *Le Magasin pittoresque*, February 9, 1833. Bibliothèque nationale de France.

All the editors and contributors to the new weekly were under forty and imbued with Saint-Simonian and republican principles, members of what the historian Alan B. Spitzer has labeled "the French Generation of 1830."[42] Lachevardière and Charton were joined by agronomist Pierre-Euryale Cazeaux (1805–1880), philosopher Jean Reynaud (1806–1863), publisher and socialist thinker Pierre Leroux (1797–1871), and literary critic Charles-Augustin de Sainte-Beuve (1804–1869).[43] None was a journalist by profession, but each had expertise in his own area.

Their choice of title, *Le Magasin pittoresque*, was significant. At the time, the French word *magasin* signified a storeroom or, by extension, a store; the word in English has preserved this meaning only in a military sense—an artillery magazine, for example. The French terms for periodicals ranged from *journal*, a newspaper, to *revue*, a serious periodical without illustrations, to *recueil*, an anthology or digest, which, like the English *miscellany*, consisted of a collection of short articles. In France the use of the English word *magazine* or the French equivalent *magasin* to designate an illustrated periodical was initiated by *Le Magasin pittoresque*; both its supporters and critics often used the English word, untranslated, to indicate precisely this type of publication. Charton played on the double meaning of the word as both store and periodical when he wrote in the first issue: "This is a real *magasin* that we have proposed to open here to all the curious, to all purses."[44] In a sociological sense, he was signaling his intention to provide the intellectual "goods" that he hoped would eradicate class conflict. Sociologist Pierre Bourdieu would have defined this endeavor as providing acquired cultural capital to those who were not fortunate enough to have inherited it through birth.[45]

The word *pittoresque* in the title of the periodical also had a double sense. Charton explained the difference between the French and English use of the term. "In England they found the word *picturesque* astonishing; a member of the Royal Society reproached me for using it. It seemed to him that I abused this term, and that it should be used only in relation to painting. But the French sense is broader and I have had the satisfaction of seeing it rapidly adopted for many illustrated works, both periodicals and other types of publications."[46] And, although *Le Magasin pittoresque* was by no means the first publication to use the word as signifying "composed of pictures"—one has only to recall the multivolume *Voyages pittoresques et romantiques dans l'ancienne France* that had begun publication in 1820—we should keep in mind that the title *Le Magasin pittoresque* carried with it the implication of a storehouse of images, Malraux's Imaginary Museum in the first age of capitalism.

The editors of *Le Magasin pittoresque* freely acknowledged their debt to Knight: "We think it is entirely appropriate to declare at the beginning of the first volume that, if we were the first to dare to import into France the idea of delivering, at the most modest price, a text varied, interspersed with engravings and divided into installments, and to do this without either patronage or advance subscription, it was only after witnessing the success of *Magazines* in England, and especially that of the digest published in London under elevated and worthy authority by Charles Knight, writer and distinguished economist, who, through his kind relations with us, has helped us to render less discouraging the early difficulties of our enterprise."[47]

The major early difficulty was the procurement of images. Other pressures of regular production could be met by mechanical presses and freelance authors, but images were produced slowly and laboriously. Charton wrote: "December 1832: Difficulties. The *Penny Magazine* intersperses its text with wood engravings. In France, this kind of engraving has long since been abandoned: there are scarcely eight engravers left."[48] He described his difficulties at length:

> When, at the beginning of 1833, I founded *Le Magasin pittoresque* with my friend Pierre Cazeaux, the greatest difficulty that I encountered was to produce each week the wood engravings that I considered useful elements for a digest intended to disseminate knowledge on a variety of subjects. Wood engraving was then an art nearly forgotten in France; engravers were rare. I addressed myself to three of them, Messrs Andrew, Best, and Leloir, who worked together and lived a rather impoverished existence in a little hovel near Place Saint-André-des-Arts.
>
> These gentlemen, if I am not mistaken, had no students. They listened to me with astonishment when I explained to them that I needed three or four engravings delivered to me each week. They protested and told me clearly that this was impossible:

> even if I had asked that many by the month instead of by the week! Obviously I had no idea of the difficulties of their profession.
>
> They were a long way from understanding that my visit to them would be the beginning of their great prosperity.
>
> I insisted, promising them that, in the early stages, they would be asked only for easy work, and even, to lighten their task, that we would combine with their engravings those ordered from London. I succeeded in persuading them to at least try to do it: they summoned assistance from students, threw themselves bravely into the work, simplified their procedures, and two years later, we no longer needed to borrow anything from England. The assistance of Messrs. Andrew, Best, and Company from then on was amply sufficient for us.[49]

In keeping with their agreement, the ABL atelier was established within a few months, in August of 1833; meanwhile, Charton had already visited London in order to procure plates from the *Penny Magazine* for his new periodical.[50]

When the first issue of *Le Magasin pittoresque* appeared, on February 9, 1833, it was almost an exact replica of the *Penny Magazine* in size, typeface, and layout. Like its English counterpart, *Le Magasin pittoresque* consisted of eight pages, published weekly, with illustrated pages alternating with pages of text, including four to six wood engravings per issue. Like the *Penny Magazine,* it aimed to be encyclopedic in content, avoiding politics and, for the most part, current events, but featuring articles of historical, industrial, scientific, or cultural interest. Neither publication offered the practical advice featured in *Journal des connaissances utiles*; instead both periodicals sought to impart a general level of familiarity with the contemporary world and its history.

While both the *Penny Magazine* and *Le Magasin pittoresque* shared a similar ideology of educating the working classes and steering readers away from "unhealthy" ideas, there were important differences between them. France had recently had still another in its series of revolutions, the "Three Glorious Days" of July 1830 that definitively ended its centuries of absolute monarchy and ushered in the constitutional monarchy of King Louis-Philippe. Article 69 of his charter promised to establish public education and to open the profession of teaching to all qualified individuals, ending the dominance of the religious orders.[51] Despite continuous opposition from the Church, the Guizot Law establishing public education was signed into law on June 28, 1833, legislating universal and obligatory primary education for boys (girls would have to wait for the Falloux Law of 1850). There would be two levels, as the law explained: "Primary elementary education includes, of necessity, moral and religious instruction, reading, writing, elements of French grammar and of arithmetic, the system of weights and measures. Advanced elementary education includes, in addition, elements of geometry and its common applications, particularly line drawing and surveying, basics of the physical sciences and natural history applied to daily life; singing, elements of history and geography, and especially the history and geography of France."[52] This was virtually the program of *Le*

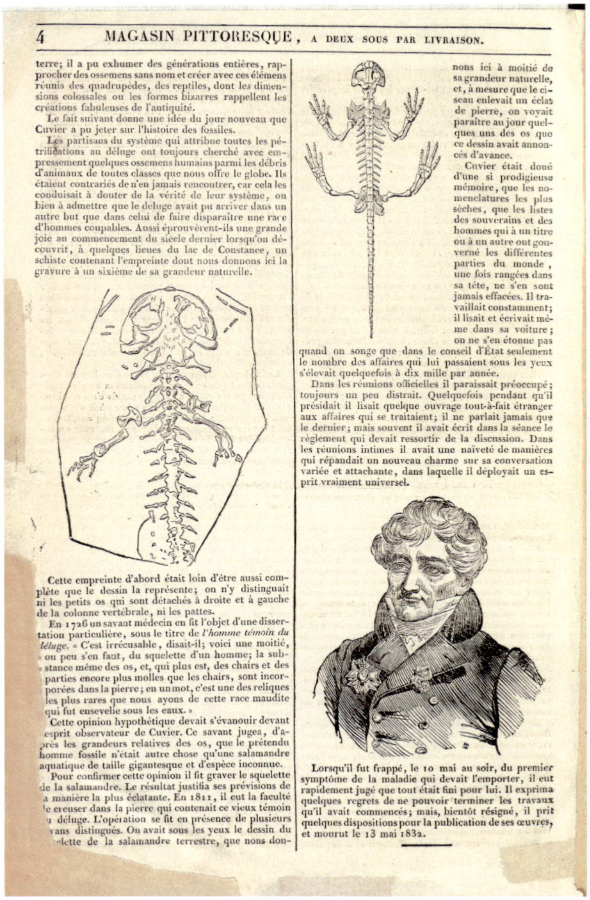

4 MAGASIN PITTORESQUE, A DEUX SOUS PAR LIVRAISON.

terre; il a pu exhumer des générations entières, rapprocher des ossemens sans nom et créer avec ces élémens réunis des quadrupèdes, des reptiles, dont les dimensions colossales ou les formes bizarres rappellent les créations fabuleuses de l'antiquité.

Le fait suivant donne une idée du jour nouveau que Cuvier a pu jeter sur l'histoire des fossiles.

Les partisans du système qui attribue toutes les pétrifications au déluge ont toujours cherché avec empressement quelques ossemens humains parmi les débris d'animaux de toutes classes que nous offre le globe. Ils étaient contrariés de n'en jamais rencontrer, car cela les conduisait à douter de la vérité de leur système, ou bien à admettre que le déluge avait pu arriver dans un autre but que dans celui de faire disparaître une race d'hommes coupables. Aussi éprouvèrent-ils une grande joie au commencement du siècle dernier lorsqu'on découvrit, à quelques lieues du lac de Constance, un schiste contenant l'empreinte dont nous donnons ici la gravure à un sixième de sa grandeur naturelle.

Cette empreinte d'abord était loin d'être aussi complète que le dessin la représente; on n'y distinguait ni les petits os qui sont détachés à droite et à gauche de la colonne vertébrale, ni les pattes.

En 1726 un savant médecin en fit l'objet d'une dissertation particulière, sous le titre de *l'homme témoin du déluge*. « C'est irrécusable, disait-il; voici une moitié, » ou peu s'en faut, du squelette d'un homme; la sub» stance même des os, et, qui plus est, des chairs et des parties encore plus molles que les chairs, sont incorporées dans la pierre; en un mot, c'est une des reliques les plus rares que nous ayons de cette race maudite qui fut ensevelie sous les eaux. »

Cette opinion hypothétique devait s'évanouir devant esprit observateur de Cuvier. Ce savant jugea, d'après les grandeurs relatives des os, que le prétendu homme fossile n'était autre chose qu'une salamandre aquatique de taille gigantesque et d'espèce inconnue.

Pour confirmer cette opinion il fit graver le squelette de la salamandre. Le résultat justifia ses prévisions de a manière la plus éclatante. En 1811, il eut la faculté le creuser dans la pierre qui contenait ce vieux témoin u déluge. L'opération se fit en présence de plusieurs vans distingués. On avait sous les yeux le dessin du elette de la salamandre terrestre, que nous donnons ici à moitié de sa grandeur naturelle, et, à mesure que le ciseau enlevait un éclat de pierre, on voyait paraître au jour quelques uns des os que ce dessin avait annoncés d'avance.

Cuvier était doué d'une si prodigieuse mémoire, que les nomenclatures les plus sèches, que les listes des souverains et des hommes qui à un titre ou à un autre ont gouverné les différentes parties du monde, une fois rangées dans sa tête, ne s'en sont jamais effacées. Il travaillait constamment; il lisait et écrivait même dans sa voiture; on ne s'en étonne pas quand on songe que dans le conseil d'État seulement le nombre des affaires qui lui passaient sous les yeux s'élevait quelquefois à dix mille par année.

Dans les réunions officielles il paraissait préoccupé; toujours un peu distrait. Quelquefois pendant qu'il présidait il lisait quelque ouvrage tout-à-fait étranger aux affaires qui se traitaient; il ne parlait jamais que le dernier; mais souvent il avait écrit dans la séance le règlement qui devait ressortir de la discussion. Dans les réunions intimes il avait une naïveté de manières qui répandait un nouveau charme sur sa conversation variée et attachante, dans laquelle il déployait un esprit vraiment universel.

Lorsqu'il fut frappé, le 10 mai au soir, du premier symptôme de la maladie qui devait l'emporter, il eut rapidement jugé que tout était fini pour lui. Il exprima quelques regrets de ne pouvoir terminer les travaux qu'il avait commencés; mais, bientôt résigné, il prit quelques dispositions pour la publication de ses œuvres, et mourut le 13 mai 1832.

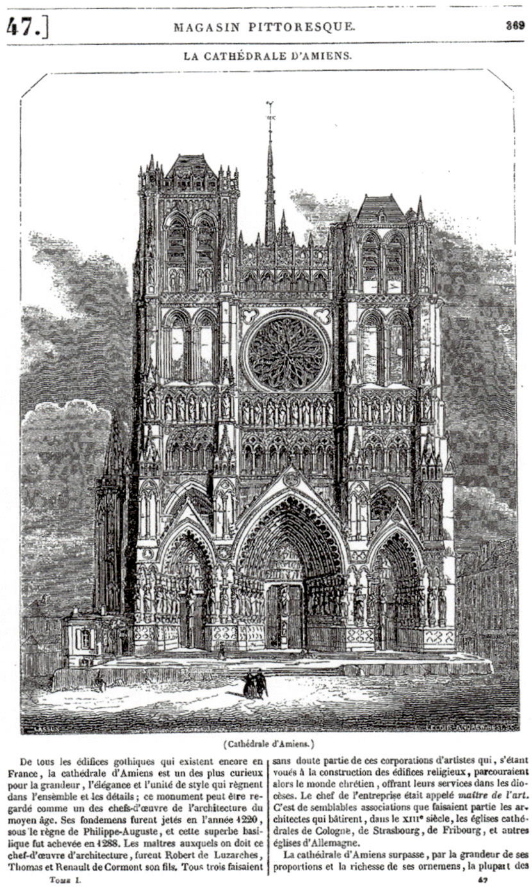

47.] MAGASIN PITTORESQUE. 369

LA CATHÉDRALE D'AMIENS.

(Cathédrale d'Amiens.)

De tous les édifices gothiques qui existent encore en France, la cathédrale d'Amiens est un des plus curieux pour la grandeur, l'élégance et l'unité de style qui règnent dans l'ensèmble et les détails; ce monument peut être regardé comme un des chefs-d'œuvre de l'architecture du moyen âge. Ses fondemens furent jetés en l'année 1220, sous le règne de Philippe-Auguste, et cette superbe basilique fut achevée en 1288. Les maîtres auxquels on doit ce chef-d'œuvre d'architecture, furent Robert de Luzarches, Thomas et Renault de Cormont son fils. Tous trois faisaient sans doute partie de ces corporations d'artistes qui, s'étant voués à la construction des édifices religieux, parcouraient alors le monde chrétien, offrant leurs services dans les diocèses. Le chef de l'entreprise était appelé *maître de l'art*. C'est de semblables associations que faisaient partie les architectes qui bâtirent, dans le XIIIe siècle, les églises cathédrales de Cologne, de Strasbourg, de Fribourg, et autres églises d'Allemagne.

La cathédrale d'Amiens surpasse, par la grandeur de ses proportions et la richesse de ses ornemens, la plupart des

Tome I. 47

FIGURE 66

Cuvier. His Life. His Works. The Fossil History of Humanity [*Cuvier. Sa vie. Ces travaux. Histoire de l'homme fossile*], *Le Magasin pittoresque* 1 (1833), no. 1. Wood engraving. Bibliothèque nationale de France.

FIGURE 67

Jean-Baptiste Lassus, *The Cathedral of Amiens* [*La Cathédrale d'Amiens*]. Engr. Andrew, Best, Leloir, *Le Magasin pittoresque* 1 (1833), no. 47. Wood engraving. Bibliothèque nationale de France.

Magasin pittoresque, whose readers had not had the benefit of the secular education proposed by Guizot that would become increasingly important in the modern world.

The idea of educating the masses was basic to the social agenda of Saint-Simonians, and so it is no accident that they were integral to the founding of this new illustrated journal. To the negative imperative of the *Penny Magazine,* of weaning the working classes away from radical politics, they added the positive Saint-Simonian imperative of promoting social and political equality. Thus Charton could write to his friend, the poet Jean-Louis Renaudot: "Let us raise the intelligence of our fellow citizens and we will witness the beginning of the reign of equality; you through your verse, me through *Le Magasin pittoresque,* we will, I hope, contribute to this great task, the only one that seems to me that can certainly lay the groundwork for the emancipation of the masses."[53] Charton's solution to class conflict was to eradicate the barriers raised by education—or lack of it. He explained his program in the first issue in an article entitled "On the Influence of Conversation." After giving several examples of how members of the same class can meet as complete strangers and yet be able to converse easily with each other, he concluded: "The difficulty in conversing divides society into two classes. Despite all efforts of good will, there is still a breach in relations between rich and poor. We speak and act in vain; there exists, nonetheless, a real line of demarcation, independent of political beliefs, one that we cannot hope to erase entirely, even through the elementary education of schooling. We can make it disappear only through the spread of basic knowledge and shared habitual and general interests, which will gradually render communications among all classes of society more pleasant, easier, and more intimate."[54]

In accordance with this objective, the very next article in the issue was an illustrated feature on the paleontologist Georges Cuvier (1769–1832) that included, along with his portrait, wood engravings of the skeleton of a salamander and a schist bearing skeletal remains (fig. 66).[55] The accompanying text explained clearly and at some length the principles of comparative anatomy that Cuvier had established.

Charton's plan to promote equality through general education formed the backbone of *Le Magasin pittoresque.* To this end, articles focused on basic historical, cultural, and scientific knowledge and dealt with the marvelous and exotic much less than did the *Penny Magazine.* In the first issue, Charton laid out the program of topics to be covered, reciting a delirious litany that evidenced the pleasures offered by this new format that would encompass the entire world in text and image: "We want to include all worthwhile subjects, for all tastes: old things, modern things, animate, inanimate, monumental, natural, civilized, savage, belonging to earth, sea, sky, from all epochs and all countries, from Indostan and China as well as Iceland, Lapland, Timbuktu, Rome, or Paris. In brief, we want to show in our engravings, describe in our articles, everything that has the merit of attracting attention and examination, everything that offers a subject of interest for contemplation, conversation, or study."[56] The tone of the magazine was unfailingly positive and optimistic, in keeping with the Saint-Simonian belief that the golden age is in the future, not the past. As a result, an article such as "How to Endure Poverty" that appeared in the *Penny Magazine* would be inconceivable in *Le Magasin pittoresque*, nor would *Le Magasin pittoresque* be likely to feature an engraving of Murillo's *The Young Beggar* on its title page, as did the *Penny Magazine.*[57]

Although the elephant, zebra, and crocodile did make their appearance in *Le Magasin pittoresque* in images borrowed from the *Penny Magazine*, and the striking cover of the boa constrictor was reused in the French journal as well, exotic animals were featured much less than in its English counterpart.[58] Both periodicals shared the intention of familiarizing the populace with their national monuments and so, if the *Penny Magazine*'s first cover illustrated the 1647 Charing Cross in London, *Le Magasin pittoresque*'s featured the 1550 Fontaine des Innocents in Paris. Within its first year, *Le Magasin pittoresque* featured cover stories with illustrations of Mont Saint-Michel, the Palais Royal in Paris, and the Palais de Justice at Dijon, clearly aiming at a national, not merely Parisian, audience. Charton's ambitions for *Le Magasin pittoresque*'s aesthetic program can be seen in the impressive cover illustration of the *Cathedral of Amiens* (fig. 67). Drawn by the prominent architect Jean-Baptiste Lassus and engraved by Andrew, Best, Leloir, it proudly displays all four signatures.[59]

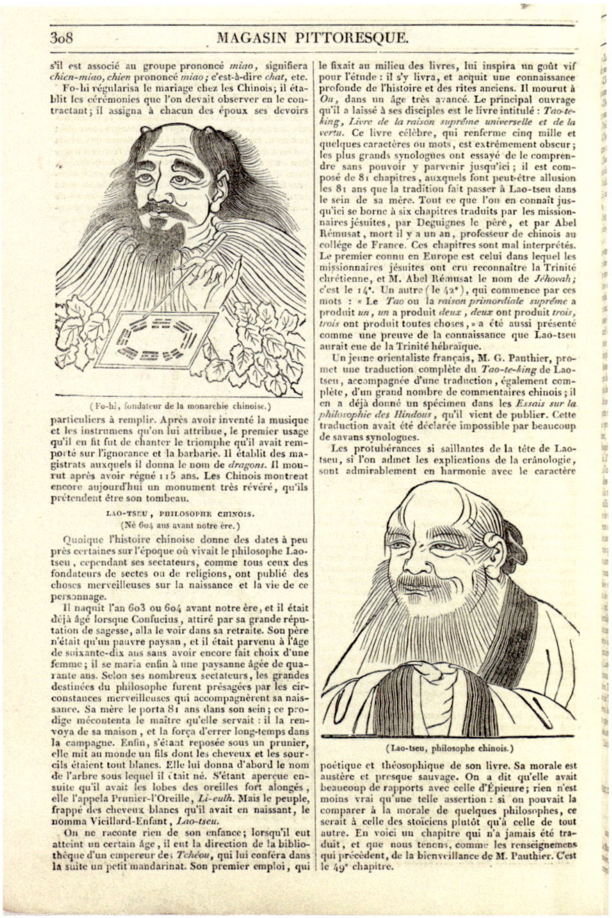

308 MAGASIN PITTORESQUE.

s'il est associé au groupe prononcé *miao*, signifiera *chien-miao, chien* prononcé *miao*; c'est-à-dire *chat*, etc.

Fo-hi régularisa le mariage chez les Chinois; il établit les cérémonies que l'on devait observer en le contractant; il assigna à chacun des époux ses devoirs particuliers à remplir. Après avoir inventé la musique et les instrumens qu'on lui attribue, le premier usage qu'il en fit fut de chanter le triomphe qu'il avait remporté sur l'ignorance et la barbarie. Il établit des magistrats auxquels il donna le nom de *dragons*. Il mourut après avoir régné 115 ans. Les Chinois montrent encore aujourd'hui un monument très révéré, qu'ils prétendent être son tombeau.

(Fo-hi, fondateur de la monarchie chinoise.)

LAO-TSEU, PHILOSOPHE CHINOIS.

(Né 604 ans avant notre ère.)

Quoique l'histoire chinoise donne des dates à peu près certaines sur l'époque où vivait le philosophe Lao-tseu, cependant ses sectateurs, comme tous ceux des fondateurs de sectes ou de religions, ont publié des choses merveilleuses sur la naissance et la vie de ce personnage.

Il naquit l'an 603 ou 604 avant notre ère, et il était déjà âgé lorsque Confucius, attiré par sa grande réputation de sagesse, alla le voir dans sa retraite. Son père n'était qu'un pauvre paysan, et il était parvenu à l'âge de soixante-dix ans sans avoir encore fait choix d'une femme; il se maria enfin à une paysanne âgée de quarante ans. Selon ses nombreux sectateurs, les grandes destinées du philosophe furent présagées par les circonstances merveilleuses qui accompagnèrent sa naissance. Sa mère le porta 81 ans dans son sein; ce prodige mécontenta le maître qu'elle servait : il la renvoya de sa maison, et la força d'errer long-temps dans la campagne. Enfin, s'étant reposée sous un prunier, elle mit au monde un fils dont les cheveux et les sourcils étaient tout blancs. Elle lui donna d'abord le nom de l'arbre sous lequel il était né. S'étant aperçue ensuite qu'il avait les lobes des oreilles fort alongés, elle l'appela Prunier-l'Oreille, *Li-eulh*. Mais le peuple, frappé des cheveux blancs qu'il avait en naissant, le nomma Vieillard-Enfant, *Lao-tseu*.

On ne raconte rien de son enfance; lorsqu'il eut atteint un certain âge, il eut la direction de la bibliothèque d'un empereur des *Tchéou*, qui lui conféra dans la suite un petit mandarinat. Son premier emploi, qui le fixait au milieu des livres, lui inspira un goût vif pour l'étude : il s'y livra, et acquit une connaissance profonde de l'histoire et des rites anciens. Il mourut à *Ou*, dans un âge très avancé. Le principal ouvrage qu'il a laissé à ses disciples est le livre intitulé : *Tao-te-king, Livre de la raison suprême universelle et de la vertu*. Ce livre célèbre, qui renferme cinq mille et quelques caractères ou mots, est extrêmement obscur; les plus grands synologues ont essayé de le comprendre sans pouvoir y parvenir jusqu'ici; il est composé de 81 chapitres, auxquels font peut-être allusion les 81 ans que la tradition fait passer à Lao-tseu dans le sein de sa mère. Tout ce que l'on en connaît jusqu'ici se borne à six chapitres traduits par les missionnaires jésuites, par Deguignes le père, et par Abel Rémusat, mort il y a un an, professeur de chinois au collége de France. Ces chapitres sont mal interprétés. Le premier connu en Europe est celui dans lequel les missionnaires jésuites ont cru reconnaître la Trinité chrétienne, et M. Abel Rémusat le nom de *Jéhovah*; c'est le 14e. Un autre (le 42e), qui commence par ces mots : « Le *Tao* ou la *raison primordiale suprême* a produit *un*, *un* a produit *deux*, *deux* ont produit *trois*, *trois* ont produit toutes choses, » a été aussi présenté comme une preuve de la connaissance que Lao-tseu aurait eue de la Trinité hébraïque.

Un jeune orientaliste français, M. G. Pauthier, promet une traduction complète du *Tao-te-king* de Lao-tseu, accompagnée d'une traduction, également complète, d'un grand nombre de commentaires chinois; il en a déjà donné un spécimen dans les *Essais sur la philosophie des Hindous*, qu'il vient de publier. Cette traduction avait été déclarée impossible par beaucoup de savans synologues.

Les protubérances si saillantes de la tête de Lao-tseu, si l'on admet les explications de la crânologie, sont admirablement en harmonie avec le caractère poétique et théosophique de son livre. Sa morale est austère et presque sauvage. On a dit qu'elle avait beaucoup de rapports avec celle d'Épicure; rien n'est moins vrai qu'une telle assertion : si on pouvait la comparer à la morale de quelques philosophes, ce serait à celle des stoïciens plutôt qu'à celle de tout autre. En voici un chapitre qui n'a jamais été traduit, et que nous tenons, comme les renseignemens qui précèdent, de la bienveillance de M. Pauthier. C'est le 49e chapitre.

(Lao-tseu, philosophe chinois.)

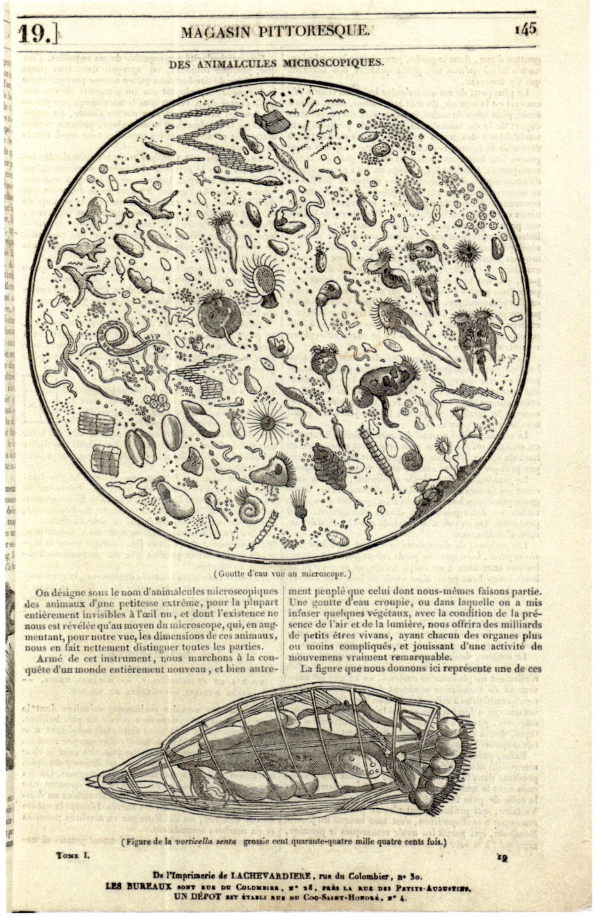

19.] MAGASIN PITTORESQUE. 145

DES ANIMALCULES MICROSCOPIQUES.

(Goutte d'eau vue au microscope.)

On désigne sous le nom d'animalcules microscopiques des animaux d'une petitesse extrême, pour la plupart entièrement invisibles à l'œil nu, et dont l'existence ne nous est révélée qu'au moyen du microscope, qui, en augmentant, pour notre vue, les dimensions de ces animaux, nous en fait nettement distinguer toutes les parties.

Armé de cet instrument, nous marchons à la conquête d'un monde entièrement nouveau, et bien autrement peuplé que celui dont nous-mêmes faisons partie. Une goutte d'eau croupie, ou dans laquelle on a mis infuser quelques végétaux, avec la condition de la présence de l'air et de la lumière, nous offrira des milliards de petits êtres vivans, ayant chacun des organes plus ou moins compliqués, et jouissant d'une activité de mouvemens vraiment remarquable.

La figure que nous donnons ici représente une de ces

(Figure de la *vorticella senta* grossie cent quarante-quatre mille quatre cents fois.)

Tome I. 19

De l'Imprimerie de LACHEVARDIERE, rue du Colombier, n° 30.
LES BUREAUX sont rue du Colombier, n° 28, près la rue des Petits-Augustins.
UN DÉPOT est établi rue du Coq-Saint-Honoré, n° 4.

FIGURE 68

Fu-Xi, Founder of the Chinese Monarchy. Lao Tse, Chinese Philosopher [*Fo-hi, fondateur de la monarchie chinoise. Lao-Tseu, philosophe chinois*], *Le Magasin pittoresque* 1 (1833), no. 39. Wood engraving. Bibliothèque nationale de France.

FIGURE 69

Drop of Water Seen through a Microscope [*Goutte d'eau vue au microscope*], *Le Magasin pittoresque* 1 (1833), no. 19. Wood engraving. Bibliothèque nationale de France.

A major difference between the magazines was the relative rarity of articles in the *Penny Magazine* on peoples and customs outside Europe; *Le Magasin pittoresque*, in contrast, demonstrated the Saint-Simonian philosophy of the equality of peoples, religions, and cultures with numerous such articles and illustrations: a feature on New Zealand included six engravings; Confucius, Fu-Xi, and Lao Tse were honored with articles and portraits (fig. 68).[60] The juxtaposition of illustrations carried its own significance of the equality of all world religions: William Penn, for example, shared a page with the many-breasted goddess Diana of Ephesus; the Egyptian zodiac calendar of Dendera was given a cover illustration, while the Catholic cemetery of the Capuchins at Palermo was relegated to an inside feature in the same issue.[61]

Science and technology, in keeping with the Saint-Simonian philosophy, were given major coverage, bringing the latest research to a general population. *Drop of Water Seen through a Microscope*, a striking cover illustration, showed a drop of water teeming with life invisible to the naked eye (fig. 69).[62] A chart illustrated the sign language alphabet for

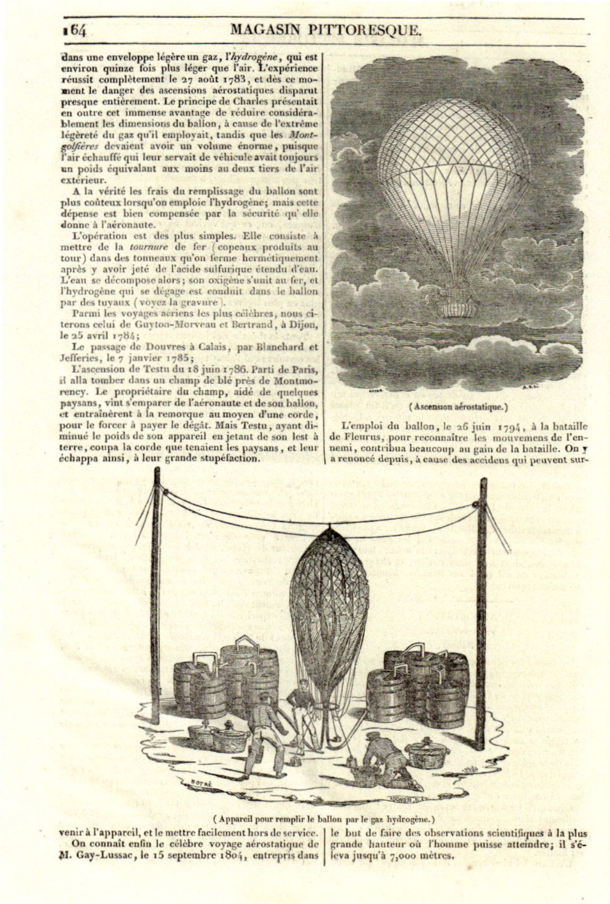
164 MAGASIN PITTORESQUE.

dans une enveloppe légère un gaz, l'*hydrogène*, qui est environ quinze fois plus léger que l'air. L'expérience réussit complètement le 27 août 1783, et dès ce moment le danger des ascensions aérostatiques disparut presque entièrement. Le principe de Charles présentait en outre cet immense avantage de réduire considérablement les dimensions du ballon, à cause de l'extrême légèreté du gaz qu'il employait, tandis que les *Montgolfières* devaient avoir un volume énorme, puisque l'air échauffé qui leur servait de véhicule avait toujours un poids équivalant aux moins au deux tiers de l'air extérieur.

A la vérité les frais du remplissage du ballon sont plus coûteux lorsqu'on emploie l'hydrogène; mais cette dépense est bien compensée par la sécurité qu'elle donne à l'aéronaute.

L'opération est des plus simples. Elle consiste à mettre de la *tournure* de fer (copeaux produits au tour) dans des tonneaux qu'on ferme hermétiquement après y avoir jeté de l'acide sulfurique étendu d'eau. L'eau se décompose alors; son oxigène s'unit au fer, et l'hydrogène qui se dégage est conduit dans le ballon par des tuyaux (voyez la gravure).

Parmi les voyages aériens les plus célèbres, nous citerons celui de Guyton-Morveau et Bertrand, à Dijon, le 25 avril 1784;

Le passage de Douvres à Calais, par Blanchard et Jefferies, le 7 janvier 1785;

L'ascension de Testu du 18 juin 1786. Parti de Paris, il alla tomber dans un champ de blé près de Montmorency. Le propriétaire du champ, aidé de quelques paysans, vint s'emparer de l'aéronaute et de son ballon, et entraînèrent à la remorque au moyen d'une corde, pour le forcer à payer le dégât. Mais Testu, ayant diminué le poids de son appareil en jetant de son lest à terre, coupa la corde que tenaient les paysans, et leur échappa ainsi, à leur grande stupéfaction.

(Ascension aérostatique.)

L'emploi du ballon, le 26 juin 1794, à la bataille de Fleurus, pour reconnaître les mouvemens de l'ennemi, contribua beaucoup au gain de la bataille. On y a renoncé depuis, à cause des accidens qui peuvent survenir à l'appareil, et le mettre facilement hors de service.

On connaît enfin le célèbre voyage aérostatique de M. Gay-Lussac, le 15 septembre 1804, entrepris dans le but de faire des observations scientifiques à la plus grande hauteur où l'homme puisse atteindre; il s'éleva jusqu'à 7,000 mètres.

(Appareil pour remplir le ballon par le gaz hydrogène.)

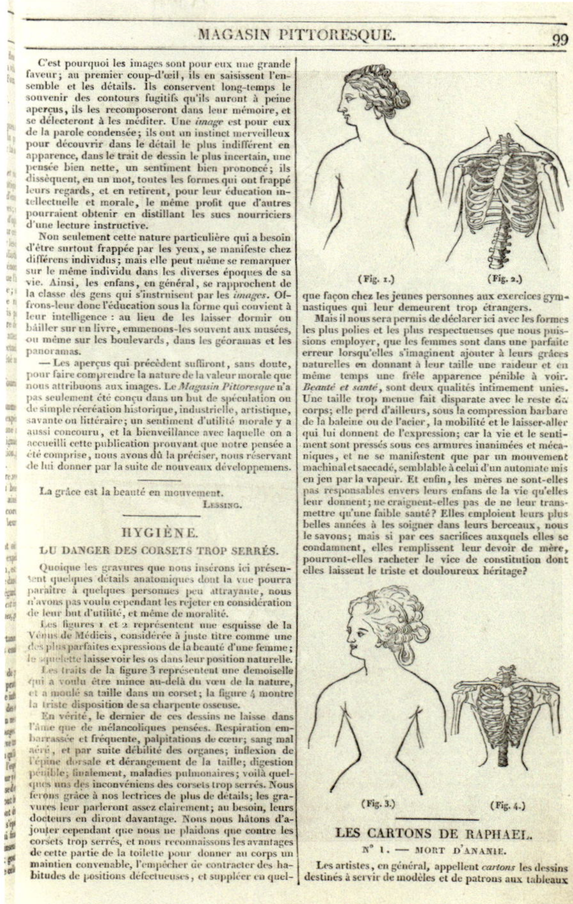
MAGASIN PITTORESQUE. 99

C'est pourquoi les images sont pour eux une grande faveur; au premier coup-d'œil, ils en saisissent l'ensemble et les détails. Ils conservent long-temps le souvenir des contours fugitifs qu'ils auront à peine aperçus, ils les recomposeront dans leur mémoire, et se délecteront à les méditer. Une *image* est pour eux de la parole condensée; ils ont un instinct merveilleux pour découvrir dans le détail le plus indifférent en apparence, dans le trait de dessin le plus incertain, une pensée bien nette, un sentiment bien prononcé; ils dissèquent, en un mot, toutes les formes qui ont frappé leurs regards, et en retirent, pour leur éducation intellectuelle et morale, le même profit que d'autres pourraient obtenir en distillant les sucs nourriciers d'une lecture instructive.

Non seulement cette nature particulière qui a besoin d'être surtout frappée par les yeux, se manifeste chez différens individus; mais elle peut même se remarquer sur le même individu dans les diverses époques de sa vie. Ainsi, les enfans, en général, se rapprochent de la classe des gens qui s'instruisent par les *images*. Offrons-leur donc l'éducation sous la forme qui convient à leur intelligence : au lieu de les laisser dormir ou bâiller sur un livre, emmenons-les souvent aux musées, ou même sur les boulevards, dans les géoramas et les panoramas.

— Les aperçus qui précèdent suffiront, sans doute, pour faire comprendre la nature de la valeur morale que nous attribuons aux images. Le *Magasin Pittoresque* n'a pas seulement été conçu dans un but de spéculation ou de simple récréation historique, industrielle, artistique, savante ou littéraire; un sentiment d'utilité morale y a aussi concouru, et la bienveillance avec laquelle on a accueilli cette publication prouvant que notre pensée a été comprise, nous avons dû la préciser, nous réservant de lui donner par la suite de nouveaux développemens.

La grâce est la beauté en mouvement.
LESSING.

HYGIÈNE.

LU DANGER DES CORSETS TROP SERRÉS.

Quoique les gravures que nous insérons ici présentent quelques détails anatomiques dont la vue pourra paraître à quelques personnes peu attrayante, nous n'avons pas voulu cependant les rejeter en considération de leur but d'utilité, et même de moralité.

Les figures 1 et 2 représentent une esquisse de la Vénus de Médicis, considérée à juste titre comme une des plus parfaites expressions de la beauté d'une femme; le squelette laisse voir les os dans leur position naturelle.

Les traits de la figure 3 représentent une demoiselle qui a voulu être mince au-delà du vœu de la nature, et a moulé sa taille dans un corset; la figure 4 montre la triste disposition de sa charpente osseuse.

En vérité, le dernier de ces dessins ne laisse dans l'âme que de mélancoliques pensées. Respiration embarrassée et fréquente, palpitations de cœur; sang mal aéré, et par suite débilité des organes; inflexion de l'épine dorsale et dérangement de la taille; digestion pénible; finalement, maladies pulmonaires; voilà quelques uns des inconvéniens des corsets trop serrés. Nous ferons grâce à nos lectrices de plus de détails; les gravures leur parleront assez clairement; au besoin, leurs docteurs en diront davantage. Nous nous hâtons d'ajouter cependant que nous ne plaidons que contre les corsets trop serrés, et nous reconnaissons les avantages de cette partie de la toilette pour donner au corps un maintien convenable, l'empêcher de contracter des habitudes de positions défectueuses, et suppléer en quelque façon chez les jeunes personnes aux exercices gymnastiques qui leur demeurent trop étrangers.

(Fig. 1.) (Fig. 2.)

Mais il nous sera permis de déclarer ici avec les formes les plus polies et les plus respectueuses que nous puissions employer, que les femmes sont dans une parfaite erreur lorsqu'elles s'imaginent ajouter à leurs grâces naturelles en donnant à leur taille une raideur et en même temps une frêle apparence pénible à voir. *Beauté et santé*, sont deux qualités intimement unies. Une taille trop menue fait disparate avec le reste du corps; elle perd d'ailleurs, sous la compression barbare de la baleine ou de l'acier, la mobilité et le laisser-aller qui lui donnent de l'expression; car la vie et le sentiment sont pressés sous ces armures inanimées et mécaniques, et ne se manifestent que par un mouvement machinal et saccadé, semblable à celui d'un automate mis en jeu par la vapeur. Et enfin, les mères ne sont-elles pas responsables envers leurs enfans de la vie qu'elles leur donnent; ne craignent-elles pas de ne leur transmettre qu'une faible santé? Elles emploient leurs plus belles années à les soigner dans leurs berceaux, nous le savons; mais si par ces sacrifices auxquels elles se condamnent, elles remplissent leur devoir de mère, pourront-elles racheter le vice de constitution dont elles laissent le triste et douloureux héritage?

(Fig. 3.) (Fig. 4.)

LES CARTONS DE RAPHAEL.

N° 1. — MORT D'ANANIE.

Les artistes, en général, appellent *cartons* les dessins destinés à servir de modèles et de patrons aux tableaux

FIGURE 70

Notré, *Ascension of an Aerostat* [*L'ascension aérostatique*]. Engr. Andrew, Best, Leloir, *Le Magasin pittoresque* 1 (1833), no. 21. Wood engraving. Bibliothèque nationale de France.

FIGURE 71

Hygiene. On the Danger of Too-Tight Corsets [*Hygiène. Du danger des corsets trop serrés*], *Le Magasin pittoresque* 1 (1833), no. 13. Wood engraving. Bibliothèque nationale de France.

the deaf and a well-illustrated article explained the technology of ballooning (fig. 70).[63] The Saint-Simonian belief in the equality of women is manifest in the article "On the Danger of Too-Tight Corsets," accompanied by four illustrations showing the damage done to the human skeleton through such compression (fig. 71).[64]

While *Le Magasin pittoresque* followed the *Penny Magazine* in featuring articles on science, industry, and natural history, it eschewed almost completely the *Penny Magazine*'s regular inclusion of literature and poetry, explaining that the English working class had a refined taste in literature, but, in France, fiction was more likely to be deleterious, seducing the popular imagination.[65] It criticized the English digest's preference for what it called "fragments" of history, biography, and the fine arts, and claimed to offer more complete coverage. Art was, in fact, always better represented in *Le Magasin pittoresque* than in the *Penny Magazine*. In addition to the engravings of major monuments of Western culture borrowed from the English periodical, *Le Magasin pittoresque* initiated a regular illustrated feature on the annual Paris Salon and published drawings or

14.] MAGASIN PITTORESQUE. 105

MUSÉES DU LOUVRE.

SALON DE 1834. — EXPOSITION DE PEINTURE.

TABLEAUX DE GENRE.

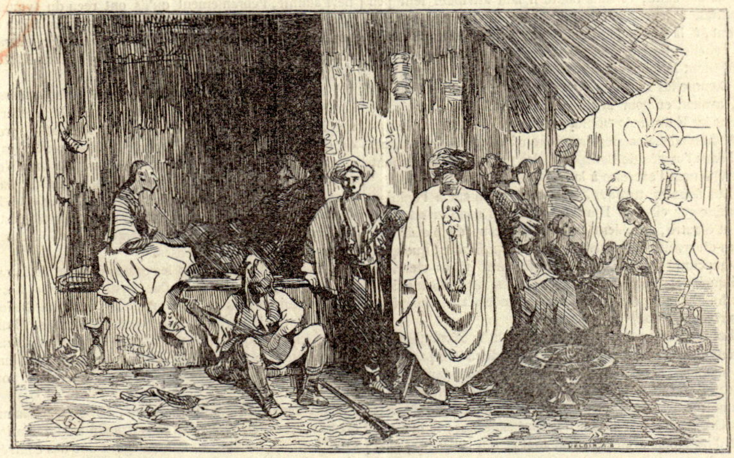

(Corps-de-garde turc sur la route de Smyrne à Magnésie, par M. Decamps.)

Des lueurs où dominent la blancheur mate et le jaune ardent éclairent l'intérieur du corps-de-garde; les soldats se reposent à l'ombre de leur baraque; au dehors, sous le soleil, on voit cheminer des voyageurs et leurs chameaux à travers la poussière qui tourbillonne.

Les armes, les pipes, la poterie, sont éparses sur le sol; au lieu du poêle de fer ou de l'âtre de nos corps-de-garde, on remarque un grand vase de cuivre où se consument quel-

pieds des bottines en maroquin, ou des souliers avec des bas. Leur manteau rouge, qui est attaché autour du cou, avec une agrafe d'argent, tombe jusqu'au genou.

M. Mac-Farlane, qui a visité la Turquie asiatique en 1828, donne des détails intéressans sur le progrès des innovations militaires. Voici comment il décrit une caserne, et les exercices des *tacticiens* à Smyrne.

« En sortant des bazars, nous traversâmes l'extrémité de

FIGURE 72

Turkish Guardhouse on the Smyrna–Magnesia Road, by Mr. Decamps. Louvre Museum. Salon of 1834. Exhibition of Painting. Genre Paintings [*Corps-de-garde turc sur la route de Smyrne à Magnésie, par M. Decamps. Musée du Louvre. Salon de 1834. Exposition de peinture. Tableaux de genre*]. Engr. Andrew, Best, Leloir, *Le Magasin pittoresque* 2 (1834), no. 14. Wood engraving. Bibliothèque nationale de France.

engravings after paintings by virtually all the major Romantic artists. Decamps, Delacroix, David d'Angers, Horace Vernet, Barye, Grandville, Boulanger, Gigoux, and Devéria, most of whom were republican in politics and Romantic in style, were all featured regularly in its pages. When covering the annual Salon, *Le Magasin pittoresque* invariably chose one of their works to illustrate, such as the wood engraving after Decamps's painting *Turkish Guardhouse on the Smyrna-Magnesia Road*, shown in the Salon of 1834 (fig. 72). The *Penny Magazine*, in contrast, much preferred the work of deceased artists.

The role of images in the project of educating the masses constitutes a debate extending back to the *Biblia pauperum*, whose pictures enabled the illiterate to "read" the Bible. Knight's wood engravings did provide elementary art appreciation for the working classes and attracted them to his publication, but Charton honestly believed that images had as much power as words, obviously an extension of the Saint-Simonian idea of the social function of art. Wood engraving, as *Le Magasin pittoresque* announced, could give "a new degree of utility heretofore unknown to the alliance of artist and writer."[66] This new emphasis on the importance of images was emphasized by an article published shortly after the magazine's inception, "On Methods of Instruction. Books and Images," undoubtedly written by Charton himself. "We note as a complementary means of instruction, still not in common use, *drawings* or *images* . . . We are so convinced of their utility that we freely state '*Without drawings, it is impossible for men, whether of high or low rank, to have a complete education.*' We attach a great moral importance to *images*, and we believe they fill the void left by books."[67]

If we can say that Knight's *Penny Magazine* was composed of text with illustrations, then Charton's *Le Magasin pittoresque* was almost the reverse: Charton described it as a "little weekly journal of engravings."[68] For him, the images were paramount, as his correspondence amply demonstrates. He traveled constantly in search of illustrations, often returning with drawings or prints for ABL to translate into engravings.[69] Once the engravings were secured, he then wrote to his contributors asking for accompanying text. To his friend Hippolyte Fortoul (1811–1856), an early Saint-Simonian who had abandoned law for a university career and was a regular contributor to *Le Magasin pittoresque*, he wrote: "I'm having some views engraved of the Midi: Toulouse, Montpellier, etc. Their museums will, I hope, furnish me with subjects for engravings. Dare I ask you for the text?"[70] To Jean Reynaud: "I'm having engraved some scenes of bears at the Jardin des Plantes. I will need *four or five columns* on the life of bears or on their history at the Jardin des Plantes."[71] To Hippolyte Fortoul again: "I'm having several views of Italy engraved. Would you like to write the text? They are general views of Florence, Naples, Pozzuoli, Castellamare, Sorrento, and a view of the piazza and fountain of Perugia. You know what liberty of digression is permitted."[72] Of his own 1845 trip to Italy, he wrote to his wife: "Lachevardière found a young draftsman who will accompany me and will make, for the *Magasin*, drawings that I will indicate to him."[73] The pressures of producing a weekly magazine were constant: "My dear friend, here is one of the engravings for which I am hoping to have some text from you. It's the one after Poussin. Unfortunately I can't send you the others, which will be finished late, for you know that I compose and lay out the page before the engravers have finished."[74] His imagination was so clearly visual that sometimes it even preceded the image itself: "I then thought that a group composed of Arab figures would make an interesting plate. But we'd have to find the subject for an article."[75]

Before being engraved, an artist's drawings first had to be redrawn on the wood plate, usually by a professional designer. In the case of well-known artists such as David d'Angers and Grandville, this plate was then returned for their approval and corrections before it went to the engravers. The process didn't always go smoothly, as the lengthy correspondence with David d'Angers demonstrates. In 1837, Charton wanted to publish an article on the Panthéon in Paris along with an engraving of David d'Angers's new sculptured pediment. The drawing was commissioned directly from the artist, as opposed to many others that were arranged between Charton and the engravers alone. At the beginning of this process, Charton suggested to the sculptor: "Perhaps you might consider that, in view of the small size of the *Magasin*, the groups of the pediment might

FIGURE 73
The Pediment of the Panthéon by Mr. David d'Angers [*Le fronton du Panthéon par M. David d'Angers*]. Engr. Andrew, Best, Leloir, *Le Magasin pittoresque* 5, no. 40 (October 1837). Wood engraving. Bibliothèque nationale de France.

FIGURE 74
Waltz, from *Music, Composed and Drawn by J. J. Grandville* [*Valse. Musique composée et dessinée par J. J. Grandville*]. Engr. Andrew, Best, Leloir, *Le Magasin pittoresque* 8, no. 31 (August 1840). Wood engraving.

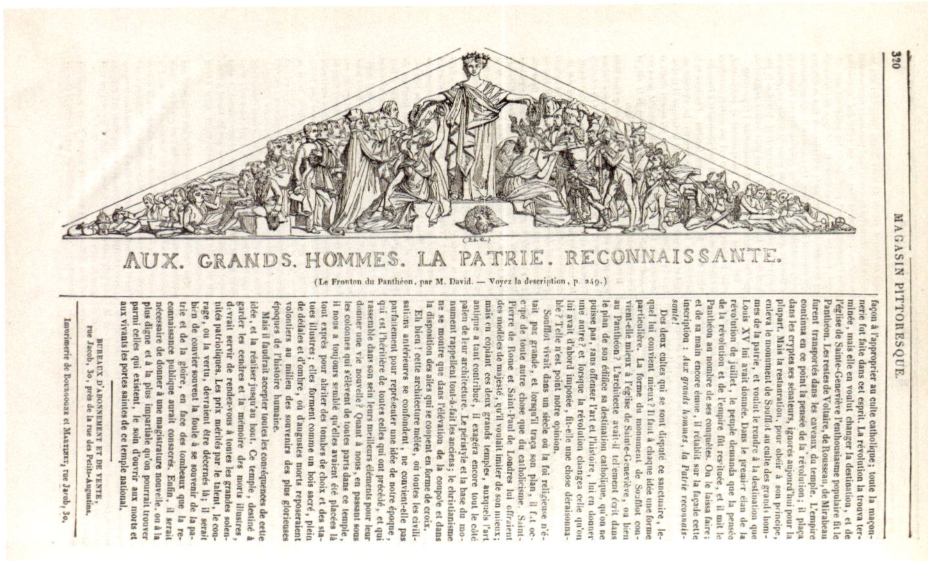

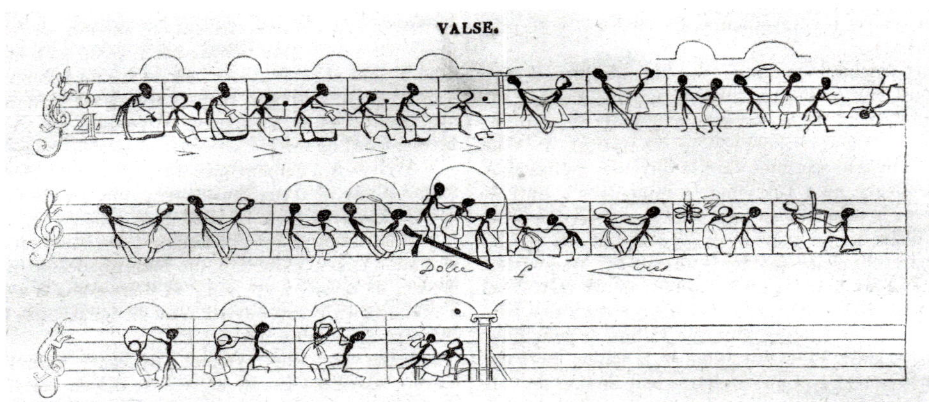

be drawn separately, along with a small representation of the whole that would give an idea of the general effect. Whatever you decide, we will do."[76] David d'Angers apparently decided against this proposal but went ahead with the project; when the first designer's work proved unsatisfactory, Charton hired another, Prosper Saint-Germain (1804–1875): "I think I've found a designer that you will like," he wrote, "and my intention is to introduce him to you Monday afternoon."[77] Then, "I'm sending you the drawing of Mr. Saint-Germain. Please let me know your opinion. The engravers will not improve upon this. Mr. Saint-Germain is coming to Paris on Monday. If you wish, I will come with him to see you, at whatever time in the afternoon is convenient to you. He will make all the corrections you desire. I am hoping for better results from this plate than the first one."[78] His hopes were dashed however, and he subsequently wrote: "I have just received the proof of the engraving from Messrs Andrew and Leloir. I find, to my great disappointment, many small errors. The figure of the Fatherland has lost an eye, Manuel is smirking, Voltaire a bit of a caricature, the arm of Bichet looks broken, etc., etc. I am writing to the engravers to ask them to do their best to correct these things. I ask you to send them your notes as well on what should be corrected. We have no time to lose. We have barely enough days left for the stereotype and the printing. I'm sorry that they didn't do better. Please accept my apologies and believe in the good will that I hold for the project, both as a friend of liberty and as your admirer."[79]

Corrections apparently continued right up until the last possible moment. The first article on the project appeared in August, but the engraving still was unfinished.[80] On October 5, Charton wrote: "The engraving of the pediment was already stereotyped and in press when your last letter to me arrived. As a result it was impossible to make the changes you desired."[81] The engraving appeared later that month with the second article on the Panthéon, and yet, despite what was evidently a less than satisfactory result, David d'Angers did continue to allow his work to be reproduced in the magazine (fig. 73).

Grandville was another case entirely. Charton clearly admired his work, and gave him a general exemption from the magazine's focus on the positive rather than the imaginative. As a result, Grandville published seventeen drawings in *Le Magasin pittoresque* within its first two years. Unlike David d'Angers, who as a sculptor had little experience with illustration, Grandville was a professional, gaining his living by supplying drawings for a variety of books and periodicals. He was accustomed to dealing with publishers, and knew how to protect the integrity of his work. He insisted on choosing the designers and engravers himself, and on seeing and correcting proofs at every stage of the process.[82] In 1840, he wrote to Charton: "I have just sent Andrew the third woodblock, which Leloir said he couldn't engrave in exact facsimile because of the short time that you are giving him. Make sure that this isn't a facile excuse to make quick work of these three plates. . . . You have to push these gentlemen (without mentioning me) to put this plate in good hands. It's as much in your interest as in mine."[83] In contrast to the work Grandville usually did for books and periodicals, his drawings for *Le Magasin pittoresque* were not illustrations; they stood alone, accompanied only by a text that he wrote himself. His 1840 series *Music, Composed and Drawn by J. J. Grandville* extends across two pages, each musical composition created from anthropomorphized notes: military music is depicted by marching soldiers, religious music by kneeling figures; a waltz is composed of black-and-white notes dancing together (fig. 74). Unlike most of the articles in *Le Magasin pittoresque,* here there is no didactic purpose; Grandville's work is included simply for visual pleasure. And, while Grandville might have been cavalier about his writing, entrusting the editing of it to Charton, he was scrupulously attentive to the fate of his drawings. In sending Charton one text, Grandville wrote: "For the rest, keep it, reject it, cut it, trim it, combine it with what I've already given you and act for the best as always. But as to the woodblock, as for the first one: it is absolutely necessary that I meet with the engraver who will never be able to complete it successfully without my advice."[84] He had no illusions about working for the periodical press. He called Charton a "picturesque ogre who makes of the best and rarest idea in the world merely a mouthful," comparing him to "Saturn who devours his artists."[85] Nonetheless, Charton's constant efforts to secure the work of all the major Romantic artists for *Le Magasin pittoresque* were largely successful. The periodical became an organ of the Romantic movement in art—albeit for a proletarian audience—just as much as *L'Artiste* fulfilled this role for the more elite and cultivated classes.

Despite the immediate popularity of the new illustrated periodicals, applause was not unanimous; attacks began virtually immediately. The strongest criticism came, as might be expected, from the entrenched art establishment. Raoul de Croy (b. 1791), writing in *Journal des artistes et des amateurs*, charged that their illustrations were "like black ink blots," and that the very idea of imparting an appreciation of art to the masses would only destroy the market for better quality images. "What will become of works of engraving, this art that is so perfect, so difficult, and so worthy of encouragement? What will become, even, of lithography, whose improvement in the last ten years is so notable, questions of morality aside?" Wood-engraved illustrations constituted "the assassination of the fine arts," he charged, proposing "Let our neighbors across the channel commercialize the productions of imagination and make art mechanically, with mechanical presses; but let us not imitate them."[86]

It wasn't only the illustrations that were criticized. Within two months, a vaudeville production entitled *Le Magasin pittoresque* opened at the Théâtre des Variétés in Paris.[87] A "Revue in fifteen issues," it lampooned everything about the new magazine, but most especially its encyclopedic content. At one point, the figure of "Competition" says to the old bookseller Basane: "Don't I see there twenty copies of the *Encyclopedia* . . . This is the mine we must exploit, this is the source we should exhaust . . . take your scissors . . . cut, edit, trim . . . all that, presented as new and accompanied by portraits of great men and huge beasts, contemporary beauties and Gothic monuments will furnish the most bizarre and varied digest of our epoch at two sous . . . there, now you have the authentic *Magasin pittoresque.*"[88]

The comment that *Le Magasin pittoresque* was merely the *Encyclopedia* cut up and reassembled does have some validity. The magnum opus published by Denis Diderot and Jean Le Rond d'Alembert in the previous century, their *Encyclopedia, or Systematic Dictionary of the Sciences, Arts, and Crafts* had, in fact, a similar intention, to collate and disseminate knowledge, but it was directed to wealthier and more literate classes.[89] And perhaps here was the real objection to Charton's new weekly, that the spread of knowledge would eradicate class barriers—as he fervently hoped.

Despite this criticism, the magazine was an indubitable success, quickly reaching a circulation of fifty thousand, which, as Charton pointed out, was greater than the major political newspapers.[90] Even Émile de Girardin was impressed; when his efforts to associate himself with *Le Magasin pittoresque* were rebuffed, he founded the rival *Musée des familles* (1833–1900).[91] Within months, there were even more illustrated general survey periodicals: *La Mosaïque* (1833–1836), *La Lanterne magique* (1833–1836), and *Le Magasin universel* (1833–1840), all advertising the number and quality of their illustrations. Clearly the public was enamored of pictures.

With numerous illustrated weeklies now publishing, all directed to the classes that had heretofore been ignored by the literary establishment, savants began to take notice. An article in *Le Temps* reflected on this new phenomenon:

> I don't know if the golden age of big books is definitively over, but it is at least clear that this moment is not favorable to them. We no longer see folio editions, no more of those thick and venerable tomes hatched after twenty years of laborious incubation. Everything is done quickly today, it seems, we write and publish quickly.... There are no more books, but on the other hand, there are newspapers. Never have there been so many of them, and among newspapers, those that are the most successful are those directed, not to the most educated minds but to the greatest numbers as well as to the smallest purses, which also constitute the greatest number. There is *Le Magasin pittoresque*, which has sixty thousand subscribers and which is sold for two sous. There are a lot of publications of the same type and the same price that circulate unnoticed in the crowd. Public knowledge doesn't gain in depth but in superficiality. True knowledge, that up to now was stoked by, nourished by, and supported by big books, sheltered in dark sanctuaries, comes out in broad daylight, runs through the streets, rubs elbows with the passersby, makes itself familiar and easily obtainable, cuts itself down to the dimensions of the smallest brains, and shrinks itself in order to go everywhere.... *Le Magasin pittoresque* and such publications that followed it have, up until now, been directed perhaps more to the amusement rather than to the education of their readers. No doubt the immense variety of facts they contain and the lovely wood engravings that accompany the articles, by reaching the mind through the eyes, are destined to spread a great deal of information and useful concepts, but no system coordinates the diverse subjects. It is an education, no doubt, but an education that is confused, anarchistic, lacking unity, and without apparent order.[92]

Much of this criticism is reminiscent of the theatrical revue of the previous year that charged that *Le Magasin pittoresque* just recycled encyclopedia entries, but there was also a grain of truth in the charge that it was disorganized. While each issue contained a variety of articles, with the major areas of science, technology, history, and the arts usually represented, they were in no particular order. The digest's strength—and perhaps its weakness as well—was that one could never be sure what one would find there. It was encyclopedic knowledge presented like a smorgasbord, with all the courses served at the same time. This situation would be rectified in the next development of the illustrated press, the illustrated weekly newspaper.

FIGURE 75

First issue of *The Illustrated London News*, May 14, 1842.

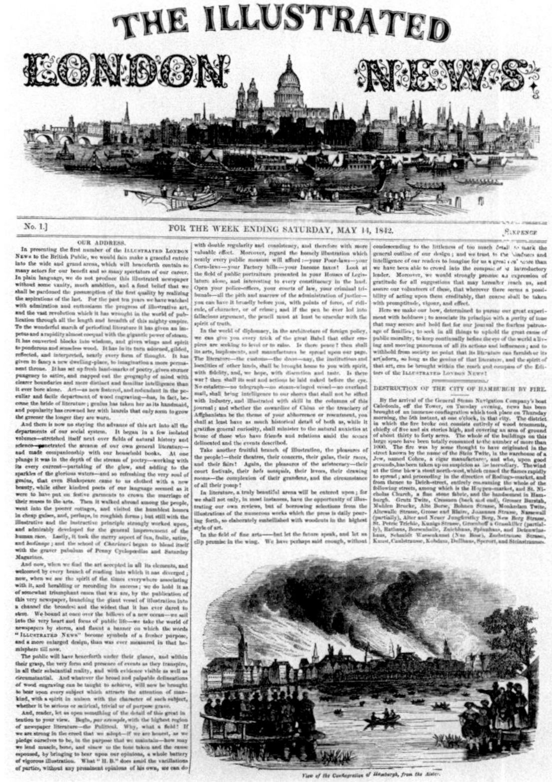
THE ILLUSTRATED LONDON NEWS

No. 1.] FOR THE WEEK ENDING SATURDAY, MAY 14, 1842. [SIXPENCE

THE ILLUSTRATED LONDON NEWS

If the first stage of illustrated universal survey periodicals was characterized by high idealism and populist reform, the second stage brought the pleasures of the illustrated press to a more prosperous audience. The British again took the lead, with the establishment of the *Illustrated London News* on May 14, 1842 (fig. 75). Four times the size of earlier periodicals like the *Penny Magazine* and *Le Magasin pittoresque*, each issue of this new illustrated weekly newspaper of contemporary events featured sixteen pages and thirty-two wood engravings for sixpence—six times the cost of the *Penny Magazine.* Its first issue announced its goals: "The public will have henceforth under their glance and within their grasp, the very form and presence of events as they transpire, in all their substantive reality, and with evidence visible as well as circumstantial." The newspaper further promised that it would "keep continually before the eye of the world a living and moving panorama of all its actions and influences."[93]

The Illustrated London News was the creation of Herbert Ingram (1811–1860), who had worked in the London print trade before opening a print shop and news agency in Nottingham in partnership with his brother-in-law, Nathaniel Cooke.[94] He had noticed that whenever a London newspaper arrived that carried an illustration—particularly if it featured an atrocity of some kind—sales increased exponentially. He became convinced that a newspaper with not one but numerous illustrations would be a great success, so he moved to London to put his ideas into practice. Capital for the enterprise came from the sale of "Parr's Life Pills," a patent medicine invented by Cooke and Ingram that became a lucrative sideline; Parr was reputed to live to the age of 152, and his portrait graced the partners' product with the bogus legend "From a Picture by Sir Peter Paul Rubens." In London, Ingram hired the engraver Henry Vizetelly (1820–1894), who had engraved the "Rubens" portrait of Parr, to oversee illustrations for his new periodical. Ingram's original idea was to produce an "illustrated criminal record," as Vizetelly noted in his memoirs, in the manner of the popular broadsheets that always sold so well. Ingram's own first attempt at illustrated journalism had been along these lines, a broadsheet entitled "The Life, Death, and Horrible Crimes of Thomas Greenacre, the Camberwell Murderer."[95] Ingram was always quite candid about acknowledging his original inspiration: "He admitted that a good murder was not to be hoped for every week or even every month, still there were police cases, and Old Bailey, and assize trials, as well as factory riots, rick-burnings, coining, sacrilege, horse-stealing, and the like. The engravings of some of which subjects, he maintained, could easily enough be prepared in advance."[96] Vizetelly took credit for having persuaded Ingram to temper his enthusiasm for crime. By the time the first issue appeared, Ingram's initial plan had been so modified that the newspaper announced that it would "associate its principle with a purity of tone that may secure and hold fast for our journal the fearless patronage of families; to seek in all things to uphold the cause of public morality."[97]

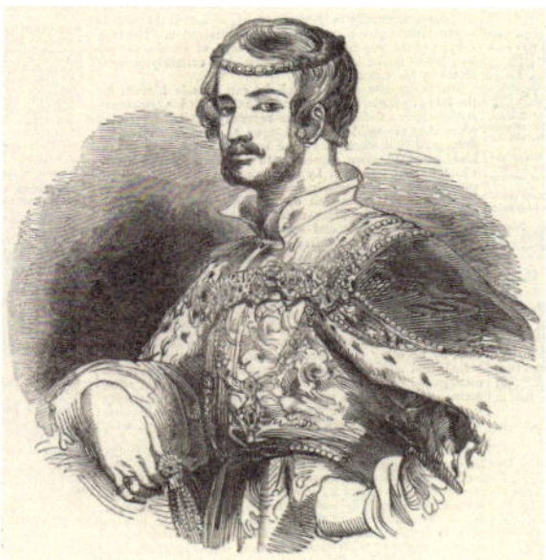

FIGURE 76
Prince Albert at Her Majesty's Bal Masqué, *The Illustrated London News*, May 14, 1842. Wood engraving. Yale Center for British Art.

Vizetelly organized the wood engravings, and Frederick William Naylor Bayley (1808–1852), previously editor of the *National Omnibus*, was named the first editor. Ingram selected the masthead engraving, a cityscape of London featuring the Thames and St. Paul's Cathedral (see fig. 75). The title, *The Illustrated London News*, was chosen simply because Ingram recalled that his Nottingham customers would come into his news agency asking only for the latest London news.[98]

The first issue appeared on May 14, 1842, two days after "Her Majesty's Bal Masqué," a fancy dress ball at Buckingham Palace; it featured a double-page spread of engravings of Queen Victoria, the prince consort, and numerous titled guests, all in historic costumes (fig. 76). Its initial run sold out immediately, and, to meet the demand, it had to be reprinted several times, totaling 24,000 copies. By the end of 1842, it was averaging 66,000 copies per issue, and by the 1851 Great Exhibition of Works of All Industry, circulation had reached 130,000.[99]

The regular contents of the publication were soon established: news from Parliament and from the foreign press, society news ("The Court and Haut Ton"), articles on fashion, horticulture, literature, sports, theater, fine arts, and (always) business news. Announcements of births, marriages, and deaths among the gentry were a regular feature. The horrible and sensational were not wholly banished, however. The first issue offered "Discovery of a Band of Murderers at Nuremburg," lengthy, detailed crime reports from the courts and the police, and a report and engraving of a disastrous fire in Hamburg (see fig. 75). As

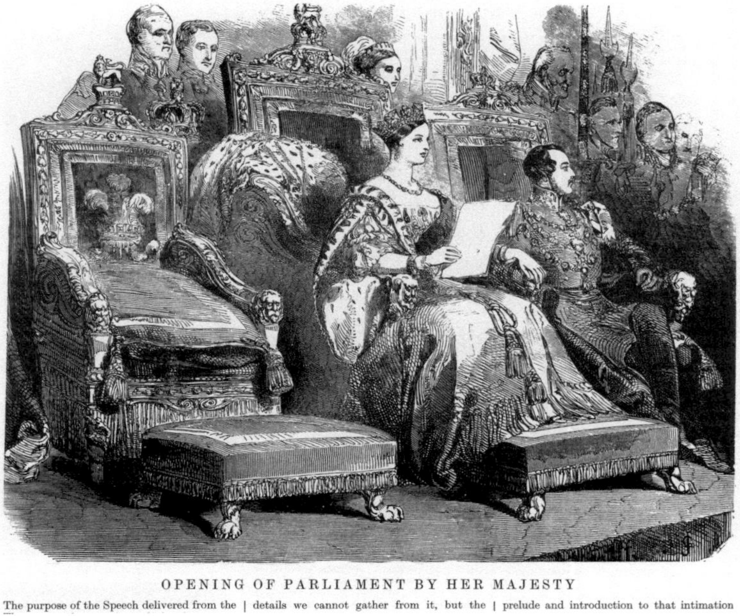

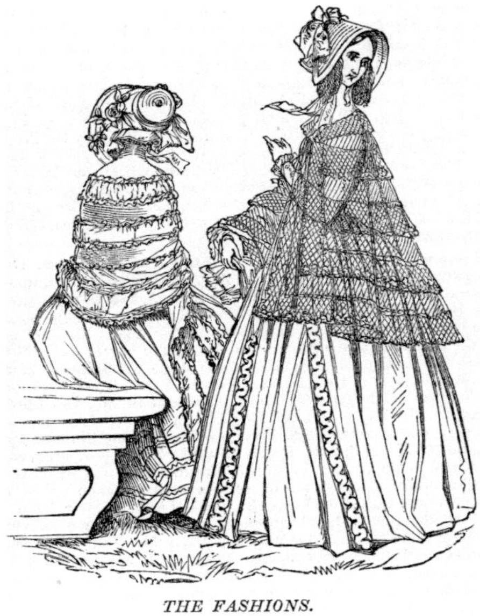

FIGURE 77
Opening of Parliament by Her Majesty, *The Illustrated London News*, January 24, 1846. Wood engraving.

FIGURE 78
The Fashions, *The Illustrated London News*, May 14, 1842. Wood engraving. Yale Center for British Art.

OPPOSITE **FIGURE 79**
Epsom Derby Day, 1844, *The Illustrated London News*, May 25, 1844. Wood engraving.

Mason Jackson, who succeeded Vizetelly as art editor, wrote, "In its pages there was something for everybody."[100] Within weeks, a certain structure developed: serious news (unless there was royal entertainment of some sort) captured the first pages, followed by more lightweight content. Queen Victoria made her appearance in virtually every issue, her activities reported in detail and almost always illustrated, such as the engraving of her with Prince Albert at the opening of Parliament in 1846 (fig. 77). Criminal news became a standard feature; there was even a column, "Coroner's Inquests," that, for a time, was accompanied by a logo of a dead body sprawled on a mortuary slab. Fashion was never absent, with regular letters from Paris, signed "Julie," or "Felicie," or "Henriette," always illustrated with the latest trends (fig. 78). In the first issue, Felicie introduces herself by writing from the rue de la Chaussée d'Antin, Paris, "I feel an inexpressible delight in inditing my first communication to your lady readers upon the fashions of this *haut ton* of this *ville de gaité*." She goes on to describe at length fashionable bonnets, beginning, "Those most decidedly *à-la-mode* are the *chapeaux paille de riz*," warns that "trimmings of every description to the bottom of dresses appear to be waning, although for the fronts they are much employed," and concludes by advising her readers that "one cannot be *comme il faut* without another charming novelty—for what is more chaste than the *capucin*, which, as it brings into use any kind of lace you may have in your *armoire*, combines at once economy and elegance."[101] Fashions, horse racing, foxhunting, and royal balls were always treated extensively, and fulsomely illustrated. *Epsom Derby Day, 1844*, for example, is one of the most ambitious illustrations of the *Illustrated London News*, a full-page engraving that clearly taxed the abilities of its production staff since the horizontal join lines in the plate remain visible (fig. 79). Social consciousness

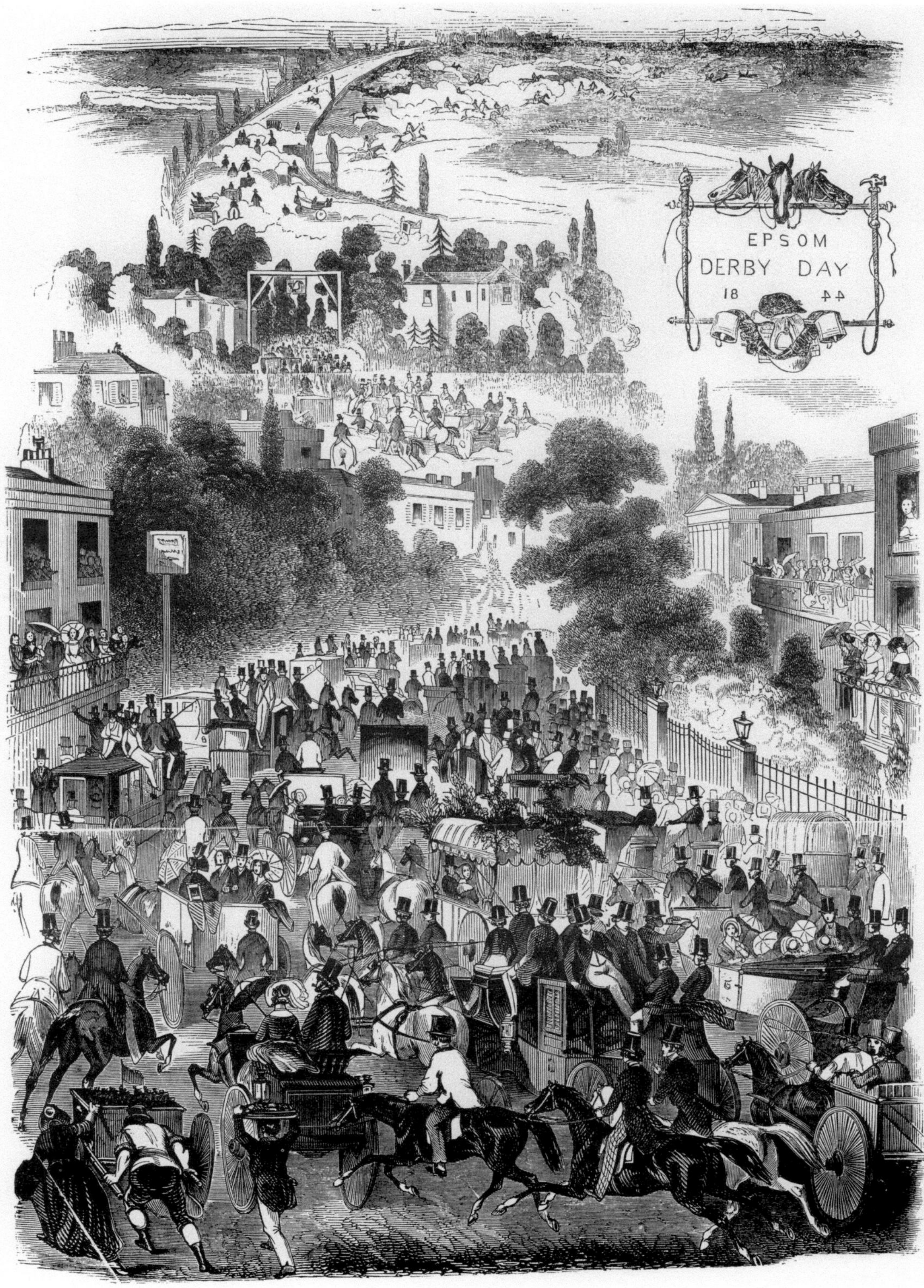
EPSOM
DERBY DAY
18 44

FIGURE 80
Preston—Attack on the Military—Two Rioters Shot. The Illustrated London News, August 20, 1842. Wood engraving.

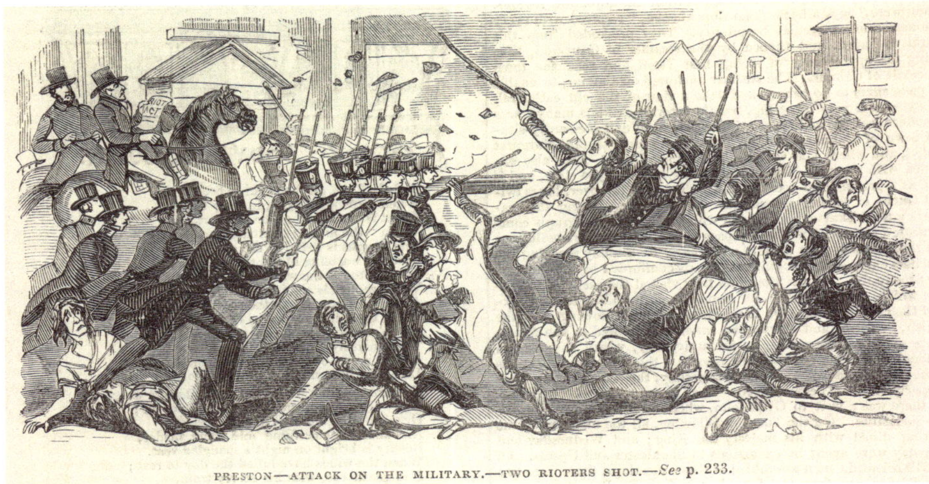

did not exist solely on this elegant plane, however. There were also articles condemning both the slave trade and workers' strikes.[102] *Preston—Attack on the Military—Two Rioters Shot* depicts the riots that took place in 1842 in the Manchester area (fig. 80). The accompanying article, "The Disturbances in the Manufacturing Districts," was written from the point of view of the wealthier classes who read this periodical, reporting at length on the violence that beset the factories, while barely mentioning the issues that had provoked it; masses of unemployed workers had attempted to shut down a factory, the army was called out, the Riot Act was read to the crowds, and clashes between workers and military followed across the entire region.[103] The *Illustrated London News* did indeed present a "living and moving panorama" of Victorian society in all its contradictions.

It is not surprising that ten years had elapsed between the establishment of the first illustrated universal survey magazines and the publication of the *Illustrated London News.* Charton's correspondence for *Le Magasin pittoresque* reveals the difficulties in obtaining wood engravings on a regular basis and in quantity; articles had to be planned months in advance in order to secure the necessary illustrations. Current events, however, demanded immediate publication. Ingram's and Vizetelly's solutions to this problem became standard practice in the news trade until the advent of photography. Vizetelly explained: "The system pursued with regard to the majority of engravings of current events—foreign, provincial, and even metropolitan when these transpired unexpectedly—was to scan the morning papers carefully, cut out such paragraphs as furnished good subjects for illustration, and send them with the necessary boxwood blocks to the draftsmen employed."[104] If the event was announced in advance, artists could be sent to make preliminary sketches of the setting, but foreign events proved more difficult. Publishers feared that the illustrated newspaper would undermine sales of their guidebooks and so refused to allow their images to be copied. Hence, editors had recourse to print collections for raw material to be copied. According to Vizetelly, the illustration of the Hamburg fire featured in the first issue had just such a genesis: "The view of the city was engraved by one of my assistants and was copied, I remember, from a print in the British Museum, the artist, in drawing the subject upon the wood, having added the necessary flames and volumes of smoke, as well as the crowd of people, in boats and on the river bank, supposed to have been attracted by the conflagration" (see fig. 75).[105] In the first issue, editor F. W. N. Bayley wrote in florid Victorian prose, "Should the pen ever be led into fallacious argument, the pencil must at least be oracular with the spirit of truth."[106] Vizetelly, in his memoirs, responded: "Not even a single engraving in the opening number derived from an authentic source!"[107] All were drawn by illustrators after written descriptions of the events—to the point where Prince Albert at "Her Majesty's Bal Masqué" was shown with a pearl headband instead of a diadem because the draftsman had misunderstood the written description (see fig. 76).[108] No matter, the authoritative promise that the *Illustrated London News* seemed to fulfill and that led to its resounding success was that "the public will have henceforth under their glance and within their grasp, the very form and presence of events as they transpire."

L'ILLUSTRATION

In the wake of the *Illustrated London News*, history repeated itself still again with the French replicating an English publication: on March 5, 1843, less than a year after the *Illustrated London News* made its debut, *L'Illustration, Journal universel* appeared (fig. 81). Where the masthead of the *Illustrated London News* showed the London skyline and the Thames, *L'Illustration*'s showed the Paris skyline and the Seine. Like the *Illustrated London News*, *L'Illustration* was, in size and format, similar to newspapers, not magazines; unlike newspapers, however, it was published weekly and was profusely illustrated.[109]

The French effort was masterminded by Édouard Charton, who continued to edit *Le Magasin pittoresque* at the same time. He wrote to his friend Hippolyte Fortoul: "Remembering an English newspaper I saw in London last summer, I proposed to Lachevardière to found an illustrated digest of news. The idea didn't appeal to him but he thought it natural that I try to do it with a different publisher. I thought of Dubochet and Paulin. We quickly agreed. Dubochet went to London to make inquiries. We started right in on costs and organization."[110] Charton became its editor while maintaining the leadership of *Le Magasin pittoresque* (Lachevardière was not pleased); its capital was advanced by Jacques-Julien Dubochet (1798–1868) and Alexandre Paulin (1793–1859), both major publishers of the period. The writer and journalist Adolphe Joanne (1813–1881) joined them; the complete editorial board of *L'Illustration* was depicted in an engraving in an 1844 issue (fig. 82). All the founding editors had extensive experience with periodicals, Charton with *Le Magasin pittoresque*, the others with the newspaper *Le National* (1830–1851), which Paulin and Dubochet founded and Paulin edited. *Le National* was so strongly opposed to the reign of King Louis-Philippe that Paulin was imprisoned—twice—for seditious writing. All four founders supported republican principles; *L'Illustration* was in the

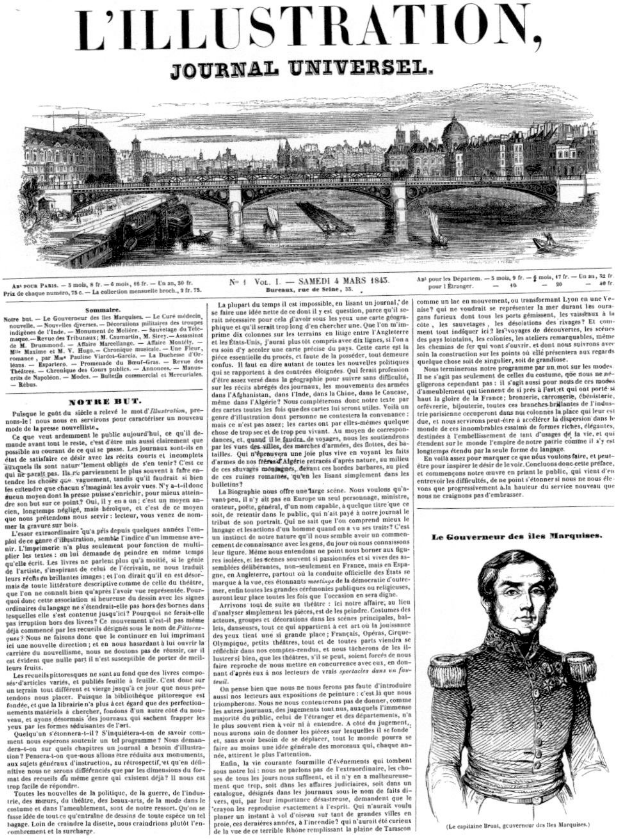

L'ILLUSTRATION,

JOURNAL UNIVERSEL.

N° 1 VOL. I. — SAMEDI 4 MARS 1843.

Bureaux, rue de Seine, 33.

Sommaire.

NOTRE BUT.

Le Gouverneur des îles Marquises.

(Le capitaine Bruat, gouverneur des îles Marquises.)

FIGURE 81

First issue of *L'Illustration*, March 5, 1843.

political opposition during most of its early decades and was even classified legally as an illustrated political journal.[111] It therefore had a genealogy very different from that of the *Illustrated London News.*

The title of the periodical carried its own message, as had *Le Magasin pittoresque* before it. In the first issue, the editorial "Our Intentions" explained: "Since contemporary taste has taken up the word *Illustration*, let's adopt it! We use it to characterize a new kind of newspaper."[112] The word *illustration* itself was new in French, adopted from the English; previously there had been no semantic distinction between drawings made for reproduction in the illustrated press and those intended to stand alone; all were drawings, *dessins.* Nonetheless, the word *illustration* was rapidly adopted in French, to the point that, after 1843, the word *pittoresque*, which had been included in the title of so many publications in the previous decade, was quickly replaced by the neologism *illustré*, and the métier of *illustrateur* came into practice as a lower-status artist, one who produced "illustrations," rather than drawings as works of art.[113] Earlier, most major artists had done at least some

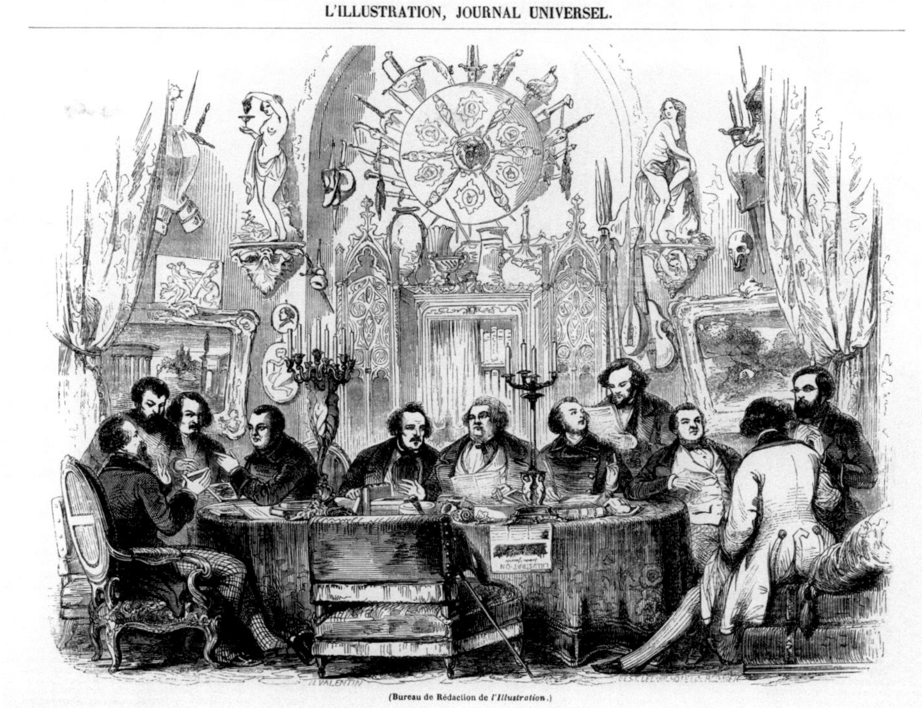

FIGURE 82
Editorial Board of L'Illustration [*Bureau de rédaction de* l'Illustration]. Engr. Valentin, *L'Illustration*, March 2, 1844. Wood engraving.

book illustration, but now individuals who specialized in drawing for periodicals increasingly did so as their principal employment and were rarely professional painters, or at least not successful ones.

Even in 1843, a decade after the foundation of *Le Magasin pittoresque*, Édouard Charton maintained his Saint-Simonian belief in the power of images. In *L'Illustration*'s first issue he wrote:

> Books only speak with half a voice if the genius of the artist, inspired by that of the writer, doesn't translate their accounts into brilliant images. From now on we can say of all descriptive writing what we say of the theater, that one doesn't really know it unless one has seen it.... What the public ardently desires today, what it demands before anything else, is to be placed as clearly as possible in the current of what is happening right now. Can newspapers do that through the short and incomplete accounts to which they are necessarily limited? It doesn't seem so. Most often they succeed in conveying a vague understanding of what they mean, while what is necessary is that it is understood so well that we think we have seen it.... Let us have newspapers that appeal to the eyes with the striking language of art.[114]

The subtitle of *L'Illustration—Journal universel*—aptly conveyed its intention of reporting on all subjects. The same kind of delirious litany that had been trumpeted by *Le*

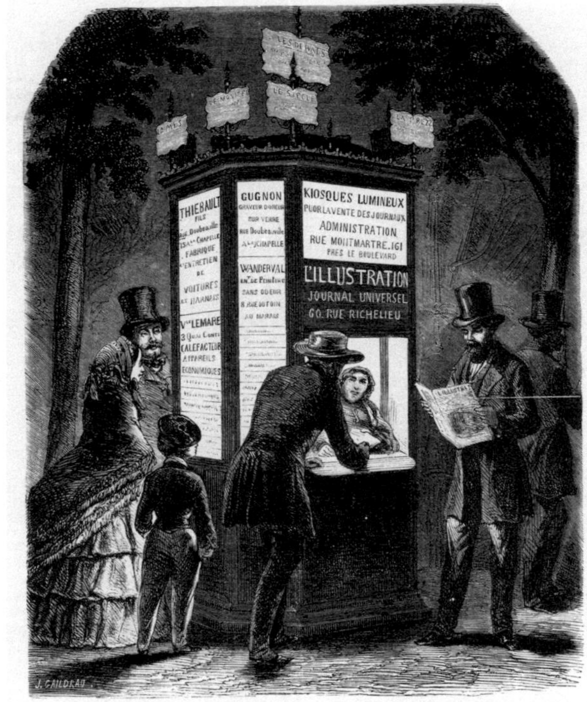

FIGURE 83
Jules Gaildrau, *Lighted Kiosks. New Stalls for the Sale of Newspapers on the Boulevards* [*Kiosques lumineux. Nouveaux bureaux pour la vente des journaux sur les boulevards*], *L'Illustration*, August 29, 1857. Wood engraving.

Magasin pittoresque again made its appearance, now transposed into the present tense. The newspaper described itself as:

> a vast annual where are recounted and illustrated at the time they occur, all the facts that contemporary history chronicles in its annals: political events, public ceremonies, great national celebrations, famous disasters, urban rumors, deaths of famous people, contemporary biographies, theatrical performances, useful discoveries, exhibitions of art and industry, new publications, glorious military exploits, fashions, types, popular scenes, etc., etc. *L'Illustration* will be, in brief, a faithful mirror that will reflect, in all its marvelous activity and all its varied movement, the life of nineteenth-century society.[115]

The analogy to a mirror resonated: In *The Red and the Black* of 1830, Stendhal had described the novel form as "a mirror walking down the road"; now that mirror would reflect the entire world.[116]

L'Illustration cost 75 centimes an issue, 30 francs annually (more than four times the price of *Le Magasin pittoresque*). Its more prosperous readership is suggested in an engraving in an 1857 issue, where an elegantly attired gentleman in a top hat is shown reading the journal (fig. 83). While its circulation was never as high as the penny press—it reached 35,000 in 1848, but usually fluctuated between 12,000 and 20,000—its readers

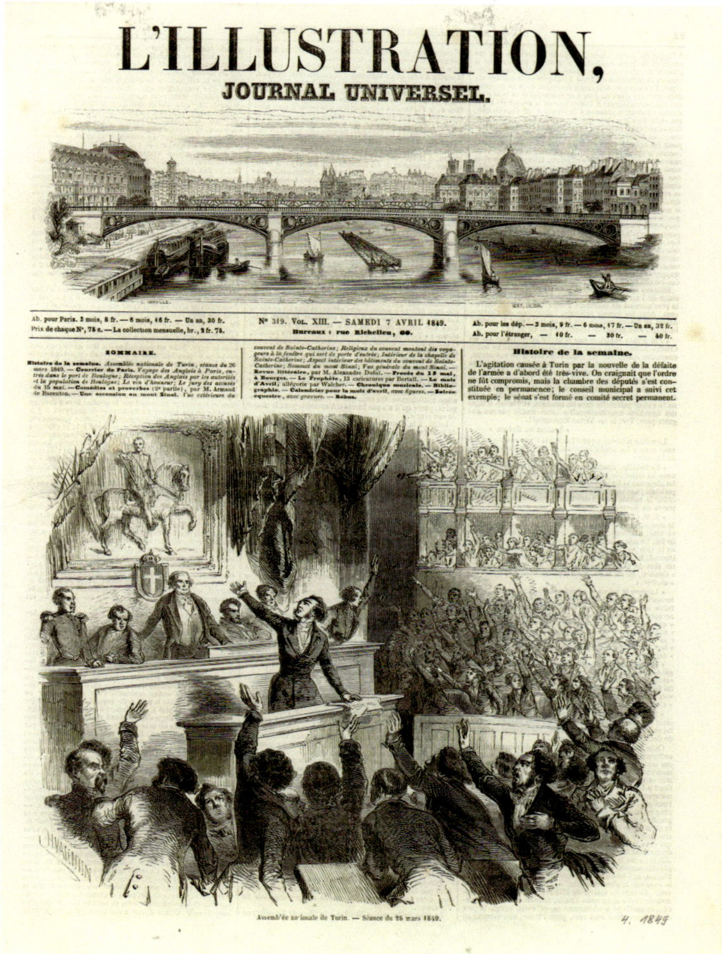
L'ILLUSTRATION,
JOURNAL UNIVERSEL.

Ab. pour Paris. 3 mois, 8 fr. — 6 mois, 16 fr. — Un an, 30 fr.
Prix de chaque N°, 75 c. — La collection mensuelle, br., 2 fr. 75.

N° 319. Vol. XIII. — SAMEDI 7 AVRIL 1849.
Bureaux : rue Richelieu, 60.

Ab. pour les dép. — 3 mois, 9 fr. — 6 mois, 17 fr. — Un an, 32 fr.
Ab. pour l'étranger, — 10 fr. — 20 fr. — 40 fr.

SOMMAIRE.

Histoire de la semaine. *Assemblée nationale de Turin, séance du 26 mars 1849.* — **Courrier de Paris.** *Voyage des Anglais à Paris, entrés dans le port de Boulogne; Réception des Anglais par les autorités et la population de Boulogne; Le vin d'honneur; Le jury des accusés du 15 mai.* — **Comédies et proverbes** (2e partie), par M. Armand de Barenton. — **Une ascension au mont Sinaï.** *Vue extérieure du couvent de Sainte-Catherine; Religieux du couvent montant des voyageurs à la fenêtre qui sert de porte d'entrée; Intérieur de la chapelle de Sainte-Catherine; Aspect intérieur des bâtiments du couvent de Sainte-Catherine; Sommet du mont Sinaï; Vue générale du mont Sinaï.* — **Revue littéraire**, par M. Alexandre Dufaï. — **Procès du 15 mai, à Bourges.** — **Le Prophète**, 13 caricatures par Bertall. — **Le mois d'Avril**, allégorie par Walcher. — **Chronique musicale.** — **Bibliographie.** — **Calendrier pour le mois d'avril**, *avec figures.* — **Soirée équestre**, *avec gravure.* — **Rébus.**

Histoire de la semaine.

L'agitation causée à Turin par la nouvelle de la défaite de l'armée a d'abord été très-vive. On craignait que l'ordre ne fût compromis, mais la chambre des députés s'est constituée en permanence; le conseil municipal a suivi cet exemple; le sénat s'est formé en comité secret permanent.

Assemblée nationale de Turin. — Séance du 26 mars 1849.

4. 1849

FIGURE 84

National Assembly at Turin—Session of March 26, 1849 [*Assemblée nationale à Turin—Séance du 26 mars 1849*], *L'Illustration*, April 7, 1849. Wood engraving. Brown University Digital Repository, David I. Kertzer Collection.

were wealthier, more educated, and more influential.[117] The bourgeoisie who could afford *L'Illustration* had no need of the basic cultural capital imparted by *Le Magasin pittoresque.* Instead, they wanted to be amused as well as informed; as a result, *L'Illustration* developed a different format for a different audience.

On its first page, there was always a striking wood engraving of topical interest, either a portrait of an important political figure or a major event, national or international; this established its currency as a serious newspaper. *National Assembly at Turin—Session of March 26, 1849,* depicts the critical moment when the abdication of Charles Albert of Sardinia was announced to the Italian parliament, after which his son, Victor Emmanuel II, succeeded him and led the Italian Risorgimento, the movement for national unification (fig. 84). Notwithstanding the republican politics of its founders, the journal supported France's colonial empire, with a regular feature, "Events in Algeria," reporting on the campaign to defeat the nationalist leader Abd-el-Kader.[118] "News of the Week,"

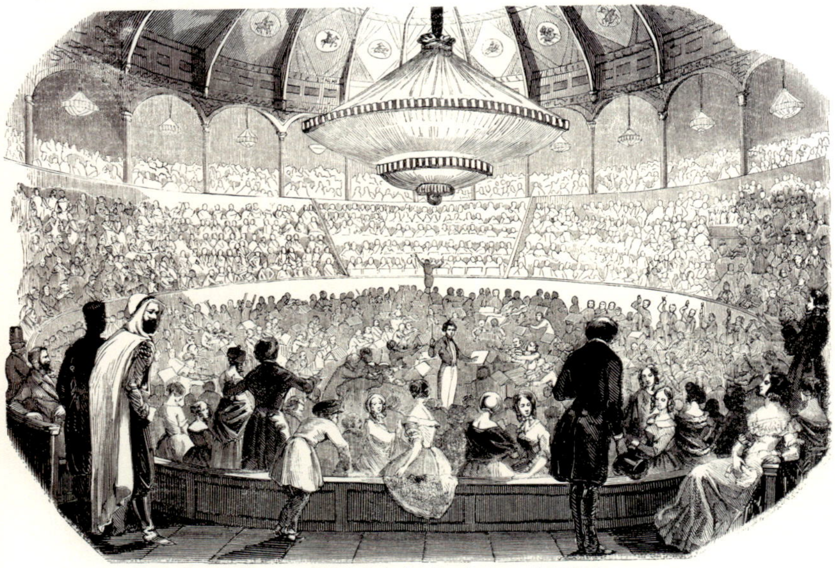
Chronique musicale.

Grand Concert des Champs-Élysées — *Concert du Conservatoire.* — *Concert* de M. Richard Mulder. — Théatre-Italien *la Rinegata,* — *Don Giovanni*

(Concert donné par M. Berlioz dans la salle du Cirque-Olympique, aux Champs-Elysées.

C'est un infatigable athlète que M. Berlioz, et un conqué- | où il met son plaisir et sa gloire. | conte d'Hoffmann. Cela vous jette dans un malaise indéfi-

FIGURE 85

Music Review. Concert Given by Mr. Berlioz in the Theater of the Cirque Olympique, at the Champs-Elysées [*Chronique musicale. Concert donné par M. Berlioz dans la salle du Cirque-Olympique, aux Champs-Elysées*], *L'Illustration*, January 25, 1845. Wood engraving.

also a regular first-page feature, provided an overview of recent national and international news; this was standard in serious political journals, but not in the *Illustrated London News,* where it was an inside feature. A summary on page one provided the table of contents for the issue, similar to that of intellectual reviews such as *Revue des deux mondes*; this feature was also lacking in the *Illustrated London News.* Inside, besides additional news articles and features, there were regular columns similar to those of the *Illustrated London News*: "Paris News" chronicled the glittering social life of the capital, much like "The Court and Haut Ton" in the *Illustrated London News*. Reviews of art exhibitions and concerts were often accompanied by illustrations, such as one of the composer Hector Berlioz conducting at the Cirque Olympique (fig. 85). Reviews of cultural events, fashion advice, and, after the example of the successful *feuilleton-roman* of the daily press, short pieces of fiction were always included, followed, on the last page, by games such as the newly fashionable rebuses (fig. 86).

While Charton was still planning *L'Illustration,* which he described as "a weekly résumé of the press, with engravings," he sought the advice of his friend, the philosopher Jean Reynaud, with whom he had started *Le Magasin pittoresque.* Reynaud wrote to him: "Beware of malicious and immoral caricatures. When one laughs at them, often one laughs without realizing that one is making fun of what one ought to respect. I've seen some Gavarni caricatures that revolt me because they attack and ridicule what one ought to respect most in the world, the love of family and the kindness of children."[119] This was clearly the opinion of

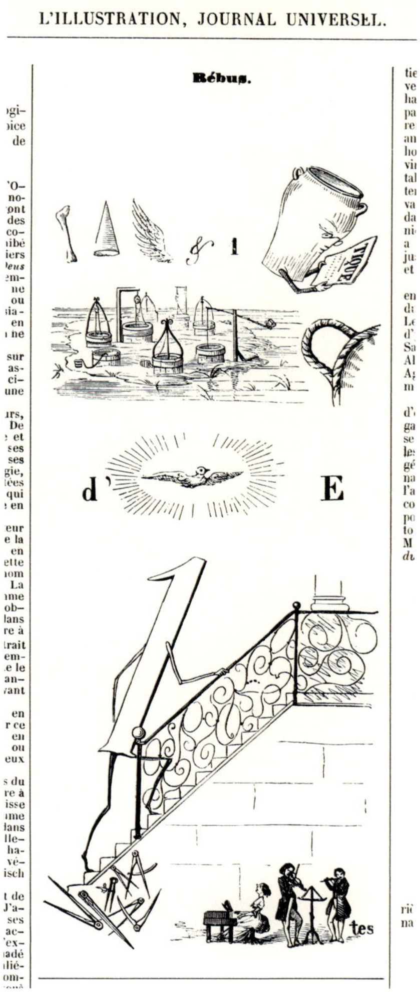

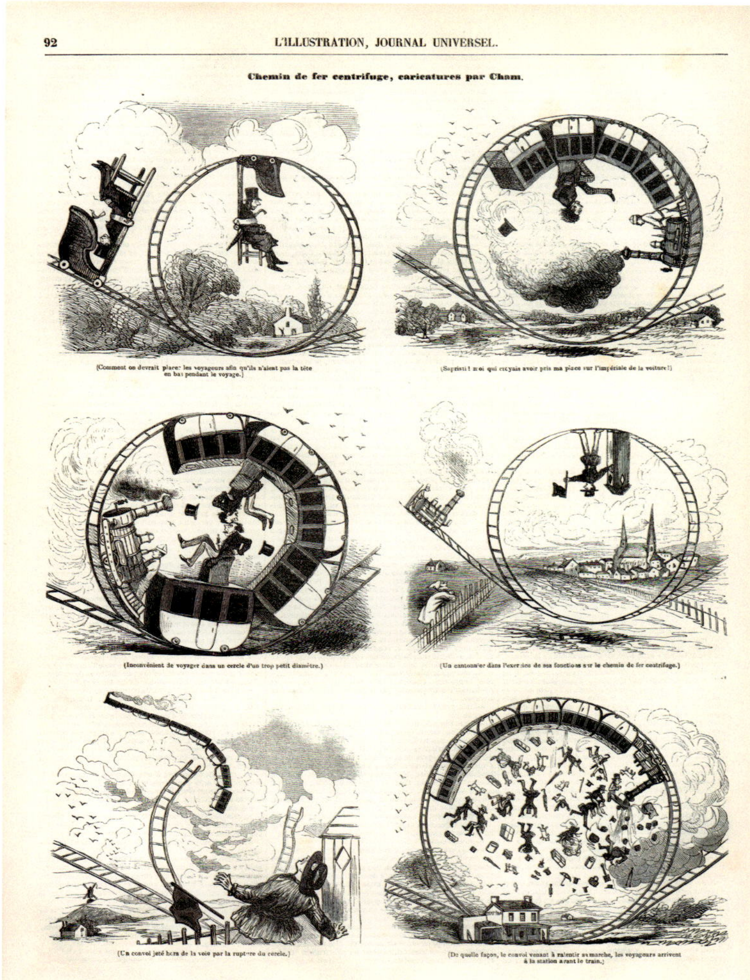

FIGURE 86
Rebus [*Rébus*], *L'Illustration*, July 26, 1845. Wood engraving.

FIGURE 87
Cham, *Railroad Roller Coaster* [*Chemin de fer centrifuge*], *L'Illustration*, October 10, 1846. Wood engraving. Brown University Digital Repository.

both gentlemen since *Le Magasin pittoresque* did not publish caricature. Regardless, graphic satire of all kinds—including the work of Gavarni (1804–1866)—would form an integral part of *L'Illustration.* Bertall (Charles-Albert d'Arnoux, 1820–1882), Cham (Amédée de Noé, 1818–1879), and Grandville worked regularly for the journal, Daumier occasionally contributed drawings, and the journal published Rodophe Töpffer's comic book, *Mr. Cryptogame*, redrawn for wood engraving by Cham.[120] Cham's drawings were a regular feature, commenting on current events in a lighter tone. In *Railroad Roller Coaster*, Cham parodies the craze for train travel, still a new experience in the 1840s, by conflating it with amusement park rides (fig. 87). While the *Penny Magazine*, *Journal des connaissances utiles*, and *Le Magasin pittoresque* earnestly gave lessons on mathematics, *L'Illustration* commissioned instead a two-page spread of caricatures by Cham, *Picturesque Arithmetic*, in which "Multiplication" was a woman with numerous children, while "Subtraction" was a street urchin robbing a

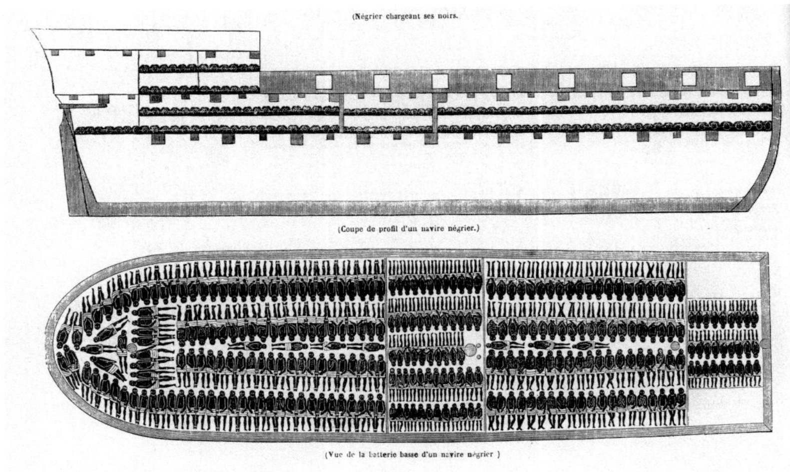

FIGURE 88
Profile View of a Slave Ship. View of the Lower Deck of a Slave Ship [Coupe du profile d'un navire négrier. Vue de la batterie basse d'un navire négrier], L'Illustration, October 21, 1843. Wood engraving.

OPPOSITE **FIGURE 89**
The Arrival of the King at the Palais Bourbon [Arrivée du roi au Palais-Bourbon], L'Illustration, December 30, 1843. Wood engraving.

rich bourgeois.[121] *L'Illustration* was clearly more entertaining than *Le Magasin pittoresque.* Devoting equal time to the problems of the world and the amusements of the leisured classes, it aspired—and succeeded—to fulfill the promise of its subtitle, *Journal universel.*

Despite the superficial aura of frivolity, however, *L'Illustration* remained serious enough in its republican politics to be in the opposition almost uninterruptedly—to the point that Napoleon III subsidized the foundation of *Le Monde illustré* (1857–1938), one of the many illustrated newspapers established in the wake of *L'Illustration,* but one that was more amenable to his policies.[122] Because censorship was so much more severe in France than in England, the politics of the *Illustrated London News* and *L'Illustration* were expressed in quite different ways. The *Illustrated London News* placed its political articles openly on its front page: "Abolition of the Slave Trade and the Right of Search," for example, was its lead article on January 21, 1843. France had not yet abolished slavery in its colonies; when *L'Illustration* ran a similar article later that same year, "On the Slave Trade and Slavery," it was positioned inside the journal, though its numerous and powerful illustrations clearly signaled its major importance.[123] The political slant of the journal was apparent, not through the placement of the article, but in the horrific engravings accompanying it and in the lavish praise it offered to the English for taking the lead on this issue. In the three-page spread, the most shocking images are the diagrams borrowed from the English of the notorious slave ship *Brookes,* showing slaves stacked in the hold like cordwood to maximize profits (fig. 88). Similarly, the obliquely expressed oppositional politics of the journal can be seen in its refusal to depict the French ruler, King Louis-Philippe, although Queen Victoria and other European royalty were often shown. When Louis-Philippe did make his appearance, he was shown as tiny and insignificant, almost impossible to identify. In *The Arrival of the King at the Palais Bourbon,* a minuscule Louis-Philippe stands far in the background between the central columns of the palace, identifiable only by his placement and the sash he wears (fig. 89).[124]

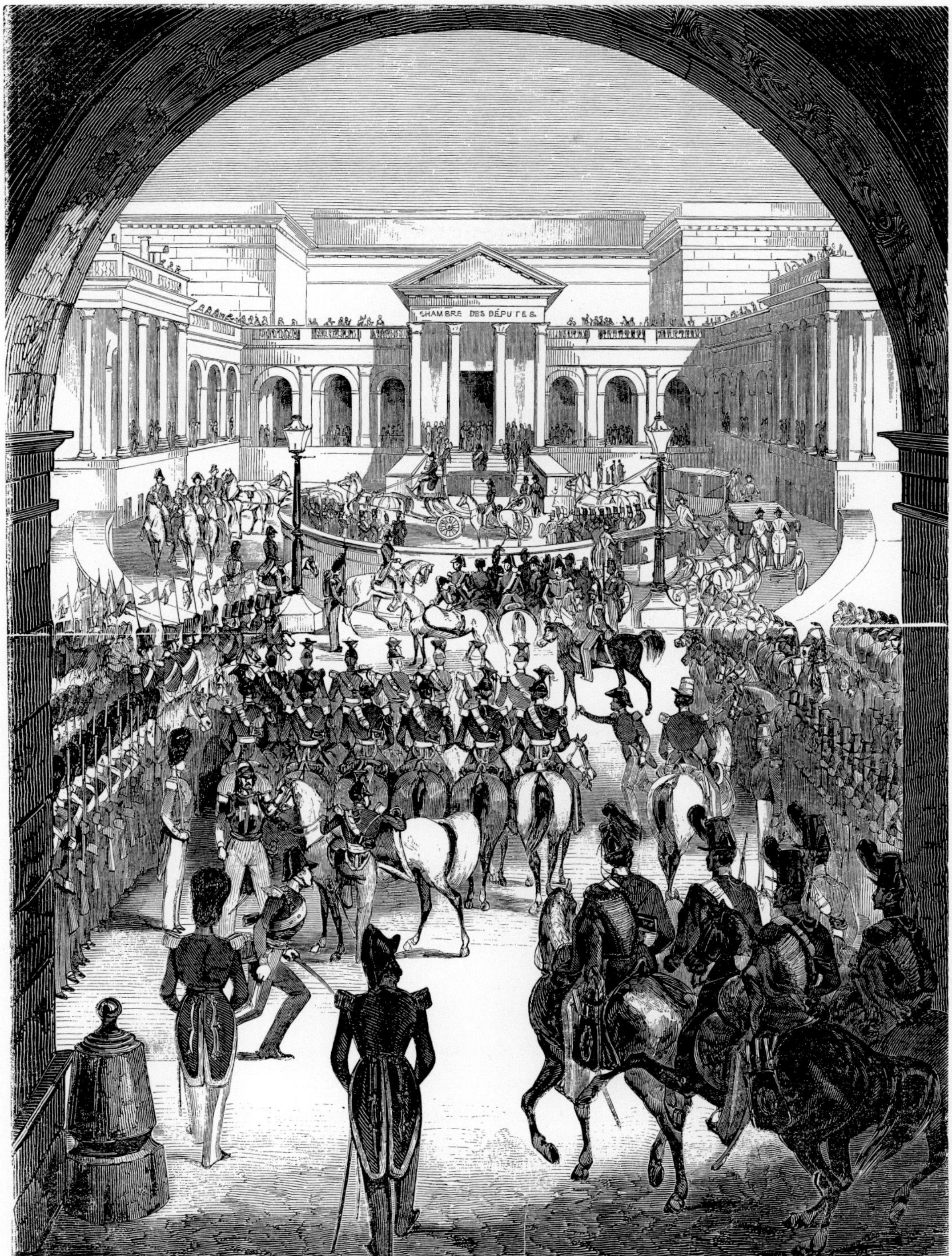

(Arrivée du roi au Palais-Bourbon.)

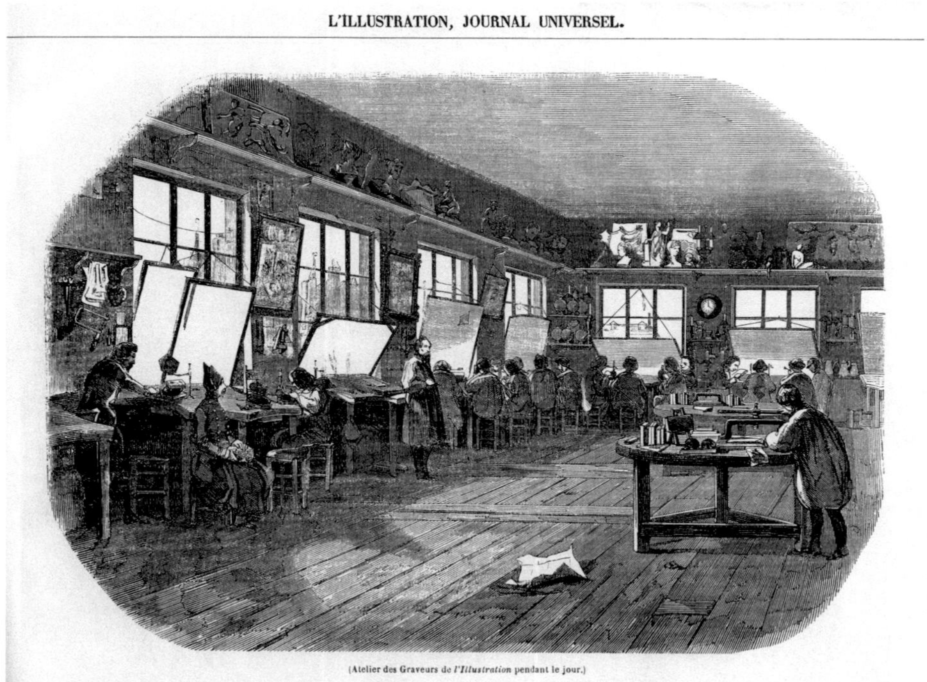

FIGURE 90
The Engraving Studio of L'Illustration *during the Day* [*Atelier des graveurs de* l'Illustration *pendant le jour*], *L'Illustration*, March 2, 1844. Wood engraving.

The *Illustrated London News* had demonstrated that a weekly newspaper could quickly produce wood engravings of current events, and *L'Illustration* followed its lead. The firm of Andrew, Best, Leloir that worked for *Le Magasin pittoresque* now worked for *L'Illustration* as well. In an obituary for his longtime colleague Jean Best, Charton wrote of the beginnings of *L'Illustration:* "But this time, no matter the difficulty, Mr. Best no longer spoke of impossibility: he gathered around him a group of skilled artists who, under his leadership, agreed to work day and night. Often the woodblocks for an entire page were cut up into little pieces and distributed among the young engravers who, especially at night, worked in two-hour shifts. This was practically a miracle: they were exhausted with fatigue, worried, they watched with anguish as the end of the week arrived, but finally, after all, they came through."[125] In *The Engraving Studio of* L'Illustration *during the Day*, we see the teams at work, baffles diffusing the glare from the windows (fig. 90). Their haste was such that join marks from the reassembled blocks often remained visible in the printed image (noticeable in fig. 89).

The journal had staked its reputation on its ability to publish images of events almost immediately after they occurred, but this was easier promised than accomplished. Charton complained: "The designers aren't any easier to deal with than the writers. No one wants to leave his studio. They draw from the accounts they receive, and they never fail to prefer the peripheral detail to the central idea."[126] Nonetheless, the journal exploited

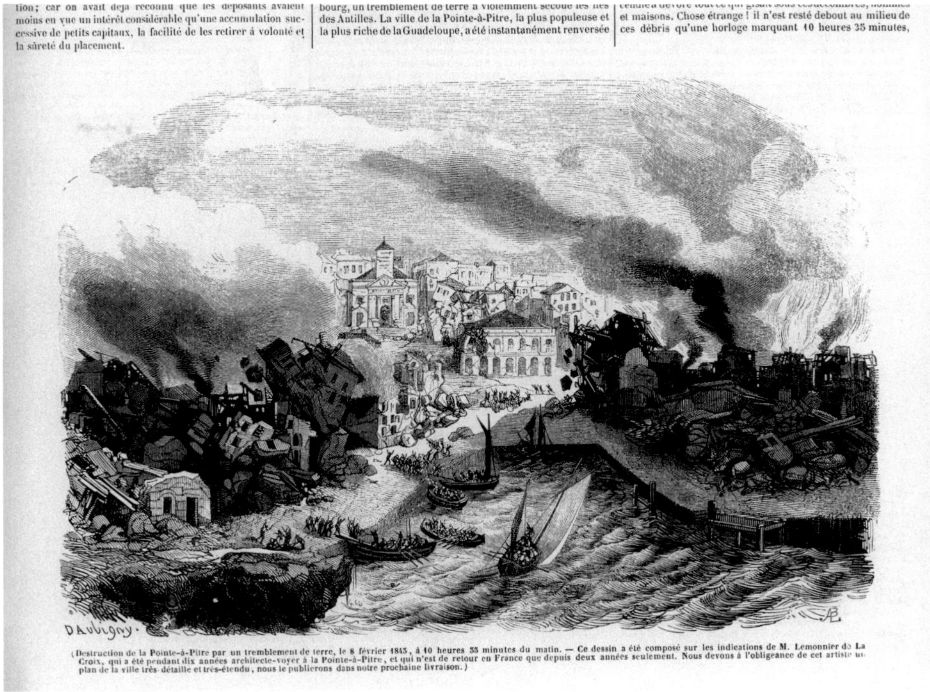

tion; car on avait déjà reconnu que les déposants avaient moins en vue un intérêt considérable qu'une accumulation successive de petits capitaux, la facilité de les retirer à volonté et la sûreté du placement.

bourg, un tremblement de terre a violemment secoué les îles des Antilles. La ville de la Pointe-à-Pitre, la plus populeuse et la plus riche de la Guadeloupe, a été instantanément renversée

et maisons. Chose étrange ! il n'est resté debout au milieu de ces débris qu'une horloge marquant 10 heures 35 minutes,

(Destruction de la Pointe-à-Pitre par un tremblement de terre, le 8 février 1843, à 10 heures 35 minutes du matin. — Ce dessin a été composé sur les indications de M. Lemonnier de La Croix, qui a été pendant dix années architecte-voyer à la Pointe-à-Pitre, et qui n'est de retour en France que depuis deux années seulement. Nous devons à l'obligeance de cet artiste un plan de la ville très-détaillé et très-étendu, nous le publierons dans notre prochaine livraison.)

FIGURE 91

Charles Daubigny, *Destruction of Pointe-à-Pitre by an Earthquake, February 8, 1843, at 10:35 in the Morning* [*Destruction de la Pointe-à-Pitre par un tremblement de terre, le 8 février 1843, à 10 heures 35 minutes du matin*]. Engr. Andrew, Best, Leloir, *L'Illustration*, March 18, 1843. Wood engraving.

whatever degree of authenticity it could muster. On March 18, 1843, *L'Illustration* published an article on the earthquake in Guadeloupe that had taken place more than five weeks earlier. Under the drawing by the well-known artist Charles Daubigny that showed the damage, the complete legend read: *Destruction of Pointe-à-Pitre by an Earthquake, February 8, 1843, at 10:35 in the Morning.—This Drawing Was Made According to Information Given by Mr. Lemonnier de La Croix, Who for Ten Years Was Municipal Architect at Pointe-À-Pitre and Who Returned to France Only Two Years Ago* (fig. 91).[127] Almost two months later, the *Illustrated London News* published the same engraving, reprinting it without the legend and removing Daubigny's signature.[128] This was not exactly the news as it was happening, nor was it eyewitness reportage, but for its time it was impressive and, considering the success that the two illustrated newspapers achieved, they obviously did manage to convince a broad readership that they were eyewitness to the great events of their time.

Nonetheless, a survey of engravings published in the first decade of *L'Illustration* reveals that a large percentage were portraits, caricatures, views, and illustrations of works of art, all of which could be prepared well in advance. The journal regularly excerpted sections of illustrated works published by its founders, Dubochet and Paulin, which not only enlivened its pages at no cost to the journal, but also served to publicize forthcoming books. There were far fewer current events. Explosions and fires were exceptionally popular, no doubt because they could be produced quickly with the judicious addition

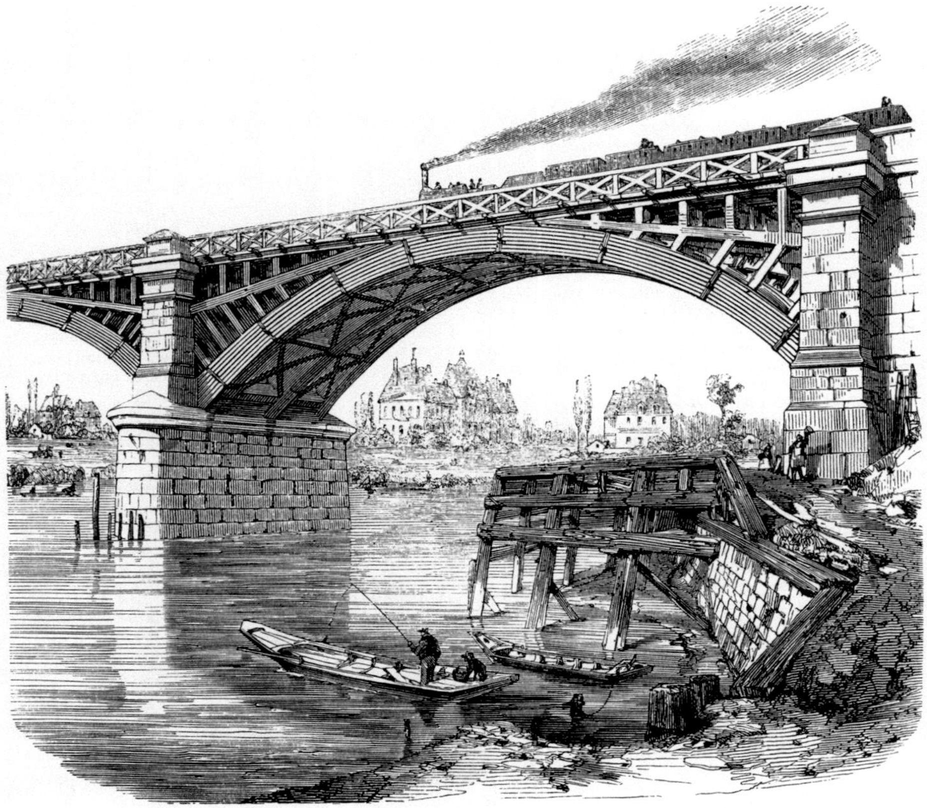

(Pont de Maisons.)

FIGURE 92
Bridge at Maisons [*Pont de Maisons*], *L'Illustration*, May 6, 1843. Wood engraving.

of flames and rubble. New bridges and railway lines were regularly featured in lavish multipage spreads attesting to the enduring attraction of modernity to a nineteenth-century audience. *Bridge at Maisons* is certainly one of the earliest images to juxtapose new and old, modern and traditional, with its cast-iron bridge and locomotive spewing steam contrasting with the peaceful rural fishermen in rowboats below (fig. 92). Though here published anonymously, it is a trope that would be adopted by subsequent generations of painters, immortalized by J. M. W. Turner in his *Rain, Steam, Speed: The Great Western Railway* of 1844 (National Gallery, London).

Gradually, the journal developed a team of illustrators who would, in fact, travel widely: Henri Valentin (1820–1855), who had worked for Charles Philipon, became the principal designer, along with Janet-Lange (Ange-Louis Janet, 1811–1872), and Pharamond Blanchard (1805–1873). The journal also called on its readers to send in sketches and articles. In its first year, it published a drawing of a bust of Napoleon in a temple in China that was captioned *Napoleon Worshipped in a Chinese Temple.—Drawing Made by an Eyewitness.*[129] Accompanying an 1845 article "Catastrophe in Algiers on March 8th" was this explanation: "The two drawings that we are publishing on this shocking catastrophe and which were sent to us from Algeria with the supplement of the newspaper *L'Abkar*, represent the explosion and the condition of the site after the explosion."[130] One way or another, the journal visualized for its readers the major events of its time.

It was fated that Édouard Charton's relationship with *L'Illustration* would not last. Charton, earnest, idealistic, dedicated to self-improvement, could not fail to condemn the frivolity that, no doubt, was a contributing factor in the enduring success of *L'Illustration.* There were signs of trouble right from the beginning, when he wrote to his friend Hippolyte Fortoul less than two months after the first issue, complaining that "Nearly my entire role insofar as the text is concerned consists in defending against the invasion of license and dishonesty. I'm treated like a prude, like a puritan."[131] The regular columns on fashion and on Parisian high society would not have pleased him, nor would the regular inclusion of caricatures skewering everything he held sacred. In addition, Lachevardière, as publisher of *Le Magasin pittoresque*, feared the competition of this new periodical, especially as Charton was now editing both publications. Charton remained at *L'Illustration* for less than a year, then wrote to his friend Fortoul, "I resigned as editor of *L'Illustration* on January 1. On the one hand, Lachevardière was harassing me; on the other hand, the character and the direction of the publication are displeasing to me."[132] Despite this, *L'Illustration* became one of the longest-running illustrated periodicals in the world, outlasting all its competitors until well into the twentieth century.

With the establishment of the great illustrated weeklies, the first era of the illustrated press drew to a close. The *Penny Magazine* published until 1845, *Journal des connaissances utiles* until 1848, both steadily losing subscribers to newer illustrated periodicals. *Le Magasin pittoresque* continued until 1938, however; *L'Illustration* to 1944, and the *Illustrated London News* to 2003. France was not the only country to follow the English lead: *Illustrirte Zeitung* was established in Leipzig in 1843, *La Ilustração. Jornal universal* in Lisbon in 1845, *La Ilustración* in Madrid in 1849, *Frank Leslie's Illustrated Newspaper* in New York in 1855, and *De Hollandsche illustratie* in Amsterdam in 1864.

CODA

The great illustrated periodicals were founded as a direct result of the development of wood engraving. The medium, however, had barely reached its apotheosis when it began to be eclipsed by the various photo processes that, in turn, would revolutionize the publishing industry just as wood engraving had done.[133] Previously, once reproductive media had outlived their technological usefulness, they were reborn as art. Etching did this successfully with international revivals in the mid-nineteenth century; lithography, epitomized by the work of Toulouse-Lautrec, followed in the 1890s. Wood engraving, however, seems to have been so identified as a proletarian medium that, aside from the illustrations of Winslow Homer and Gustave Doré, it never benefited from a successful transition to the status of art. Not that the effort wasn't made. In 1862, the first (and last) attempt in this direction appeared: *L'Album du Magasin pittoresque. Cent gravures choisies dans la collection.* With a preface by Alfred Michiels, art critic and librarian of the École des beaux-arts in Paris, it was indeed a luxury production. Its full-page plates, printed on thick paper, were signed in the block by both the artist who made the drawing and the wood engraver who cut the plate. Michiels wrote that, in choosing the plates, the editors of *Le Magasin pittoresque* "gave preference to works engraved from an original drawing, transferred to the plate by the artist who signed it."[134] In other words, these were not reproductive prints but original works of art. The publication included works by Grandville, Daubigny, Meissonier, Delacroix, Constable, and numerous other artists. Nonetheless, it does not appear to have enjoyed a great success, as it was never reprinted and the experiment was not repeated. Wood engraving would soon become extinct, revived only in the twentieth century by Max Ernst, whose *La femme 100 têtes* (1929) and *La semaine de bonté* (1933) utilized cut-up wood engravings in Surrealist collages, playing up the "retro" look of this medium that had been, by then, completely eclipsed by photography.

In the course of the nineteenth century, photo processes slowly made their way into the illustrated press.[135] The *Illustrated London News* made some use of daguerreotypes right from the beginning. *The Colosseum Print*, a three-by-four-foot view of London given to subscribers in 1843 was based on a series of daguerreotypes by Antoine Claudet (1797–1867) transferred in sections to wooden blocks for engraving. The collodion process, invented in 1851, allowed the artist's drawing to be transferred directly to the plate, instead of being traced or redrawn; engravers then reworked it for printing, a procedure used for the journal's illustrations of the 1851 Great Exhibition of Works of Industry of All Nations. *L'Illustration*, from 1848 on, featured images with the caption "after a daguerreotype," or "after a photograph," although such images were always reworked by wood engravers, who added figures and details that film was too slow to capture. The earliest published news "photographs" were images of the June 25–26 uprising during the 1848 revolution that appeared in *L'Illustration* in its July 1–8 issue (fig. 93).[136] *The Barricades on Rue Saint-Maur-Popincourt*, shown here twice, are labeled *Sunday Morning* and *Monday after the Attack*; the principal difference seems to be the addition of crowds that filled the streets after the event rather than any editorial comment about the effects of the uprising. Both images carry the legend *from a Daguerreotype by Mr. Thibault.* Despite the desirable sense of authenticity conveyed by photographic images such as these, and especially by their legends that assure readers of their veracity, their use proved problematic for mass production. As is obvious here, figures, details, clouds, all had to be added or augmented by the engraver who translated the image so that it could be printed along with the text. As a result of this double effort, only 214 photographs were published in *L'Illustration* from 1843 to 1859, as against 19,800 to 26,400 wood engravings. Essentially they were all wood engravings; the printed images were so similar that often the only way they can be differentiated is through their captions.[137] By the 1880s, however, halftone processes had come into general usage, and photographs could be utilized more easily, although they continued to coexist in the periodical press with wood engraving until late in the century.

Color also entered the periodical press, albeit slowly. Hand-colored lithographs had been included in periodicals such as *La Silhouette* and *La Caricature* since the early decades of the century, and hand-colored etchings and engravings had been included even earlier. Chromolithography, however, made its appearance first as single-sheet images before it was economically or technologically feasible to utilize in the periodical press.[138] *L'Illustration* published its first chromos in the July 12, 1884 issue, although they were hardly full-color productions as they included only a few tones. Smaller periodicals led the way, often printing from zinc instead of stone plates in a process called gillotage. Satiric publications such as *L'Éclipse* (1868–1919), *Don Quichotte* (1873–1893), and *Gil Blas* (1891–1903), chose a direction diametrically opposed to the eye-witness verisimilitude that was always the ideal of mass-circulation newspapers like *L'Illustration* and the *Illustrated London News* and that led them to photography. These smaller periodicals preferred instead to commission major artists to create striking hand-drawn images with editorial content.[139] These drawings were topical and often caricatured, the late-century heirs of the hand-colored caricatures that appeared in *La Caricature* in the 1830s.[140] André Gill (1840–1885) was a major artist in this medium, working for a variety of periodicals.[141] *The Victor* was one of a supplemental portfolio of five prints that Gill drew for the weekly *L'Eclipse* during the Franco-Prussian War of 1870 (fig. 94). Here there is no pretense to the seemingly dispassionate reportage of photography apparent in Thibault's photographs of the 1848 revolution (see fig. 93); Gill clearly considered the war a military debacle. His verdict is clear: the only hero in this war is the skeletal figure of Death, laurels crowning its head and medals pinned to its fleshless ribs. On a scrap of paper near its amputated legs is written "19 juillet"—the date that Napoleon III declared war against the Prussians—and "Metz," the French city that was the site of the Prussian siege that lasted until the French army was defeated at Sedan on September 2, bringing the Second Empire to an inglorious close. *L'Éclipse*, as well as the numerous other illustrated satirical periodicals established in the wake of *La Caricature*, continuing to the twenty-first century's *Charlie Hebdo*, should remind us that even in the periodical press there would always be a place for the hand-drawn image.[142]

Unlike these smaller publications, however, universal survey periodicals like the *Illustrated London News* and *L'Illustration* established the format that took root in modern culture and that still governs everything from our daily newspapers to the television news: serious events first, followed by culture and then entertainment. The early illustrated journals seized on whatever technological improvements would bring them ever closer

FIGURE 93

The Barricades on Rue Saint-Maur-Popincourt, Sunday Morning, from a Daguerreotype by Mr. Thibault. The Barricades of Rue Saint-Maur-Popincourt, Monday after the Attack, from a Daguerreotype by Mr. Thibault [*La barricade de la rue Saint-Maur-Popincourt, le dimanche matin, d'après une planche daguerréotypée par M. Thibault. La barricade de la rue Saint-Maur-Popincourt, le lundi après l'attaque, d'après une planche daguerréotypée par M. Thibault*], *L'Illustration*, July 1–8, 1848. Wood engraving.

OPPOSITE **FIGURE 94**

André Gill, *The Victor* [*Le vainqueur*], *L'Éclipse*, supplement no. 5, August 28, 1870. Gillotype. Heidelberg University Library.

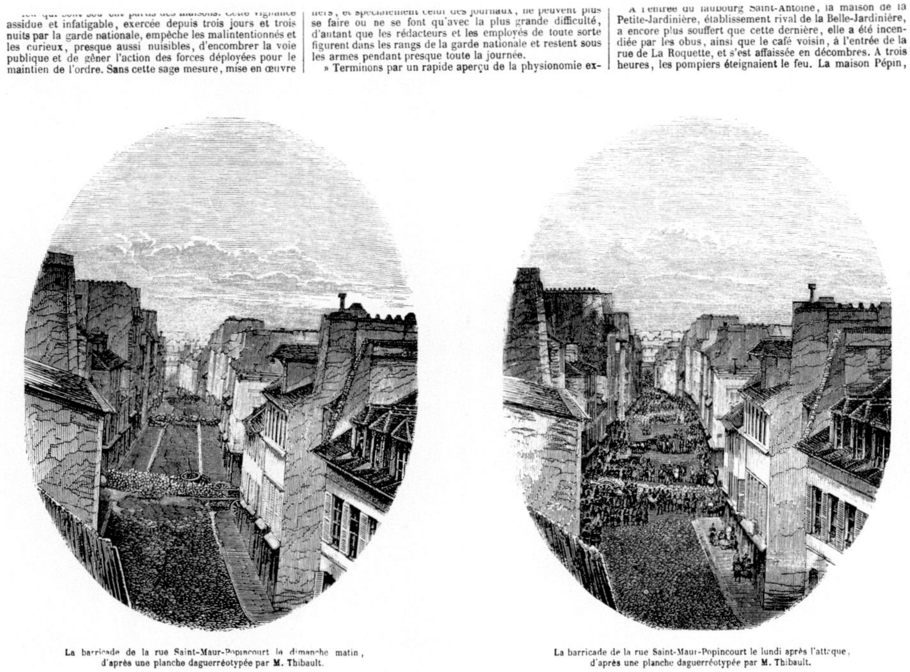

assidue et infatigable, exercée depuis trois jours et trois nuits par la garde nationale, empêche les malintentionnés et les curieux, presque aussi nuisibles, d'encombrer la voie publique et de gêner l'action des forces déployées pour le maintien de l'ordre. Sans cette sage mesure, mise en œuvre

se faire ou ne se font qu'avec la plus grande difficulté, d'autant que les rédacteurs et les employés de toute sorte figurent dans les rangs de la garde nationale et restent sous les armes pendant presque toute la journée.

» Terminons par un rapide aperçu de la physionomie ex-

A l'entrée du faubourg Saint-Antoine, la maison de la Petite-Jardinière, établissement rival de la Belle-Jardinière, a encore plus souffert que cette dernière, elle a été incendiée par les obus, ainsi que le café voisin, à l'entrée de la rue de La Roquette, et s'est affaissée en décombres. A trois heures, les pompiers éteignaient le feu. La maison Pépin,

La barricade de la rue Saint-Maur-Popincourt le dimanche matin, d'après une planche daguerréotypée par M. Thibault.

La barricade de la rue Saint-Maur-Popincourt le lundi après l'attaque, d'après une planche daguerréotypée par M. Thibault.

to their ideal of veracity, from images after written descriptions of events, to eyewitness drawings of these events, to photography that seemed to offer a dispassionate and accurate transcription of the events. In the course of the century, beginning when illustrated periodicals had their origins in an ideal of improving the lot of the working classes, they multiplied many times over into agents of every political, cultural, and social opinion. Looking back on the idealism characteristic of the early illustrated press, however, Charles Knight wrote in his memoirs: "All the cheap literature was not good at the period of this triumphant retrospect. This was a circumstance that was particularly mortifying to those who, like myself, had formed an over sanguine estimate of the benefit that was likely to result from the general diffusion of the ability to read."[143] What had happened was that, as the illustrated press developed and achieved hegemony, it had become a medium that, like all media, was intrinsically neither good nor bad but was shaped by the vision of those who used it.

Supplément N° 5. DIX CENTIMES

RÉDACTEUR EN CHEF
F. POLO

ABONNEMENTS
PARIS
Un an. . . . . 5 fr. »
Six mois. . . . . . 3 »
Trois mois. 1 50

Rue du Croissant, 16

L'ECLIPSE

DIRECTEUR
F. POLO

ABONNEMENTS
DÉPARTEMENTS
Un an. . . . . 6 fr. »
Six mois. 3 50
Trois mois. . . . . 2 »

Rue du Croissant, 16

LE VAINQUEUR, PAR GILL

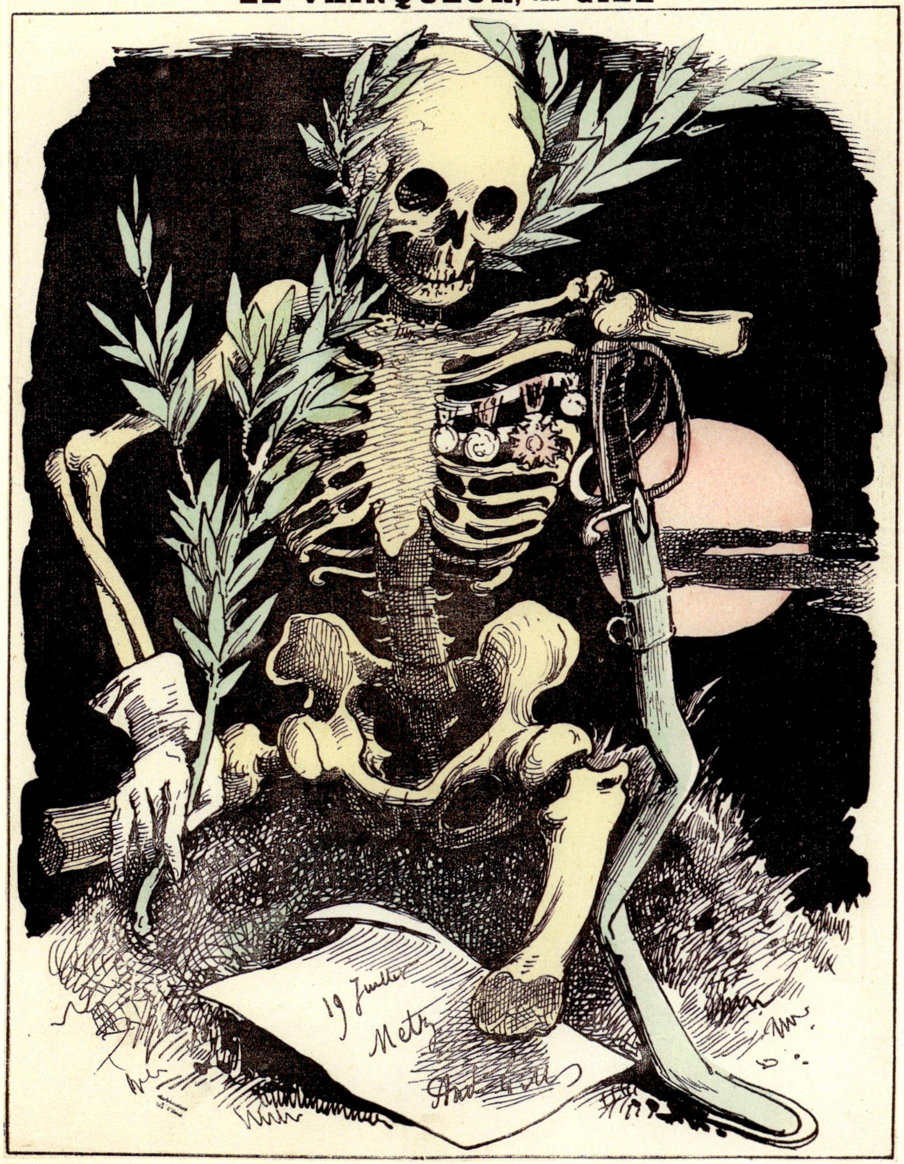

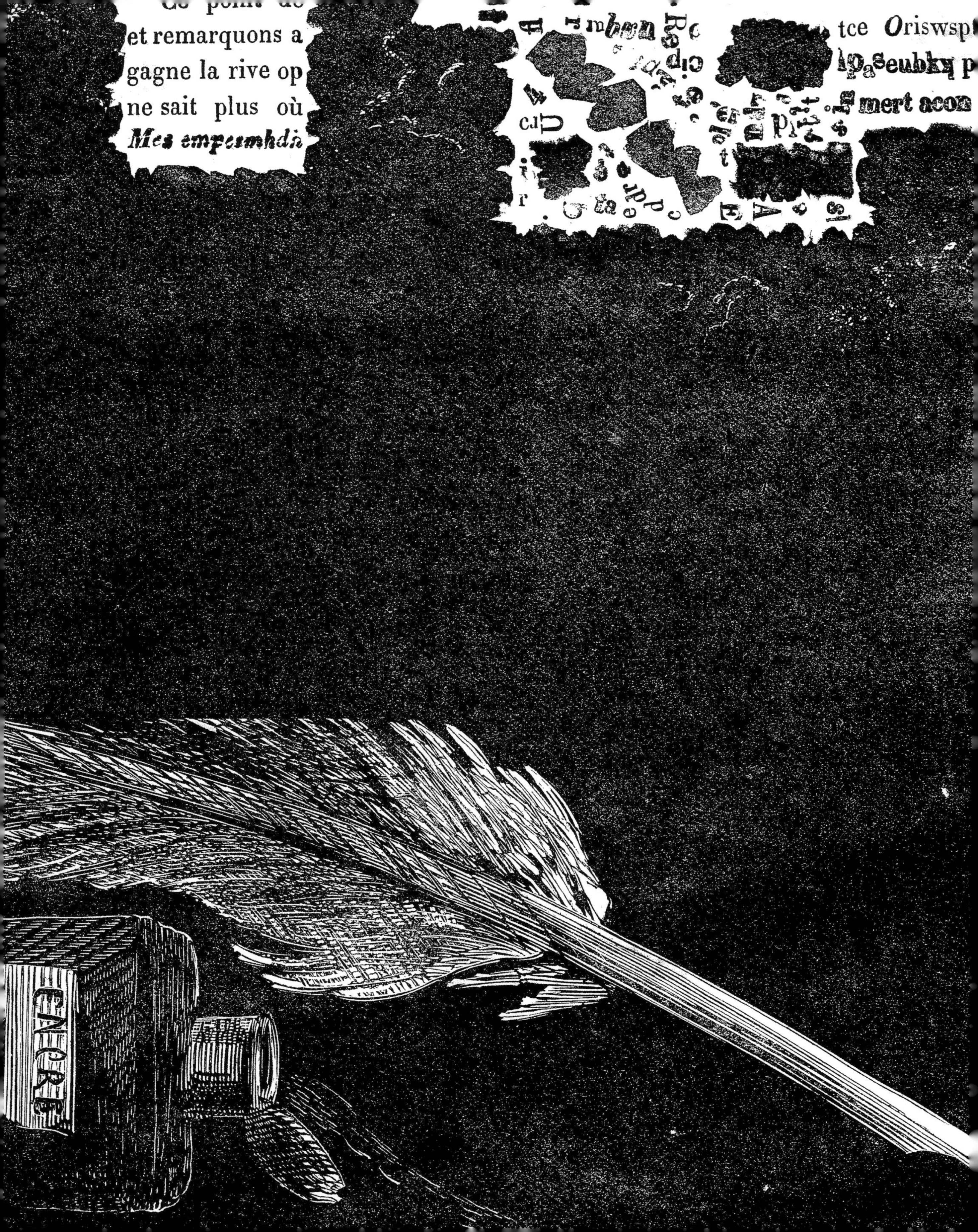
et remarquons a
gagne la rive op
ne sait plus où
tce Oriswspt
mert acon
ENCRE

3 The Invention of Comics

Stories in Pictures

Today the comic strip is the most familiar form of caricature, yet this was not always the case. Comics are relative newcomers to the field of visual humor, only gradually displacing the single-sheet popular prints and caricatures popular across Europe for centuries.[1] While sharing with caricature a humorous drawing style, comics also share with two earlier art forms, history painting and book illustration, the intention of narrating a story. But whether called comic strip or comic book, *bande dessinée*, or graphic novel, comics have become one of the most widely read examples of visual print culture in the modern era, and as such merit serious attention. While French artists did not invent the comic strip—that honor belongs to the Swiss schoolmaster Rodolphe Töpffer (1799–1846), who published his first, *The Story of Mr. Jabot*, in 1833—they played an important though largely unacknowledged role in its development (fig. 95).[2] By examining precedents and early development, we can better understand how disparate elements were united into a new form, the comic book, and a new visual language of drawing.

To understand the transformation of single-sheet imagery into the sequential narration of comics we must follow two separate routes. We have already seen, in chapter 1, that albums began as miscellaneous collections of images without narrative development. Visual narrative had, however, long existed in the form of stories told through series of separate images, each of which depicts a single episode. Biblical texts were often treated in narrative cycles adorning churches, but in the graphic arts the print series of William Hogarth are, perhaps, the best-known examples. Closer to our definition of comics are the single-sheet prints that include multiple frames arranged in more than one register on the page. These often had a single theme, although they were not always intended to be read sequentially. The many versions of *The World Upside-Down* that appeared throughout Europe over several centuries are good examples of this.[3] In the last of an 1829 set of four sheets, where each frame of the color woodcut depicts a scene from an "upside-down world" where things are the reverse of what was then considered normal, we see the ass who leads his master; children who beat their parents; a woman carrying a rifle who is going off hunting while leaving her husband to care for their child; and a horse who grooms the stableman (fig. 96). These and numerous

DETAIL OF FIGURE 125

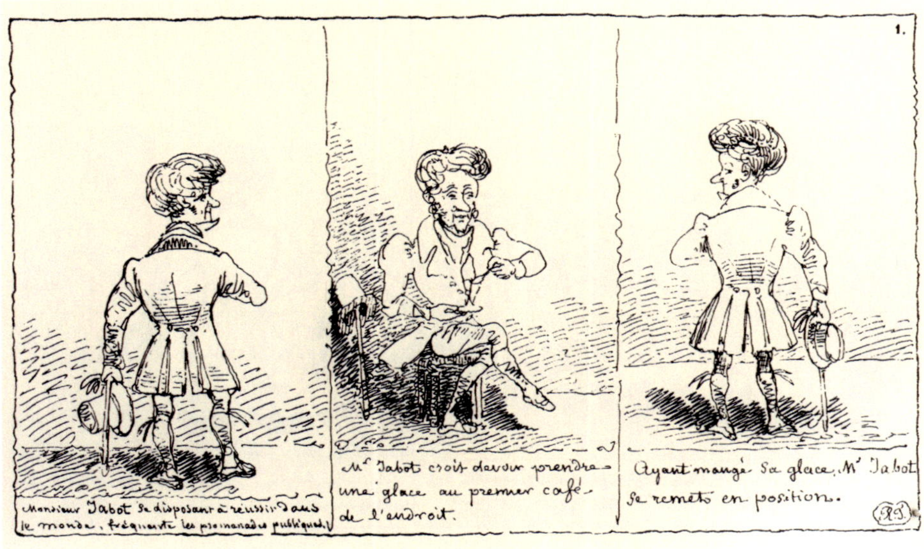

FIGURE 95

Rodolphe Töpffer, *The Story of Mr. Jabot* [*Histoire de M. Jabot*]. Geneva, 1833, pl. 1. Lithograph. Bibliothèque nationale de France.

other such images can be read either separately or serially, although there is no narrative development.

Hogarth is usually cited as the progenitor of both caricature and comics. His *Characters and Caricaturas*, discussed in chapter 1 (see fig. 17), made a sharp distinction between drawings revealing character and those whose distortions and exaggerations relegated them to the realm of caricature. Despite this, his picture stories, each composed of six to eight scenes, combined aspects of both character and caricature. Although they are not what we today would call comic books, since they were unbound with each engraved sheet carrying only a single image, they exercised a major influence on the subsequent development of the comic book.[4] It might be argued that cycles such as Rubens's twenty-four paintings that each depicted an episode in the life of Marie de' Medici (1622–1625) functioned as sequential narratives in much the same way, but Hogarth's work differs because he depicted the lives of fictional, historically insignificant, individuals and events, as opposed to, for example, the queen Marie de' Medici. Hogarth produced numerous narrative print series, providing his original stories with rich imagery laden with symbolism and topical references. Their captions are brief, their narrative advanced through images rather than text. His stories are always moralizing—"modern moral subjects," he called them—and there is little that is "comic" about them. His first, *A Harlot's Progress* (1732), traces the downfall of an innocent country lass who arrives in London and is soon led astray into a life of sin. *The Rake's Progress* (1735) tells a parallel tale, depicting the life of a young man who, having inherited wealth, gives himself over to pleasures and debauchery and ends up in the insane asylum of Bedlam. In *Marriage A-la-Mode* (1745) we see in six "acts" the sad results of arranged and loveless marriages (figs. 97a–f). In the first engraving,

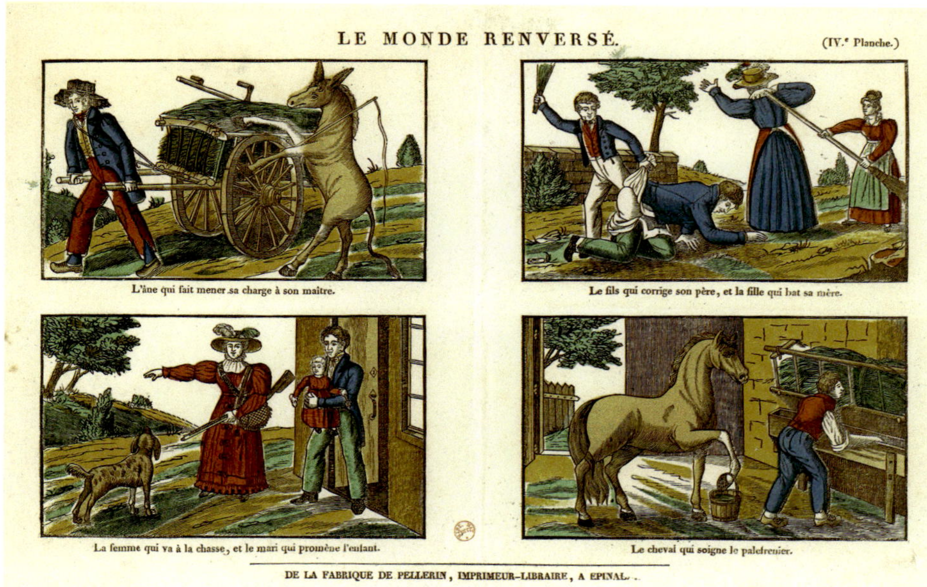

FIGURE 96

The World Upside-Down [*Le monde renversé*], pl. 3, 1829. Pub. Pellerin, Épinal. Stencil-colored woodcut.

the fathers of the luckless couple arrange their union with little regard for the happiness of their children; not surprisingly, both spouses stray to find lovers, as the next three engravings reveal. In the denouement, image five, the husband discovers the wife with her lover, who stabs him and then escapes out the window, and the series concludes with the wife, her husband dead, her lover hanged for murder, taking poison and expiring. Moral, yes; comic, no.

Hogarth's followers, James Gillray, Thomas Rowlandson, and George Cruikshank, developed these beginnings into a golden age of English caricature.[5] They produced thousands of prints that lampooned everyone and everything and served as the immediate precedent for comics. These artists invented a style of caricature that, like Hogarth's, was crowded with visual and symbolic allusions, but, unlike Hogarth's, was rapidly executed and often featured grotesque exaggerations. With a cursory style of etching and with hand-coloring replacing the labor-intensive modeling of crosshatched engraving, such images could be produced cheaply and sold in quantity.

Some of their prints had the superficial appearance of comics, in that, like *The World Upside-Down*, they contained multiple frames on a single page and were thematically linked. Nonetheless, they lacked narrative development. This can be seen in the two-part series *The Secret History of Crim Con* (1812), drawn by George Moutard Woodward (1760–1809) but etched by Thomas Rowlandson (fig. 98). Its subject, adultery ("Criminal Conversation"), was certainly not new, but here it is depicted with wit and cynicism and with none of the moralizing of Hogarth. Each of the twelve frames of the Woodward–Rowlandson work features different protagonists and can be read as a separate cartoon. Hogarth would have called them *caricatura* to emphasize their

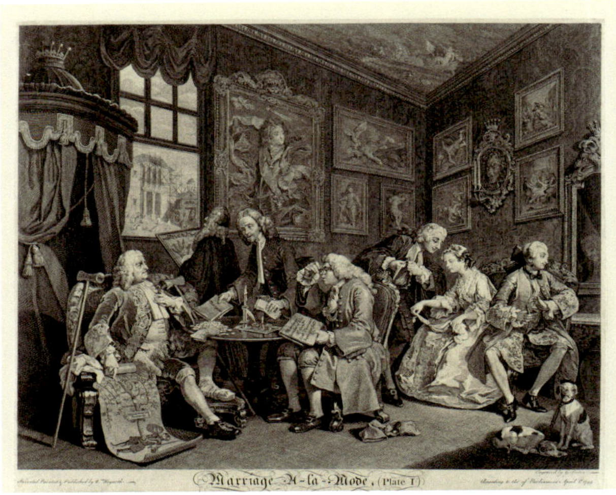

A

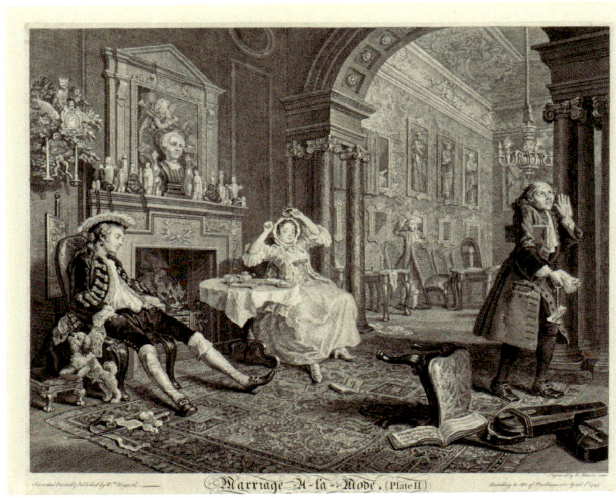

B

FIGURES 97A–F
William Hogarth, *Marriage-A-la-Mode*, 1745. Engr. G. Scotin. Yale University, Lewis Walpole Library.

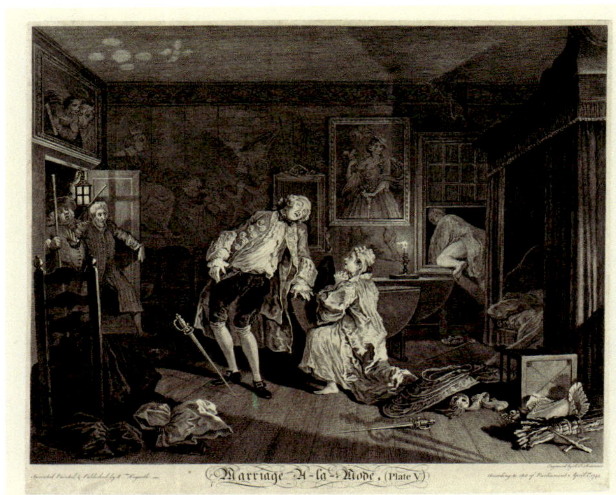

E

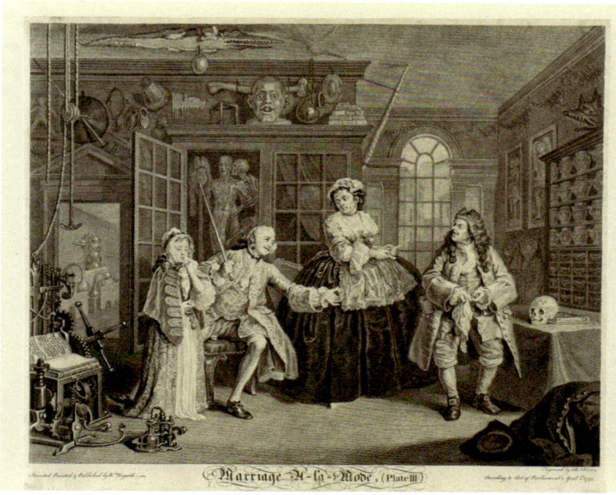
Marriage A-la-Mode, (Plate III)

C

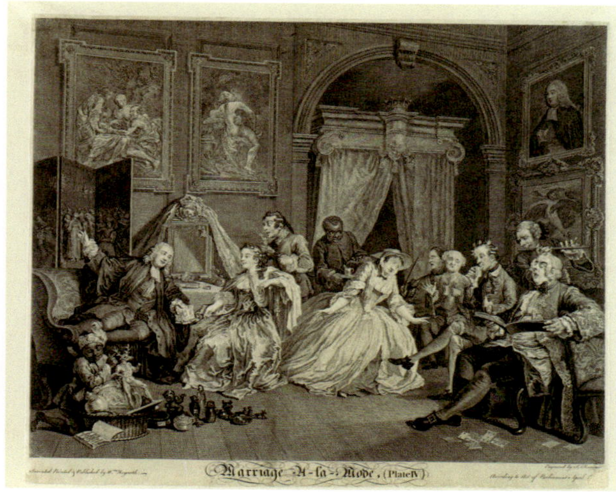
Marriage A-la-Mode, (Plate IV)

D

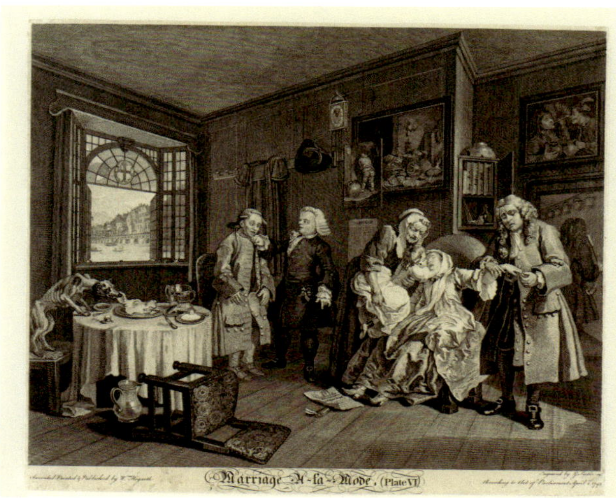
Marriage A-la-Mode, (Plate VI)

F

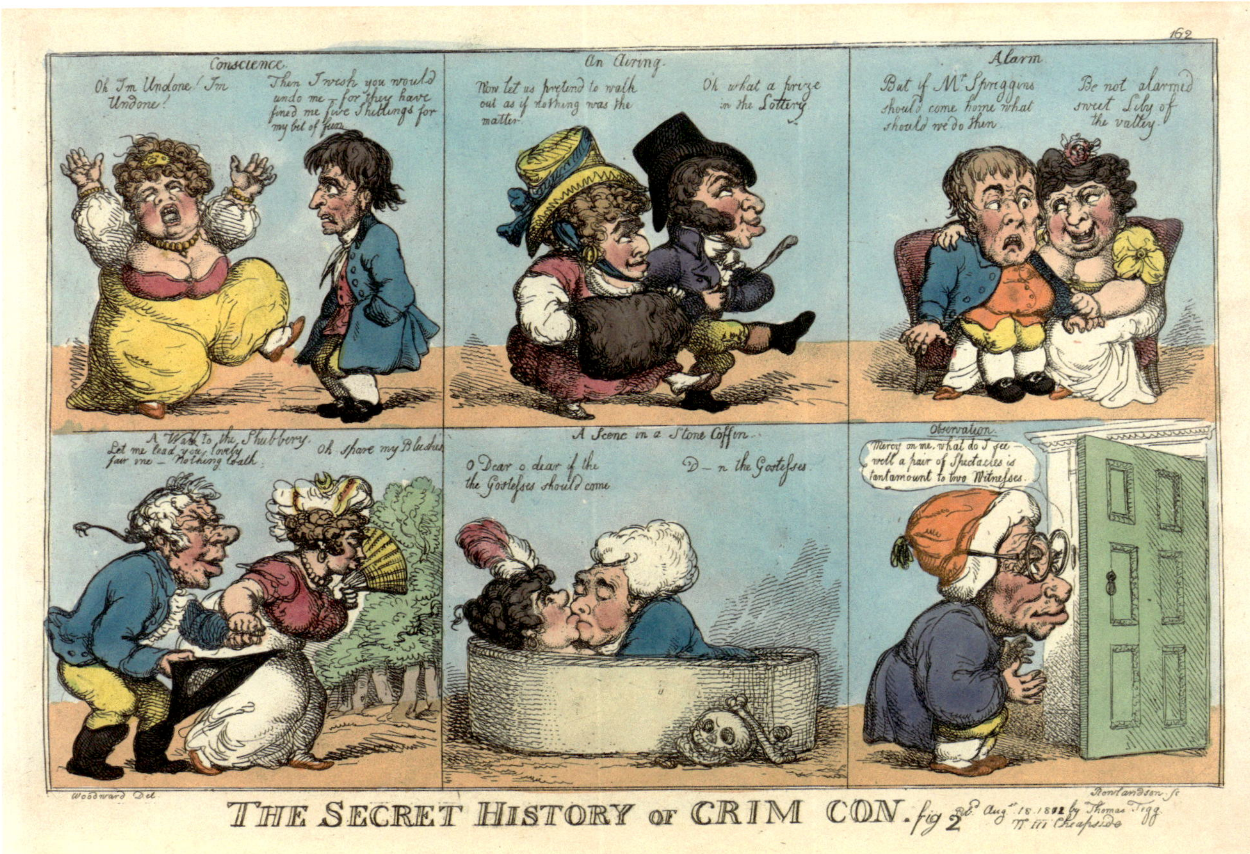

FIGURE 98

Thomas Rowlandson, after George Moutard Woodward, *The Secret History of Crim Con*, no. 2, from *The Caricature Magazine*, or *Hudibrastic Mirror.* London: Thomas Tegg, 1812. Hand-colored etching. Yale Center for British Art, Paul Mellon Collection.

exaggerations and distortions, as opposed to the more elevated depiction of character that he preferred. And in truth, all the figures in *Crim Con* are caricatured in ways that the French—and Hogarth—would consider grotesque; big bellies and misshapen noses abound. The organization of the frames is more like vaudeville than literature; each frame—we could call them "skits"—has a title and is accompanied by a brief verbal exchange. In the opening frame of the first print, its title, "Morality," sets the stage with two gents talking: "Sad times, sad times, friend Nicodemus. This crim con business is quite shocking." Nicodemus responds: "Ah it is of no use talking to them—they will have their own way. Shocking doings indeed." Then follow frames depicting with relish all these shocking doings. In the second print of the series, the illicit lovers are shown under titles such as "Conscience," where the woman cries out "Oh I'm Undone, I'm Undone!" (fig. 98). In "An Airing," she tells her consort, "Let us pretend to walk out as if nothing was the matter." In "Alarm," he worries "but if Mr. Spriggins should come home, what should we do then?" There is even a touch of gothic humor, the lovers embracing in "A Scene in a Stone Coffin," where the terrified woman shrieks "O Dear, o dear, if the gostesses should come." The series culminates with "Observation," where a betrayed husband peers through a doorway and, clearly considering some legal action, concludes "Mercy on me, what do I see—well a pair of spectacles is tantamount to two witnesses." The two prints strike a subtle balance between, on the one hand, working as a set of individual cartoons united by a theme, with each of the principal characters of husband, wife, and lover redrawn from frame to frame with no repeated physiognomies, and, on the other, as a mildly narrative sequence leading to the "discovery" in the last frame. While Hogarth had expressed his subtle form of caricature through gesture and symbolism, his followers' work was raucous and bawdy in form and content. Although still some distance from the comic book, these prints have begun to approach its rowdy, irreverent tone.

The characters in these prints all represent types: Hogarth's naïve young woman and wastrel young man, Rowlandson's unattractive, cuckolded husbands. Most individuals in caricature were anonymous, good for a laugh, but they were not recurring characters who captured the public's imagination. Thomas Rowlandson changed this with his *Tour of Doctor Syntax, in Search of the Picturesque*; it depicts the misadventures of a country parson who wants to make his fortune by producing one of the travel books then fashionable. First published as installments in Rudolph Ackermann's *Poetical Magazine*, with text by William Combe (1742–1823) and illustrations by Rowlandson, *Doctor Syntax* was then published in book format in 1812, sold out five editions in twelve months, and was reissued numerous times throughout the century.[6] What is surprising is that Combe produced each chapter only after Rowlandson had provided the drawing for it, thus reversing what is normally accepted as the logical relationship between artist and writer. In this sense, Rowlandson was the creator of *Doctor Syntax*, Combe the illustrator.

Rowlandson's images all show the humorous exaggerations and deformations of caricature, as well as the sequential organization of narrative comics. In *Doctor Syntax*

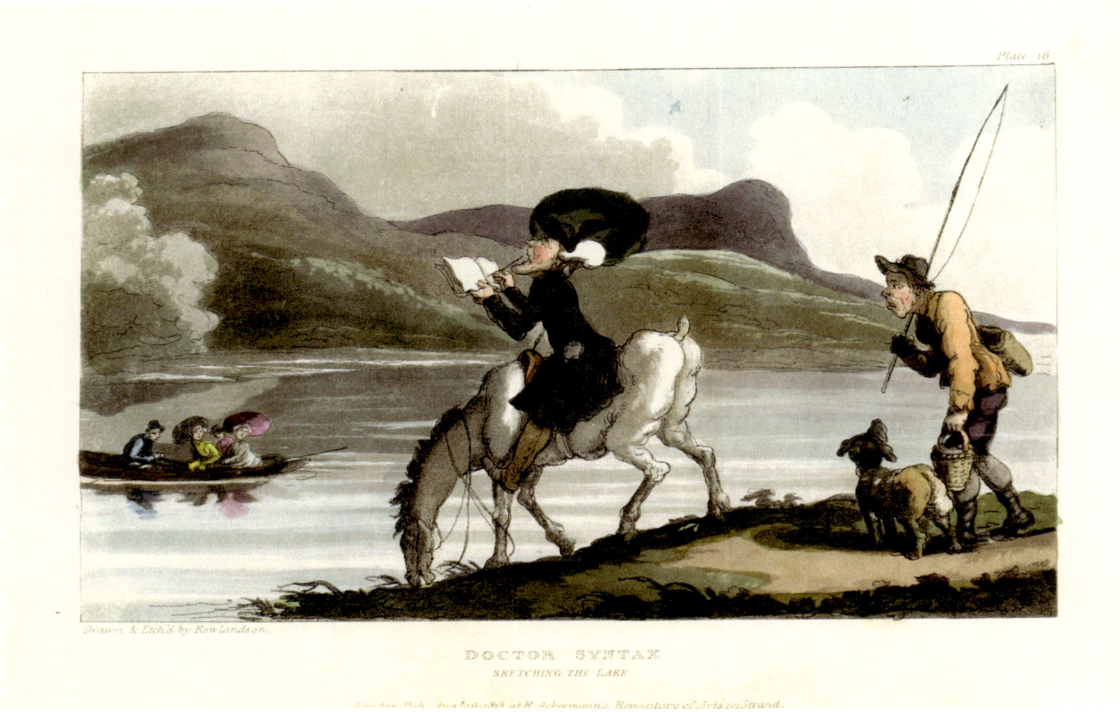
Plate 16
Drawn & Etch'd by Rowlandson.
DOCTOR SYNTAX
SKETCHING THE LAKE

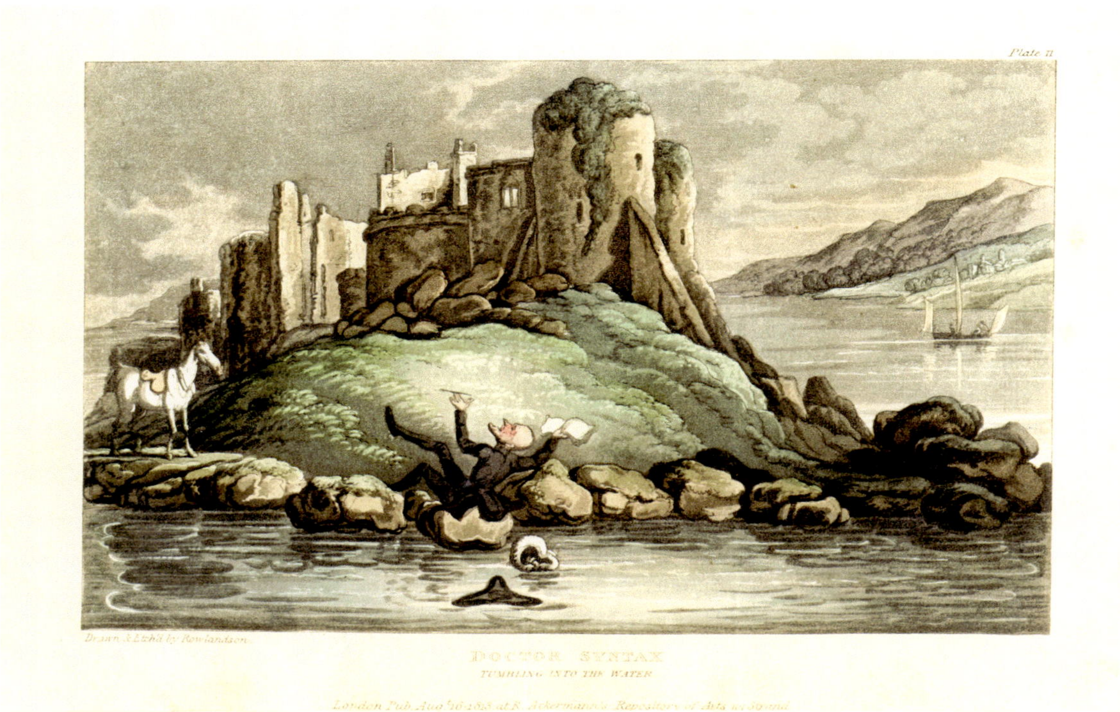
Plate 11
Drawn & Etch'd by Rowlandson.
DOCTOR SYNTAX
TUMBLING INTO THE WATER

FIGURE 99
Thomas Rowlandson, *Doctor Syntax Sketching the Lake, The Tour of Doctor Syntax, in Search of the Picturesque.* London: R. Ackermann, 1812, pl. 16. Hand-colored aquatint.

FIGURE 100
Thomas Rowlandson, *Doctor Syntax Tumbling into the Water, The Tour of Doctor Syntax, in Search of the Picturesque.* London: R. Ackermann, 1812, pl. 11. Hand-colored aquatint.

Sketching the Lake, the parson attempts to draw picturesque scenery, as the travel books say he must (fig. 99). When he backs up to get a good view of a crumbling Gothic castle—a most desirable subject according to travel accounts—he loses his balance, with the result seen in *Doctor Syntax Tumbling into the Water* (fig. 100). His pointy chin, beaklike nose, and ungainly limbs only increase our amusement at his pratfall. In the end, despite cutting a ridiculous figure wherever he goes, Doctor Syntax does, in fact, write his book, and it becomes an enormous success—as did Rowlandson and Combe's *Tour of Doctor Syntax, in Search of the Picturesque.*

Through thirty hand-colored aquatints in *Doctor Syntax,* many times the number in any of Hogarth's narrative series, Rowlandson created not just a contemporary type, but a unique individual, the protagonist of a narrative tale, well in advance of the French characters discussed in chapter 1—Mayeux, Robert Macaire, and Mr. Prudhomme—whose adventures never coalesced into narratives. *Doctor Syntax* became an immediate sensation, inspiring Rowlandson and Combe to produce two additional tours for the good parson, *The Second Tour of Doctor Syntax, in Search of Consolation* and *The Third Tour of Doctor Syntax, in Search of a Wife.*[7] Other writers and artists shamelessly borrowed the character and sent him roaming the world having adventures. There was even a plagiarized French version titled *The Romantic Don Quixote, or The Tour of Doctor Syntax in Search of the Picturesque and Romantic.*[8]

Unlike the series by Hogarth, there was no moral message in the *Doctor Syntax* stories. We have most definitely left behind the morality of the Age of Enlightenment and entered the Age of Individualism where the hero—or antihero—reigns supreme. Beginning with *Doctor Syntax,* the history of comic art can be read through the adventures (or misadventures) of fictitious individuals who, in the course of the nineteenth century, took their place in the public imagination much as historical and religious figures had done previously.

Doctor Syntax's successors in terms of public acclaim were Tom and Jerry—not the familiar American cat and mouse cartoon characters, but the original Tom and Jerry created by George and Isaac Cruikshank for Pierce Egan's *Life in London; or, The Day and Night Scenes of Jerry Hawthorn, Esq., and His Elegant Friend Corinthian Tom, Accompanied by Bob Logic, the Oxonian, in Their Rambles and Sprees through the Metropolis.* Pierce Egan (1722–1849), a prolific journalist and sportswriter (see fig. 16) launched *Life in London* in 1821 as a monthly publication, recounting the three chums' adventures through all the social strata of London from high to low; like *Doctor Syntax* it had images and text interspersed with an average of one hand-colored etching per chapter, thirty-six in all. *A Shilling Well Laid Out. Tom and Jerry at the Exhibition of Pictures at the Royal Academy* shows the two chums, elegantly dressed, hobnobbing with the elite of society (fig. 101). *Tom and Jerry "Masquerading it" among the Cadgers, in the "Back Slums," in the Holy Land* depicts them cavorting in a low tavern in St. Giles parish, one of the most notorious slums

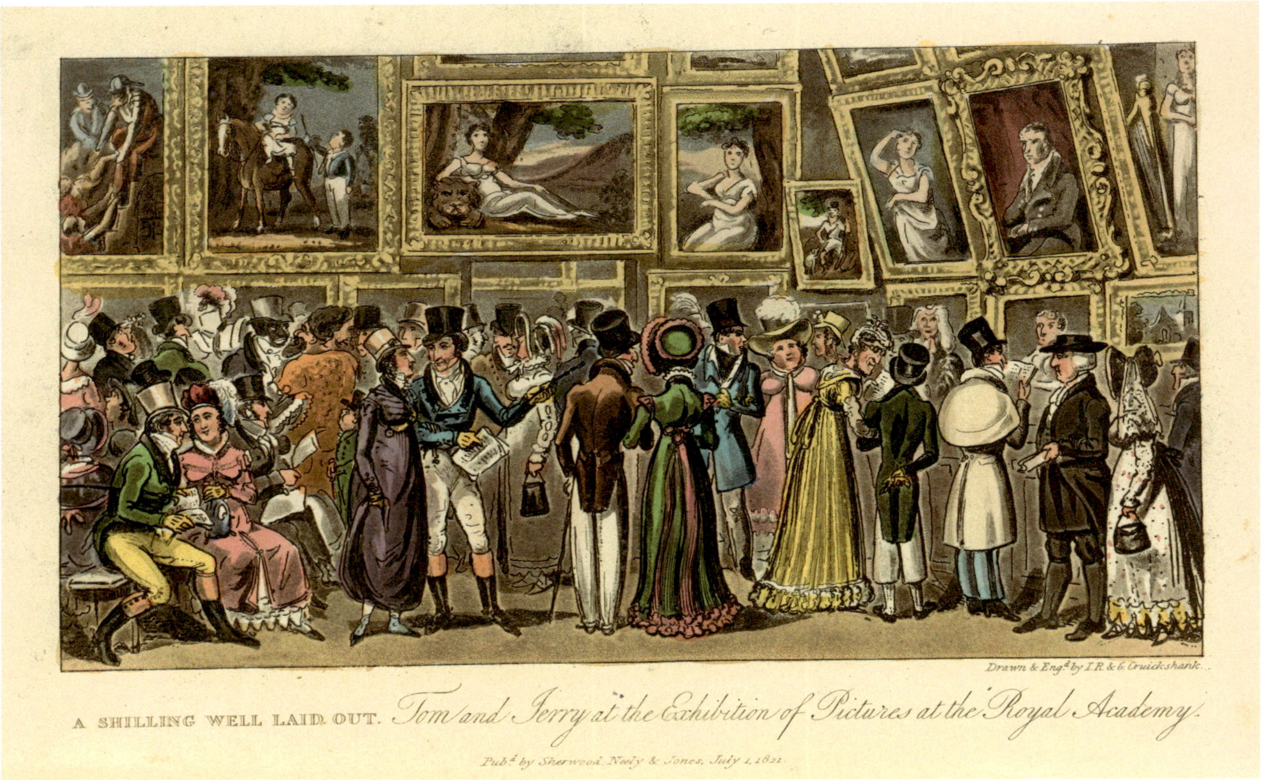

FIGURE 101

Isaac Robert Cruikshank and George Cruikshank, *A Shilling Well Laid Out. Tom and Jerry at the Exhibition of Pictures at the Royal Academy*, from Pierce Egan, *Life in London.* London: Sherwood, Neely & Jones, 1821. Hand-colored aquatint. Yale Center for British Art, Paul Mellon Collection.

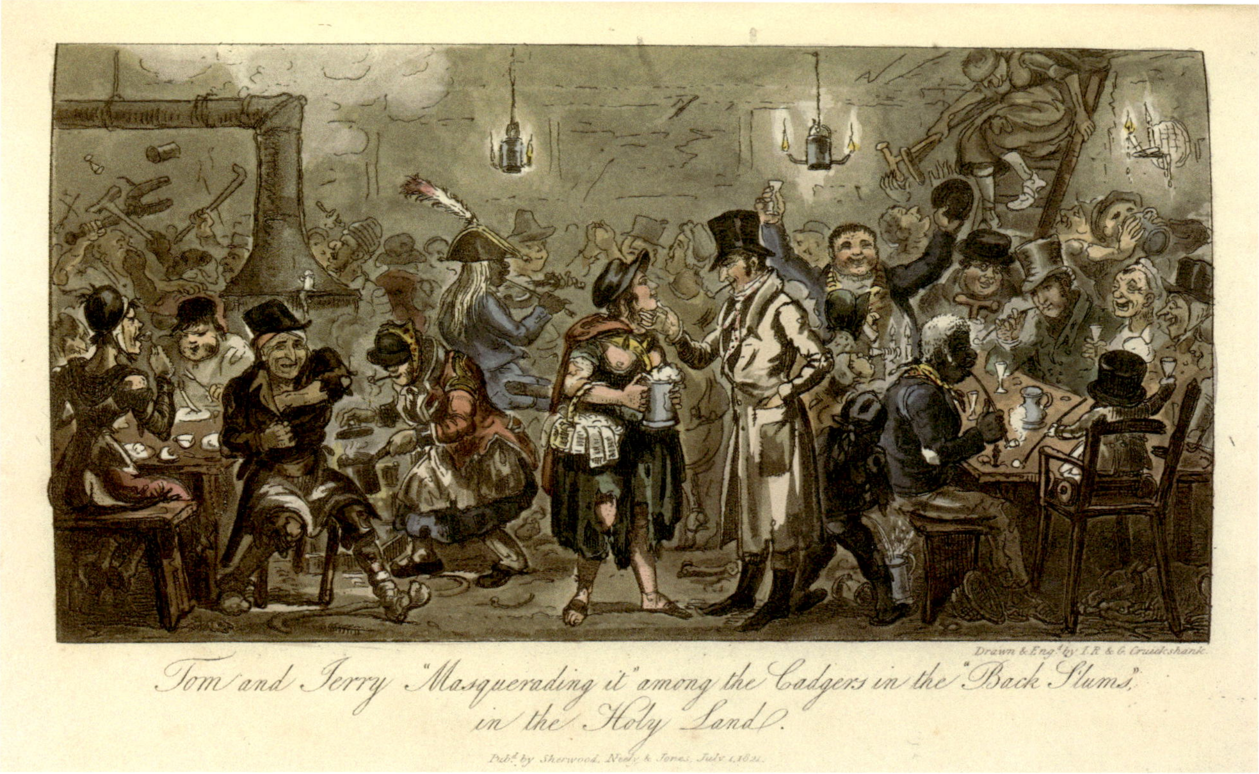

FIGURE 102

Isaac Robert Cruikshank and George Cruikshank, *Tom and Jerry "Masquerading it" among the Cadgers, in the "Back Slums," in the Holy Land*, from Pierce Egan, *Life in London.* London: Sherwood, Neely & Jones, 1821. Hand-colored aquatint. Yale Center for British Art, Paul Mellon Collection.

in nineteenth-century London (fig. 102). These depictions of high life and low proved irresistible to an audience that existed, no doubt, somewhere in-between. What seems distasteful to us today—the vision of the haves slumming among the have-nots—proved so popular in nineteenth-century England that *Life in London* was reprinted numerous times, including a cheap black-and-white edition that is more like a comic book than the quality hand-colored publication of the original.[9] Like Doctor Syntax, the three chums were quickly adopted and sent on adventures by other artists and writers; *Life in London* was immediately plagiarized in France, where it was translated and reissued as *English Diorama, or, Picturesque Promenades in London, Including the Most Accurate Notes on the Character, Manners, and Customs of the English Nation.*[10]

Although the French had initially lagged behind the British in the production of caricature, their trajectory was essentially the same. In chapter 1 we made the acquaintance of the clerk Calicot (see fig. 30) who, from 1817, had many adventures pretending to be a courageous hero; the hunchback Mayeux (see fig. 31) made his first appearance in the 1820s, drawn by several artists, including Cham, Daumier, and, principally, Traviès. After 1830, Daumier's Ratapoil and Robert Macaire (see fig. 32) as well as Monnier's Mr. Prudhomme (see fig. 33) became as widely known in France as Doctor Syntax or Tom and Jerry in England. These recurring French characters' adventures, however, rarely continued past a single frame; confined to the realm of caricature, they never developed into comics.

PICTURE STORIES TO COMICS

In the history of ideas, it is always difficult to establish why something happens when and why it does. During the years between Hogarth's work in the eighteenth century and the publication of the first comic book in the 1830s, it is easy to see—especially in hindsight—all the ways in which caricaturists were desirous of telling longer stories, working their way toward multisheet narratives either by cramming their individual prints with visual and verbal allusions, or by actually expanding their program to include additional prints on the same theme. While it is clear that the genre of comics was established incrementally over several centuries, with precedents ranging from broadsheets to the print series of Hogarth, probably the most important single factor was the new technology that made mass-produced illustrated books and periodicals feasible. Improved literacy, which guaranteed a wider audience for printed material, was an essential factor, but when we realize that a comic book has—not the six to eight illustrations of William Hogarth, nor even the twenty or more of Thomas Rowlandson or George and Isaac Cruikshank—but literally hundreds of frames, it should be obvious that the invention of comics had to wait, not just for cheaper paper, faster presses, and wider audiences, but also for an illustration medium faster and less expensive than etching or engraving. That medium was lithography. As we saw in chapter 1, its ability to reproduce images cheaply and quickly encouraged the transformation of drawing into caricature; within a few decades it accelerated the transformation of caricature into comics.

If we identify a comic strip as a single page or series of pages, each containing multiple frames of images narrating an original story, then Rodolphe Töpffer undoubtedly created the first such work with his 1833 *Story of Mr. Jabot,* the tale of a maladroit social climber whom we meet in the first frame (see fig. 95) striking the pose that he hopes will distinguish him as a man of the world.[11] Töpffer was the son of a prominent Swiss painter, and was himself a would-be artist who had given up painting because of his weak eyesight; instead, he turned to a teaching career in Geneva where he ran a boarding school for boys. So fearful was he of damaging his reputation as a serious author by publishing a work as frivolous as *Mr. Jabot* that he waited two years after it was printed before allowing it to be distributed. *Mr. Jabot* was not his first—he had drawn *The Loves of Mr. Vieux Bois* earlier, in 1827, but did not publish it for ten years. In plot, *Mr. Vieux Bois* veers close to slapstick, with all kinds of improbable adventures befalling its aging protagonist in pursuit of his lady love. In this satire on romantic fiction, Mr. Vieux Bois (literally Mr. Old Wood—English plagiaries called him Mr. Oldbuck) attempts suicide repeatedly but unsuccessfully until, in the end, he succeeds in winning the hand of his beloved. In his fourth suicide attempt he jumps out a window, only to get his shirt snagged on the sundial below, after which he returns home, changes his shirt, and sinks into romantic melancholy (fig. 103).

Töpffer died young, at forty-seven, but in his short life, in addition to his "serious" writing, he published seven comic albums and left several more incomplete.[12] Like *The Tour of Doctor Syntax,* each of Töpffer's stories is organized around the adventures of an eponymous protagonist. As literature, they should be understood within the tradition of the picaresque novel, which recounts one adventure after another. The form was widespread in Europe: Cervantes's *Don Quixote* (1605–1615) is perhaps the best-known example of the genre, which includes Alain-René Lesage's *Gil Blas* (1715–1735) and Henry Fielding's *Tom Jones* (1749). Although Töpffer claimed that his comics were created to amuse his students, his stories are too sophisticated for children. Mr. Jabot, for example, postures and poses, attempting to flatter his way into elite society. Despite many mishaps, eventually he, like Doctor Syntax, gains his heart's desire: Jabot manages to marry a marquise, just as Syntax does eventually publish his travel book. *Mr. Crépin* (1837) tells the story of a family that searches for the best education for its badly behaved children; this gives Töpffer the opportunity to satirize the leading educational theories of his day. Two pedagogical methods—each clearly unsuccessful—are depicted in figure 104. In the left frame, Mr. Bonichon introduces the younger children to his method of studying physics by reading the *Adventures of Telemachus,* a heroic poem first published by Fénelon in 1699 and a standard part of classical education; on the right, Mr. Fadet attempts to lecture the older boys on the system he has invented of reducing everything to fractional numbers. Other tutors come and go, each with his own pedagogical system—Mr. Craniose teaches

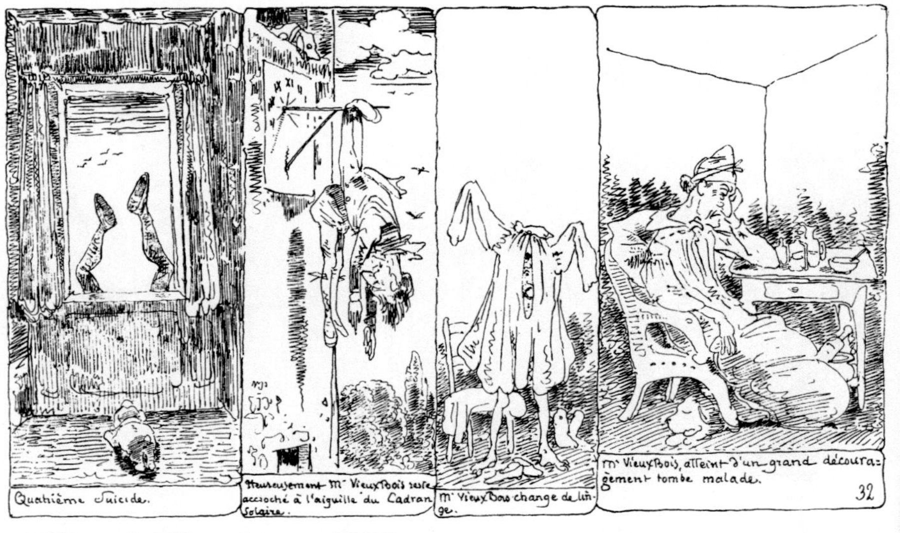

FIGURE 103
Rodolphe Töpffer, *The Loves of Mr. Vieux Bois* [*Les amours de Mr Vieux Bois*]. Geneva: Frutiger, 1837, pl. 32. Lithograph.

"phrenologically" and Mr. Gribouille believes in giving prizes to all his pupils. After several additional experiments, Crépin discovers Mr. Bonnefoi (Goodfaith) whose method is simply to have each student do his best according to his abilities. The story ends happily with all of Crépin's children having excellent careers. In *Doctor Festus* (1840), Töpffer's target is the pretensions of academicians, in *Mr. Pencil* (1845), government bureaucrats. In *The Story of Albert* (1845), the young Albert joins a band of revolutionaries, but as soon as the police arrive he abandons them, shaves off his beard, and flees (fig. 105). Each album satirizes issues and foibles of the day, much like caricature, but now in the form of lengthy, highly developed narratives.

Töpffer was an admirer of Rowlandson, so the resemblances between Doctor Syntax and characters like Mr. Jabot or Mr. Vieux Bois are not coincidental. All are picaresque, satiric tales of ordinary men who are ambitious beyond their talents, but who succeed in the end. This type of story would appeal to a new, postrevolutionary public where, in principle at least, all things were possible to those not fortunate enough to be born into the elite. Because we today are so familiar with every variety of comic art, it is difficult for us to appreciate the impact of Töpffer's work on a public that had never yet seen a comic book and where literature was rarely directed to this segment of the populace. The charm of Töpffer's albums, however, extended, like the tales of Tom and Jerry, to a spectrum of classes; the great German writer Goethe (1749–1832), to whom Töpffer sent a copy of *The Story of Mr. Cryptogame*, claimed that he rationed himself to "looking at only ten pages or so at a time, resting afterward, because, he said, he risked getting an indigestion of ideas."[13] *Mr. Cryptogame*, the last of Topffer's comics published in his lifetime, was one of the longest. The plot revolves around the unrequited love of Elvire for the butterfly collector Mr. Cryptogame. His repeatedly unsuccessful attempts to flee result in numerous

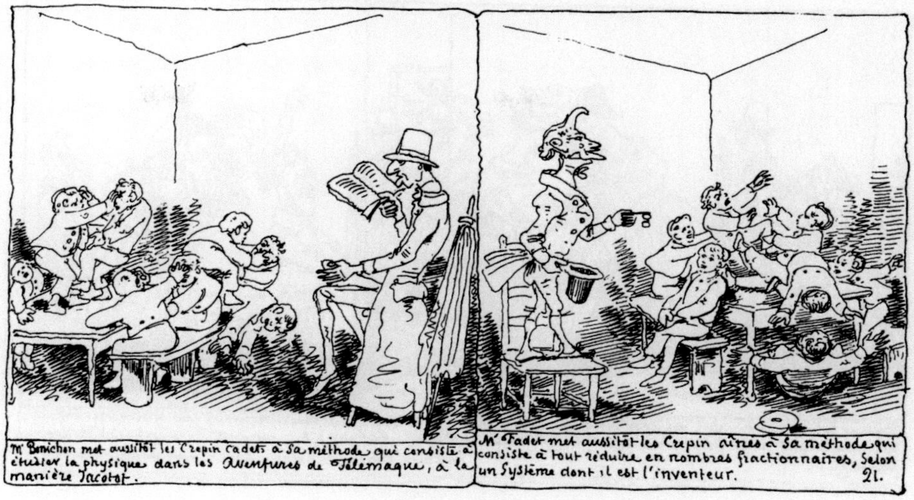

FIGURE 104
Rodolphe Töpffer, *Mr. Crépin*. Geneva: Frutiger, 1837, pl. 21. Lithograph.

adventures in distant climes, taking them both to polar regions and to North Africa as well. He sets off for the freedom of the New World with exhilaration, when, much to his dismay, he discovers Elvire hiding on the boat (fig. 106).

In all these stories, Töpffer created a hybrid medium encompassing word and image simultaneously; he recognized this by calling his works "literature in prints," stories in pictures.[14] While their literary subjects can be discussed in terms of contemporaneous social history or popular culture, their formal aspects must be situated within the arena of drawing. Visual conventions invented by early comic strip artists, still in use today, created a graphic language parallel to, but distinct from, that of painting, illustration, or even caricature. In terms of narrative strategies, these works differ markedly from the albums of lithographs that had been popular for over a decade.[15] As discussed in chapter 1, earlier albums presented each image individually, not as part of a sequence. Even when linked by a common theme, each image could be read and enjoyed in isolation from its neighbors. Each of Töpffer's albums, in contrast, contains hundreds of frames filling from forty to more than ninety pages. They demand to be read in a linear, horizontal mode from the beginning, like a book; they cannot be dipped into like earlier albums.

Töpffer produced his albums using transfer paper and pen lithography, a method that did not reverse drawings and captions as would standard lithography or engraving. He maintained a boxed grid format throughout, though the width of the frames does vary. The pages are in horizontal format (the album style called *à l'italienne*), with only one tier of frames per page. As a result, he could write his captions below the frames on each page, giving the published work the appearance of a handwritten journal; this "handmade" aspect even extended to the wavery outlines of the boxed grid. Although English artists often used the "speech balloons" characteristic of modern caricature and comics, or at

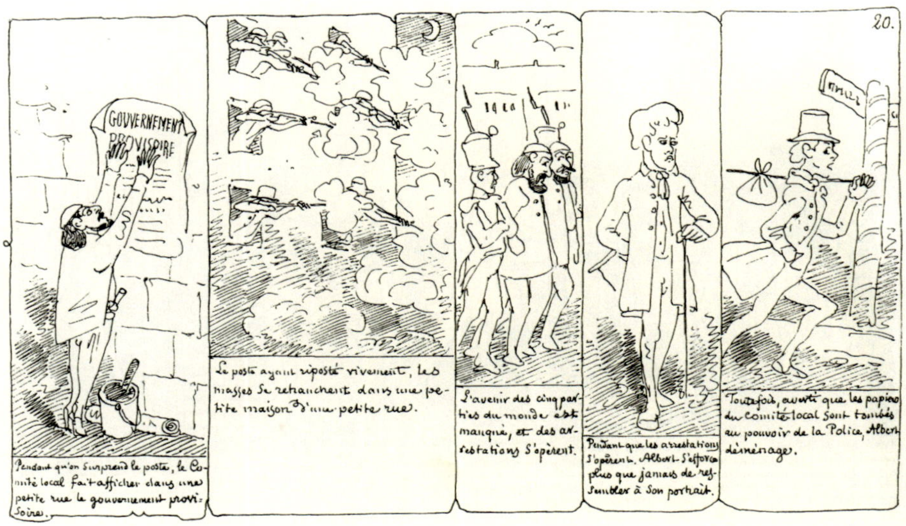

FIGURE 105
Rodolphe Töpffer, *The Story of Albert by Simon of Nantua* [*Histoire d'Albert par Simon de Nantua*]. Geneva, 1845, pl. 20. Lithograph.

least wrote the text in the same frame as the images (see fig. 18), Töpffer always preserved the integrity of his image by rigidly separating text and drawing.

Töpffer's drawing style differed notably from traditional illustration or even caricature. It was more abstract, almost cursory, with volume and space indicated, if at all, by a scribbled line. Had he attempted to render his images like those of Hogarth, it is unlikely that he would ever have been able to complete even one story in his lifetime. *The Story of Mr. Jabot,* for example, has one hundred fifty-five separate frames—and that is one of his shorter works. Despite the differences in rendering, Töpffer's treatment of events was very like that of illustration of the period, ultimately based on conventions adapted from painting. The viewer is situated at a middle distance, as a neutral observer of the events depicted. The stories are told through human agency, as in a play where the events depicted are seen from the vantage point of the audience, not that of the protagonist. There are no distortions of scale, no close-up views, no departures from his set drawing style; these visual strategies, typical of modern comic books, would be introduced by his French followers.

THE SECOND GENERATION: FRENCH COMIC BOOKS

Following Töpffer's paramount achievement in creating the first comic books, French artists adopted his new narrative mode and completed its transformation from humorous illustration based on painting to the madcap modern language of comics. Although two of France's most distinguished caricaturists, Daumier and Grandville, had no interest in this new genre of comics, younger artists, such as Cham and Doré, adopted it immediately.

The French contribution to the comic-strip genre can be seen in even a brief survey from Töpffer's albums to the publication in 1854 of Gustave Doré's *Dramatic and Picturesque*

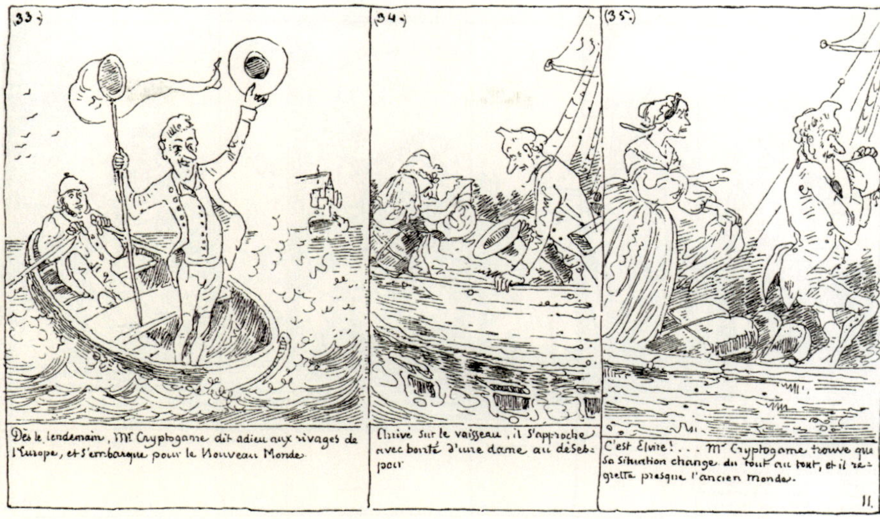

FIGURE 106

Rodolphe Töpffer, *The Story of Mr. Cryptogame* [*Histoire de monsieur Cryptogame*]. Geneva, 1845, pl. 11, nos. 33–35. Lith. Schmidt.

History of Holy Russia in Caricature, the longest comic book of the nineteenth century.[16] *Holy Russia* contains five hundred drawings arranged on more than two hundred pages, and can be seen as the culmination of the second chapter in the creation of modern comic books, Töpffer himself having written the first.

Soon after their publication Töpffer's innovative comic books circulated throughout Europe and the United States, in pirated as well as authorized editions. They had their earliest influence in Paris when, in 1839, Charles Philipon's Maison Aubert, which had already established the journal *Le Charivari,* published plagiarized editions of three of Töpffer's albums, *Mr. Vieux Bois, Mr. Jabot,* and *Mr. Crépin.*[17] Judging by the number of additional comic books that Aubert published subsequently, eight in the following four years, this must have seemed a promising new genre. Aubert titled these works *Collection of Jabots,* after Töpffer's first published album, a play on the meaning of the word *jabot* as something frivolous.[18] The albums probably did not sell very well, however, because between 1842, the year when Aubert published the tenth and eleventh (*The Story of Prince Colibri and the Fairy Caperdulaboula* and *The Adventures of Telemachus, Son of Ulysses,* both by Cham), and the very end of 1847, when the last of the collection—Gustave Doré's *Labors of Hercules*—was published, there were no additional albums in this series.[19] Aubert did continue to publish single-sheet caricatures both in albums and portfolios, however, often on a set theme but with no continuous narrative.

It is not difficult to understand why these early comics had only a limited success. Though they were ephemeral literature, they needed to be read at one sitting, demanding too much concentration from the reader; albums of single-sheet caricatures, in contrast, could be enjoyed even in a few spare moments, and could be viewed in no particular order. Unlike Töpffer's albums, which, by satirizing contemporary society were intended more

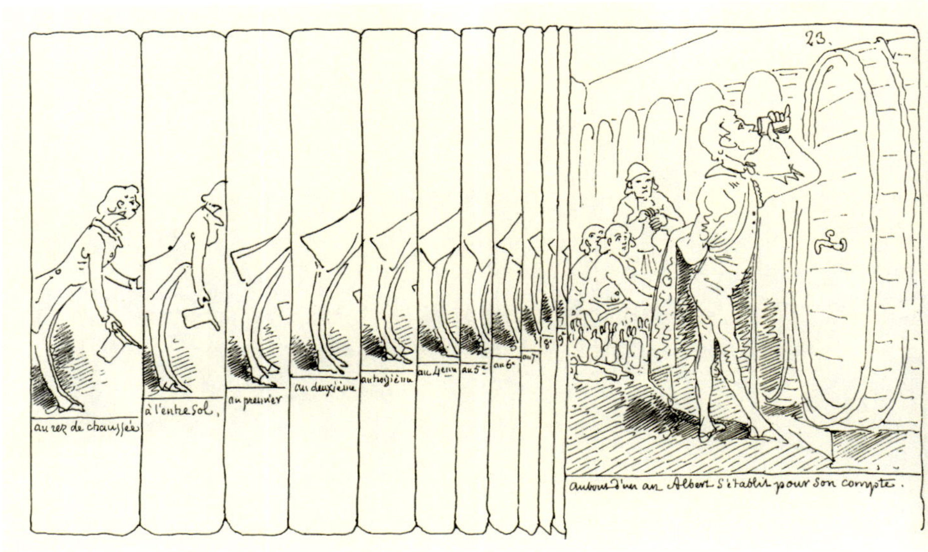

FIGURE 107
Rodolphe Töpffer, *The Story of Albert by Simon of Nantua* [*Histoire d'Albert par Simon de Nantua*]. Geneva, 1845, pl. 23. Lithograph.

for adults than for children, the early French comic book seemed to be an art form unsure of its audience. Too expensive and too long for children, their stories were not interesting enough to attract adults. Albums of individual caricatures were more sophisticated and topical and, as a result, proved more attractive to an adult audience. Indeed they were specifically advertised as "intended to be displayed on salon tables or to decorate the library of the book lover," as one contemporaneous publisher advised.[20]

Each of the twelve comic books published by Aubert in the *Collection of Jabots* has the same size and format, identical with those of Töpffer and no doubt dictated by the publisher. Seven were written by the caricaturist Cham, who wrote his first two, *The Story of Mr. Lajaunisse* and *Mr. Lamélasse,* in 1839, when he was twenty-one years old.[21] An admirer of Töpffer, Cham later redrew the illustrations for Töpffer's *Mr. Cryptogame* when it was published in the journal *L'Illustration* in 1845 in wood-engraved format.[22] While it is beyond question that Töpffer influenced these French artists, it is also possible that the French artists influenced Töpffer as well. In the accordionlike image that Töpffer drew in his 1845 *Story of Albert,* Albert ascends the staircase as a traveling wine salesman (fig. 107). The progressively narrower frames show Albert, hat in hand, meekly approaching his potential customers; they are captioned *On the Ground Floor, On the Mezzanine, On the First Floor, On the Second Floor,* and so on to the ninth floor.[23] The last frame shows the result of his efforts, Albert happily tippling a year later, master of his own business. The entire page sets up an insistent staccato rhythm that Töpffer had not used previously, which is then resolved into the relaxed cadenza at the end. It represents one of the few times he departs from his naturalistic mode of drawing. And yet this trope had appeared earlier in one of Aubert's *Collection of Jabots,* the 1840 *Story of Mr. de Vertpré and His Housekeeper Too* that is often mistakenly attributed to Cham but is listed in Aubert's

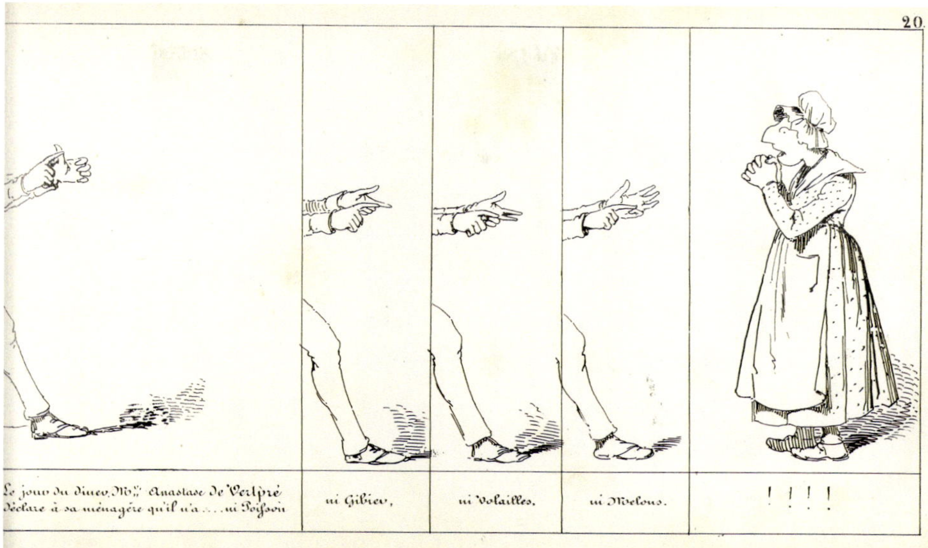

FIGURE 108
Eugène Forest, *The Story of Mr. Vertpré and His Housekeeper Too* [*Histoire de Mr de Vertpré et de sa ménagère aussi*]. Paris: Aubert, 1840, pl. 20. Lithograph. Getty Research Institute, Los Angeles (84-B21489).

catalogues as by E. [Eugène] Forest (1808–1891), an artist who drew regularly for *Le Charivari*.[24] Here, Forest has drawn his protagonist in a series of frames showing only anatomical fragments. The caption explains: *The Day of the Dinner, Mr. Anastase de Vertpré Tells His Housekeeper That He Has… neither Fish, / nor Fowl, / nor Poultry, / nor Melons* (fig. 108).[25] The staccato set up by the frames depicting his fingers ticking off the essentials he lacks is then resolved in the last frame that shows his housekeeper's shocked reaction: she wrings her hands, her mouth agape, and her only response is *!!!!*

All the albums in the *Collection of Jabots* utilized Töpffer's horizontal format containing one or more boxed frames in a single tier with captions handwritten below. The artists even used pen lithography in emulation of Töpffer. Cham's graphic style from the beginning, however, was more varied than Töpffer's, showing more volumetric drawing and shading, as well as the beginnings of the abstract language familiar to us in modern comics. In his albums, Cham varied the scale of objects to show their importance, as when Mr. Lamélasse's servant carries a huge letter of invitation (fig. 109). Cham was also the first to combine different drawing styles. While Töpffer always drew like Töpffer, Cham could, for example, draw like a child, as he did in *A Misunderstood Genius*, which recounts the life of the incompetent artist Barnabé Gogo. *At Ten Years of Age*, the caption tells us, *Mr. Barnabé Had Made So Much Progress That His Drawings of Horses Could Be Mistaken for Those of Géricault*.[26] The accompanying frame shows a child's stick-figure drawing of men on horseback, leaving the reader wondering if it is the doting parents or the inept art critics who are being satirized here (fig. 110). Cham was also the first to expand the subjects of comics from contemporary life to include parodies of literature: his *Adventures of Telemachus* retells the classic text by Fenélon obliquely caricatured by Töpffer in his *Mr. Crépin* (see fig. 104). Perhaps the most influential of Cham's innovations was his

FIGURE 109
Cham, *Mr. Lamélasse*. Paris: Aubert, 1839. Lithograph. Bibliothèque nationale de France.

BELOW FIGURE 110
Cham, *A Misunderstood Genius* [*Un génie incompris*]. Paris: Aubert, 1841, pl. 5. Lithograph. Bibliothèque nationale de France.

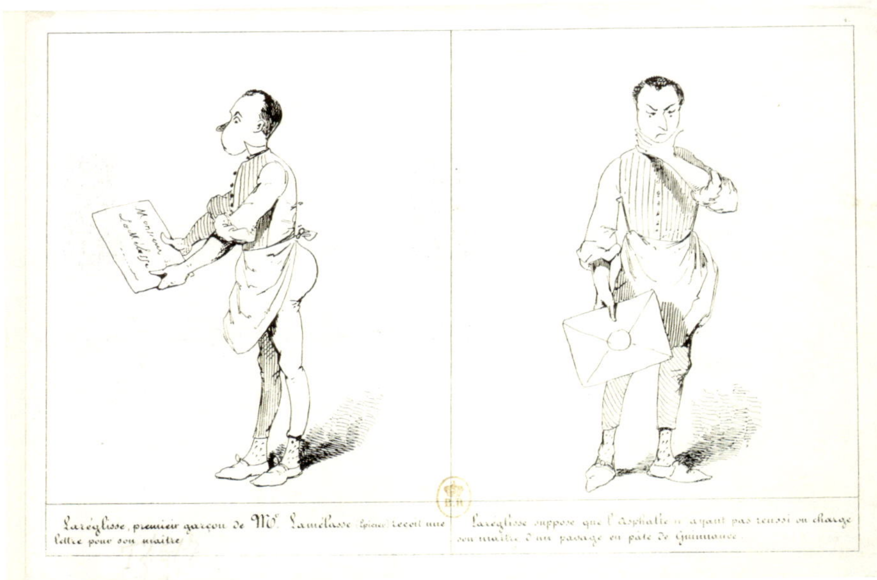

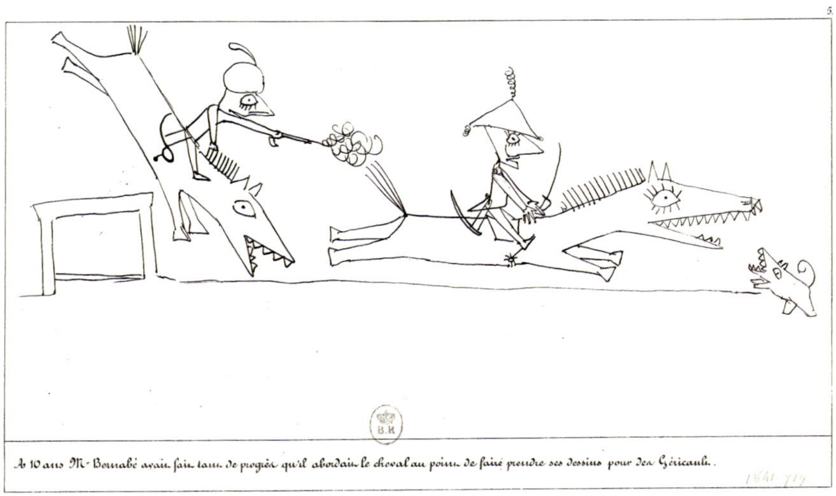

inclusion of an all-black vignette, where *Mr. Lajaunisse* blows out his candle and is left in darkness (fig. 111), and an all-white vignette, where Cham states that he is much too discreet to show what transpires in *Two Marriageable Vaccinated Spinsters,* a satire on the marriage "market" (fig. 112). Although both had precedents in Laurence Sterne's novel *The Life and Opinions of Tristram Shandy, Gentleman* (1759–1767), Cham's publication of these panels disseminated Sterne's visual tropes to both caricature and comic books where they remain in standard usage even today.[27]

In Cham's first seven comic books he adhered to the boxed grid format of the Töpffer–Aubert layout, but during the five years when Aubert had ceased publishing this series, 1842 to 1847, Cham created *Travel Impressions of Mr. Boniface,* published in 1844

FIGURE 111

Cham, *Mr. Lajaunisse.* Paris: Aubert, 1839, pl. 8. Lithograph. Bibliothèque nationale de France.

FIGURE 112

Cham, *Two Marriageable Vaccinated Spinsters* [*Deux vieilles filles vaccinées à marier*]. Paris: Aubert, 1839, pl. 13. Lithograph. Bibliothèque nationale de France.

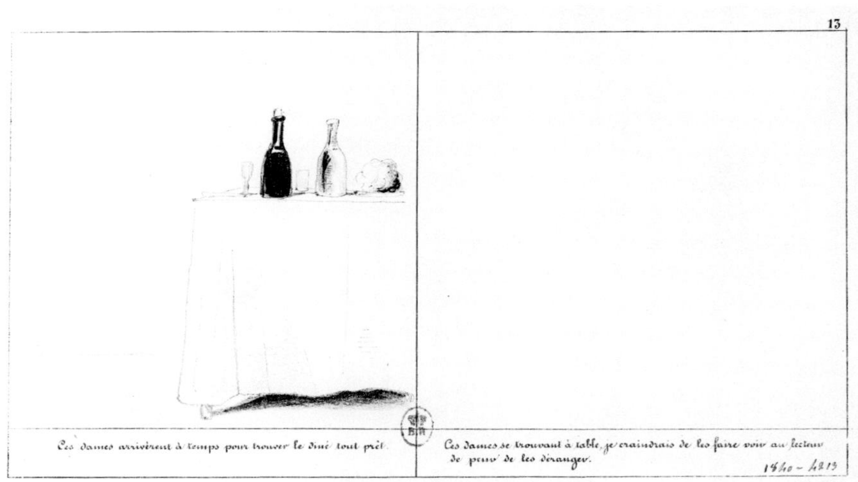

by Alexandre Paulin, a major publisher and one of the founders of *L'Illustration* the previous year.[28] In *Mr. Boniface* Cham maintained the horizontal format of previous albums, but enlarged the dimensions. Artistically, it was the most ambitious comic book after Töpffer's own work, surpassed only later by Doré's *Holy Russia* of 1854.

There are dramatic differences between *Mr. Boniface* and Cham's earlier albums. *Mr. Boniface* was produced through wood engraving, not lithography, and the earlier convention of handwritten text was replaced here by standard letterpress typeface. With these changes, Cham took the comic album out of the realm of the "artist's book" and brought it into the arena of the mass-produced printed book. The most striking change is that the frames are no longer arranged horizontally in strip format. Cham replaced

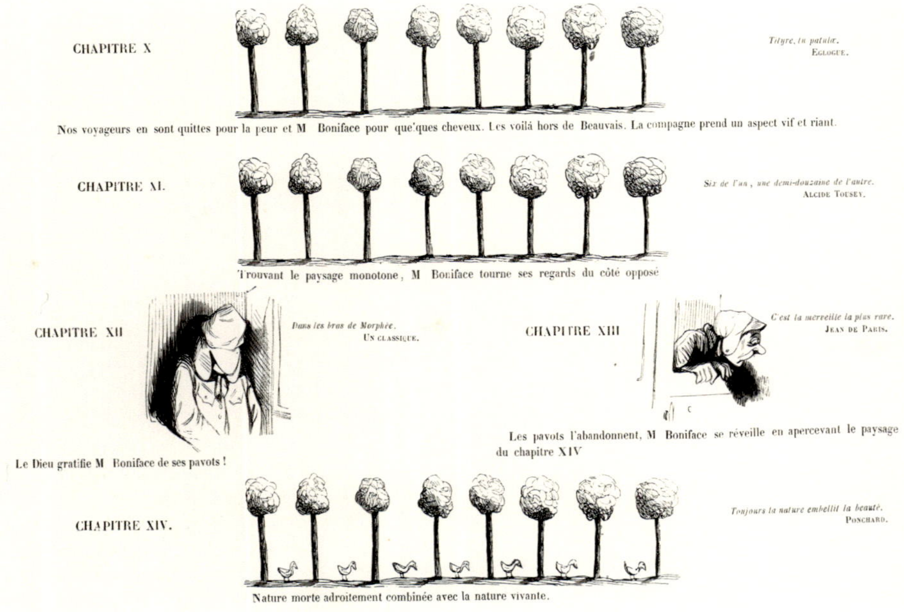

FIGURE 113
Cham, *Travel Impressions of Mr. Boniface* [*Impressions de voyage de monsieur Boniface*]. Paris: Paulin, 1844, pl. 7. Wood engraving.

Töpffer's single horizontal tier of boxed frames with a free-form arrangement of vignettes to be read either top to bottom, left to right, or in some combination thereof, depending on the events depicted. This innovation allowed him to create for the reader a kinesthetic equivalent to the narrative. When Mr. Boniface is on a boring coach voyage, we are forced to follow the monotonous landscape across the horizontal page (fig. 113). When he is seasick, we reenact the pitching of the boat by reading up and down, up and down, up and down, until we share his misery (fig. 114). Cham even added to each chapter heading a short apocryphal epigram of commentary, satirizing the pomposity of "serious" books (especially those of the Romantics) and underscoring their distance from this new irreverent comic genre. Since each "chapter" of *Mr. Boniface* consists of only one image with a brief epigraph, he was also parodying the contemporaneous literary mode for long, pretentious, multichapter books, particularly travel memoirs. Chapter LXX depicts only Mr. Boniface's barking dog Azor, accompanied by the epigraph, "'*Do, re, mi, fa, sol, la, si, do*.'—Rossini." Chapter XL's vignette of a ship's deck strewn with seasick passengers is accompanied by the epigraph, "'I am suffering.'—A Romantic."[29]

The frenetic rhythms familiar to us in classic comic strips make their first appearance in *Mr. Boniface* in a series of what in film technology are called "jump cuts," abrupt shifts of viewpoint. This narrative technique did not make its appearance in cinema until the twentieth century. One vignette shows Mr. Boniface's view through his coach window where he—and we—can see only the coachman's boots (fig. 115). This image becomes all the more poignant through the inclusion of the epigram, "'Ah, Nature, how beautiful you are.'

FIGURE 114

Cham, *Travel Impressions of Mr. Boniface* [*Impressions de voyage de monsieur Boniface*]. Paris: Paulin, 1844, pl. 18. Wood engraving.

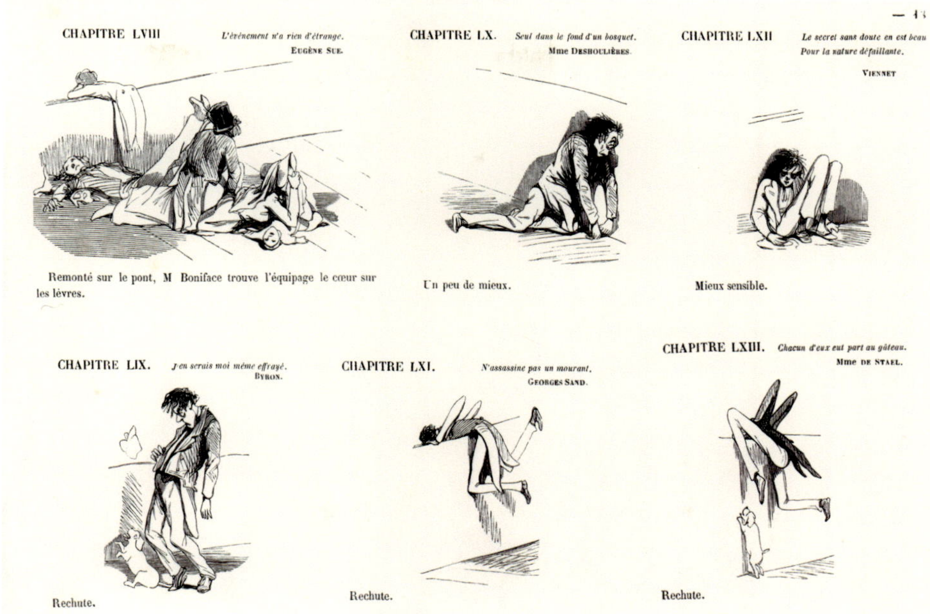

FIGURE 115

Cham, *Travel Impressions of Mr. Boniface* [*Impressions de voyage de monsieur Boniface*]. Paris: Paulin, 1844, pl. 8. Wood engraving.

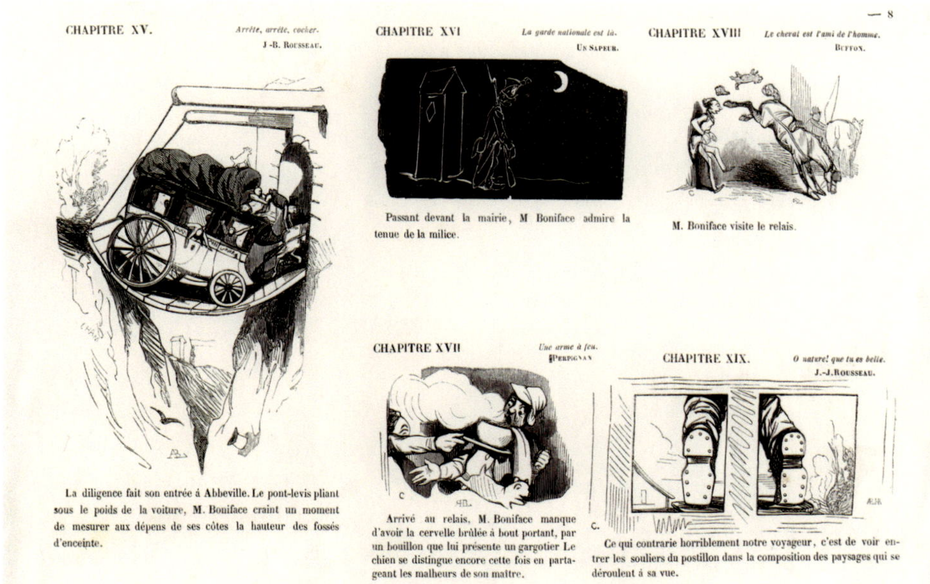

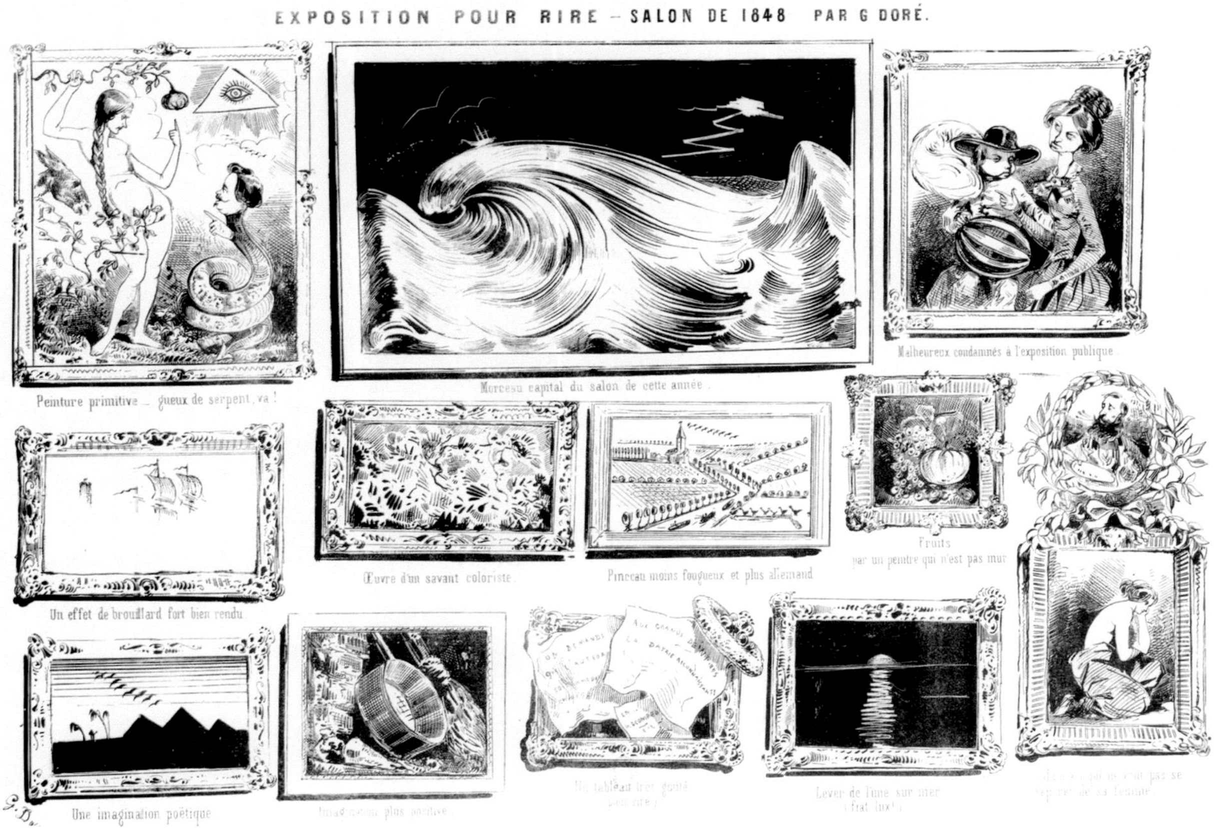

FIGURE 116

Gustave Doré, *Comic Exhibition.—Salon of 1848* [*Exposition pour rire.—Salon de 1848*], *Le Journal pour rire*, March 25, 1848. Wood engraving. Bibliothèque nationale de France.

J.-J. Rousseau."[30] By abandoning Töpffer's viewpoint of the neutral observer in favor of the highly personal viewpoint of the protagonist, Cham has created the visual equivalent of the transformation that was taking place in literature during these same years. This transformation gave us the modern novel, which is less a collection of picaresque adventures than an exploration of personal experience and psychological verism. Cham's *Mr. Boniface* and Stendhal's *Charterhouse of Parma* (1839) have this in common: both recount their protagonist's unique and individual private responses to public events.

An inventory of the formal innovations that Cham introduced into the comic book would be extensive and would include: the elimination of the boxed format, the substitution of typography for the handwritten text, the use of multiple tiers of frames in irregular formats, combinations of disparate styles, fragments, out-of-scale objects, different points of view, close-ups as well as distant views, and all-black or all-white panels.

While it is indisputable that many of these visual conventions made their first appearance in the work of Cham, there is still another precedent that must be cited, a particularly French phenomenon called "The Salon in Caricature" briefly mentioned in chapter 1.[31] Well before the arrival of the comic strip, there had been caricatures of individual paintings, but only in the 1840s, and only in France, were these images combined into entire pages, often entire "comic books" devoted to that year's Salon installation. The first were published almost simultaneously in 1843 by Raimond Pelez in *Le Charivari* (see fig. 55), and Bertall in *Les Omnibus* (see fig. 162), one year before Cham's groundbreaking *Travel Impressions of Mr. Boniface.*[32] The Salon in Caricature became a staple of art exhibitions, produced by Bertall, Cham, Nadar, Doré, and numerous others, and continued to be published in France throughout the century, waning only after the government abandoned its sponsorship of the Salon in 1880. On a single page, artists could caricature a variety of paintings—history, portraiture, genre, landscape, still life—each with its own scale, subject, and style, and then combine the images into the free-form layouts soon to be familiar in comics. These caricatured Salons gave artists the freedom to break from the Töpffer–Aubert grid format since here there were no strictures as to scale, style, or narrative continuity, and the large pages of journals permitted an anarchistic layout of multiple frames. When the pages were hand-colored, as many were, their resemblance to modern comic books is even more striking. Gustave Doré's *Comic Exhibition. The Salon of 1848* (fig. 116) caricatures each work separately, as was the custom (see fig. 55), but now he juxtaposes them in a way that invites a sequential reading. A nocturnal view of pyramids, for example, is captioned *A Poetical Imagination,* but next to it is a still life of kitchen pots labeled *More Positivist Imagination.* What looks like a wildly overgrown forest is identified as *Work of a Brilliant Colorist,* juxtaposed with a geometrically regular landscape captioned *Less Impassioned, More Germanic Brushwork.*[33] And, in an example of the cross-fertilization that would become standard across the various print media, Doré includes a virtually all-black image, *Moonrise at Sea (Let There Be Light!)* and a virtually all-white one, *Mist Very Well Rendered.*[34] As it developed, the early Salon in Caricature presented a clear precedent for the freewheeling combinations of imagery that soon would appear in French comics.

Although Cham continued to produce comic books after *Travel Impressions of Mr. Boniface,* none surpassed this album. His later works, such as *Soulouque and His Court* (1850) or *The Adventures of Mr. Beaucoq* (1856), are more conventional in both drawing and format.[35] Most of his production after 1844 consisted of individual illustrations and caricatures, often published first in *Le Charivari* or *L'Illustration* (see fig. 87) and then reissued as albums. He created over a hundred such albums as well as regular Salons in Caricature, producing them until his death in 1879.

It was the young Gustave Doré (1832–1883) who took up the challenge of the comic book, which, by 1847 had been abandoned by both Cham and the Maison Aubert. Doré is the best-known illustrator of the nineteenth century, and yet his contributions to the development of the comic book remain relatively unknown and unacknowledged, probably because his career as a caricaturist and comic-book artist lasted fewer than eight years, from late 1847 to 1854, from when he was fifteen to twenty-two years old.

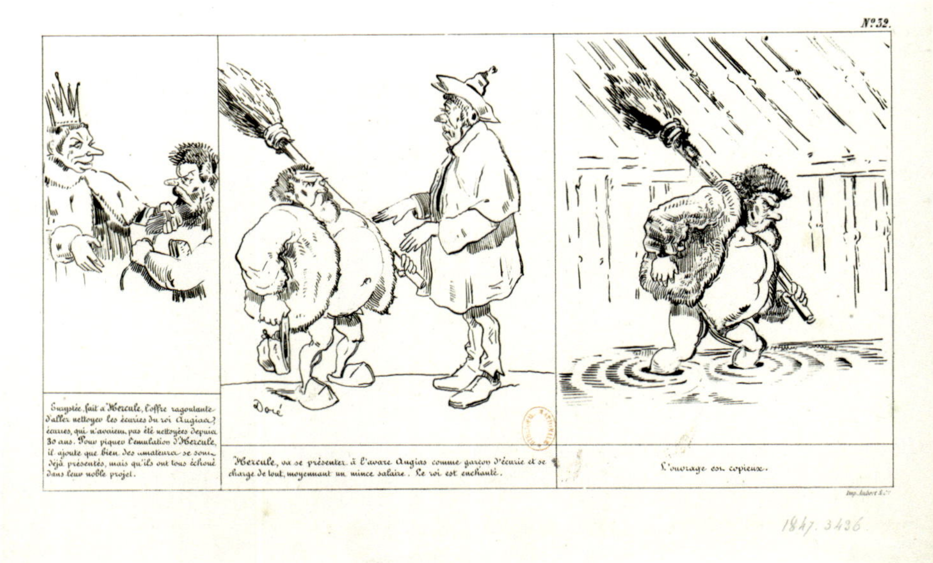

FIGURE 117
Gustave Doré, *The Labors of Hercules* [*Les travaux d'Hercule*]. Paris: Aubert, 1847, pl. 32. Lithograph. Bibliothèque nationale de France.

During this time he produced thousands of images and four full-length comic books.[36] When Doré wrote and drew his first, *The Labors of Hercules*, he was only fifteen, but he was already negotiating a contract to publish regularly in Philipon's new weekly, *Le Journal pour rire. Hercules* was the twelfth and last album in the *Collection of Jabots*; after that the project ceased, although the Maison Aubert later published two comic books by Doré in formats very different from the earlier series.

In many ways, Doré's *Labors of Hercules* was outmoded even before it was published because it followed closely the Töpffer model of a wavery linear style of drawing, executed in pen lithography with a single tier of boxed frames. Cham had already abandoned this format three years earlier. Like Töpffer, Doré positioned the handwritten text beneath the frame, and his album, like Töpffer's, is intended to be read horizontally. Töpffer's subjects, however, had been contemporary, often politically charged, and he had never attempted a literary theme even in parody. In subject, Doré's *Labors of Hercules* recalls Cham's *Adventures of Telemachus* of 1842. Since the French educational system placed such great emphasis on the classics, it is not surprising that both young men chose to parody Greek mythology in the style of a lycée student, probably their intended audience.

Even in his first youthful effort, Doré was already showing his narrative gifts. Artists do not often write interesting fiction, nor do many writers draw remarkable images. Rare is the individual like Töpffer or Doré who could do both. Despite Cham's fourteen years seniority over Doré, the younger man's creation is more unified and wittier. Doré exploited the rhythm of turning pages with the element of surprise that a new page

FIGURE 118
Gustave Doré, *The Labors of Hercules* [*Les travaux d'Hercule*]. Paris: Aubert, 1847, pl. 33. Lithograph. Bibliothèque nationale de France.

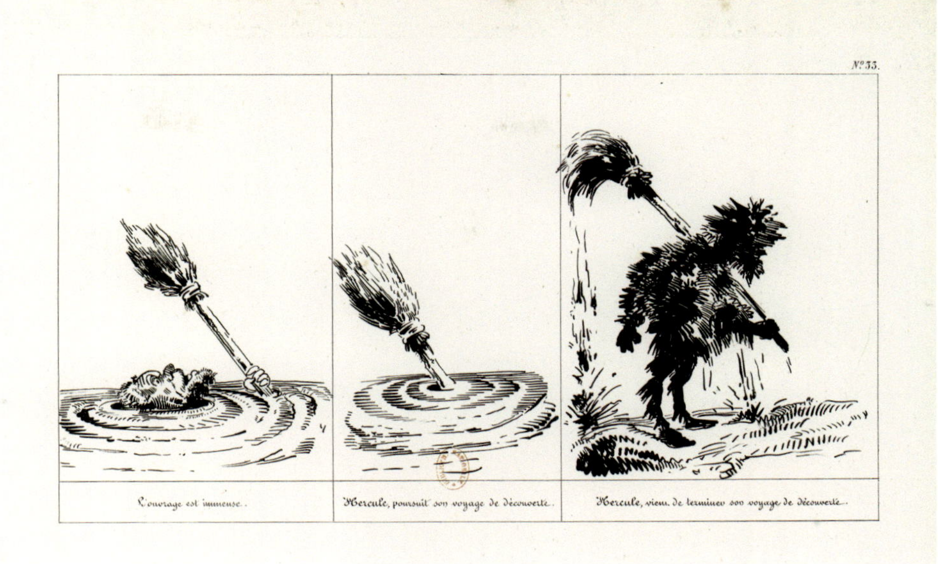

could bring. Thus, when Hercules attempts to clean the Augean stables, a turn of the page shows him sinking into the mire (figs. 117 and 118). What is missing in all the early comic strips, however, are the "speed lines" that have since become the standard visual symbol of motion. In Töpffer's work, and even in the early pages of the *Labors of Hercules*, figures do not look like they are moving at all, but seem to be frozen in space: Töpffer's handwritten caption might tell us, in figure 119, that *Mr. Cryptogame Circles the Deck Nine Times without Finding the Way Out*, but there is little visual indication of speed.[37] Doré actually invented what has since become the conventional symbol of speed later in *The Labors of Hercules* where Cacus, the fire-breathing giant, vomits forth whirlwinds of flame and smoke (fig. 120). These speed lines have since become one of the most characteristic graphic signs in the language of comics.

Doré published three additional comic books after *The Labors of Hercules. Three Artists Misunderstood and Malcontent* (1850) and *(Dis)Pleasures of a Pleasure Trip* (1851) were both published by Aubert, although their style and formats differed markedly from the by-then defunct *Collection of Jabots*.[38] Both were executed in lithographic crayon instead of pen lithography, resulting in more illusionistic rendering. In *Three Artists*, Doré abandoned boxed frames and handwritten legends, as Cham had done earlier in his *Mr. Boniface*. But it was in his *(Dis)pleasures of a Pleasure Trip* where Doré first displayed the imaginative leaps that would define his comic style and that of future graphic artists.

(Dis)pleasures, reprinted three times and still in print today, was the most successful of Doré's first three albums. In it, he exploited shifts in scale in an unprecedented way. In its most striking frame, he recounts how César Plumet, his artist-protagonist, was drawing

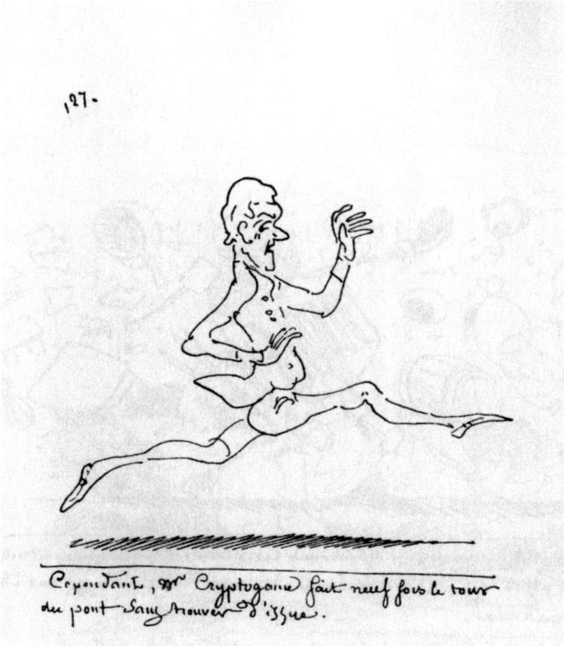

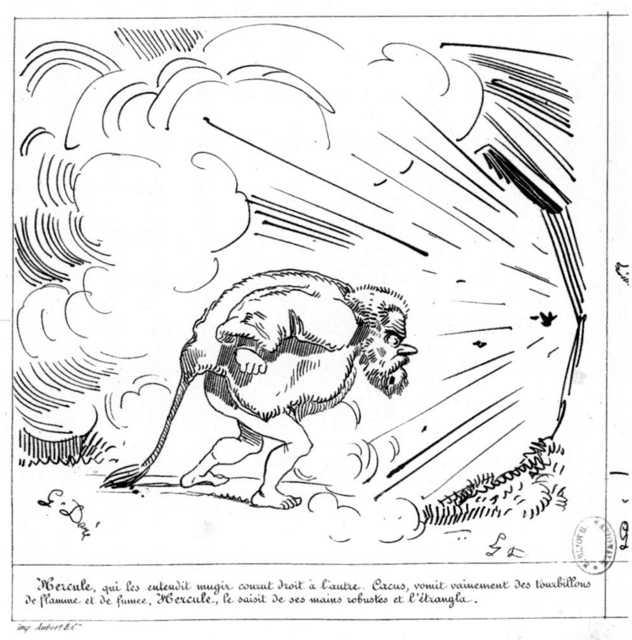

FIGURE 119
Rodolphe Töpffer, *The Story of Mr. Cryptogame* [*Histoire de monsieur Cryptogame*]. Geneva, 1845, pl. 127. Lith. Schmidt.

FIGURE 120
Gustave Doré, *The Labors of Hercules* [*Les travaux d'Hercule*]. Paris: Aubert, 1847, p . 36. Lithograph. Bibliothèque nationale de France.

an inhabitant of the Swiss Alps when his model slipped and stepped on his drawing. The result of this mishap is shown in Doré's drawing of Plumet's drawing marred by a huge footprint across its surface (fig. 121). Later, when a cow licks Plumet's drawing, Doré draws the cow's huge muzzle and tongue looming over the tiny artist and drawing (fig. 122). In both panels, he shatters the illusionistic window and normative perspective that had governed art production from the Renaissance. Nor are Doré's stories told only through figures; now all kinds of visual imagery, objects, or even designs carry the storyline forward. He draws frames that are not even "art," showing a close-up of Mr. Plumet's account book (fig. 123) or his crossed-out sketchbook drawing, visible in figure 121. When Mme Plumet looks through her telescope, Doré shows us what she sees, in an extraordinary plate of six circular drawings, one even including the fly that has landed on her telescope lens (fig. 124). With *(Dis)Pleasures of a Pleasure Trip*, Doré severed the last ties to traditional figure-centered illustration in order to advance the narrative by any means available. Images expand and contract across the page; on turning the page one cannot be certain what one will find, sometimes a full-page drawing, sometimes up to eighteen tiny ones. On occasion, readers cannot even be certain of the order in which the frames should be read. Doré's work becomes self-referential in terms of formal innovations and begins to establish a historiography of caricature and the comic book. He quotes Cham's all-black panels (the fifth drawing in figure 124, when Mme Plumet looks through her telescope at night), as well as Grandville's figurative musical compositions, as seen in figure 123, instantly recognizable to his audience (see fig. 74).

FIGURE 121
Gustave Doré, *(Dis)pleasures of a Pleasure Trip* [*Dés-agréments d'un voyage d'agrément*]. Paris: Aubert, 1851, pl. 10. Lithograph.

FIGURE 122
Gustave Doré, *(Dis)pleasures of a Pleasure Trip* [*Dés-agréments d'un voyage d'agrément*]. Paris: Aubert, 1851, pl. 21. Lithograph.

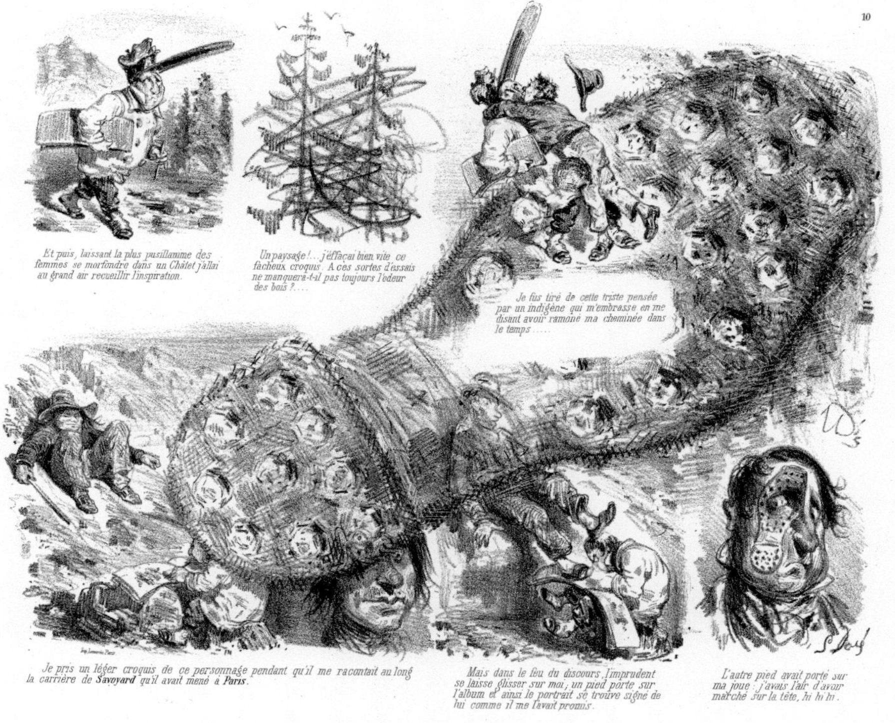

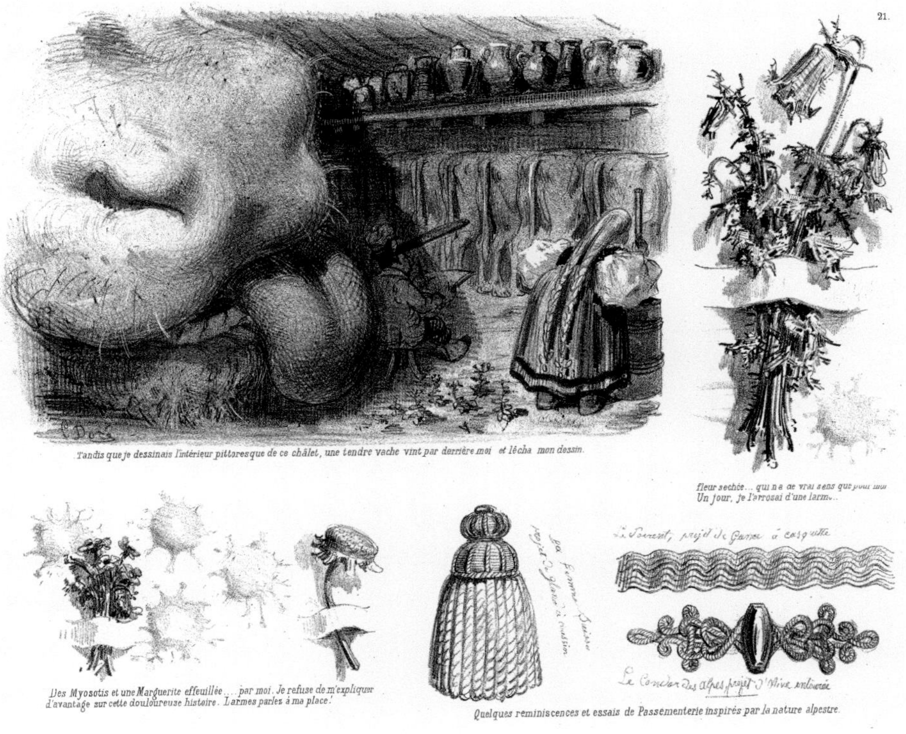

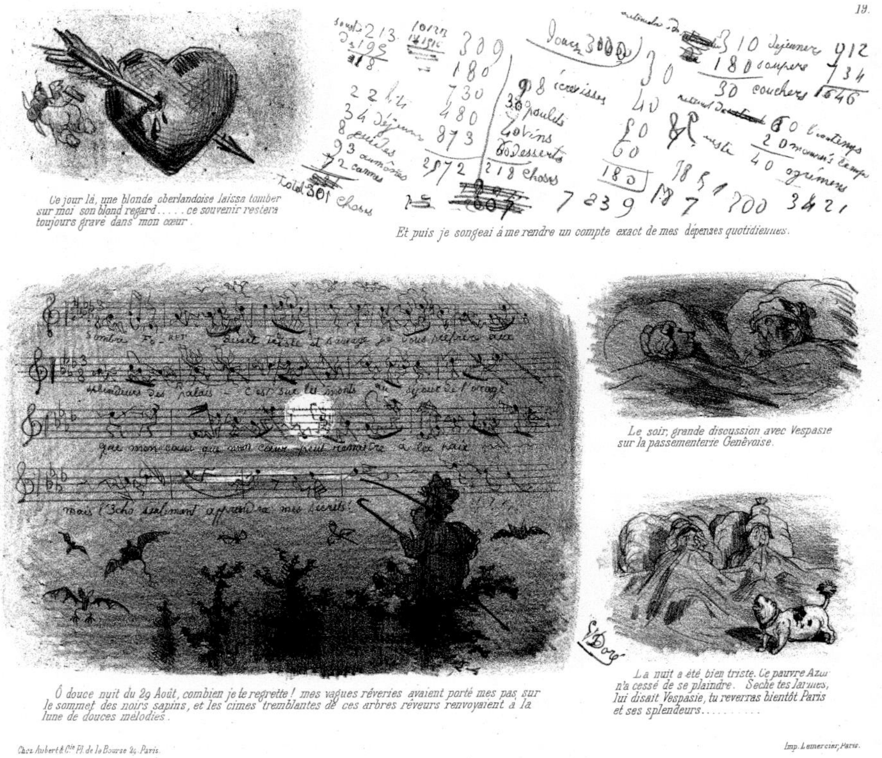

FIGURE 123
Gustave Doré, *(Dis)pleasures of a Pleasure Trip* [*Dés-agréments d'un voyage d'agrément*]. Paris: Aubert, 1851, pl. 19. Lithograph.

Nothing, however, prepares us for Doré's last foray into the world of comics, his 1854 *Dramatic and Picturesque History of Holy Russia in Caricature.* Doré was then still a young man, only twenty-two, when he wrote and drew this work, whose exuberance and sheer anarchy set it off from all previous examples of the genre. It was published not by Aubert, but by J. Bry aîné, who probably subsidized it, since despite its being the most ambitious comic book of the nineteenth century, with five hundred frames spread over two hundred seven pages, it sold for only four francs, cheaper even than the much more modest albums of the *Collection of Jabots. Holy Russia* represents not only the culmination of Doré's work in this medium, but also the culmination of the comic book's development by the post-Töpffer generation.

In *Holy Russia,* Doré rejects the horizontal album format that had characterized comic books since their inception, adopting instead the vertical format that brought his publication further into alignment with "real" books. He abandoned the lithography in which he had drawn all his previous comic books in favor of wood engraving, a medium in which he would work for the rest of his life. His team of wood engravers was led by Eugène-Noël Sotain (1816–1874), to whom he gave star billing on *Holy Russia*'s cover.[39]

Holy Russia was published during the Crimean War, and took full advantage of the contemporary interest in all things Russian.[40] Purporting to narrate the history of Russia, Doré takes the reader from the beginnings of recorded history (which he indicates with

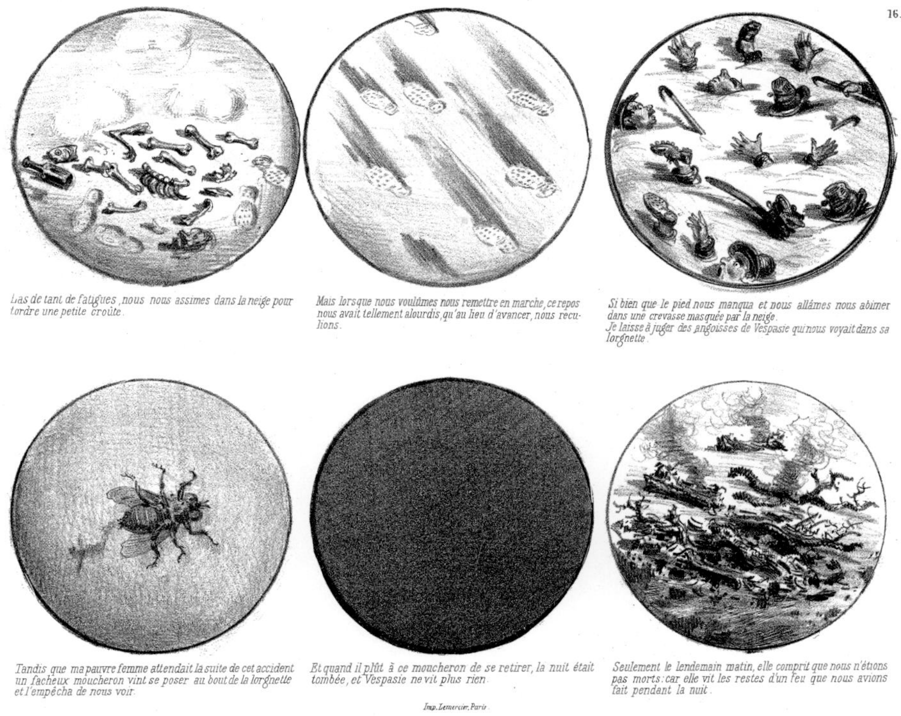

FIGURE 124
Gustave Doré, *(Dis)pleasures of a Pleasure Trip* [*Dés-agréments d'un voyage d'agrément*]. Paris: Aubert, 1851, pl. 16. Lithograph.

an all-black frame), to 1854 where he predicts French success in the Crimean War. His text is in the tradition of English humorous "histories" such as Gilbert Abbott à Beckett's *The Comic History of England*, which had illustrations by John Leech (1817–1764).[41] Doré's text follows closely actual events in the history of Russia, utilizing a complex word-image counterpoint where images comment on or even contradict the text. For example, a progressively more boring account of the origins of the Russian people is shown slowly being swallowed up by spilled ink, a brilliant graphic symbol of the medium overwhelming the message (fig. 125). No one earlier had so closely imbricated word and image.

Although much could, and should, be written on this masterpiece of comic art, Doré's formal innovations present its most striking aspect. By juxtaposing different styles of drawing, intercutting figural scenes with inanimate objects, and setting his frames against a counterpoint of textual commentary, Doré succeeded in defining a new form of comic art. He recounts the history of Ivan the Terrible "who stimulated Russian industry by immediately encouraging the branch of ironwork that he had himself studied extensively."[42] The page of accompanying images forms an ironic commentary, showing a variety of lethal weapons and instruments of torture that Ivan devised, identifying each as to its function (fig. 126): for the advancement of civilization, for the implementation of justice, as an aid in interrogation, and—presented as the pinnacle of civilization—the

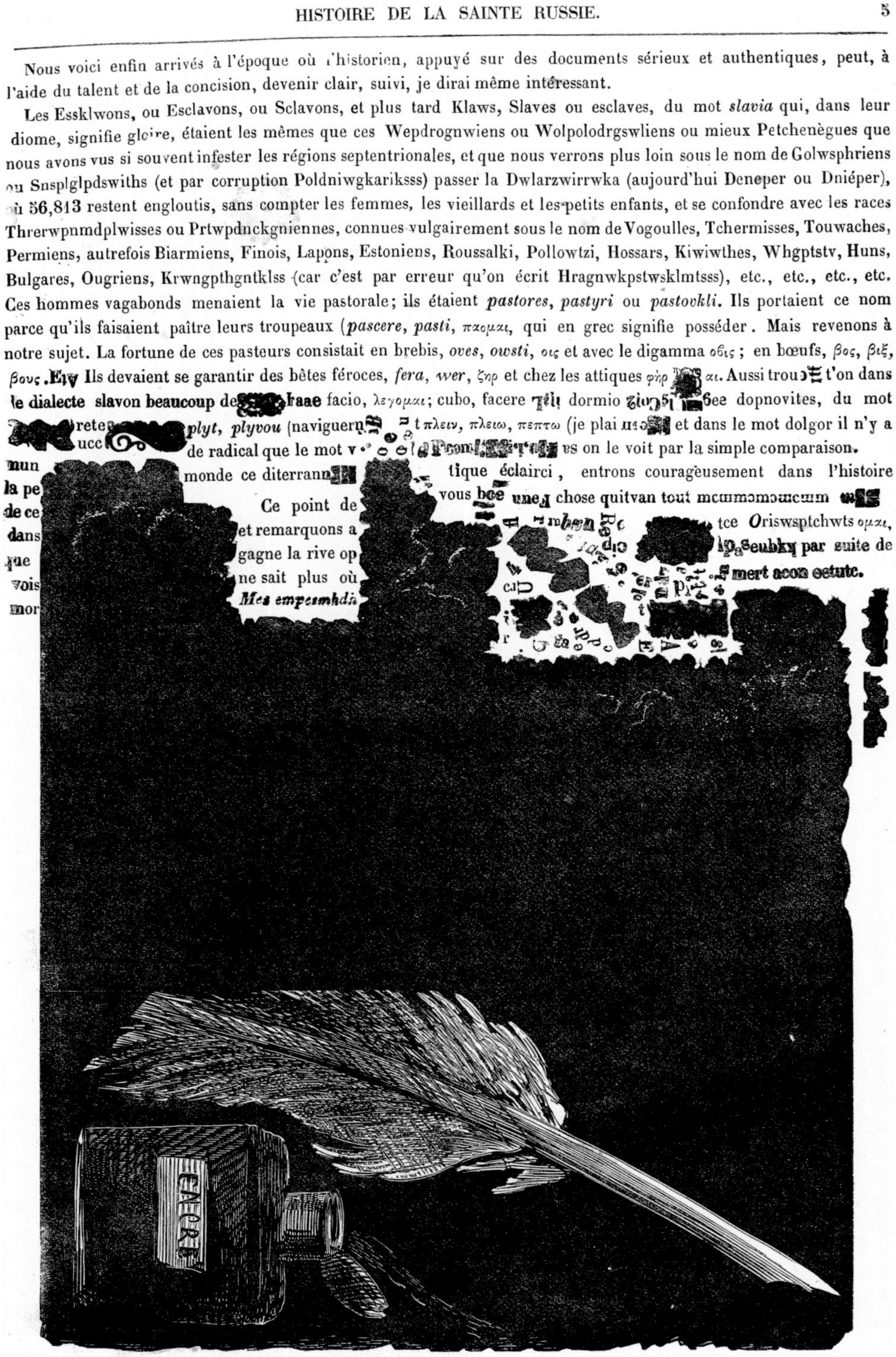
HISTOIRE DE LA SAINTE RUSSIE. 5

Nous voici enfin arrivés à l'époque où l'historien, appuyé sur des documents sérieux et authentiques, peut, à l'aide du talent et de la concision, devenir clair, suivi, je dirai même intéressant.

Les Essklwons, ou Esclavons, ou Sclavons, et plus tard Klaws, Slaves ou esclaves, du mot *slavia* qui, dans leur diome, signifie gloire, étaient les mêmes que ces Wepdrognwiens ou Wolpolodrgswliens ou mieux Petchenègues que nous avons vus si souvent infester les régions septentrionales, et que nous verrons plus loin sous le nom de Golwsphriens ou Snsplglpdswiths (et par corruption Poldniwgkariksss) passer la Dwlarzwirrwka (aujourd'hui Deneper ou Dniéper), où 56,813 restent engloutis, sans compter les femmes, les vieillards et les petits enfants, et se confondre avec les races Threrwpnmdplwisses ou Prtwpdnckgniennes, connues vulgairement sous le nom de Vogoulles, Tchermisses, Touwaches, Permiens, autrefois Biarmiens, Finois, Lapons, Estoniens, Roussalki, Pollowtzi, Hossars, Kiwiwthes, Whgptstv, Huns, Bulgares, Ougriens, Krwngpthgntklss (car c'est par erreur qu'on écrit Hragnwkpstwsklmtsss), etc., etc., etc., etc. Ces hommes vagabonds menaient la vie pastorale; ils étaient *pastores*, *pastyri* ou *pastovkli*. Ils portaient ce nom parce qu'ils faisaient paître leurs troupeaux (*pascere*, *pasti*, πασμαι, qui en grec signifie posséder. Mais revenons à notre sujet. La fortune de ces pasteurs consistait en brebis, *oves*, *owsti*, οις et avec le digamma οϜις ; en bœufs, βος, βιξ, βους. Ils devaient se garantir des bêtes féroces, *fera*, *wer*, ζηρ et chez les attiques φὴρ ... αι. Aussi trou... t'on dans le dialecte slavon beaucoup de ... facio, λεγομαι; cubo, facere ... dormio ... dopnovites, du mot ... *plyt*, *plyvou* (naviguer ... πλειν, πλειω, πεπτω (je plai ... et dans le mot dolgor il n'y a ... de radical que le mot ... on le voit par la simple comparaison. ... monde ce diterran ... tique éclairci, entrons courageusement dans l'histoire ... vous ... chose quitvan tout ...

Ce point de ... et remarquons a ... gagne la rive op ... ne sait plus où ... *Mes empesmhdh* ... tce Oriswsptchwts ... par suite de ... mert acon estute.

FIGURE 125

Gustave Doré, *Dramatic and Picturesque History of Holy Russia in Caricature* [*Histoire dramatique, pittoresque et caricaturale de la sainte Russie*]. Engr. Eugène-Noël Sotain. Paris: J. Bry aîné, 1854, pl. 5. Wood engraving.

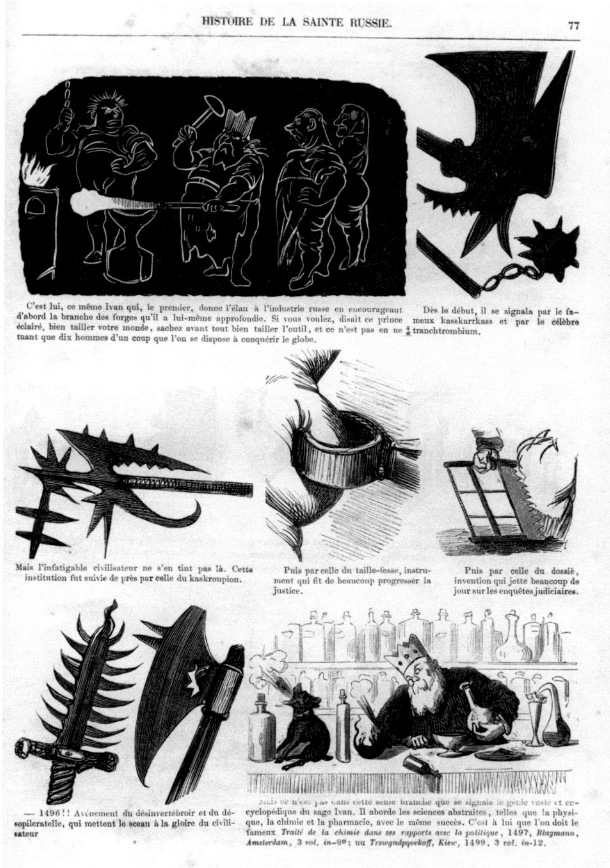

FIGURE 126
Gustave Doré, *Dramatic and Picturesque History of Holy Russia in Caricature* [*Histoire dramatique, pittoresque et caricaturale de la sainte Russie*]. Engr. Eugène-Noël Sotain. Paris: J. Bry aîné, 1854, pl. 77. Wood engraving.

backboner (*désinvertébroir*) and the obstruction clearer (*désopileratelle*). Doré's homage to Cham includes not only all-black frames, but also a page of all-white ones where, he tells us, the events are too boring to depict.[43] There are several citations from Grandville, whom he obviously admired.[44] The distinctive all-over contour style of the English artist Richard Doyle (1824–1883), who drew for *Punch*, also appears in several panels, often juxtaposed with other styles (fig. 127).[45] Doré also parodied the mid-century interest in folk art by drawing in the crude woodcut style of the *image d'Épinal*, as such works were called; he even adds a facetious legend *Drawn from the Collection of Popular Russian Prints (Facsimile)* (fig. 128)[46] He borrows from Töpffer, recreating the map in *Doctor Festus* of the fictitious land of Ginvernais (fig. 129), remaking it into a map of Russia (fig. 130) and giving each region a satirical name such as Brutaslaw (Uglyland) or Vanitéislaw (Vanityland).[47] Interestingly, the only occasions in *Holy Russia* on which Doré utilized the standard grid format were where it demands to be read as satire. On one page, for example, the uniformity of the frames and captions underscores the boring, repetitive nature

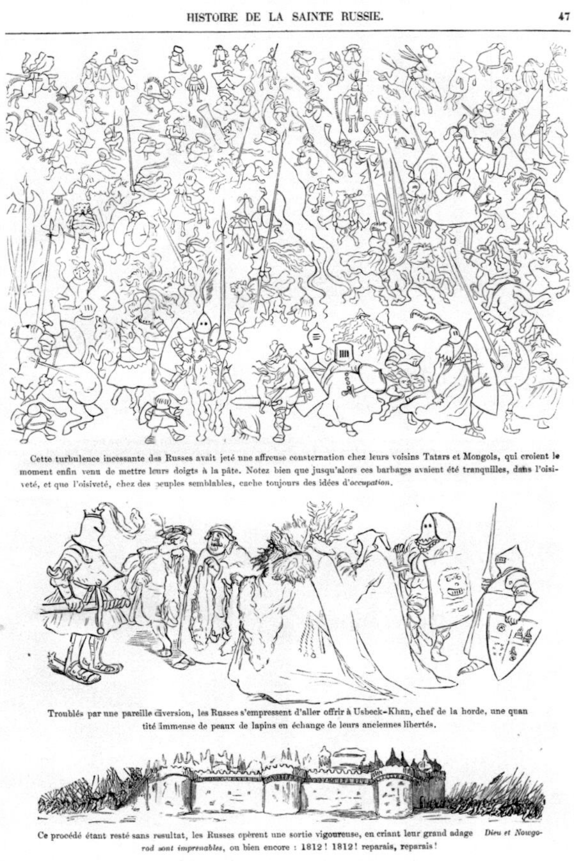

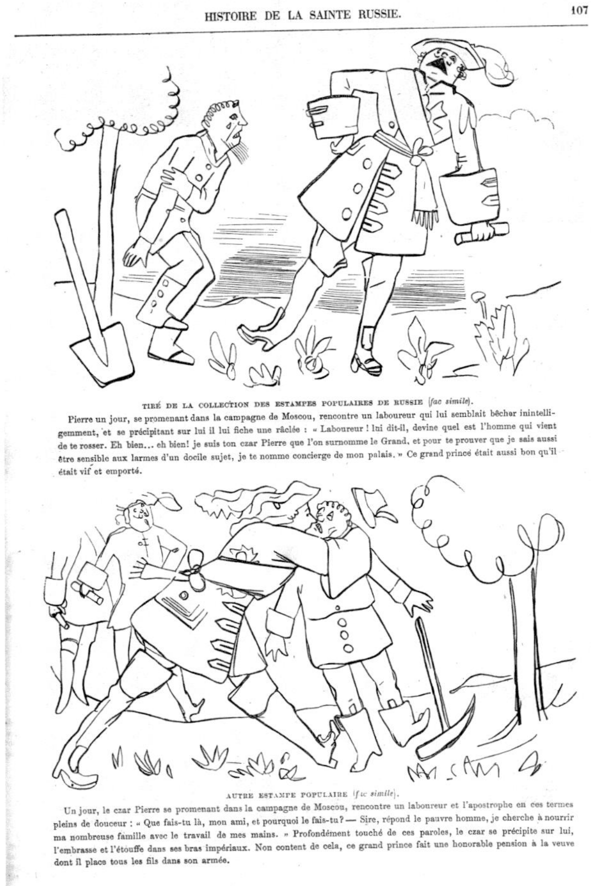

FIGURE 127
Gustave Doré, *Dramatic and Picturesque History of Holy Russia in Caricature* [*Histoire dramatique, pittoresque et caricaturale de la sainte Russie*]. Engr. Eugène-Noël Sotain. Paris: J. Bry aîné, 1854, pl. 47. Wood engraving.

FIGURE 128
Gustave Doré, *Dramatic and Picturesque History of Holy Russia in Caricature* [*Histoire dramatique, pittoresque et caricaturale de la sainte Russie*]. Engr. Eugène-Noël Sotain. Paris: J. Bry aîné, 1854, pl. 107. Wood engraving.

of the recitation of Russia's continual civil wars narrated by the text (fig. 131). The most dramatic pages in this album, however, appeared only in a small, more expensive edition. There the relentless black and white of the text and illustrations are shattered by two pages stained red to symbolize the violence and bloodshed of the czars (fig. 132).

With his *Dramatic and Picturesque History of Holy Russia in Caricature,* Doré created a tour de force of the graphic arts. It is a summation of the comic book art as invented by Rodolphe Töpffer and transformed by French artists in subsequent decades; its trajectory leads directly to the modern graphic novel.

SERIALIZED COMICS

A less dramatic contribution than Doré's that, nonetheless, had even more far-reaching consequences for the history of comics is their serialization in the periodical press. This was something that neither Töpffer nor the artists of the *Collection of Jabots* accomplished.

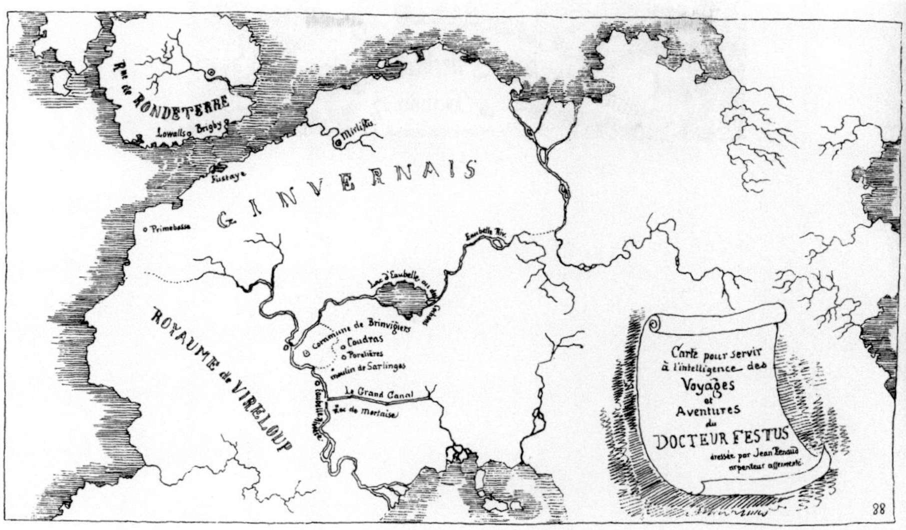

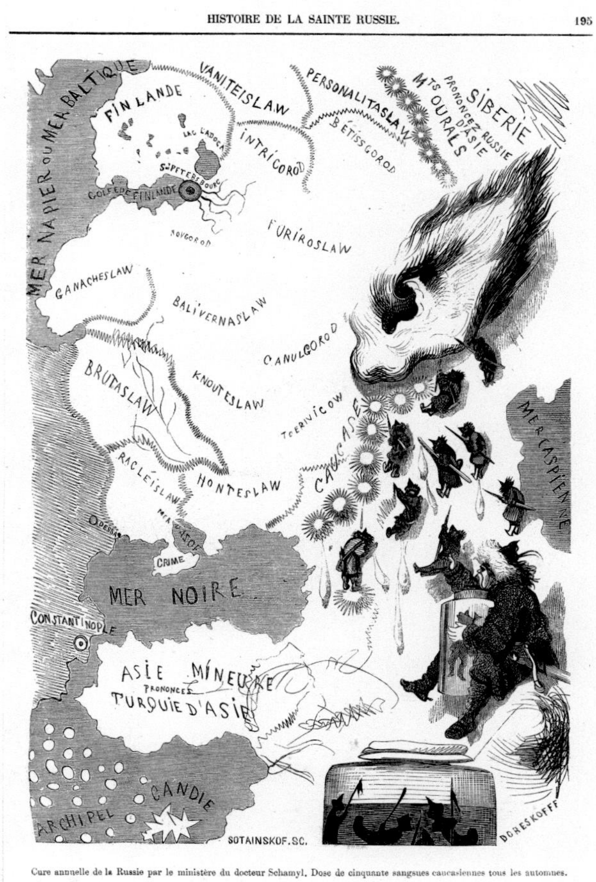

FIGURE 129
Rodolphe Töpffer, *Doctor Festus* [*Le docteur Festus*]. Paris: Abraham Cherbuliez, 1840, pl. 88. Lithograph.

FIGURE 130
Gustave Doré, *Dramatic and Picturesque History of Holy Russia in Caricature* [*Histoire dramatique, pittoresque et caricaturale de la sainte Russie*]. Engr. Eugène-Noël Sotain. Paris: J. Bry aîné, 1854, pl. 195. Wood engraving.

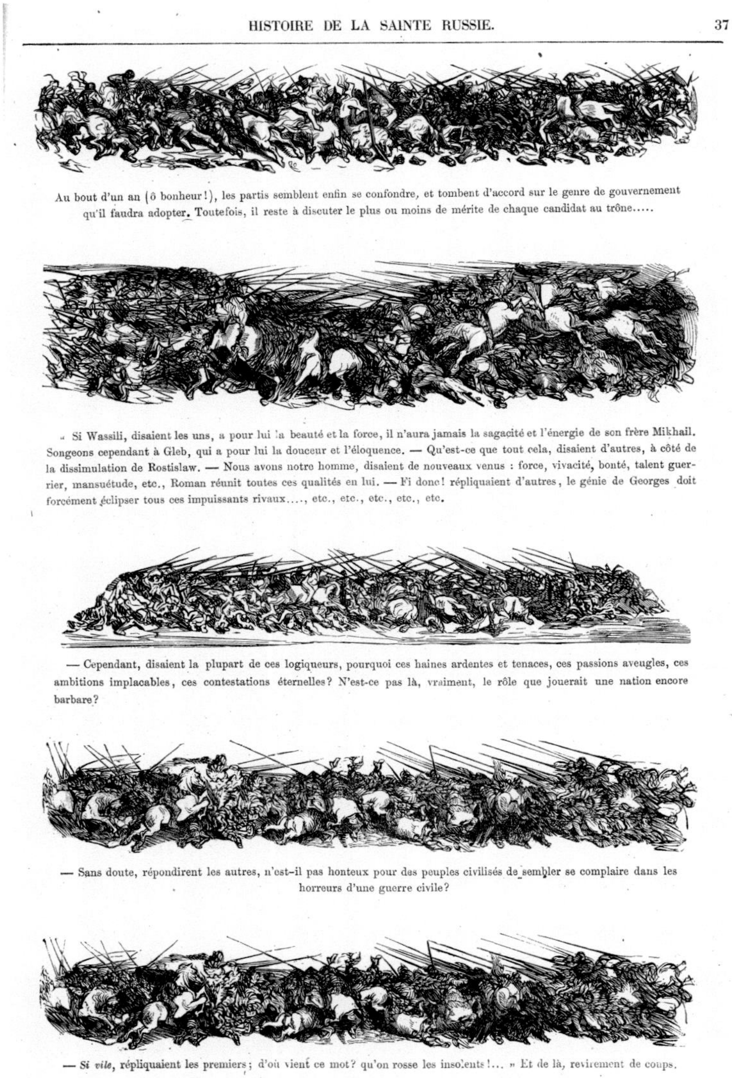

FIGURE 131
Gustave Doré, *Dramatic and Picturesque History of Holy Russia in Caricature* [*Histoire dramatique, pittoresque et caricaturale de la sainte Russie*], . Engr. Eugène-Noël Sotain. Paris: J. Bry aîné, 1854, pl. 37. Wood engraving.

OPPOSITE FIGURE 132
Gustave Doré, *Dramatic and Picturesque History of Holy Russia in Caricature* [*Histoire dramatique, pittoresque et caricaturale de la sainte Russie*]. Engr. Eugène-Noël Sotain. Paris: J. Bry aîné, 1854, pl. 89. Hand-colored wood engraving. Bibliothèque nationale de France.

Although *L'Illustration* reprinted Rodolphe Töpffer's *Story of Mr. Cryptogame* in 1845, and often published full-page comic strip stories, the first ongoing serial actually created for a periodical was by Nadar (Gaspard-Félix Tournachon, 1820–1910), whose *Public and Private Life of Mister Reactionary* (fig. 133) appeared in 1849 in eleven weekly installments in the *Revue comique à l'usage des gens sérieux*, "the comic revue for serious people."[48] Nadar's storyline follows the exploits of "Mossieu Réac," a hustler and con artist reminiscent of Daumier's Robert Macaire (see fig. 32). Daumier, however, had drawn a single image for each episode, while Nadar published a narrative strip along the bottom margin of the weekly journal, continuing his story in this format for several months and concluding

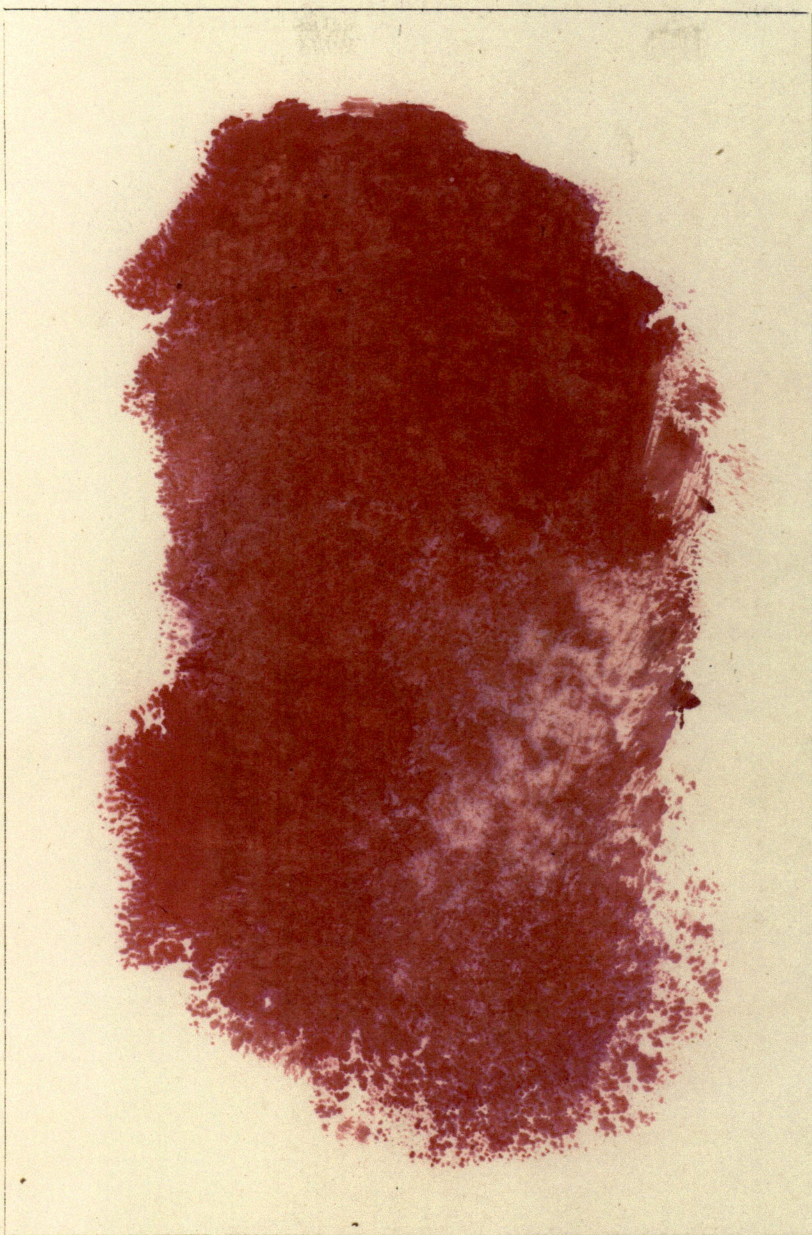

1542-1580. — Suite du règne d'Ivan-le-Terrible. Devant tant de crimes, clignons l'œil pour n'en rien voir que l'aspect général.

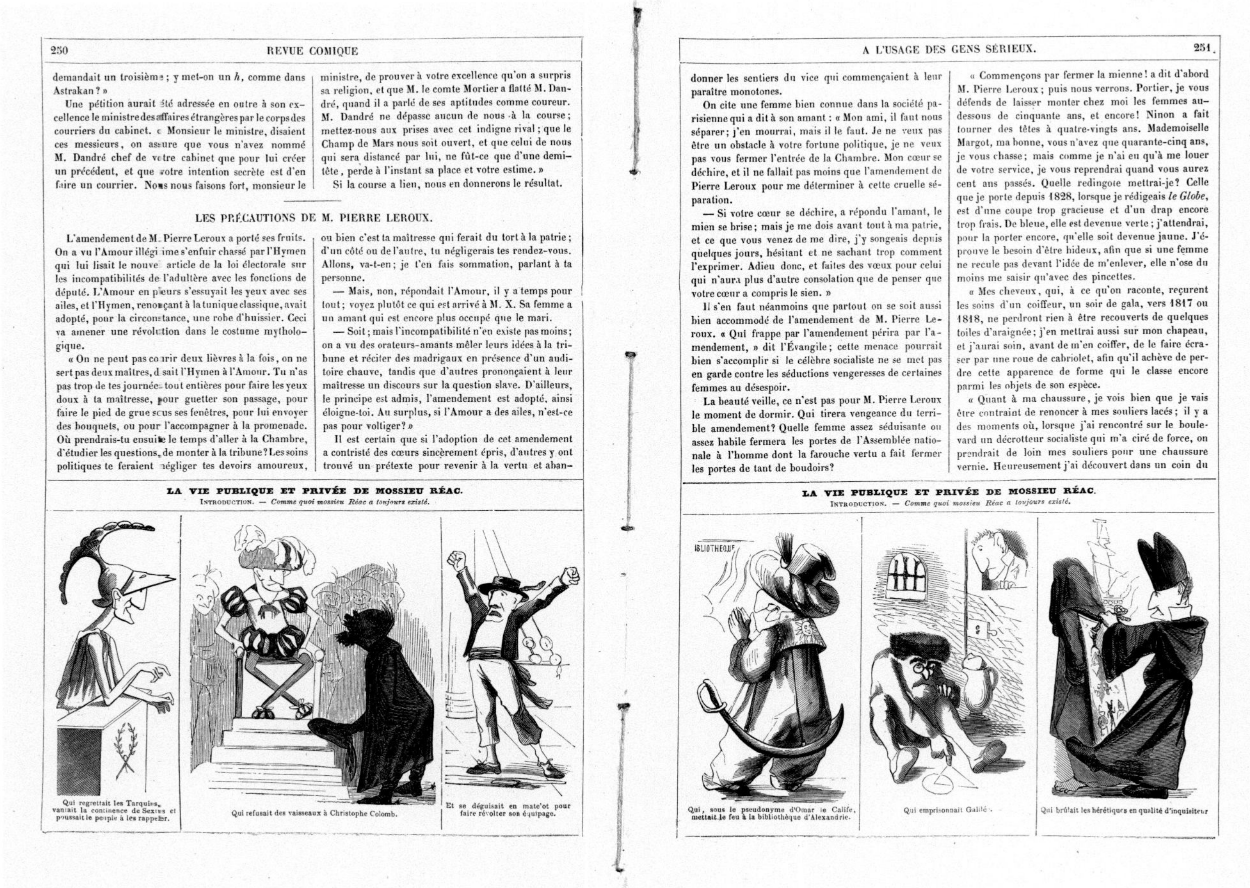

250 REVUE COMIQUE

demandait un troisième; y met-on un *h*, comme dans Astrakan? »

Une pétition aurait été adressée en outre à son excellence le ministre des affaires étrangères par le corps des courriers du cabinet. « Monsieur le ministre, disaient ces messieurs, on assure que vous n'avez nommé M. Dandré chef de votre cabinet que pour lui créer un précédent, et que votre intention secrète est d'en faire un courrier. Nous nous faisons fort, monsieur le ministre, de prouver à votre excellence qu'on a surpris sa religion, et que M. le comte Mortier a flatté M. Dandré, quand il a parlé de ses aptitudes comme coureur. M. Dandré ne dépasse aucun de nous à la course; mettez-nous aux prises avec cet indigne rival; que le Champ de Mars nous soit ouvert, et que celui de nous qui sera distancé par lui, ne fût-ce que d'une demi-tête, perde à l'instant sa place et votre estime. »

Si la course a lieu, nous en donnerons le résultat.

LES PRÉCAUTIONS DE M. PIERRE LEROUX.

L'amendement de M. Pierre Leroux a porté ses fruits. On a vu l'Amour illégitime s'enfuir chassé par l'Hymen qui lui lisait le nouvel article de la loi électorale sur les incompatibilités de l'adultère avec les fonctions de député. L'Amour en pleurs s'essuyait les yeux avec ses ailes, et l'Hymen, renonçant à la tunique classique, avait adopté, pour la circonstance, une robe d'huissier. Ceci va amener une révolution dans le costume mythologique.

« On ne peut pas courir deux lièvres à la fois, on ne sert pas deux maîtres, disait l'Hymen à l'Amour. Tu n'as pas trop de tes journées tout entières pour faire les yeux doux à ta maîtresse, pour guetter son passage, pour faire le pied de grue sous ses fenêtres, pour lui envoyer des bouquets, ou pour l'accompagner à la promenade. Où prendrais-tu ensuite le temps d'aller à la Chambre, d'étudier les questions, de monter à la tribune? Les soins politiques te feraient négliger tes devoirs amoureux, ou bien c'est ta maîtresse qui ferait du tort à la patrie; d'un côté ou de l'autre, tu négligerais tes rendez-vous. Allons, va-t-en; je t'en fais sommation, parlant à ta personne.

— Mais, non, répondait l'Amour, il y a temps pour tout; voyez plutôt ce qui est arrivé à M. X. Sa femme a un amant qui est encore plus occupé que le mari.

— Soit; mais l'incompatibilité n'en existe pas moins; on a vu des orateurs-amants mêler leurs idées à la tribune et réciter des madrigaux en présence d'un auditoire chauve, tandis que d'autres prononçaient à leur maîtresse un discours sur la question slave. D'ailleurs, le principe est admis, l'amendement est adopté, ainsi éloigne-toi. Au surplus, si l'Amour a des ailes, n'est-ce pas pour voltiger? »

Il est certain que si l'adoption de cet amendement a contristé des cœurs sincèrement épris, d'autres y ont trouvé un prétexte pour revenir à la vertu et abandonner les sentiers du vice qui commençaient à leur paraître monotones.

On cite une femme bien connue dans la société parisienne qui a dit à son amant: « Mon ami, il faut nous séparer; j'en mourrai, mais il le faut. Je ne veux pas être un obstacle à votre fortune politique, je ne veux pas vous fermer l'entrée de la Chambre. Mon cœur se déchire, et il ne fallait pas moins que l'amendement de Pierre Leroux pour me déterminer à cette cruelle séparation.

— Si votre cœur se déchire, a répondu l'amant, le mien se brise; mais je me dois avant tout à ma patrie, et ce que vous venez de me dire, j'y songeais depuis quelques jours, hésitant et ne sachant trop comment l'exprimer. Adieu donc, et faites des vœux pour celui qui n'aura plus d'autre consolation que de penser que votre cœur a compris le sien. »

Il s'en faut néanmoins que partout on se soit aussi bien accommodé de l'amendement de M. Pierre Leroux. « Qui frappe par l'amendement périra par l'amendement, » dit l'Évangile; cette menace pourrait bien s'accomplir si le célèbre socialiste ne se met pas en garde contre les séductions vengeresses de certaines femmes au désespoir.

La beauté veille, ce n'est pas pour M. Pierre Leroux le moment de dormir. Qui tirera vengeance du terrible amendement? Quelle femme assez séduisante ou assez habile fermera les portes de l'Assemblée nationale à l'homme dont la farouche vertu a fait fermer les portes de tant de boudoirs?

« Commençons par fermer la mienne! a dit d'abord M. Pierre Leroux; puis nous verrons. Portier, je vous défends de laisser monter chez moi les femmes au-dessous de cinquante ans, et encore! Ninon a fait tourner des têtes à quatre-vingts ans. Mademoiselle Margot, ma bonne, vous n'avez que quarante-cinq ans, je vous chasse; mais comme je n'ai eu qu'à me louer de votre service, je vous reprendrai quand vous aurez cent ans passés. Quelle redingote mettrai-je? Celle que je porte depuis 1828, lorsque je rédigeais *le Globe*, est d'une coupe trop gracieuse et d'un drap encore trop frais. De bleue, elle est devenue verte; j'attendrai, pour la porter encore, qu'elle soit devenue jaune. J'éprouve le besoin d'être hideux, afin que si une femme ne recule pas devant l'idée de m'enlever, elle n'ose du moins me saisir qu'avec des pincettes.

« Mes cheveux, qui, à ce qu'on raconte, reçurent les soins d'un coiffeur, un soir de gala, vers 1817 ou 1818, ne perdront rien à être recouverts de quelques toiles d'araignée; j'en mettrai aussi sur mon chapeau, et j'aurai soin, avant de m'en coiffer, de le faire écraser par une roue de cabriolet, afin qu'il achève de perdre cette apparence de forme qui le classe encore parmi les objets de son espèce.

« Quant à ma chaussure, je vois bien que je vais être contraint de renoncer à mes souliers lacés; il y a des moments où, lorsque j'ai rencontré sur le boulevard un décrotteur socialiste qui m'a ciré de force, on prendrait de loin mes souliers pour une chaussure vernie. Heureusement j'ai découvert dans un coin du

LA VIE PUBLIQUE ET PRIVÉE DE MOSSIEU RÉAC.
INTRODUCTION. — *Comme quoi mossieu Réac a toujours existé.*

Qui regrettait les Tarquins, vantait la continence de Sextus et poussait le peuple à les rappeler.

Qui refusait des vaisseaux à Christophe Colomb.

Et se déguisait en matelot pour faire révolter son équipage.

LA VIE PUBLIQUE ET PRIVÉE DE MOSSIEU RÉAC.
INTRODUCTION. — *Comme quoi mossieu Réac a toujours existé.*

Qui, sous le pseudonyme d'Omar le Calife, mettait le feu à la bibliothèque d'Alexandrie.

Qui emprisonnait Galilée.

Qui brûlait les hérétiques en qualité d'inquisiteur

251

FIGURE 133

Nadar, *Introduction. The Public and Private Life of Mister Reactionary* [*Introduction. La Vie publique et privée de mossieu Réac*], *Revue comique à l'usage des gens sérieux*, March 3, 1849, 250–51. Wood engraving.

with the con man conned into marriage. While *Mister Reactionary* adds little to the formal language of comics, it opened an important venue for its further development into the ubiquitous serialized strips of modern newspapers.

Unfortunately for the history of comics after its brilliant beginnings, Rodolph Töpffer died prematurely and his French followers soon abandoned the medium. Their work was ill paid and undervalued, so it is no surprise that they so often took their talents elsewhere. Cham went back to creating single-sheet images, Nadar turned to the nascent medium of photography, and Gustave Doré, who soon after the publication of *Holy Russia* enjoyed a major success with his illustrations to Rabelais, decided to devote himself to the more prestigious media of painting, sculpture, and book illustration. After this initial period of innovation, comic books actually became more conservative, and the next truly original comic artist would be the German Wilhelm Busch (1832–1908), who began publishing in the 1860s. And yet each of these early comic book artists—Rodolphe Töpffer, Cham, Nadar, and Gustave Doré—contributed to the establishment of the language of comics: from Töpffer who invented "literature in prints," to Cham, who created many of its familiar visual conventions, to Nadar, who gave us our weekly comics, to Doré, who, by deliberately transgressing all the newly established comic book traditions, established it as a subversive medium that obeys no rules, a definition still applicable in the twenty-first century.

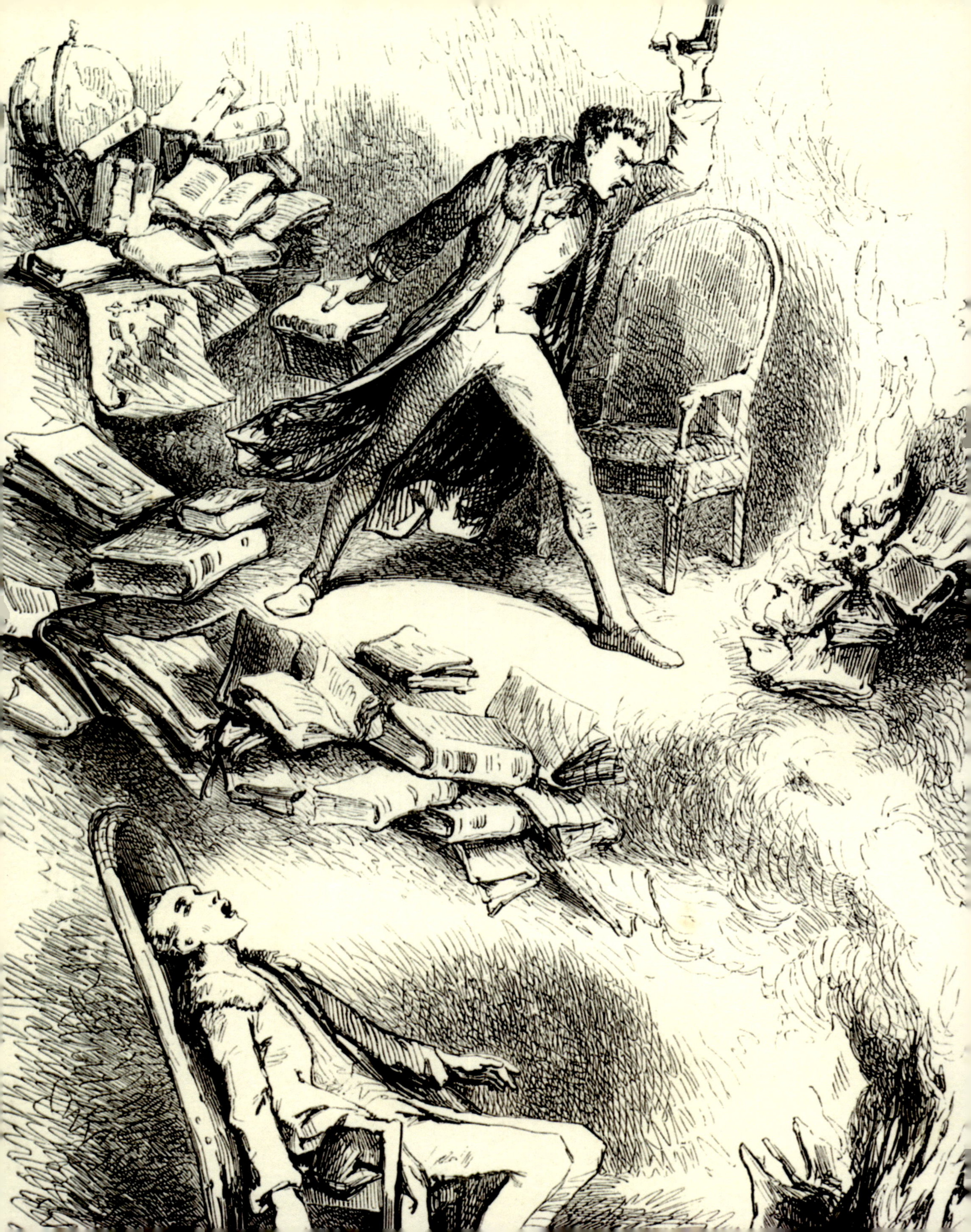

4 Paths Forgotten, Calls Unheard

Pictures in Stories

When we talk about sequential visual narration, if we are not discussing film we usually mean comics. But this is only one form of visual narration. The great age of book illustration in the nineteenth century saw the creation of more illustrated volumes than ever before; yet while the genres of illustrated print culture discussed so far—caricature, the illustrated press, and comics—have survived, and in the case of comics have even gained in importance, by 1900 illustrated books had largely disappeared, surviving mainly as luxury goods or as children's literature. As a result, the prevalent conception today is that comic strips and the modern graphic novel are the inevitable outcome of nineteenth-century illustration. There were, however, other contemporaneous developments. I take as my model the thoughts of the novelist Milan Kundera (b. 1929), who wrote: "I often hear it said that the novel has exhausted all its possibilities. I have the opposite impression: during its 400-year history, the novel has missed many of its possibilities: it has left many great opportunities unexplored, many paths forgotten, calls unheard."[1] In the course of its development, visual narration has also left many paths forgotten and calls unheard. In this chapter I propose to explore several of them.

STORIES IN PICTURES TO PICTURES IN STORIES

Comics are perhaps the illustrated books most familiar to us today; the preceding chapter described the works and influence of the Swiss schoolmaster Rodolphe Töpffer, who published his first, *The Story of Mr. Jabot*, in 1833 and his last, *Mr. Cryptogame*, in 1845. Töpffer called his creations "literature in prints," stories in pictures; in this chapter, we will reverse that word-image relationship and look at the contemporaneous proliferation of illustrated books, pictures in stories.[2]

While many artists both wrote the text and drew the images for their comic albums, for book illustration as a whole it was more common for artists to collaborate with writers rather than to produce the entire work alone. But collaboration has always been a problematic concept in art history, which privileges originality above all else; since here the images are illustrative of someone else's text, illustrated books have received little attention. Nonetheless, we should consider the possibility that the provision of a text might

DETAIL OF FIGURE 156

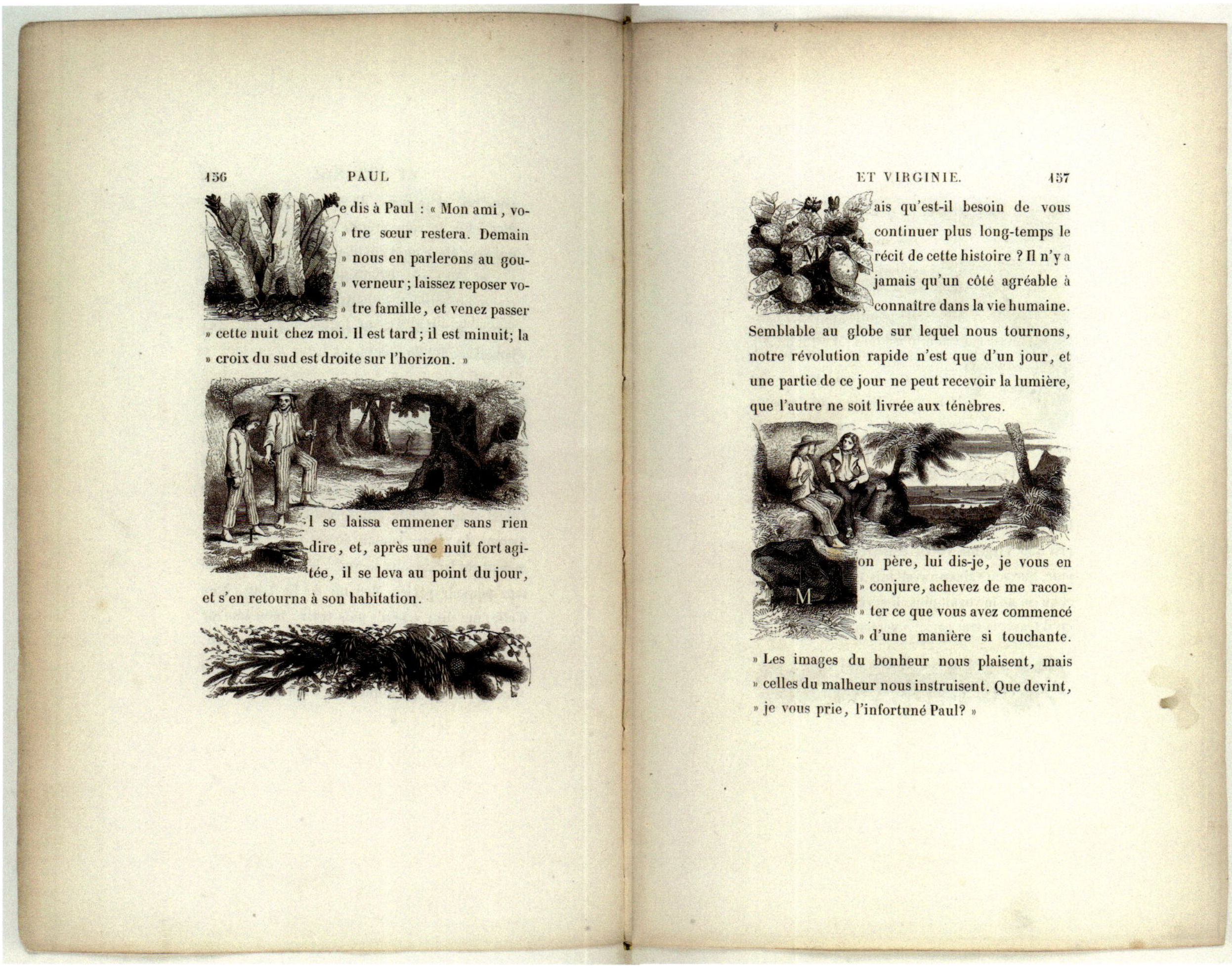

156 PAUL

e dis à Paul : « Mon ami, vo-
» tre sœur restera. Demain
» nous en parlerons au gou-
» verneur ; laissez reposer vo-
» tre famille, et venez passer
» cette nuit chez moi. Il est tard ; il est minuit; la
» croix du sud est droite sur l'horizon. »

I se laissa emmener sans rien
dire, et, après une nuit fort agi-
tée, il se leva au point du jour,
et s'en retourna à son habitation.

ET VIRGINIE. 157

ais qu'est-il besoin de vous
continuer plus long-temps le
récit de cette histoire ? Il n'y a
jamais qu'un côté agréable à
connaître dans la vie humaine.
Semblable au globe sur lequel nous tournons,
notre révolution rapide n'est que d'un jour, et
une partie de ce jour ne peut recevoir la lumière,
que l'autre ne soit livrée aux ténèbres.

on père, lui dis-je, je vous en
» conjure, achevez de me racon-
» ter ce que vous avez commencé
» d'une manière si touchante.
» Les images du bonheur nous plaisent, mais
» celles du malheur nous instruisent. Que devint,
» je vous prie, l'infortuné Paul? »

134A–B
Vignettes by François-Louis Français, Tony Johannot, Louis-Charles-Auguste Steinheil. Engr. Robert Hart, Orrin Smith, Charles Gray, Hippolyte Lavoignat, from Bernardin de Saint-Pierre, *Paul and Virginia* [*Paul et Virginie*] (1788). Paris: Curmer, 1838, 156–57. Wood engravings. Bibliothèque nationale de France.

well be liberating to artists and might allow them a greater degree of creative freedom. It might even result in a different medium altogether. By exploring nineteenth-century varieties of word-image collaboration, we can better understand the richness of the century's visual culture, whole continents of which are unfamiliar to us today.

Probably the best-known example of an original collaboration between an artist and writer in the comic genre is the 1812 English publication, *The Tour of Doctor Syntax, in Search of the Picturesque*, with images by Thomas Rowlandson and text by William Combe (see figs. 99 and 100). This was discussed in the previous chapter as an important precursor of modern comics, but here I would like to consider it in a different context—as an illustrated book. In Combe's own account of his collaboration with Rowlandson, he stated that each month he received from the artist a colored sketch, around which he composed the chapter, "the Artist and the Writer having no personal communication with, or knowledge of each other."[3] Aside from the sheer brilliance of Rowlandson's images and general hilarity of Combe's text, what is most significant is that Rowlandson created his images *before* William Combe wrote the text; in other words, the writer illustrated the artist, the reverse of what we today think is the normal order of precedence. Rowlandson's innovation was repeated often enough in subsequent decades that there is a nineteenth-century subgenre of works conceived by artists and illustrated by writers.

The innovative precedence of artist over writer in *Doctor Syntax* was offset by its traditional organization, in which a full-page, hand-colored etching by Rowlandson accompanied each canto by Combe. In this sense, *Doctor Syntax* is typical of traditional book illustration, which rarely had more than one illustration per chapter and often far fewer. The primary reason for this was technological. Because in the early part of the nineteenth century the same reproductive processes could not be used for both word and image, text had to be printed in letterpress and images printed separately in etching, engraving, or lithography. The same technological problem retarded the development of the illustrated press, as we saw in chapter 2. Text and images in books were bound together subsequent to printing, as was the case with *Doctor Syntax.* Because of this extra labor and expense, illustrations were sparse and books were costly. The technological processes that developed toward midcentury brought illustrated books into an economic range more feasible for an increasingly literate public. As we have seen, wood engraving allowed word and image to be combined on a single page, and lithography produced images cheaply and quickly; both encouraged the proliferation of illustrations. Periodicals, comics, and illustrated books, although vastly different in content, all utilized the same reproductive processes. The result was a great age of illustrated publications of all types, peaking in France in the 1840s, in England in the 1860s, and in the United States by the late nineteenth century. It is no coincidence that both comic books and illustrated books of all kinds had their golden age during this period.

Mass-produced illustrated books were, at first, quite similar to earlier publications. *Doctor Syntax,* for example, has only twenty-nine illustrations in twenty-six cantos, essentially one illustration per chapter. Until the nineteenth century, books—and therefore their illustrations—were small because paper was handmade and expensive. As the century wore on, however, paper became cheaper and books became larger. When illustrations proved a major commercial attraction, publishers began increasing their number and using that as a point of advertisement. The 1838 Curmer edition of Bernardin de Saint-Pierre's novel *Paul and Virginia* (1788) advertised thirty full-page illustrations and four hundred vignettes, smaller images inserted within the text. Sometimes there were so many vignettes interrupting the text that it begins to look like a modern graphic novel: on one page we see the forest and Paul walking through it, and, on the facing page, the narrator who is recounting the story (figs. 134a–b). This page layout had not been possible with metal-plate engraving or lithography, where text and images had to be printed separately. This new feature delighted readers, although it was condemned by literary critics and even by writers who resented the interruptions to their text.

The increase in the number of illustrations was not only quantitative, however. It affected the conceptual organization of the artist's task, which, in turn, changed the way the reader-viewer experienced the work. The earlier practice of book illustration had much in common with history painting, where the artist chose to depict an event that in some way presented the crux of the situation: a moment of high drama, a crossroads of action, a moral decision. This can readily be seen in Anne-Louis Girodet's illustration for Racine's play *Andromache*, where he depicts the moment when Pyrrhus, who has captured Andromache's son Astyanax, offers her a choice: her son will live if she abandons her fidelity to her dead husband Hector and marries him; otherwise her son will die (fig. 135). Girodet has depicted Andromache at this crucial moment of moral decision. Girodet was a renowned history painter, and in this image the rectilinear frame, minimal setting, dramatic moment, and highly finished execution are all characteristic of his work and of Neoclassical art in general. Until the modern period, most European painting, or at least its most prestigious category of history painting, was basically illustrative of a text, whether biblical, mythological, or historical; artists provided drawings to be engraved for luxury editions of these texts. And so, as proposed in chapter 3, it should not be surprising that early illustrated comic books such as *Doctor Syntax* adopted a narrative strategy of depicting pivotal moments that was not so different from that of painted historical cycles like, as noted in the previous chapter, Peter Paul Rubens's narrative cycle of the life of Marie de' Medici, or even Michelangelo's Sistine Chapel ceiling. The focus on key moments of high drama was the lingua franca of sequential narration in all media. But—and here is the crux of the nineteenth-century transformation—what happens to that concept if the artist is asked to produce not twenty-four images, as in Rubens's cycle, but several hundred? If we look to the work of Töpffer to answer that question, we will find a narrative strategy that has become so familiar to us in both comic books and cinema that it seems transparently "natural." Since Töpffer told his stories almost entirely through images, with minimal text, his frames are figure-centered like a film storyboard and move the narrative along rapidly, based on actions and events. The French comic artists who followed Töpffer made many formal innovations and expanded his semiotic vocabulary into the formal language familiar to us today in comics. Gustave Doré's invention of speed lines to indicate velocity (see fig. 120) and Cham's close-ups showing the protagonist's point of view (see fig. 115) are just two of the innovations by comic artists that departed from the standard narrative strategies that had governed both illustration and history painting. There were, however, other visual narrative strategies that played no role

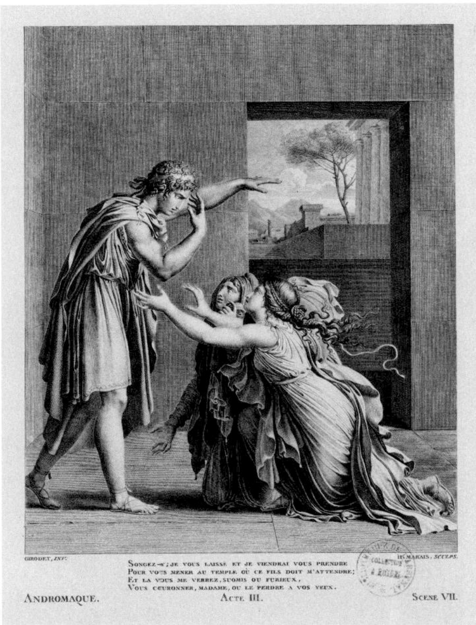

FIGURE 135
Anne-Louis Girodet, *Andromache [Andromaque], Act III, Scene VII.* Engr. Henri Marais, from Jean Racine, *Oeuvres de Jean Racine.* Paris: P. Didot l'aîné, an IX-1801. Bibliothèque nationale de France.

in early comics but were invented instead by illustrators of midcentury novels, and these inventions also led to differences in our experience of the work.

Jean Gigoux (1806–1894), a painter as well as a major nineteenth-century illustrator, recounted in his autobiographical collection of essays how in 1835 (the year that Töpffer's first comic album was sold) he was approached by the publishers Jacques-Julien Dubochet and Alexandre Paulin, who would later found the periodical *L'Illustration.* They wanted to bring out a new edition of Alain-René Lesage's classic novel *Gil Blas* (1715–1735): "One day I was asked to do a hundred vignettes for a new edition of this marvelous book. I swear I nearly had an attack of terror. It seemed to me that I would never find a hundred subjects to illustrate. Nonetheless, I did it. Some days later, the publishers asked me to do three hundred more. Then I had to reread the book and to draw more illustrations as I went along. The following week, the publishers, recognizing the pleasure that these vignettes gave to subscribers, asked me for two hundred additional drawings. In all I made six hundred, and I think that I would have been able to continue indefinitely."[4] Clearly the experience resulted in a conceptual shift in Gigoux's attitude toward illustration, from the time when he fears that he "would never find a hundred subjects to illustrate"—and here he was clearly thinking of illustration in the traditional sense of depicting major events in the text—to when, after six hundred, he feels that he "would have been able to continue indefinitely." The importance of this shift should not be underestimated: in the process, Gigoux reconceptualized his previous practice of visual narration. Instead of choosing a single image to represent a book, or a chapter of a book, or even an event in a chapter of a book, as Rowlandson and earlier artists had done, Gigoux established a parallel narrative in which the story unfolds through both word and image. While, by 1835, Töpffer was already embarked on a similar endeavor, there are differences that can be seen as paradigmatic of paths forgotten. Although Töpffer's drawing style was caricature, his stories are told in a straightforward illustrational manner with a minimum of text accompanying each frame. His images lead the narrative, and that narrative unfolds through the agency of the figures depicted. Gigoux, and all the artists who provided thousands of drawings for midcentury illustrated novels, tell the stories with a much greater variety of narrative strategies. Earlier book illustration, as well as Töpffer's comic albums, preserved the rectilinear frame around the image, subsuming it into the tradition of a history painting translated into graphic form. This parallel to painting is also evident in the preservation of the integrity of this rectilinear image, into which only English caricaturists allowed text to intrude. Midcentury illustrators discarded these conventions, often creating vignettes that fade out at the edges like dream imagery.[5] Since the text was bearing the major burden of the narrative, these artists could—and did—respond with flights of imagination, using the writer's words as a point of departure. In addition, the chameleonlike nature of illustration, through which the artist adapts his style to the task at hand instead of always working in his own signature style (as Töpffer did), might also have encouraged a variety

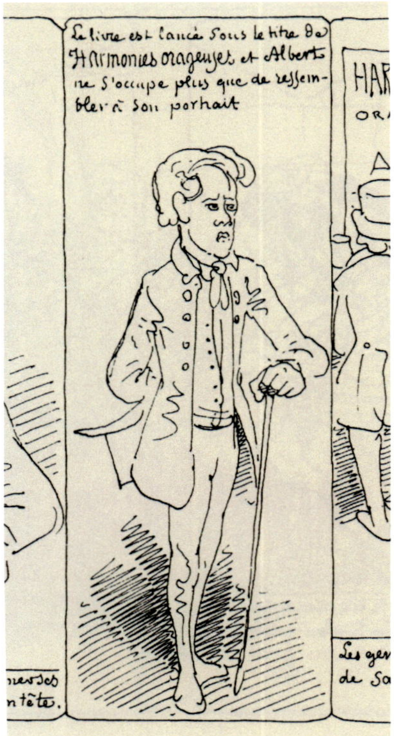

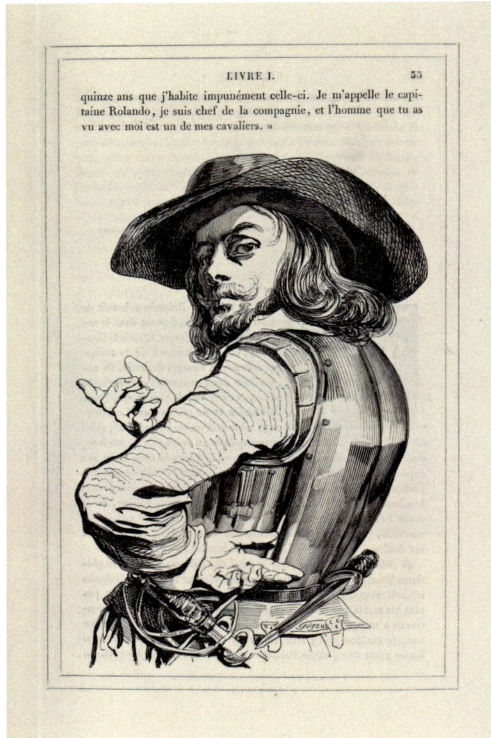
LIVRE I. 55

quinze ans que j'habite impunément celle-ci. Je m'appelle le capitaine Rolando, je suis chef de la compagnie, et l'homme que tu as vu avec moi est un de mes cavaliers. »

FIGURE 136
Rodolphe Töpffer, *Albert*, from *The Story of Albert by Simon of Nantua* [*Histoire d'Albert par Simon de Nantua*]. Geneva, 1845, pl. 13. 1845. Lithograph.

FIGURE 137
Jean Gigoux, *I Am Called Captain Rolando*, from Alain-René Lesage, *The Story of Gil Blas of Santillana* (1715–1735) [*Je m'appelle le capitaine Rolando / Histoire de Gil Blas de Santillane*]. Paris: Paulin, 1835, 35. Wood engraving. Bibliothèque nationale de France.

of means. And so, instead of focusing on the limitations of the artist's working with a writer's text—that is, on the negative aspect of the collaboration—it might be more valuable to understand what advantages an artist might derive from this process.

The great print historian Henri Béraldi (1849–1935) called the 1835 edition of *Gil Blas* "one of the five or six principal illustrated books of the nineteenth century," because it was the first to bring wood engraving into common use, enabling numerous illustrations at a price cheaper than etching or metal-plate engraving.[6] Many of Gigoux's images are typical of book illustration, although uncommon in Töpffer's new medium of comic books: portraiture of the main characters, for example, is something that Töpffer's drawing style, which was more caricature than illustration, scarcely could allow. The drawing style of caricature inevitably distances the viewer from the image, like an editorial comment, and prompts the response of ridicule—and indeed, in Töpffer's *Story of Albert*, the protagonist, whom we see trying to look like his formal portrait, is very much an object of ridicule (fig. 136). Illustrational images offer a wider latitude for the reader's response: we laugh at Albert, but Gigoux's Captain Rolando in *Gil Blas* inspires fear or admiration, even curiosity about his dress and weapons, which Gigoux attentively depicts (fig. 137). Illustrators can also present pictorial allusions that would be disruptive in a comic album. For example, when, in *Gil Blas*, Don Raphael says *We Resembled, as Homer Might Have Said,*

FIGURE 138

Jean Gigoux, *We Resembled, as Homer Might Have Said, Two Birds of Prey*, from Alain-René Lesage, *The Story of Gil Blas of Santillana* (1715–1735) [*Nous ressemblions, comme auroit dit Homère, à deux milans / Histoire de Gil Blas de Santillane*]. Paris: Paulin, 1835, 394. Wood engraving. Bibliothèque nationale de France.

FIGURE 139

Jean Gigoux, *I Took the Almanza Road and from There Continued on My Way, Going from City to City until I Reached Granada*, from Alain-René Lesage, *The Story of Gil Blas of Santillana* (1715–1735) [*Je pris le chemin d'Almanza, d'où, poursuivant ma route, j'allai de ville en ville jusqu'à celle de Grenade / Histoire de Gil Blas de Santillane*]. Paris: Paulin, 1835, 497. Wood engraving. Bibliothèque nationale de France.

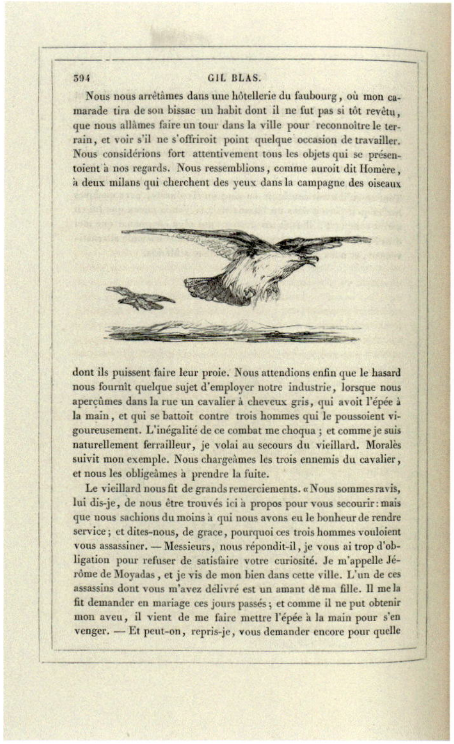

594 GIL BLAS.

Nous nous arrêtâmes dans une hôtellerie du faubourg, où mon camarade tira de son bissac un habit dont il ne fut pas si tôt revêtu, que nous allâmes faire un tour dans la ville pour reconnoître le terrain, et voir s'il ne s'offriroit point quelque occasion de travailler. Nous considérions fort attentivement tous les objets qui se présentoient à nos regards. Nous ressemblions, comme auroit dit Homère, à deux milans qui cherchent des yeux dans la campagne des oiseaux dont ils puissent faire leur proie. Nous attendions enfin que le hasard nous fournît quelque sujet d'employer notre industrie, lorsque nous aperçûmes dans la rue un cavalier à cheveux gris, qui avoit l'épée à la main, et qui se battoit contre trois hommes qui le poussoient vigoureusement. L'inégalité de ce combat me choqua ; et comme je suis naturellement ferrailleur, je volai au secours du vieillard. Moralès suivit mon exemple. Nous chargeâmes les trois ennemis du cavalier, et nous les obligeâmes à prendre la fuite.

Le vieillard nous fit de grands remerciements. « Nous sommes ravis, lui dis-je, de nous être trouvés ici à propos pour vous secourir : mais que nous sachions du moins à qui nous avons eu le bonheur de rendre service ; et dites-nous, de grace, pourquoi ces trois hommes vouloient vous assassiner. — Messieurs, nous répondit-il, je vous ai trop d'obligation pour refuser de satisfaire votre curiosité. Je m'appelle Jérôme de Moyadas, et je vis de mon bien dans cette ville. L'un de ces assassins dont vous m'avez délivré est un amant de ma fille. Il me la fit demander en mariage ces jours passés ; et comme il ne put obtenir mon aveu, il vient de me faire mettre l'épée à la main pour s'en venger. — Et peut-on, repris-je, vous demander encore pour quelle

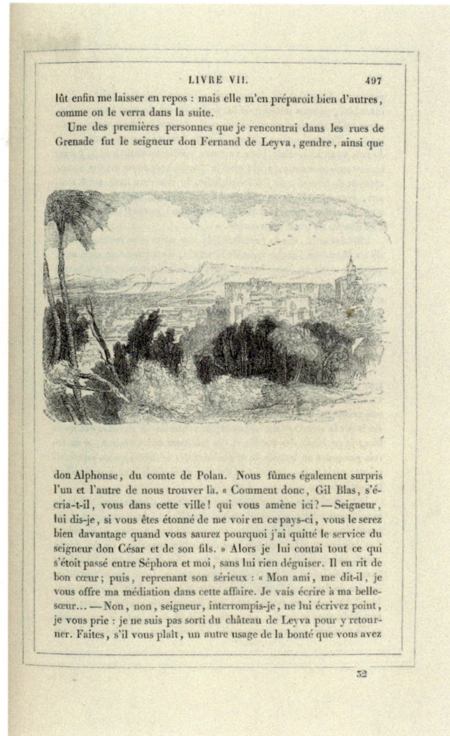

LIVRE VII. 497

lût enfin me laisser en repos : mais elle m'en préparoit bien d'autres, comme on le verra dans la suite.

Une des premières personnes que je rencontrai dans les rues de Grenade fut le seigneur don Fernand de Leyva, gendre, ainsi que don Alphonse, du comte de Polan. Nous fûmes également surpris l'un et l'autre de nous trouver là. « Comment donc, Gil Blas, s'écria-t-il, vous dans cette ville ! qui vous amène ici ? — Seigneur, lui dis-je, si vous êtes étonné de me voir en ce pays-ci, vous le serez bien davantage quand vous saurez pourquoi j'ai quitté le service du seigneur don César et de son fils. » Alors je lui contai tout ce qui s'étoit passé entre Séphora et moi, sans lui rien déguiser. Il en rit de bon cœur ; puis, reprenant son sérieux : « Mon ami, me dit-il, je vous offre ma médiation dans cette affaire. Je vais écrire à ma belle-sœur... — Non, non, seigneur, interrompis-je, ne lui écrivez point, je vous prie : je ne suis pas sorti du château de Leyva pour y retourner. Faites, s'il vous plaît, un autre usage de la bonté que vous avez

32

Two Birds of Prey, the reader immediately *sees* the birds of prey soaring off the page as metaphors (fig. 138). In a comic book they would have been "real" birds somehow implicated in the plot. When Gil Blas arrives at Granada, we see the city at the same time as he does, as he stands on a hillside surveying it laid out before him—and us (fig. 139). Neither Töpffer nor later comic book artists ever bothered to depict a landscape other than as a schematic setting for their comic narratives; but with book illustration we can linger over the images, much as a real traveler might, or as an armchair tourist looking at the travel imagery that was ubiquitous in the nineteenth century. We do not just read the text, we reenact it by looking at the illustrations simultaneously through our own eyes and through those of the protagonists. What is salient here is the variety of visual experiences that Gigoux provides and the various moods his images evoke. When Gil Blas visits a mosque in Algiers, its shadowy presence and exaggerated perspective cast a spell over the reader far in excess of a simple delineation of locale (fig. 140). Later, on successive pages, we follow the fate of Gil Blas, where first his carriage is stopped at night and he is arrested. The dramatic contrasts of light and dark and turbulent sky communicate the violence of the scene: *I Was Scarcely Two Hundred Feet from the House of Lord Gabriel When Fifteen or Twenty Men, Some on Foot, Others on Horseback, All Armed with Swords and Rifles, Surrounded My Carriage and Halted It…* (fig. 141). When we turn the page, we read: *We Changed Horses at Cormenar*

FIGURE 140
Jean Gigoux, *You Can Well Imagine That, If I Attended Prayers with Muslims in Their Mosques and Followed the Rites of Their Religion, It Was Only by Necessity*, from Alain-René Lesage, *The Story of Gil Blas of Santillana* (1715–1735) [*Vous vous imaginez bien que si j'assistois aux prières que les muselmans font dans leurs mosquées, et remplissois les autres devoirs de leur religion, ce n'étoit que par grimace / Histoire de Gil Blas de Santillane*]. Paris: Paulin, 1835, 421. Wood engraving. Bibliothèque nationale de France.

FIGURE 141
Jean Gigoux, *I Was Scarcely Two Hundred Feet from the House of Lord Gabriel When Fifteen or Twenty Men, Some on Foot, Others on Horseback, All Armed with Swords and Rifles, Surrounded My Carriage and Halted It…*, from Alain-René Lesage, *The Story of Gil Blas of Santillana* (1715–1735) [*J'étois à peine à deux cents pas de la maison du seigneur Gabriel, que quinze ou vingt hommes, les uns à pied, les autres à cheval, tous armés d'épées et de carabines, entourèrent mon carrosse et l'ar-rêtèrent… / Histoire de Gil Blas de Santillane*]. Paris: Paulin, 1835, 687. Wood engraving. Bibliothèque nationale de France

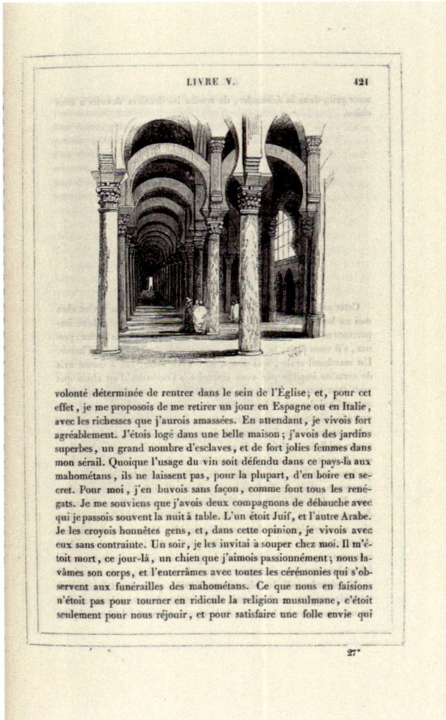
LIVRE V. 421

volonté déterminée de rentrer dans le sein de l'Église; et, pour cet effet, je me proposois de me retirer un jour en Espagne ou en Italie, avec les richesses que j'aurois amassées. En attendant, je vivois fort agréablement. J'étois logé dans une belle maison; j'avois des jardins superbes, un grand nombre d'esclaves, et de fort jolies femmes dans mon sérail. Quoique l'usage du vin soit défendu dans ce pays-là aux mahométans, ils ne laissent pas, pour la plupart, d'en boire en secret. Pour moi, j'en buvois sans façon, comme font tous les renégats. Je me souviens que j'avois deux compagnons de débauche avec qui je passois souvent la nuit à table. L'un étoit Juif, et l'autre Arabe. Je les croyois honnêtes gens, et, dans cette opinion, je vivois avec eux sans contrainte. Un soir, je les invitai à souper chez moi. Il m'étoit mort, ce jour-là, un chien que j'aimois passionnément; nous lavâmes son corps, et l'enterrâmes avec toutes les cérémonies qui s'observent aux funérailles des mahométans. Ce que nous en faisions n'étoit pas pour tourner en ridicule la religion musulmane, c'étoit seulement pour nous réjouir, et pour satisfaire une folle envie qui

27*

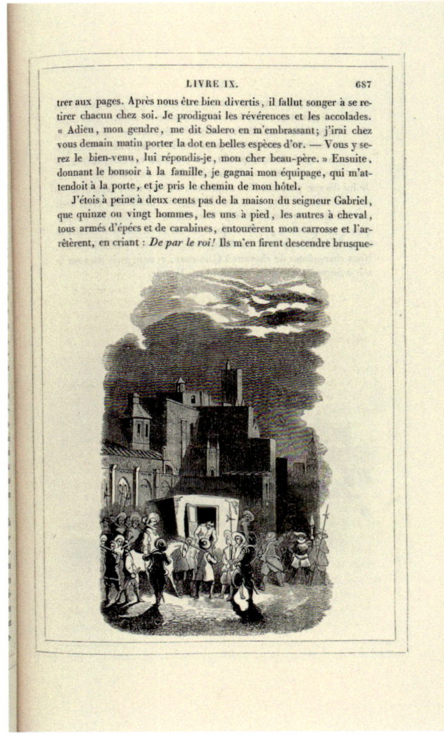
LIVRE IX. 687

trer aux pages. Après nous être bien divertis, il fallut songer à se retirer chacun chez soi. Je prodiguai les révérences et les accolades. « Adieu, mon gendre, me dit Salero en m'embrassant; j'irai chez vous demain matin porter la dot en belles espèces d'or. — Vous y serez le bien-venu, lui répondis-je, mon cher beau-père. » Ensuite, donnant le bonsoir à la famille, je gagnai mon équipage, qui m'attendoit à la porte, et je pris le chemin de mon hôtel.

J'étois à peine à deux cents pas de la maison du seigneur Gabriel, que quinze ou vingt hommes, les uns à pied, les autres à cheval, tous armés d'épées et de carabines, entourèrent mon carrosse et l'arrêtèrent, en criant : *De par le roi!* Ils m'en firent descendre brusque-

and in the Evening We Arrived at Segovia, Where They Imprisoned Me in the Tower. We see his prison high on a cliff in Segovia at the same time as he does, the distant perspective accentuating his isolation (fig. 142). Finally, we see him in his miserable cell, his body constrained within the narrow frame of the drawing, conveying his anguish: *I Spent the Whole Day Cursing My Fate* (fig. 143) All the drawings are vignettes, visionary, fading out at the edges, focused inward. Unlike the amusing images of Töpffer, Gigoux's, like all illustrational imagery, offer a much broader spectrum of emotional affect. We are caught up in the events they depict, we share them, and, in our reading and looking, we reenact them both visually and imaginatively.

ILLUSTRATION AS LITERARY DECADENCE

The illustrator's art was not appreciated universally, however. Writers often complained bitterly about the growing fashion for illustrated books. Many authors, Flaubert among them, refused to allow their works to be illustrated, and tradition-minded literary critics loathed the new fashion for illustration. Opposition grew until, in 1843, the literary critic Henri Blaze de Bury (1813–1888), writing under his pseudonym F. de Lagenevais, published a twenty-six-page jeremiad in the venerable conservative monthly *Revue des deux mondes* attacking illustrated literature of all kinds.[7] Since his article appeared at the

FIGURE 142
Jean Gigoux, *We Changed Horses at Cormenar and in the Evening Arrived at Segovia, Where They Imprisoned Me in The Tower*, from Alain-René Lesage, *The Story of Gil Blas of Santillana* (1715–1735) [*Nous changeâmes de chevaux à Cormenar, et nous arrivâmes sur le soir à Ségovie, où l'on m'enferma dans la tour / Histoire de Gil Blas de Santillane*]. Paris: Paulin, 1835, 688. Wood engraving. Bibliothèque nationale de France.

FIGURE 143
Jean Gigoux, *I Spent the Whole Day Cursing My Fate*, from Alain-René Lesage, *The Story of Gil Blas of Santillana* (1715–1735) [*J'employai tout le jour à maudire mon étoile... / Histoire de Gil Blas de Santillane*]. Paris: Paulin, 1835, 690. Wood engraving. Bibliothèque nationale de France.

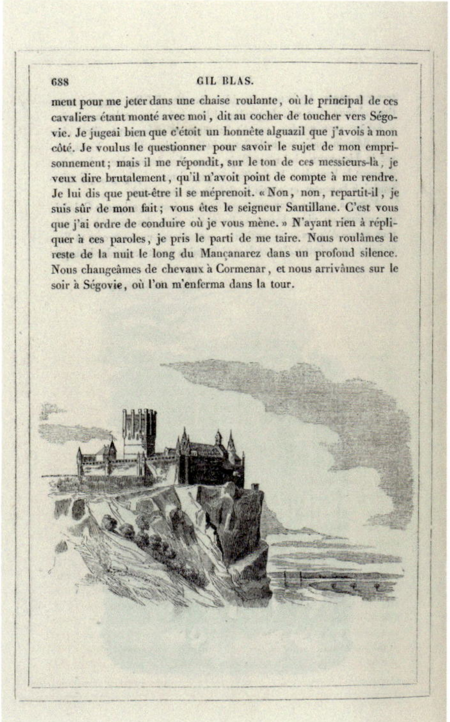
688 GIL BLAS.

ment pour me jeter dans une chaise roulante, où le principal de ces cavaliers étant monté avec moi, dit au cocher de toucher vers Ségovie. Je jugeai bien que c'étoit un honnête alguazil que j'avois à mon côté. Je voulus le questionner pour savoir le sujet de mon emprisonnement; mais il me répondit, sur le ton de ces messieurs-là, je veux dire brutalement, qu'il n'avoit point de compte à me rendre. Je lui dis que peut-être il se méprenoit. « Non, non, repartit-il, je suis sûr de mon fait; vous êtes le seigneur Santillane. C'est vous que j'ai ordre de conduire où je vous mène. » N'ayant rien à répliquer à ces paroles, je pris le parti de me taire. Nous roulâmes le reste de la nuit le long du Mançanarez dans un profond silence. Nous changeâmes de chevaux à Cormenar, et nous arrivâmes sur le soir à Ségovie, où l'on m'enferma dans la tour.

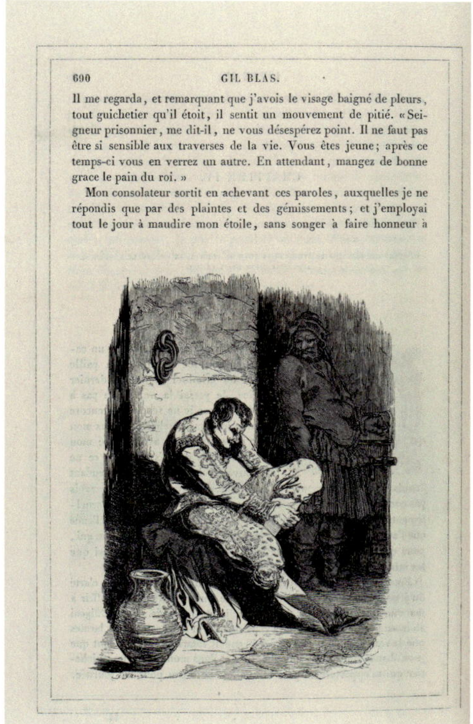
690 GIL BLAS.

Il me regarda, et remarquant que j'avois le visage baigné de pleurs, tout guichetier qu'il étoit, il sentit un mouvement de pitié. « Seigneur prisonnier, me dit-il, ne vous désespérez point. Il ne faut pas être si sensible aux traverses de la vie. Vous êtes jeune; après ce temps-ci vous en verrez un autre. En attendant, mangez de bonne grace le pain du roi. »

Mon consolateur sortit en achevant ces paroles, auxquelles je ne répondis que par des plaintes et des gémissements; et j'employai tout le jour à maudire mon étoile, sans songer à faire honneur à

height of production of illustrated literature and in the most respected French periodical, it can be seen as paradigmatic of this negative point of view. In it he identified and pilloried all the transformations that had been taking place, none of which met with his approval. "Illustration is a symptom of literary decadence," he thundered, and listed all the reasons for this.[8] Sounding very much like a Modernist formalist critic, he pronounced that "each art has its own specific kind of beauty."[9] He saw the combination of word and image as the worst kind of miscegenation, a bastardization of the high aspirations of both media, each of which, he felt, should compete only in its own sphere. The new technology of wood engraving allowed vignettes to be inserted directly into the text, instead of being segregated on separate pages (see figs. 134a–b). This was, for him, intolerable, for while the earlier practice of inserting full-page metal-plate engravings between pages "did not break the unity of impression so necessary for reading," wood-engraved vignettes "throw the pages into disorder, upset the regular harmony of lines to which the eye is accustomed," and—the ultimate transgression—"substitute the artist's vision for that of the poet."[10] In other words, the traditional hierarchy of word over image was being challenged. What followed was an itemized denunciation of all the major illustrated works of his period, beginning with illustrated novels and ending with illustrated periodicals. Comics were omitted from his tirade, no doubt because they were beneath contempt, ranking even lower than illustration. "Picturesque literature" he labeled it all, a pejorative

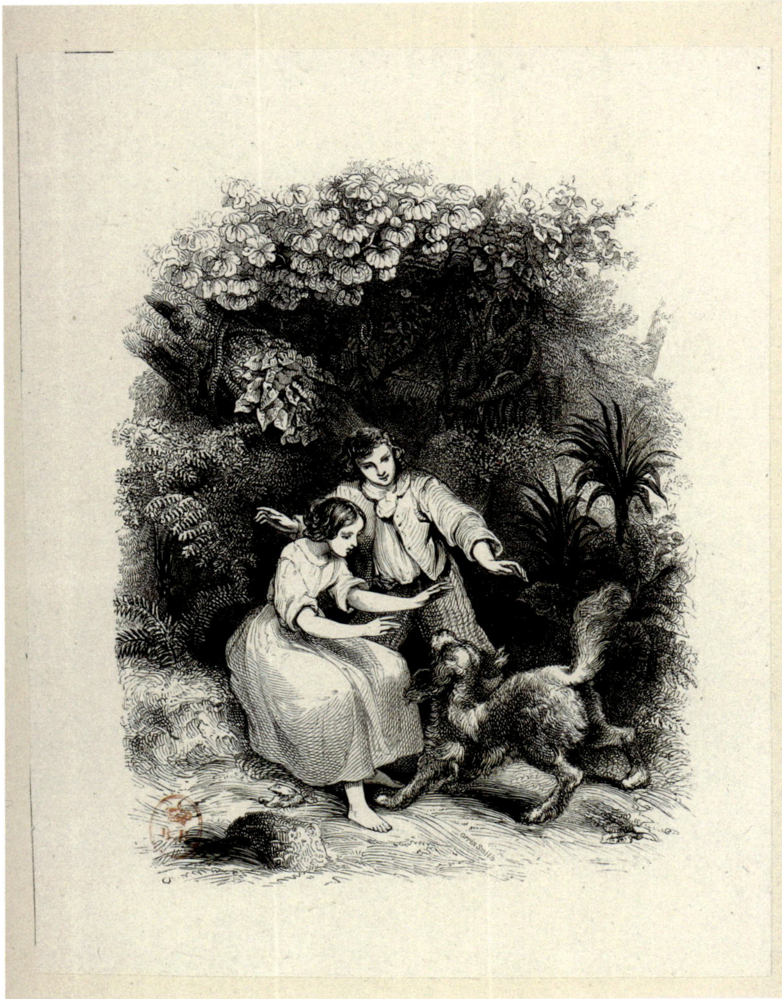

FIGURE 144
Tony Johannot and François-Louis Français, *Paul and Virginia Found in the Forest by Fidèle*. Engr. Orrin Smith, from Bernardin de Saint-Pierre, *Paul and Virginia* (1788) [*Paul et Virginie retrouvé dans la forêt par Fidèle/Paul et Virginie*]. Paris: Curmer, 1838, 54–55. Wood engraving. Bibliothèque nationale de France.

pun relating to its picture-laden content and to the titles of numerous publications such as *Le Magasin pittoresque.*[11] Indeed, the Picturesque, the lowly third realm of art that lagged far behind the Beautiful and the Sublime in traditional aesthetic theory, had been for decades under attack by high-minded critics. For Lagenevais, the current prominence of lithography in particular seemed to be symptomatic of the appeal of the picturesque to the semiliterate lower classes.[12]

Lagenevais's most acerbic criticism, however, was reserved for the 1838 Curmer edition of Bernardin de Saint-Pierre's *Paul and Virginia* (1788), now recognized as the most extraordinary illustrated book of the nineteenth century (see figs. 134a and b). Béraldi called it "one of the most remarkable ever published."[13] The most profusely illustrated printed book since the *Nuremburg Chronicle* of 1493, it boasted twenty-nine full-page plates, four hundred fifty vignettes, and seven steel-engraved portraits. In *Paul and*

Virginia Found in the Forest by Fidèle, a collaboration between the landscape artist François-Louis Français (1814–1897) and the leading illustrator of figures, Tony Johannot (1803–1852), we see Paul and Virginia in a lush woodland setting on the island of Mauritius in the Indian Ocean, where they lead an idyllic life away from the corruptions of society (fig. 144). Even a brief comparison with Girodet's *Andromache* (see fig. 135), shows the changes that have transpired in the language of illustration within a few decades. *Andromache* demands to be considered as a painting translated into the print medium, a "window on the world" in the Renaissance tradition, while the centrifugal focus and fade-out perimeters of *Paul and Virginia Found in the Forest by Fidèle* give it a subjective aura that underscores at the same time its distance from the "real world" and its fragmentary vision of it. We saw, in the previous chapter, a parallel transformation in the format of comic books, which also began with rectilinear frames, but soon abandoned them in favor of more subjective layouts. In the vignettes of *Paul and Virginia*, the staccato alternation of text and image further drives the imagery into the realm of the subjective. The double-page layout of the novel's pages 156 and 157 is paradigmatic; here the illustrators show us the narrator on the recto page, but, at the same time, we see on the preceding verso the events that he is recounting taking place in the past (see figs. 134a–b). These events are thrice removed, recounted by the narrator, envisioned by the artists, observed by the reader. In Eugène Isabey's *Stormy Coast* (fig. 145), the images actually seem to take precedence over the text; the literary functions here as an interruption of the visual and not the reverse.

All the leading illustrators of the period—Tony Johannot, Louis-François Français, Eugène Isabey, Paul Huet (1803–1869), and Ernest Meissonier (1815–1891)—contributed drawings to the Curmer *Paul and Virginia*, resulting in a tour de force of the art of illustration. Nonetheless, according to Lagenevais, illustrated books addressed themselves only to women, children, and the barely literate, who leafed through their pages looking at the pictures instead of actually reading them—as opposed to the male audience for serious literature who, to be sure, constituted the readership of *Revue des deux mondes*.[14]

For Lagenevais, the only publications worse than illustrated books were the illustrated periodicals newly established in France discussed here in chapters 1 and 2: *Le Charivari* (1832), *Le Magasin pittoresque* (1833), and, concurrently with his article, *L'Illustration* (1843). He and other conservatives blamed England—quite correctly—for the wave of illustration that they feared was inundating serious literature in France: "Just like wood engraving and mechanical printing, like all innovations that tend to seduce the purchaser through cheap price, *picturesque magazines* [magasins pittoresques] were born in England, the natural homeland of all commercial ideas."[15] The *Illustrated London News* had begun publishing less than a year earlier, in May 1842, inspiring the foundation of *L'Illustration*, whose first issue was published just after Lagenevais's diatribe appeared—and might even have provoked it as it was much publicized and eagerly anticipated.[16]

The 1830s and 1840s had seen an explosion of illustrated publications of all kinds, many of which have already been discussed here. Lagenevais cited those he found

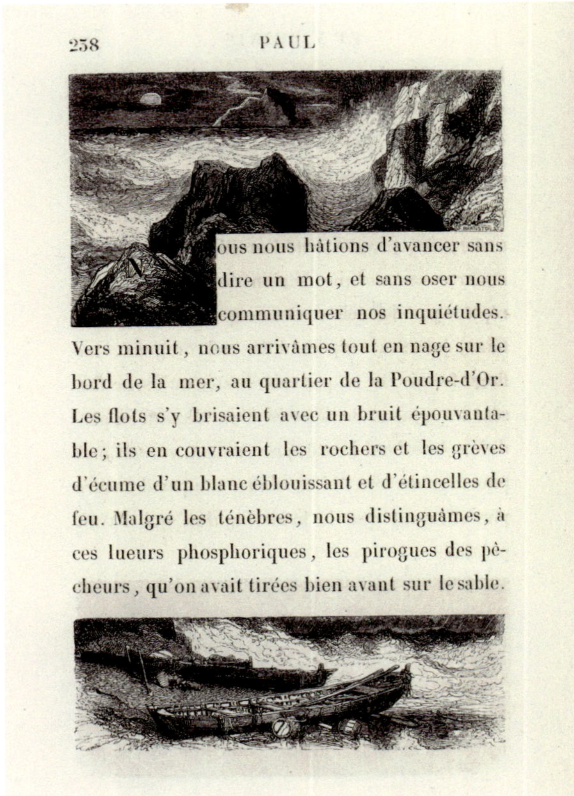

ous nous hâtions d'avancer sans dire un mot, et sans oser nous communiquer nos inquiétudes. Vers minuit, nous arrivâmes tout en nage sur le bord de la mer, au quartier de la Poudre-d'Or. Les flots s'y brisaient avec un bruit épouvantable; ils en couvraient les rochers et les grèves d'écume d'un blanc éblouissant et d'étincelles de feu. Malgré les ténèbres, nous distinguâmes, à ces lueurs phosphoriques, les pirogues des pêcheurs, qu'on avait tirées bien avant sur le sable.

FIGURE 145
Eugène Isabey, *The Stormy Coast*. Engr. Branston, from Bernardin de Saint-Pierre, *Paul and Virginia* (1788) [*Bords de la mer orageuse / Paul et Virginie*]. Paris: Curmer, 1838, 238. Wood engraving. Bibliothèque nationale de France.

OPPOSITE **FIGURE 146**
Honoré Daumier, *Attorney for the Defense*. Engr. Birouste, from *The French Depicted by Themselves* [*Le Défenseur officieux en justice de paix/Les Français peints par eux-mêmes*]. Paris: Curmer, 1840–1842, vol. 2, 309. Wood engraving. Bibliothèque nationale de France.

particularly objectionable: albums, travel books, anthologies, landscape views, caricatures, physiologies, and almanacs.[17] What he found particularly offensive was the popularity of what Walter Benjamin called "panoramic literature," entertaining lightweight publications that surveyed in word and image contemporary manners, morals, customs, and characters.[18] Following the publication of the fifteen volumes of *Paris, or The Book of a Hundred and One* (1831–1834), there had been numerous compilations of literary scenes and sketches of modern life; these publications had few if any illustrations, however, and so remained safely within the genre of light literature minus what was for Lagenevais the miscegenation of imagery.[19] This was not the case, however, with the ten-volume, profusely illustrated series *The French Depicted by Themselves* (1840–1842), which Lagenevais singled out for condemnation, no doubt because the installments had just been completed, it was immensely popular, and it juxtaposed word and image on the same page.[20] An anthology combining wood engravings, many hand-colored, with brief essays on contemporary social types, it was produced by the period's most prominent graphic artists, including Gavarni, Grandville, and Daumier, alongside writers such as Honoré de Balzac, Théophile Gautier (1811–1872), and Alexandre Dumas (1802–1870). Daumier's *Attorney for the Defense* (fig. 146) accompanied an essay by Émile Dufour that detailed all

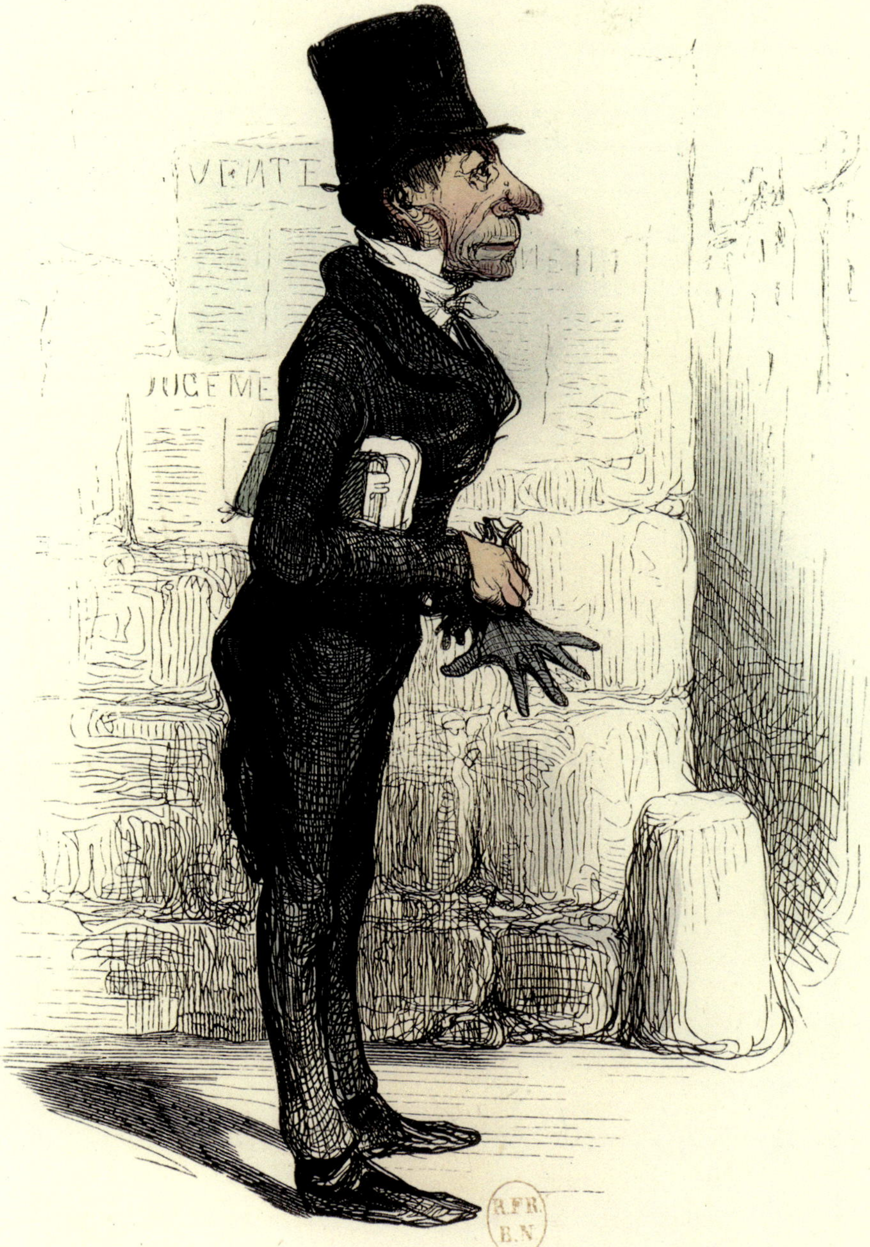

LE DEFENSEUR OFFICIEUX

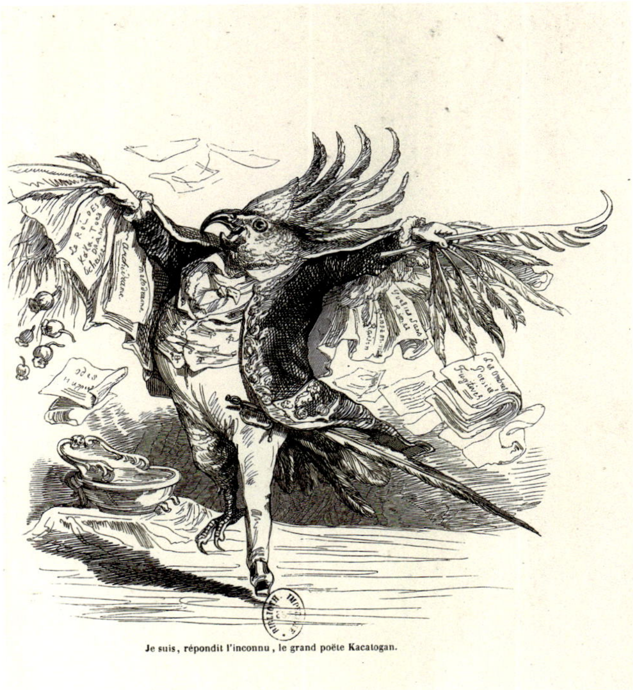

FIGURE 147

Grandville, *I Am, Answered the Unknown Man, the Great Poet Cockatoo*. Engr. John Andrew, for "The Story of a White Blackbird" by Alfred de Musset, from Grandville, *Scenes of the Public and Private Life of Animals: Studies of Contemporary Manners and Morals* [*Je suis, répondit l'inconnu, le grand poëte Kacatogan* / "Histoire d'un merle blanc" / *Scènes de la vie privée et publique des animaux. Études de moeurs contemporains*]. Paris: Hetzel and Paulin, 1842, vol. 2, 252–53. Wood engraving. Bibliothèque nationale de France.

the cunning employed by this lowest class of lawyers as they trolled the courts on the lookout for clients. Here we see him standing before official notices reading *Vente* (Sale) and *Jugement* (Judgment), his red nose signaling his fondness for drink, his pose evenly balanced between aggression and obsequiousness.[21] *The French Depicted by Themselves* was so successful that it was reissued several times throughout the century and spawned numerous imitations. Following close behind both in popularity and in Lagenevais's wrath was another recent publication, Grandville's *Scenes of the Public and Private Life of Animals: Studies of Contemporary Manners and Morals* (1842), a two-volume anthology with drawings by Grandville and text by popular authors such as Balzac, Charles Nodier, Alfred de Musset (1810–1857), and George Sand (Amantine-Aurore-Lucie Dupin, 1804–1876). Lagenevais referred to it sarcastically as *Animals Depicted by Themselves*.[22] Here there is the same delight in skewering representatives of all classes and professions in both word and image, but Grandville now depicts them in the guise of the animal whose physiognomic traits they share, as he had done previously in his series *The Metamorphoses of the Day* (see fig. 21). Here he draws the self-proclaimed *Great Poet Cockatoo*, not only flamboyant and vain, but clad in the robes of an academician, a detail that, no doubt, did not escape the literati class (fig. 147).

These anthologies had been inspired by *Heads of the People* (1838–1840), an English publication with drawings by Kenny Meadows (1790–1874) and texts by Douglas Jerrold (1803–1857) and William Makepeace Thackeray (1811–1863), among others. It described

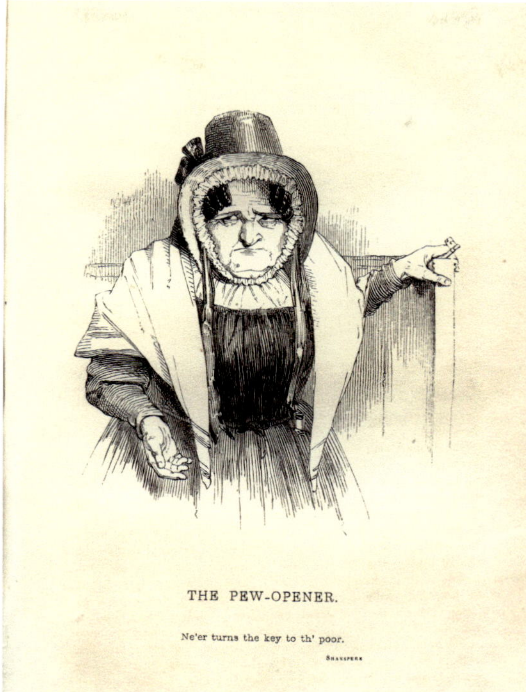

FIGURE 148
Kenny Meadows, *The Pew-Opener.* Engr. Orrin Smith, from *Heads of the People: Portraits of the English.* London: Robert Tyas, 1840, 224–25. Wood engraving.

and illustrated various social types, liberally mixing humor with acerbic social criticism.[23] Kenny Meadows's wood engraving *The Pew-Opener* (fig. 148), for example, carries the epigraph "Ne'er turns the key to th' poor. Shakspere"; Douglas Jerrold's accompanying text begins with criticism of this "hopeful looker-out for sixpences" whom Meadows has depicted with her hand ever-outstretched for an offering, but soon expands into an attack on religious hypocrisy that encompasses clergy and worshippers alike.[24] Quickly plagiarized and translated into French, *Heads of the People* spawned not only *The French Depicted by Themselves*, but also hundreds of similar "physiologies," as they were called, small inexpensive books combining text and image, each focusing on one particular social type, amusing, but never rising to the acerbic level of their English progenitor.[25] The student, the society woman, the bourgeois, all the denizens of modern Paris, were each described in a satirical pseudosociological style and illustrated with witty wood-engravings interspersed with text, a format that was impossible with lithography or metal-plate engraving. The illustrators Lagenevais singled out for especially harsh criticism—Tony Johannot, Gavarni, and Grandville—were precisely those who were the most celebrated at the time and who contributed most actively to all these various genres of illustrated literature.

While some forms of ephemeral literature, such as almanacs, had been popular in France since at least the eighteenth century, as had travel literature, landscapes, and views, the new technologies of lithography and wood engraving now brought them into an economic range affordable to a vast middle and artisanal class, rather than being limited to

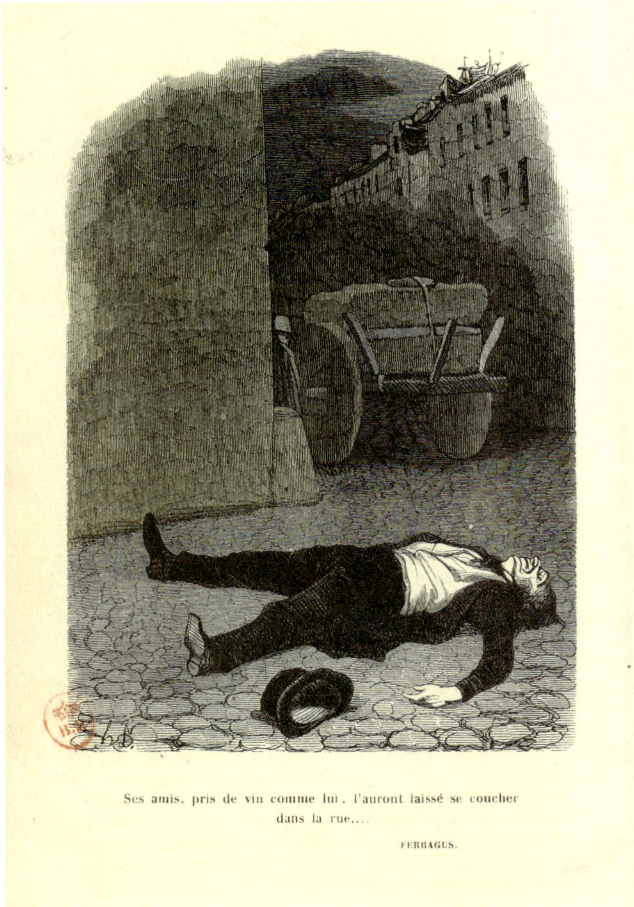

FIGURE 149

Honoré Daumier, *His Friends, Just as Drunk as He Was, Left Him Sleeping in the Street* from Honoré de Balzac, *Ferragus* (1833) [*Ses amis, pris de vin comme lui, l'auront laissé se coucher dans la rue*], frontispiece, *Oeuvres complètes de M. de Balzac.* Paris: Furne, 1843, vol. 9, pt. 1. Wood engraving. Bibliothèque nationale de France.

the elite audiences of the previous century.[26] As a result, their numbers multiplied. The major compendium of topographical images, *Picturesque and Romantic Travels through Old France* (see fig. 26), had begun publication in 1820 and was so successful that, by 1878, over twenty volumes had been published, marshaling several generations of writers and artists to describe, in word and image, the different regions of France; similar publications multiplied throughout the Western world.[27] In addition to these focused publications, there were literary potpourris assembled by publishers that featured the combined efforts of artists and writers. Modeled on the English gift books known as "keepsakes," which combined short texts and poems with high-quality steel-engraved illustrations, these had become popular New Year's gifts in France.[28]

Among literary figures, Lagenevais singled out Balzac (he called him "the ubiquitous Balzac") for disapproval because he so often collaborated with illustrators.[29] Balzac had always been interested in illustration and, unlike the authors whom Lagenevais preferred, he actually wanted his novels published in illustrated editions.

The Furne illustrated edition of Balzac's *Human Comedy* had just begun to appear the previous year, 1842, and would grow to seventeen volumes by 1848. Their illustrations demonstrate Balzac's ability to envision his writing enhanced by a broad spectrum of artistic styles. In the frontispiece for the Furne edition of Balzac's *Ferragus*, Daumier has drawn a dramatic image of the faithful retainer Justin, abandoned in the street, about to be run over and killed; the stark realism and ominous lighting of the scene represent a grim departure from Daumier's familiar wit and caricature (fig. 149). A conceptual abstraction by Bertall for Balzac's *Little Miseries of Conjugal Life* is worlds apart: the artist "draws" a blank sheet of paper, whose caption on this imagined page is half drawn above (*Draw for yourself, please*), and half typeset below (*the one you dream of!*) (fig. 150). Across the drawn page is another drawing, of an artist's pencil—but without the artist's hand, the imagined dream is only a black blot. Bertall has given us an ingenious emblem of the collaboration between author, artist, and reader. Balzac could coordinate this spectrum of illustration in his novels because he knew all the artists of his time; he was a regular contributor to the very periodicals and anthologies that Lagenevais condemned, and had served as editor of the satirical journal *La Caricature* in its early years.[30] But Balzac was by no means the only author enamored of illustration. His English counterpart Charles Dickens (1812–1870) considered it an integral aspect of his novels and had them illustrated by major artists such as George Cruikshank, John Leech, and Phiz (Hablot Knight Browne, 1815–1882).[31] Phiz drew the illustrations for *The Posthumous Papers of the Pickwick Club* (1837), including a full-page etching of *The First Appearance of Mr. Samuel Weller*, the cockney valet who became Dickens's first popular success (fig. 151). The novelist William Makepeace Thackeray (1811–1863) had a second career as an illustrator, providing drawings to be etched or wood engraved for *Punch* and for his own works as well.[32] *Vanity Fair* (1848) is considered his greatest novel, a scathing portrait of Victorian society. He shows us its protagonist, the antihero Becky Sharp in the guise of a Greek tragic hero, Clytemnestra; here she is all in shadow, the classic graphic sign of guilt (fig. 152). Like Balzac, these writers had an imagination that was visual as well as literary.

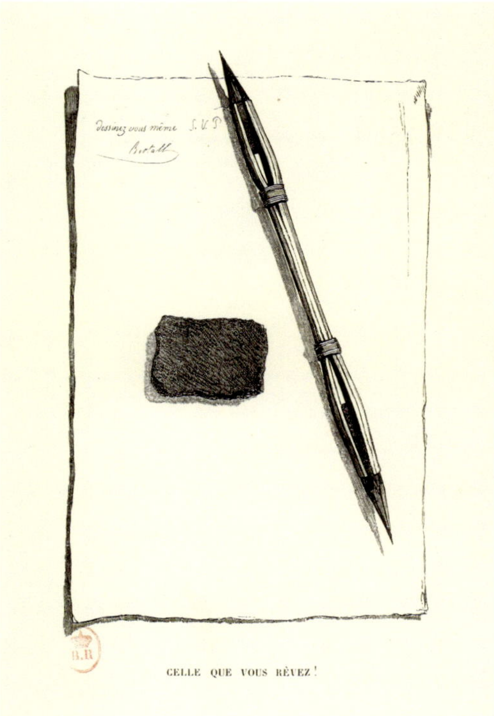

FIGURE 150
Bertall, *Draw for Yourself, Please… the One You Dream of!* Engr. André Castan, from Honoré de Balzac, *Little Miseries of Conjugal Life* [*Dessinez vous même S.V.P. celle que vous rêvez! / Petites misères de la vie conjugale*]. Paris: Chlendowski, 1846, 90–91. Wood engraving. Bibliothèque nationale de France.

Lagenevais did not comprehend this, however. He thought it deplorable when artists usurped the author's prerogative by superimposing their own conception of events over those of the writer or even the reader. Even worse, he saw Armageddon arriving when artists "no longer wanted to translate the text, but to dictate it."[33] For him, this was truly the world upside-down, when writers became subservient to artists. It was the very innovation introduced in England by *Doctor Syntax* and soon to be taken up in France.

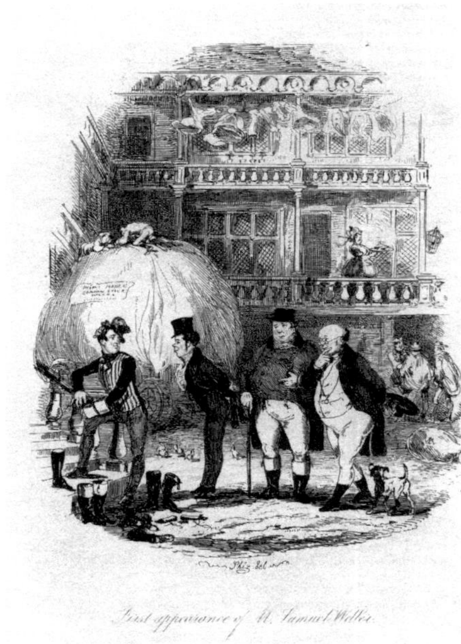

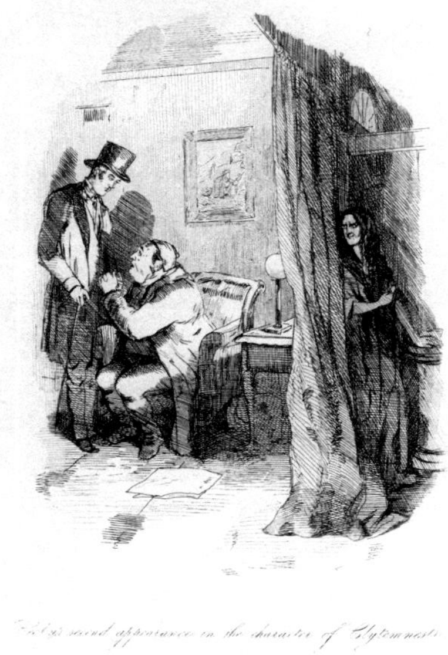

FIGURE 151
Phiz [Hablot Knight Browne], *First Appearance of Mr. Samuel Weller*, from Charles Dickens, *The Posthumous Papers of the Pickwick Club*. London: Chapman and Hall, 1837. Etching.

FIGURE 152
William Makepeace Thackeray, *Becky's Second Appearance in the Character of Clytemnestra*, from *Vanity Fair. A Novel without a Hero.* London: Bradbury & Evans, 1848. Etching.

The artist-driven illustrated text that Lagenevais feared would completely destroy literature became especially prevalent in France in the 1840s. By then not only were Töpffer's comic books available (the first plagiarized French editions had appeared in 1839), but, as was noted in the previous chapter, by 1843 when Lagenevais was writing, the publishing house of Aubert, where Charles Philipon was the guiding force, had already brought out eight original comic books by French authors. These comic books represent one aspect of the word-image continuum, certainly the predominant survival into the twentieth century, when illustrated novels had declined in numbers and illustrated periodicals had moved on to photography. But for an understanding of the richness and variety of artist-driven publications, there are other examples of artist-driven texts that should be cited. The two preeminent ones are Tony Johannot's *Travel Where You Will* (1843) and Grandville's *Another World* (1844). Both artists enjoyed huge audiences for their work, and so it was only logical that both were singled out for condemnation by Lagenevais.

TRAVEL WHERE YOU WILL: "A WORK OF FANTASY" BY TONY JOHANNOT

Tony Johannot, whom Théophile Gautier called "the king of illustration," was one of the most prolific illustrators of the nineteenth century, surpassed only later by Gustave Doré.[34] Johannot was credited in the epilogue of *Travel Where You Will* as the book's guiding force and even received precedence on its title page over his collaborators, the well-known writer Alfred de Musset, and P.-J. Stahl, the pseudonym of the book's publisher Jules Hetzel (1814–1886).[35] Like most expensive publications, the book was sold on subscription and published, usually weekly, in installments called *livraisons*. A prospectus was issued in advance to generate these subscriptions; the one for *Travel Where You Will* describes the collaborators' working method and an included "Notice" forewarned subscribers of the unusual nature of the installments, thirty-three in all, that they would receive: "*Travel Where You Will* being essentially and by its very nature a work of fantasy, complete freedom is given to the authors and to the illustrator. Their project is to write and draw *alternately* according to whether it would be better to write or to draw and whether the pen or the pencil would be more appropriate to render their intentions. Since there will be a considerable number of vignettes relative to the length of the text, subscribers will receive sometimes *both* text and vignettes, sometimes *only* vignettes."[36] And indeed, the book's subtitle announces its departure from the norm, stating that "the book was written with pen and pencil" and would contain "vignettes, legends, episodes, commentaries, incidents, notes, and poetry."[37]

While Johannot's work has received little attention either from literary critics or from art historians, in many ways *Travel Where You Will* is the most adventurous of the word-image texts of the nineteenth century. Its illustrations were drawn in a variety of styles, ranging from topographic to fantastic. The bizarre subject and shifting scale of *And So, in the Guise of Friendship, the Rogue Stole My Brain* is unlike anything that had gone before (fig. 153). What set it apart from other artist-inspired projects, however, was Johannot's departure from traditional strategies of illustration, and even from Töpffer's picture stories, five of which had been published by 1842 when Johannot began work on this project.[38] The book's conception provided more of a relay than previous collaborative projects, for here the artist and writers actually took turns advancing the plot, as its advance publicity had promised. Most innovative was the insertion of several pages of illustration as a series instead of as single pages. In this way, Johannot repeatedly took over to tell the story himself, interrupting the literary narrative in order to present a visual narrative unfolding in real time. These visual interludes break into the literary experience of the text and replace it with another order of experience altogether, akin to the difference between reading a book and watching a film. For example, when the narrator Franz describes how he was obliged to flee for his life on a wild horse, at the line "Our horses ran like the wind," Johannot inserts a vignette depicting the horses and riders, a standard illustrational strategy.[39] The narration continues below the vignette: "To leave the city, we had to pass in front of my fiancée's house. Her bedroom window was half open and I saw the sweet girl seated before a harpsichord that her uncle, the organist, had left her. She was singing:"—the page ends with the colon.[40] Then follows five pages of images alone. The first full-page drawing immediately following changes the medium from literary to visual and the reader's focus from Franz to Marie (fig. 154). The vignette of horsemen on the preceding page is repeated, much diminished but still visible through the open window, like a cinematic fade-out in the background while in the foreground Marie sings, unaware. The caption offers a synopsis of the preceding text: *Her Bedroom Window Was Half Open and She*

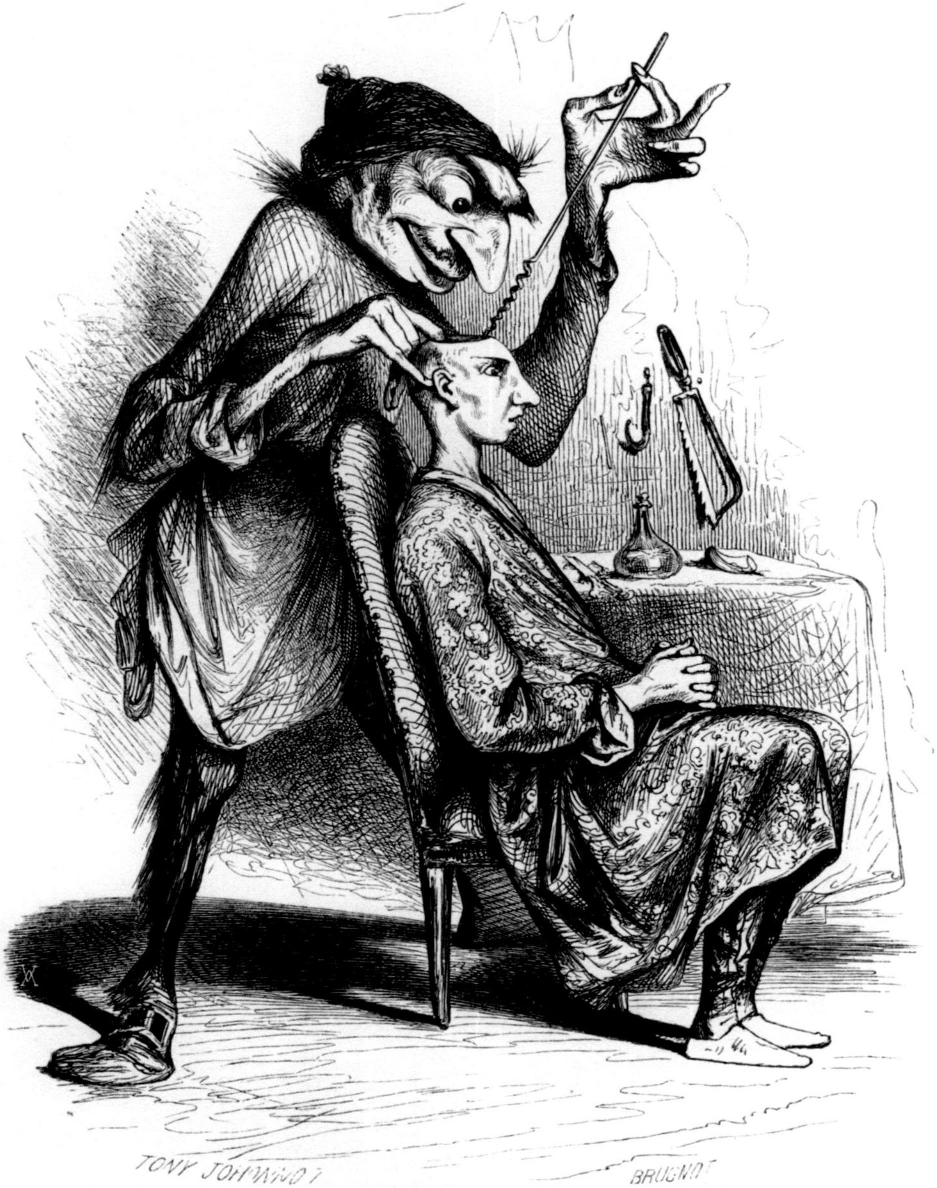

C'est ainsi que, sous le voile de l'amitié, le scélérat vint
à bout de s'emparer de ma cervelle, dont il fit son profit, car, à mesure que ma tête
diminuait de volume, la sienne grossissait.. ..

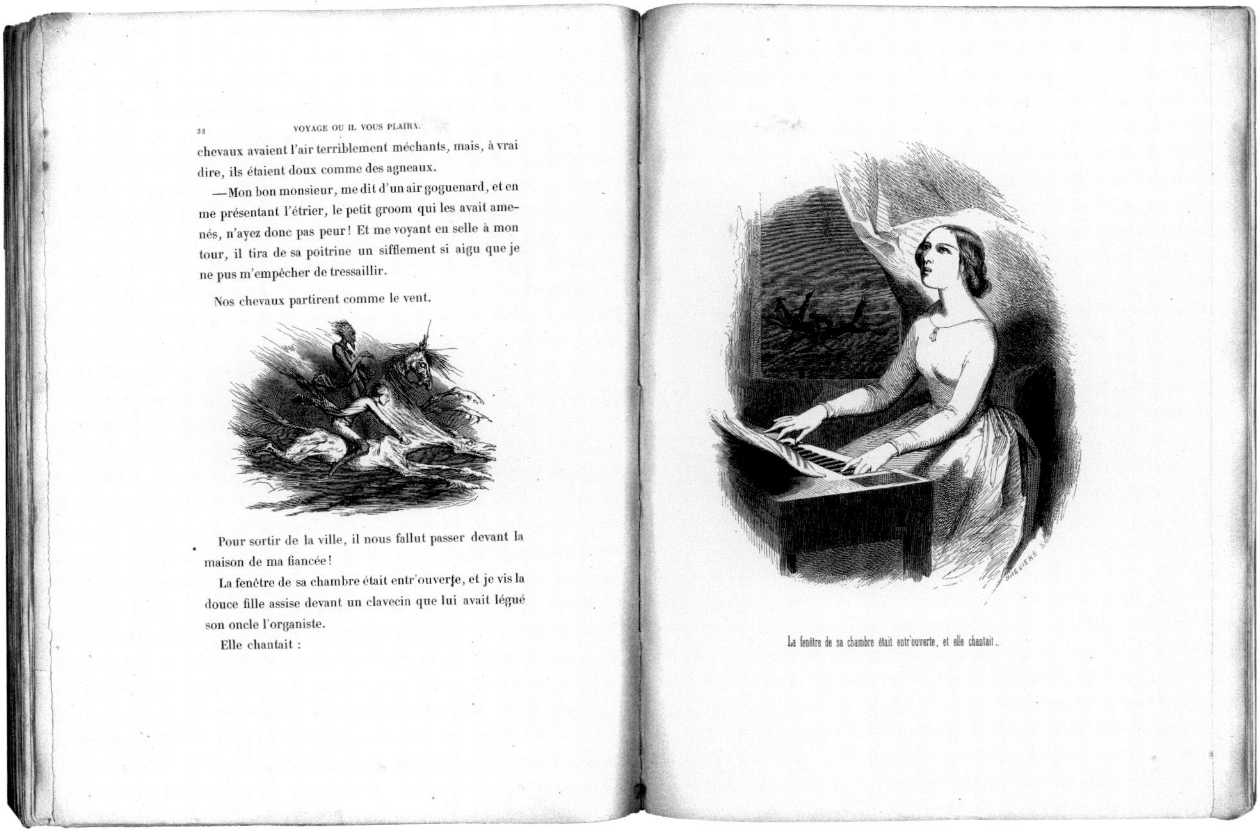
32 VOYAGE OU IL VOUS PLAIRA.

chevaux avaient l'air terriblement méchants, mais, à vrai dire, ils étaient doux comme des agneaux.

—Mon bon monsieur, me dit d'un air goguenard, et en me présentant l'étrier, le petit groom qui les avait amenés, n'ayez donc pas peur! Et me voyant en selle à mon tour, il tira de sa poitrine un sifflement si aigu que je ne pus m'empêcher de tressaillir.

Nos chevaux partirent comme le vent.

Pour sortir de la ville, il nous fallut passer devant la maison de ma fiancée!

La fenêtre de sa chambre était entr'ouverte, et je vis la douce fille assise devant un clavecin que lui avait légué son oncle l'organiste.

Elle chantait :

La fenêtre de sa chambre était entr'ouverte, et elle chantait..

OPPOSITE **FIGURE 153**

Tony Johannot, *And So, in the Guise of Friendship, the Rogue Stole My Brain*. Engr. Brugnot, from *Travel Where You Will* [*C'est ainsi que, sous le voile de l'amitié, le scélérat vint à bout de s'emparer de ma cervelle / Voyage où il vous plaira*]. Paris: J. Hetzel, 1843, 108–9. Wood engraving.

FIGURE 154

Tony Johannot, *Her Bedroom Window Was Half Open, and She Was Singing*... Engr. Brevière, from *Travel Where You Will* [*La fenêtre de sa chambre était entr'ouverte, et elle chantait... / Voyage où il vous plaira*]. Paris: J. Hetzel, 1843, 32–33. Wood engraving.

Was Singing.[41] Only now we actually *see* her singing. This page marks the transition between word and image. A turn of the page changes the medium again, showing an imaginary Marie in the dark hostile night and presenting us with the words of the song she sings, Mozart's melancholy love song "Forget Me Not" (Vergiss mein nicht), rendered in French as "Remember" (Rappelle-toi). In it a departed lover begs his love not to forget him when he is dead and buried. Here Johannot illustrates text with text, a tour de force in using words to illustrate words. Then follows a triple-page spread showing the actual sheet music of the song Marie is singing and playing. Johannot has now illustrated text with music, so we have gone from reading that Marie is singing, to seeing Marie singing, to reading the words she sings, to actually seeing the sheet music that she is playing and singing. As the music fades away, the section ends with a final vignette of the young woman silently mourning over her lover's grave. This interlude concludes on a verso page, with the following recto returning us to the text, its first line echoing the last text page: "She was singing... and I was leaving!"[42] Abruptly the vision and music vanish, to be replaced by text alone. The interlude seems to unfold in real time as the reader turns the pages, with the music actually present, not simply described, nor even simply illustrated by an image of Marie singing.

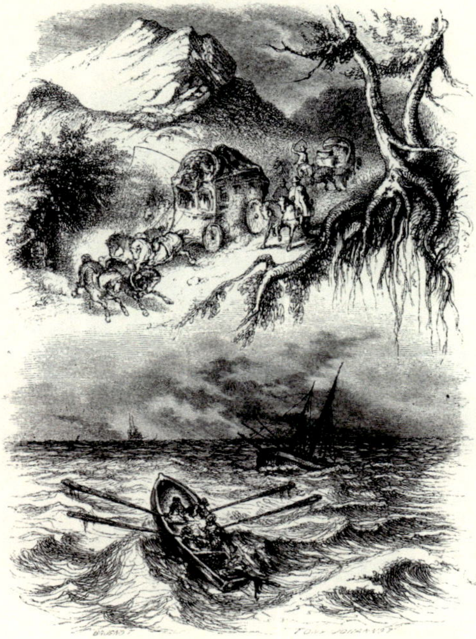

FIGURE 155
Tony Johannot, *I See Only Ships Crossing Immense Oceans, Horses and Carriages Lumbering toward All Corners of the World...* Engr. Brugnot, from *Travel Where You Will* [*Je ne vois plus que navires qui se croisent sur l'immensité des mers, que chevaux et voitures qui roulent pesamment sur tous les points du globe... / Voyage où il vous plaira*]. Paris: J. Hetzel, 1843, 14–15. Wood engraving.

While Johannot's inventiveness has some precedents in earlier works, he was the first to include a sequential series of illustrations that break into the text and usurp its storytelling function. The innovative aspect of this narrative device would have been all the more striking to readers since subscribers received their installments at intervals and thus, on receiving this one, were forced to wait another week before returning to the text. We can well understand the distress that many writers felt in the face of this kind of competition from artists.

A further example of Johannot's inventiveness occurs when Franz is distraught because his beloved Marie is weeping; to distract himself, he opens the book *Travels Around the World by Captain Cook*, a popular publication with many French editions.[43] The subsequent pages show the illustrations Franz is looking at—or is he imagining them? Are we, the readers, seeing the book illustrations that Franz sees or only what he is imagining? In the drawings, Cook and his explorers traverse mountains, deserts, and seas. *I See Only Ships Crossing Immense Oceans, Horses and Carriages Lumbering toward All Corners of the World...* (fig. 155). In this drawing the image itself is bisected by the horizon of the seascape, the upper half presenting an alpine landscape with, improbably, tree roots snaking downward through the stormy sky to the turbulent ocean of the lower vignette. There a defenseless

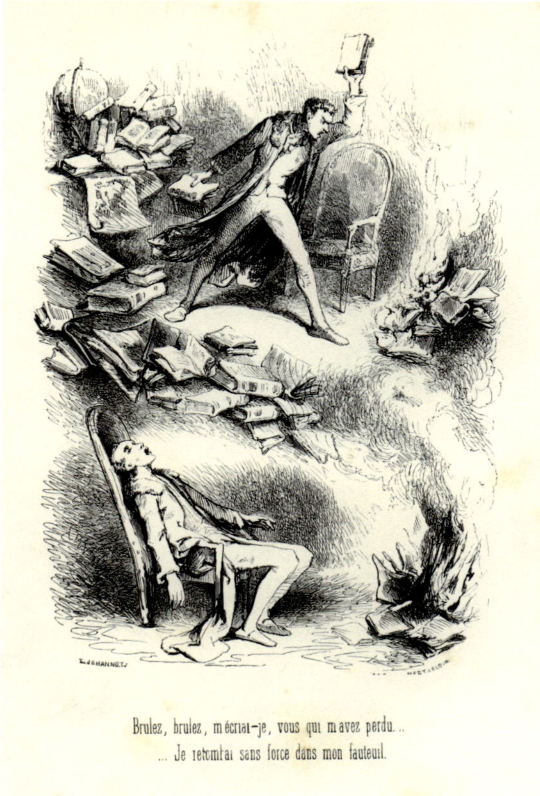

FIGURE 156
Tony Johannot, *Burn, Burn, I Cried, You Who Ruined Me*... Engr. Andrew, Best, Leloir, from *Travel Where You Will* [*Brûlez, brûlez, m'écriai-je, vous qui m'avez perdu*... / *Voyage où il vous plaira*]. Paris: J. Hetzel, 1843, 20–21. Wood engraving.

crew escaping a shipwreck rows frantically in an attempt to reach the distant ship on the horizon. On the next page there is a similar set of interpenetrating images: ... *Brave and Hardy Travelers Who Cross the Burning Sands of Deserts and Who Climb the Rugged Peaks of the Alps and the Cordillera.*[44] The instructions to the binder specify that these two plates should follow each other but should precede the corresponding passage in the novel, with the result that the reader encounters first the images and only later the text that "illustrates" them.[45] This strategy, situating the writer as illustrator of the artist's preceding images, is repeated throughout the novel, with the artist regularly "upstaging" the writer, presenting situations and dialogue before they occur in the text. In many of these illustrations, Johannot has adapted for his purposes the macédoine or medley print discussed in chapter 1, the composite image made up of several vignettes. Here he extends the macédoine's reach from single-sheet prints and albums into the illustrated book, where a page of illustration could contain a sequence of images both united and separated by a related motif. In *Burn, Burn, I Cried, You Who Ruined Me*... the whirlwind of books being thrown into the fire divides the image into two separate moments, with its caption reflecting both past and present (fig. 156). Above, Franz tosses his travel books into the fire; below, we see the result, he is exhausted by his efforts: *I Fell Back Weakly into My Armchair.*[46] In these images,

Johannot has invented a complex visual strategy to subvert narrative time and suggest how events interweave in our consciousness instead of unfolding in a linear manner.

In looking for precedents for these imaginative types of literary illustration, the obvious example is that of the English author Laurence Sterne, whose novel *The Life and Opinions of Tristram Shandy, Gentleman* (1759–1767) contained the all-black and all-white pages discussed in the previous chapter in relation to the comic albums of Cham. Sterne included numerous additional graphic interventions, such as blank pages, doodles, and a marbled endpaper inserted in the text.[47] His work was translated almost immediately into most European languages and found an appreciative audience in France, where he was called "the Rabelais of England" and praised by Voltaire and Diderot.[48] Among Sterne's later admirers was the French author Charles Nodier, who, besides being instrumental in the production of *Picturesque and Romantic Travels through Old France*, published the novel *The Story of the King of Bohemia and His Seven Castles* (1830)—the title is borrowed from a tale mentioned in *Tristram Shandy* that never quite gets told in either work.[49] Nodier imitated Sterne's typographical fancies, varying the typeface and font and occasionally composing the text in pictographic format, but he also did something that Sterne did not do: he engaged the services of an artist to add vignettes interspersed throughout the text. Though he was not credited in the book, that artist was Tony Johannot in his first major commission. The resultant work, a collaboration between artist and writer, is credited with having introduced wood engraving to French book illustration, an innovation that Johannot would develop more fully later, in his *Voyage Where You Will.*[50] Perhaps it was through following in the footsteps of Laurence Sterne and Charles Nodier that Johannot found the inspiration to create his own novel as a vehicle to display the full range of his graphic imagination. Neither of the two earlier works, however, can compare with Johannot's *Travel Where You Will* as an illustrated book: *Tristram Shandy* has very few images; *The Story of the King of Bohemia* is more conventional in illustrative strategy. *Travel Where You Will* stands alone as a free-form illustrated novel, equaled in imaginative structure only by Grandville's *Another World.*

GRANDVILLE'S *ANOTHER WORLD*

Grandville published *Another World*, a loosely organized novel in pictures, in 1844, after Johannot's *Travel Where You Will*, although he had completed it earlier, in 1843. While the two works are quite dissimilar, Grandville claimed that Johannot had plagiarized his idea, provoking a bitter quarrel and narrowly avoiding a duel.[51] Johannot's project had been actively collaborative between the artist and two writers: its final illustration depicts the two writers' pens and the artist's pencil lying together in peaceful repose. Grandville's *Another World*, however, shows the writer and artist in opposition: in its opening pages the artist's pencil declares its independence, and, as it sets off on its own adventures, the writer's pen thumbs its nose in derision.

Grandville actually attempted to write the text himself but gave up, stating "the pen rebels in my hands at forming sentences."[52] His publisher, Henri Fournier, then engaged Taxile Delord (1815–1877), the editor of the journal *Le Charivari,* to write the text, although the contract stipulated that it would be written according to Grandville's own notes.[53] Probably it was in deference to Grandville's decisive role in the conception of the work that Delord's name appears nowhere except on the final page—a gesture that would have infuriated Lagenevais. Grandville's nonlinear concept of narration is demonstrated by the work's complete title: *Another World: Transformations, Visions, Incarnations, Ascensions, Locomotions, Explorations, Peregrinations, Excursions, Vacations, Cosmogonies, Phantasmagorias, Reveries, Gaiety, Jokes, Whimsies, Metamorphoses, Zoomorphoses, Lithomorphoses, Metempsychoses, Apotheoses, and Other Things.*[54] The organization of Tony Johannot's *Travel Where You Will* had been similar, although not quite so radical. In this way, both artists demonstrate a logic that is peculiarly visual, rather than linear and narrative, and one that would resurface only later in the experimental novels and poetry of the twentieth century.

In *Another World,* neither Grandville's images nor Delord's text present the kind of narrative that nineteenth-century audiences had come to expect, and so it is no surprise that there was only one edition, although it now ranks among the most sought-after of nineteenth-century illustrated books. Grandville's contemporaries much preferred his earlier work, *Scenes of the Public and Private Life of Animals,* which was more conventional in narrative structure. Indeed, Théophile Gautier, in his 1847 obituary for Grandville published in *La Presse,* singled out this earlier work as Grandville's "true masterpiece," while *Another World* was not even mentioned.[55] In *Another World,* Grandville is thinking in pictures, in the kind of free association characteristic of the dream state, of Surrealism, or of later experimental literature. With these works, the very definition of narrative is challenged. The linear and logical development of plot in nineteenth-century novels was itself new and had barely replaced the earlier picaresque formula that was basically a compilation of adventures, such as Lesage's *Gil Blas* or Cervantes's *Don Quixote*—or even early comics. Grandville basically rejected both structures, the tight narrative form of contemporary novels as well as the earlier loose picaresque model, in favor of presenting his images in series, each organized around a fantastic idea and illustrated by the text. For example, in chapters 3 and 4, "A Steam Concert" and its sequel "The Rhubarb and the Senna," Grandville was inspired by the contemporary interest in steam power, the basis for much of nineteenth-century industry. He starts from the notion that if steam can power industrial machinery, trains, and boats, why not a steam orchestra? He then imagines what that would look like and how it would function—or malfunction, as the case may be. Trombones could be powered by steam, but so could singers and their accompanists. In a manner that would become familiar in the twentieth century through the calligrams of Guillaume Apollinaire, music is represented by words scattered across the page: *ha ha ha ha ha, ah ah ah ah, oh la la la la!* (fig. 157). There could

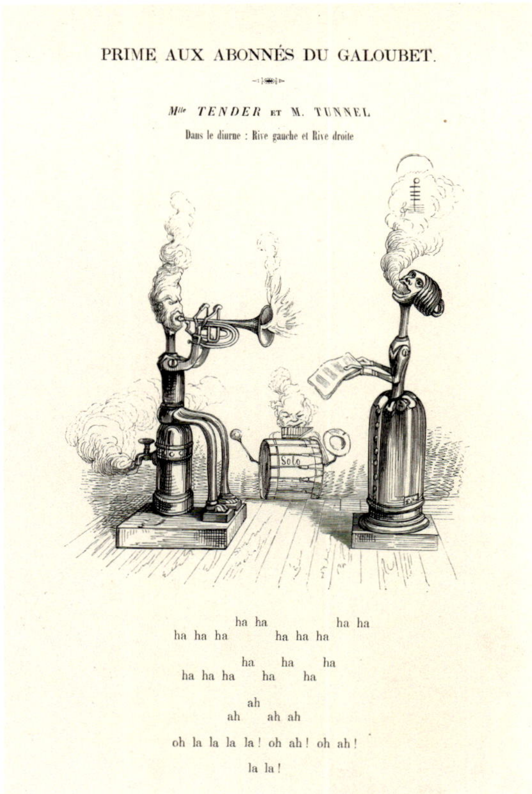

FIGURE 157
Grandville, *Ms. Tender and Mr. Tunnel in the Duet "Left Bank and Right Bank,"* from *Another World* [*Mlle Tender et M. Tunnel dans le diurne "Rive gauche et Rive droite" / Un autre monde*]. Paris: H. Fournier, 1844, 20. Wood engraving. Bibliothèque nationale de France.

OPPOSITE **FIGURE 158**
Grandville, *At the Fireworks in D*, from *Another World* [*Dans le feu d'artifice en ré / Un autre monde*]. Paris: H. Fournier, 1844, 24. Wood engraving. Bibliothèque nationale de France.

even be a steam-driven child-prodigy musician, who, as Grandville tells us in a caption cleverly mimicking a child's handwriting, *Has Barely Left His Nursemaid, at Most Twenty-Two Months Old I Would Guess.*[56] Too much steam, however, could be disastrous, as noted in the text accompanying one of the work's most imaginative images: "At the fireworks in *D*, at the moment when the fugue is ending *smorzando* with a sweet and dreamy melody, an ophicleide suddenly burst from too much harmony like a bomb hurling forth black notes and white notes, screeching trills, quavers, and semiquavers. clouds of musical smoke and flames of melody filled the air. the ears of many music lovers were damaged, others were hurt by the clashing notes of the key of *F* with the key of *G*" (fig. 158).[57] In Grandville's image, the ophicleide, a large tuba-like instrument, crouches on the concert stage like a beast of prey, spewing out anthropomorphic notes that either attack the audience, represented as large ears, or collapse, stricken, onto the stage.

Each chapter in *Another World* begins with a different concept. In chapter 6, "In Flight, a Bird's-Eye View," the protagonist's ability to fly through the air results in a series of images seen from above. This theme is revisited in chapter 20, "Aerial Locomotion," which shows the bizarre possibilities of human flying machines. Chapter 24, ""The High and the Low," depicts a society where the rank of the inhabitants is indicated by their dans l'atmosphère. Plusieurs dilettanti ont eu les oreilles déchirées, d'autres ont été blessés par les éclats de la clé de *fa* et de la clé de *sol*.

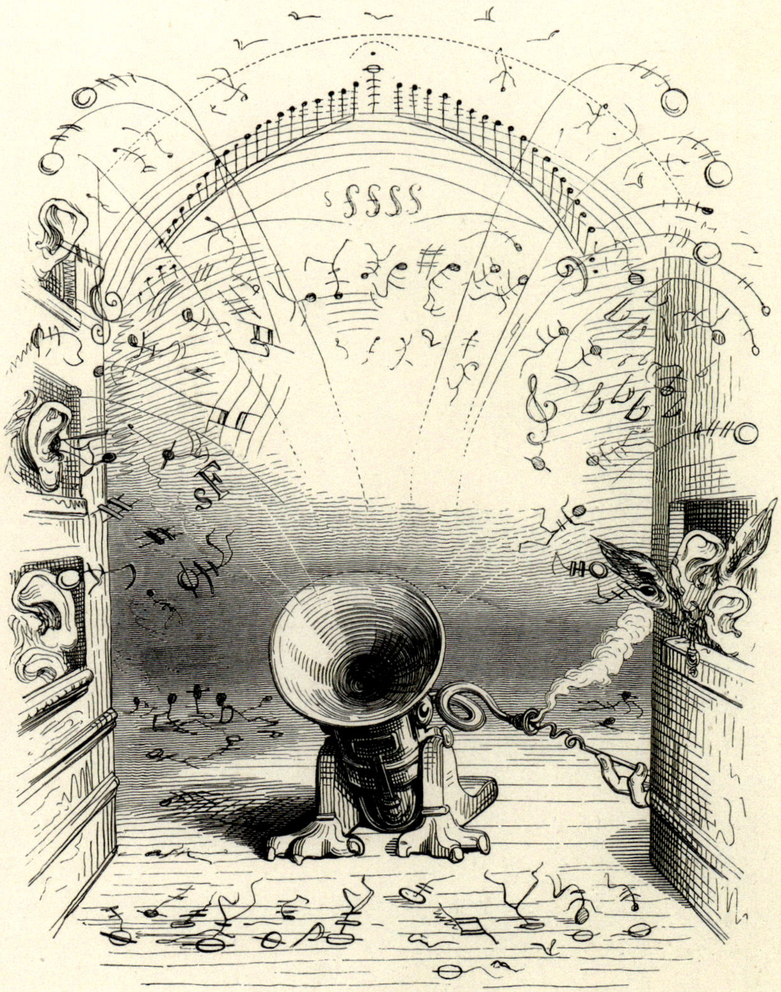

Des mesures sont prises pour qu'un pareil accident ne se renouvelle plus.

Le Dr Puff, qui prouve par son double caractère qu'Esculape est fils d'Apollon, a prodigué à tous les assistants les soins de son art avec un désintéressement au-dessus de tout éloge.

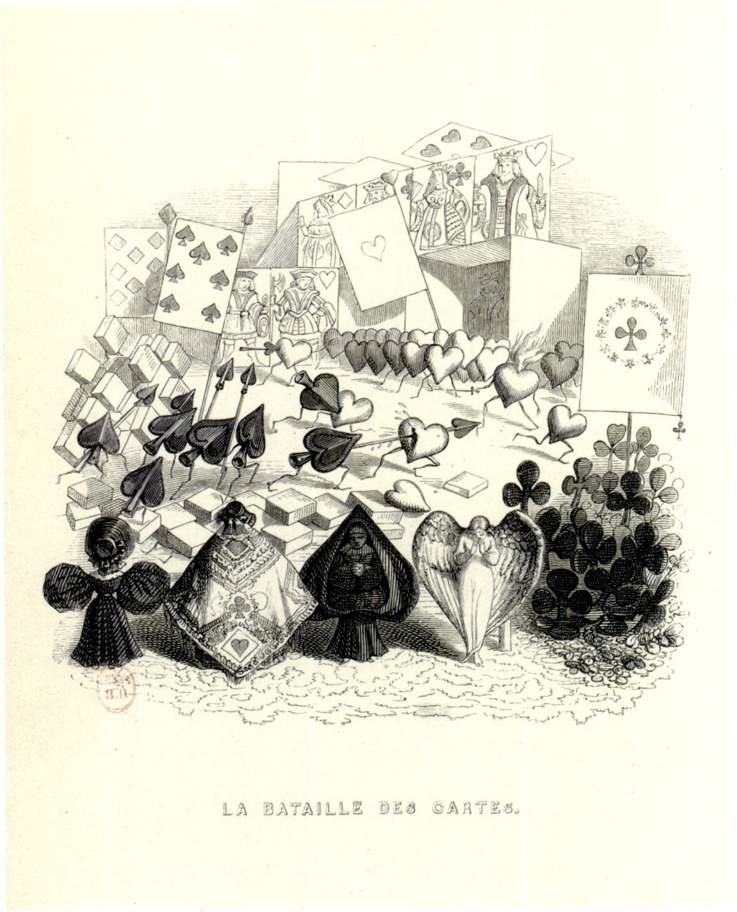

FIGURE 159

Grandville, *The Battle of Playing Cards*, from *Another World* [*La bataille des cartes / Un autre monde*]. Paris: H. Fournier, 1844, 246–47. Wood engraving. Bibliothèque nationale de France.

relative height. Probably his most influential image occurred in chapter 32, "The Metamorphoses of Sleep," where *The Battle of Playing Cards* (fig. 159) clearly provided inspiration for John Tenniel's drawings for Lewis Carroll's *Alice's Adventures in Wonderland* (1866).[58] Tenniel borrowed Grandville's trope of playing cards for *The King and Queen Inspect the Tarts* where the entire royal court is made up of playing-card figures (fig. 160).

In *Another World*, Grandville shifts easily between modes, from fantasy to satire. In chapter 14, "The Louvre of the Marionettes," the protagonist Hahblle makes a balloon flight to an exotic country peopled with marionettes instead of humans. He is surprised to discover that this strange country has a national museum very like the Louvre, and even, like Paris, has an annual Salon exhibition of art that he can visit. Grandville then takes us through this imaginary exhibition, a thinly disguised satire on the French art establishment. Paintings here are so big that the museum entrance must be demolished to bring them inside. Upon entering, Hahblle finds a catalogue that looks very like the *livrets* sold at the Paris Salon that list all the works in the exhibition. With the aid of this catalogue,

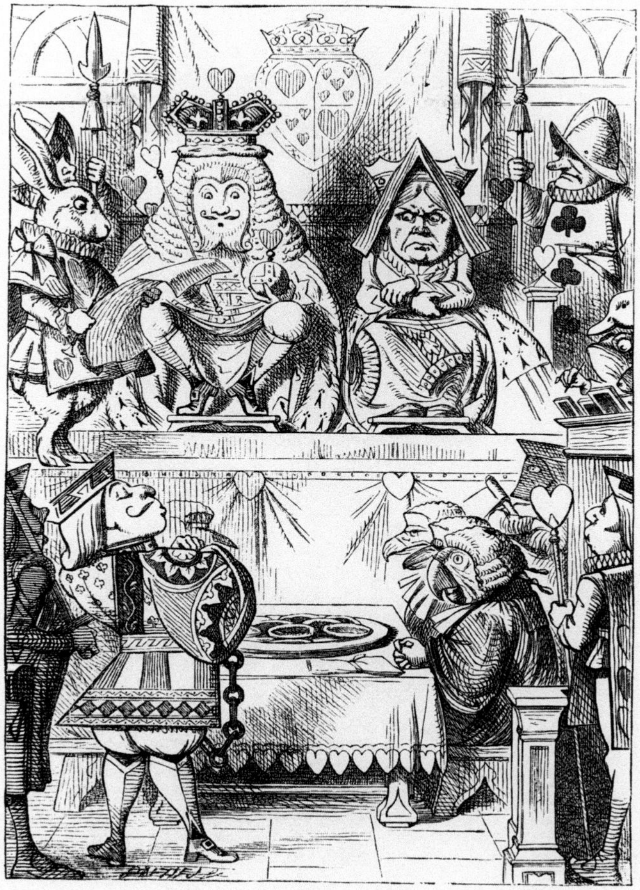

FIGURE 160
John Tenniel, *The King and Queen Inspect the Tarts*, from Lewis Carroll, *Alice's Adventures in Wonderland*. London: Macmillan, 1866. Wood engraving.

Grandville then depicts these works (fig. 161). Prominent among them and listed first in the catalogue is *The Angel of Painting Imploring Divine Mercy for the Jury.*[59] Above it is *The Crossing of the Red Sea*, which is indeed very red, and surrounding it are sendups of other Salon favorites: *An Eclogue from Virgil* (three trees, two cows), *Portrait of Mme P. de L. with Her Dog and Her Diamonds* (seen from the back), *The Angry Waves* (which emerge from the frame and threaten to swamp the gallery), and finally *A Frame Worth Two Thousand Francs Based on the Process of Ruolz and Elkington*, a frame so large that we can't even see the painting within it.[60] All of these works are being admired by mechanical marionettes, who, like Parisian audiences, clearly lack any aesthetic discrimination.

Hahblle proceeds to another gallery where he reads the critics' reviews aloud, each a satire on the pomposity and vacuity of art critics and each accompanied by Grandville's literal rendering of the critics' words: "Nothing is comparable to the battle scene of our great painter Jérôme Tulipier. What combat! What shock! What turbulence! What tumult! What an eruption! Angry heads, menacing arms, sabers and swords, everything is alive,

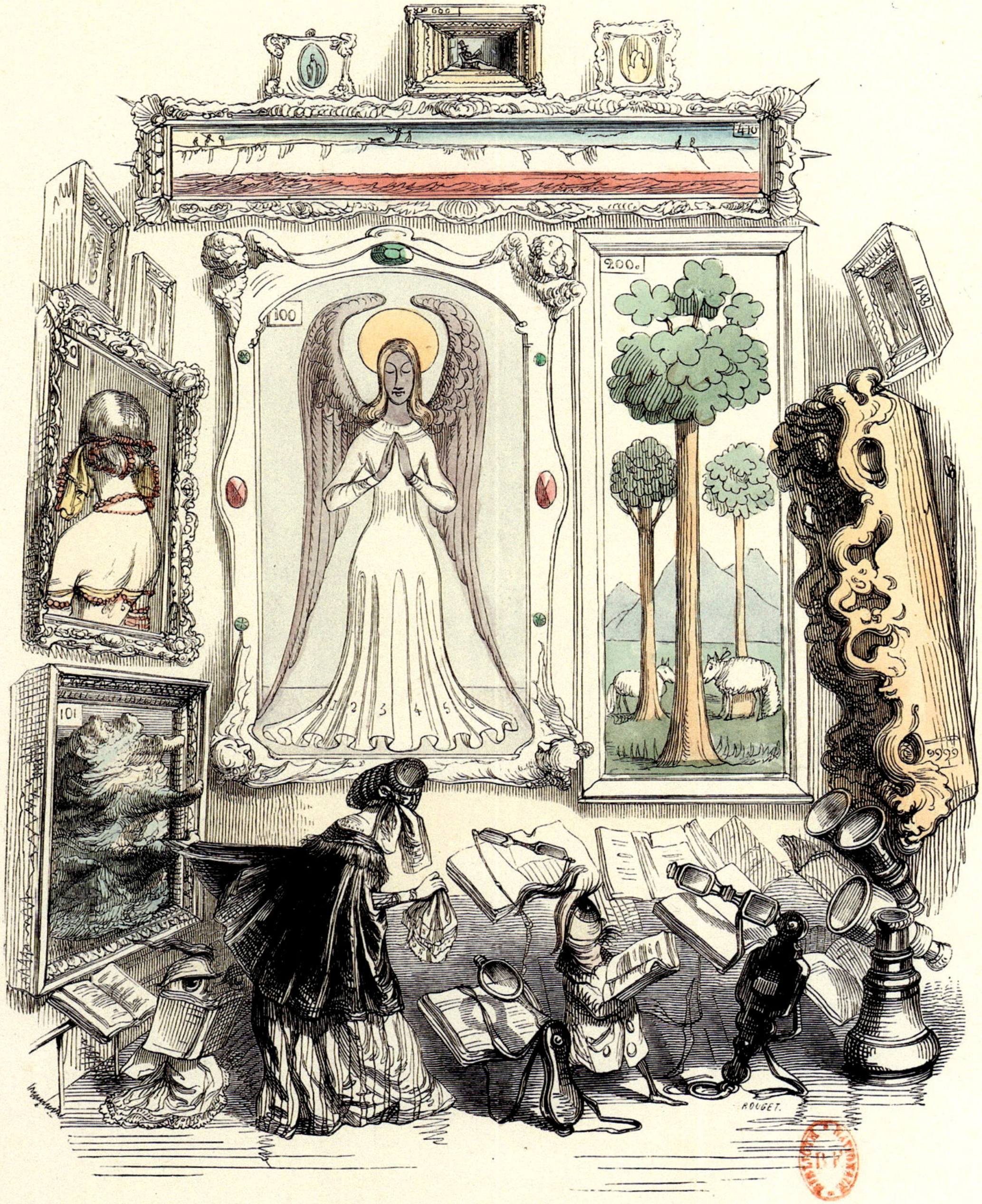

LE LOUVRE DES MARIONNETTES.

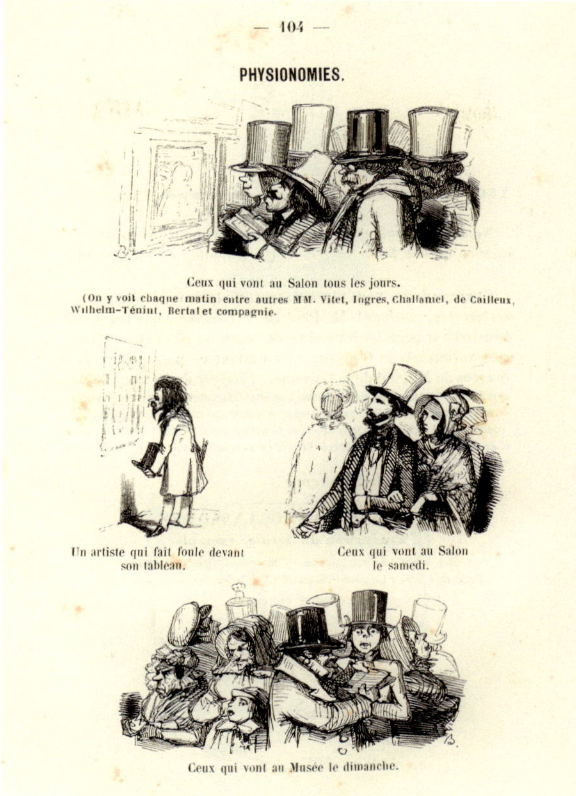

OPPOSITE FIGURE 161
Grandville, *The Louvre of the Marionettes*. Engr. Rouget, from *Another World* [*Le Louvre des marionettes / Un autre monde*]. Paris: H. Fournier, 1844, 84–85. Wood engraving. Bibliothèque nationale de France.

FIGURE 162
Bertall, *Physiognomies*, from *The Salon of 1843* [*Physionomies / Le salon de 1843*], from *Les Omnibus*, 7me livraison, 1843, 104. Wood engraving. Houghton Library, Harvard University. Typ 815.43.2040 no. 7.

emerges right out of the canvas, and does battle"—and indeed, the horses, warriors, and weapons depicted in this painting do emerge directly into the gallery space.[61] "This morning, when the windows were opened to let in a breath of air for a lady who had just fainted from the heat, birds entered and threw themselves on the landscape by our famous Thomas Gorju representing an orchard in Normandy"—here Grandville shows us live birds circling the painting.[62] "The brilliance of *A Sunrise* has dazzled everyone, even a blind mole who happened to wander into the gallery"—in Grandville's rendering, the canvas has disappeared into a white blaze.[63]

In form, "The Louvre of the Marionettes" follows the Salon in Caricature discussed in the preceding chapters, a genre that first appeared in 1843 (see fig. 55), the year that Grandville began working on *Another World*. Although caricatures of individual works of art had already become familiar features in the periodical press, Bertall had, in 1843, begun the practice of publishing a series of images satirizing the installation and the art-viewing public as well as individual works of art. In *Physiognomies* from his *Salon of 1843*, Bertall caricatures artists who come to the Salon to see their own work, bourgeois who can afford to pay the admission fee to visit on Saturdays, and the working classes who are obliged to brave the crowds on Sunday, the only free day (fig. 162).[64] The Salon in Caricature quickly

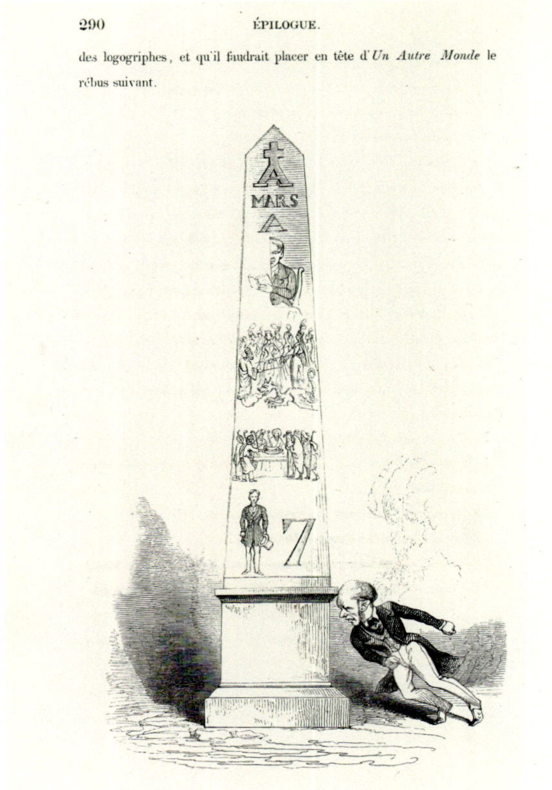

FIGURE 163
Grandville, *Ah, Believe Me, Dear Reader, Do Not Behave Like This!*, from *Another World* [*Ah! Crois-moi, ami lecteur, ne fais pas comme cet / Un autre monde*]. Paris: H. Fournier, 1844, 290. Wood engraving. Bibliothèque nationale de France.

became a standard feature of the annual art season, published in the illustrated press and sold as inexpensive booklets that resembled and parodied the exhibition catalogue. Nonetheless, no Salon in Caricature ever proved as radical as Grandville's "The Louvre of the Marionettes." Individual images in the caricatured Salons were often witty, and caricatures of major paintings are still familiar to us today, but as series they lacked the unifying narrative premise and biting satire of Grandville.

Grandville ended *Another World* with a rebus cautioning the reader against attempting to too closely decipher his images (fig. 163). *Ah, Believe Me, Dear Reader, Do Not Behave Like This!* the caption warns, as a man butts his head against stone. In other words, we should not attempt to read images the way we read text. Grandville's choice of a rebus was significant because rebuses had just become all the rage in Paris, introduced that very year in the journals *Le Charivari* and *L'Illustration* (see fig. 86).[65] Rebuses present the very essence of the word-image relationship, where the images literally incarnate the words and vice versa. This returns us to Thomas Rowlandson, who began *The Tour of Doctor Syntax, in Search of the Picturesque* with a rebus, the first three letters of the word *picturesque* being formed by picturesque ruins (fig. 164). These images well illustrate the kind of nonlinear imaginative structure and broad spectrum of illustrational strategies

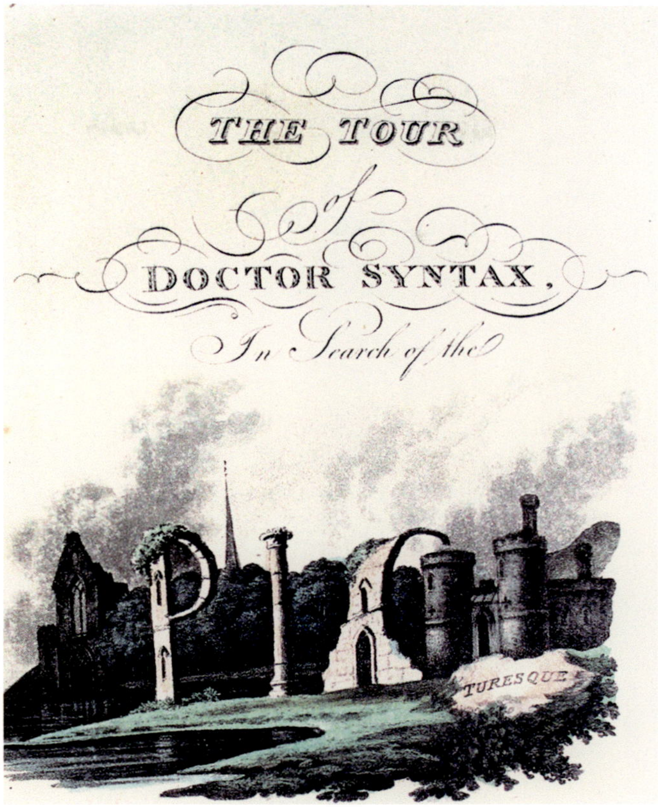

FIGURE 164

Thomas Rowlandson, frontispiece, *The Tour of Doctor Syntax, in Search of the Picturesque.* London: R. Ackermann, 1812. Hand-colored aquatint.

characteristic of the work of nineteenth-century artists. Those I have discussed here, and many of their colleagues, were among the first to seek alternative forms of narration, introducing new formal languages into the visual arts.

The introduction of a new medium often results in a period of fertile experimentation before it settles into a tradition. After the decades of the mid-nineteenth century, when book illustration was defining itself, there were many renowned artists devoted to the medium, Gustave Doré chief among them. The efflorescence of the early decades gradually subsided, however, and a division of labor became, once again, the norm: the writer provided the text and only then did the artist draw. Although the illustrated book scarcely outlasted the century that saw its birth, surviving today principally in the realms of the artist's book, children's literature, and the graphic novel, it had its golden age in the nineteenth century, when more illustrated books were published than in the entire world's history until then. But if we continue to think of illustration merely as the servant of literary texts, or as somehow inferior to painting and sculpture, we will not only fail to understand much of nineteenth-century visual culture, but we will also deprive ourselves of the singular pleasure that comes from the simultaneous enjoyment of word and image.

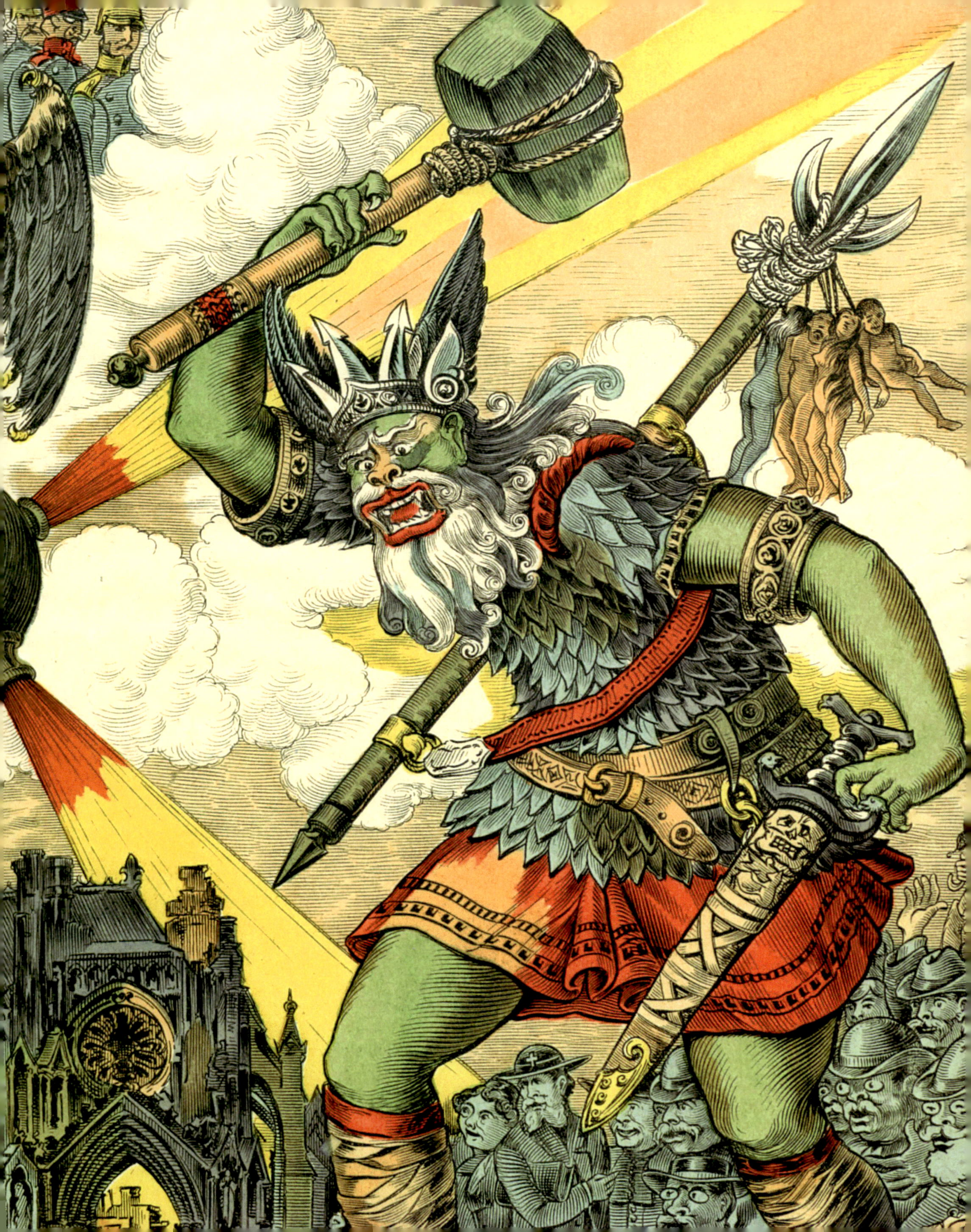

5 The Curious History of Popular Imagery in France

This study of illustrated print culture would not be complete without a discussion of popular prints, whose history of innovation begins earlier but then interweaves with that of caricature, the illustrated press, comics, and illustrated books. Earlier chapters focused on each medium's early decades of innovation and expansion, but popular prints followed a different trajectory and continued to mutate throughout the century. Chameleon-like, they absorbed influences from the full spectrum of printed imagery, responsive to every shift in taste, style, and markets. As a result, this chapter will chart an arc both different from and longer than the others.

The earliest popular prints were broadsheets, single-sheet woodblock prints, crudely cut and garishly colored by stencils. Created by artisan-designers, they were cheap and ephemeral, intended for a semiliterate, mostly rural, audience.[1] Such prints, depicting saints, rulers, legends, and current events, had been produced for centuries all over Europe. Early printmaking centers existed throughout France, in Orléans, Lille, and Beauvais, for example. *Our Lady of Liesse, Pray for Us,* published in Orléans by the printer Jean-Baptiste Letourmy (1747–1800), is typical of the genre with its simplified contours and coloring (fig. 165). In the nineteenth century such printmaking became concentrated in eastern France, in Épinal, south of Nancy. This small town eventually became synonymous with French popular print production and, as a result, these prints are often called *images d'Épinal,* regardless of where they were actually produced.

In many ways these images can be understood as the rural parallel to the lithographs discussed in chapter 1, although their point of view rarely encompassed the sly, cynical commentary of the urban lithographic imagery that we saw in chapter 1. Popular prints spread news of current events well before the advent of the illustrated press, they portrayed heroes and homilies and religious figures, and, like illustrated books, they depicted stories and legends. They provided both entertainment and household decoration; in poorer households, they were often tacked up on walls. Like all ephemeral art, they were discarded and replaced often, and for much of the populace, they constituted the only art ever seen. By the twentieth century, however, modern technology and advanced literacy had narrowed their audience to children, whose affection for picture books and comics persists even today.

DETAIL OF FIGURE 200

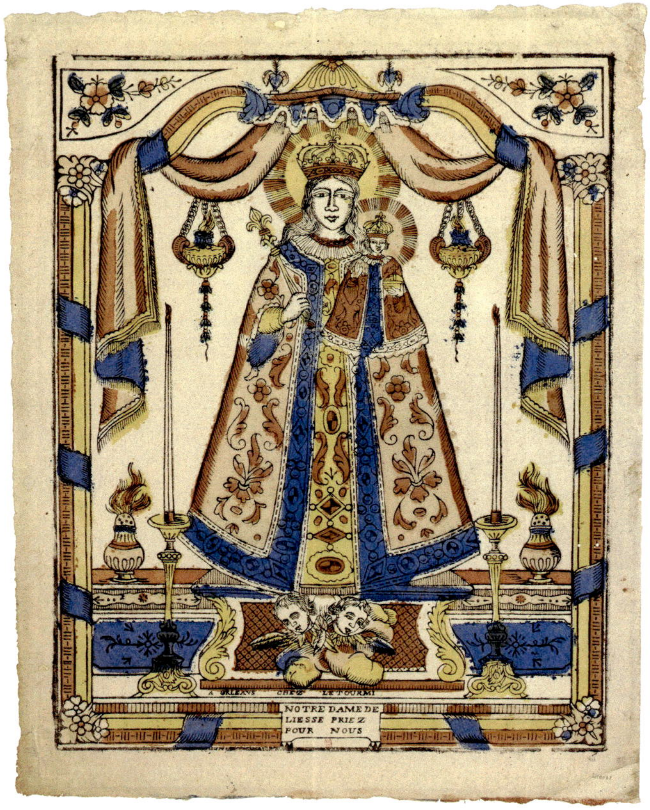

FIGURE 165
Our Lady of Liesse, Pray for Us [*Notre Dame de Liesse, priez pour nous*], 1774–1800. Pub. Letourmy, Orléans. Stencil-colored woodcut. Bibliothèque nationale de France.

In Épinal, the firm of Pellerin, established in 1796, was the largest publisher and has remained in business for over two centuries. Renamed Imagerie d'Épinal in 1984, it still operates, located on the outskirts of the city where it occupies the factory built in 1897 after its original workshop was destroyed by fire.[2] Pellerin's earliest productions were playing cards, a lucrative business since they wore out with repeated use and had to be regularly replaced (fig. 166). By the late nineteenth century, the firm had moved on to comics, producing amusing strips such as Dous Y'Nell's *Amazing Fertilizer Potion*, about the great scientist Poussdru, who invents a fertilizer so powerful that his houseplant eventually engulfs his house (fig. 167). And yet, despite this long history, the longest in the French print trade, the trajectory of the *image d'Épinal* from the playing cards of the eighteenth century to the comic strips of the twentieth has yet to be conceptualized in its entirety.

A BRIEF HISTORY OF *IMAGES D'ÉPINAL*

The playing cards and religious imagery that constituted the earliest popular prints appealed, one might say, to both ends of the consumer spectrum. Such prints were usually sold by traveling salesmen, peddlers known as *colporteurs*, who also sold the cheap books called *bibliothèque bleue* (for their blue covers), as well as almanacs, catechisms, and broadsheets.[3] Colporteurs also sold a good deal of pornography and politically suspect images and texts, all of which gave them an aura of ill repute that rubbed off on their wares as well.

The earliest prints produced in the town of Épinal date from the seventeenth century, but Nicolas Pellerin (1703–1773), who gave his name to the firm that came to dominate Épinal printmaking, did not appear there until 1735–1740.[4] By trade, he was a designer of playing cards; by politics, he and his family were Freemasons, which rendered them politically suspect across several regimes. His brother Gabriel Pellerin also designed playing cards, as did Gabriel's son Jean-Charles Pellerin (1756–1836), who was the real force in the establishment of the Maison Pellerin in 1796.[5]

By 1800, four years after the establishment of the Pellerin firm in Épinal, Jean-Charles Pellerin had installed four printing presses and had expanded production from playing cards to include the full range of colporteur printed wares, both books and images.[6] Throughout the nineteenth century, the Maison Pellerin continued to design, print, and market a variety of books and images, prospering and expanding. In 1810 it produced 16,000 prints; in 1823, 102,000; by 1842 it was averaging several million annually, and in the period from 1870 to 1914 the average had increased to 10 to 15 million images a year, and Pellerin prints were being shipped throughout the world.[7] Even as early as 1845 the firm employed eighty to one hundred workers, an enormous enterprise by the standards of the time. In 1876, the playing-card branch of the family business was sold to the Parisian firm of Grimaud, and, after this, Pellerin focused on the broadsheets and albums that, by this time, had a wide international market. It is difficult to establish an accurate chronology of imagery because Pellerin, like other printers of his day, continued to reproduce earlier works and borrowed freely from wherever he found them; since international copyright protection was not established until the 1886 Berne Convention, this kind of "borrowing" was common practice. As a result, there are family resemblances of basic print types throughout Europe and the Americas, and untangling who did what first is a Herculean task.[8] The earliest Pellerin catalogue dates from 1814, when *The Stages of Life* had already been published (fig. 168). Pellerin reissued it, with minor changes, in 1826, and again around 1850; Armand-François Hurez published virtually the same image sometime around 1817 in Cambrai, as did Frédéric Dupont-Diot in Beauvais in the late 1820s.[9] There were numerous later variations, and there were, no doubt, earlier ones as well. This simple graphic image, symbolizing the inevitable progression of human life as an ascending and descending staircase from cradle to grave was evidently a source of comfort over many generations.

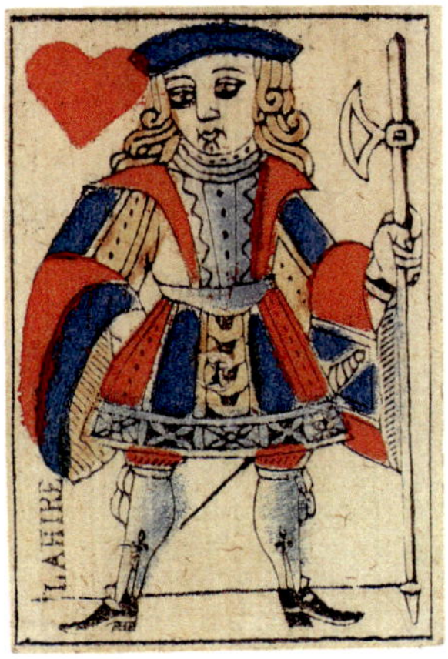

FIGURE 166
Jack of Hearts [*Valet de coeur*], playing card, late eighteenth to early nineteenth century. Pub. Pellerin, Épinal. Stencil-colored woodcut.

What is particularly noteworthy about the trajectory of the Pellerin firm is that, despite its almost total identification with what historian Peter Laslett has called "the world we have lost"—in other words, with a nostalgic view of a simpler rural past—from the very beginning the firm adopted industrial methods and constantly updated its

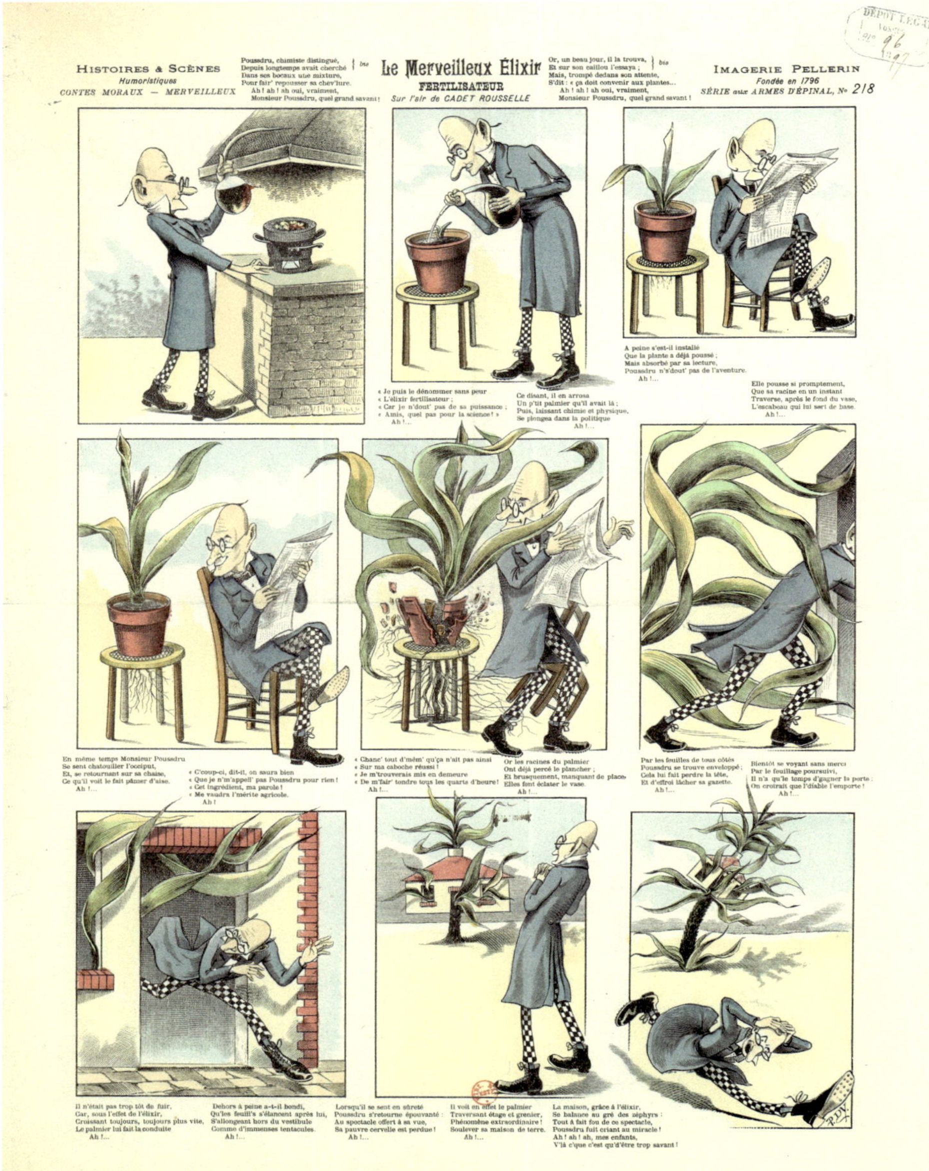

FIGURE 167

P. Dous Y'Nell, *The Amazing Fertilizer Potion* [*Le merveilleux élixir fertilisateur*], *Série aux armes d'Épinal*, no. 218, 1897. Pub. Pellerin, Épinal. Color lithograph. Bibliothèque nationale de France.

subject matter.[10] In 1809 Pellerin added stereotype production to his establishment.[11] We encountered this process in chapter 2, where it facilitated the expansion of the illustrated press. These metal plates, produced from a plaster mold of the original woodblock and used for the actual printing, had two major advantages: since numerous stereotypes could be cast from the same woodblock, quantities of identical prints could be produced by numerous presses working simultaneously, and, if damaged, the stereotype could be easily replaced, while the woodblock would have to be recut. For Pellerin this date, 1809,

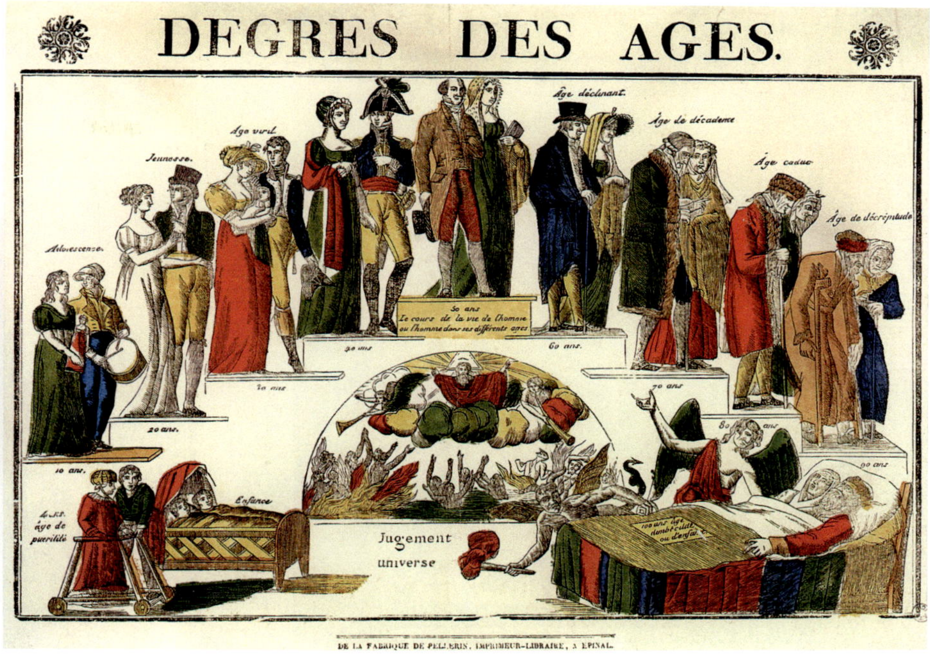

FIGURE 168
The Stages of Life [*Degrés des âges*], before 1814. Pub. Pellerin, Épinal. Stencil-colored woodcut.

marked the beginnings of industrialization, although histories of the publishing house prefer to date industrialization to later in the century, thus recouping most of his production for an idealized preindustrial past. Indeed, the two-volume catalogue of French popular prints published by the Bibliothèque nationale de France and the Musée national des arts et traditions populaires terminates in the 1860s and includes only the two thousand or so prints originally cut or engraved on wood or metal plates, not the later lithographic or photographic processes.[12] Since it is customary to cite these earlier prints by their original medium (a practice observed here as well), it is easy to forget that they were actually printed from metal stereotypes, a more technologically advanced industrial process.

Before the Revolutionary period of the late eighteenth century, religious imagery was predominant. Such images, known as *images de préservation* after their intended purpose of protection, were most often tacked up on the walls of dwellings to invoke the intercession of the depicted saint. St. Agatha, for example, protects against fire (fig. 169). St. Roch protects against the plague; St. Nicolas is the patron saint of sailors, while St. Geneviève protects the city of Paris. By the early decades of the nineteenth century, the bulk of popular print imagery included religious subjects, broadsheets showing contemporaneous events, portraits of rulers, military imagery, or well-known themes such as *The Stages of Life* (see fig. 168) or *The World Upside-Down* (see fig. 96). These images, cheap and ephemeral, formed the stock-in-trade of the colporteur.

Censorship of one kind or another was always a problem. Because pictures make a direct appeal even to the illiterate, censorship of imagery in France was always more

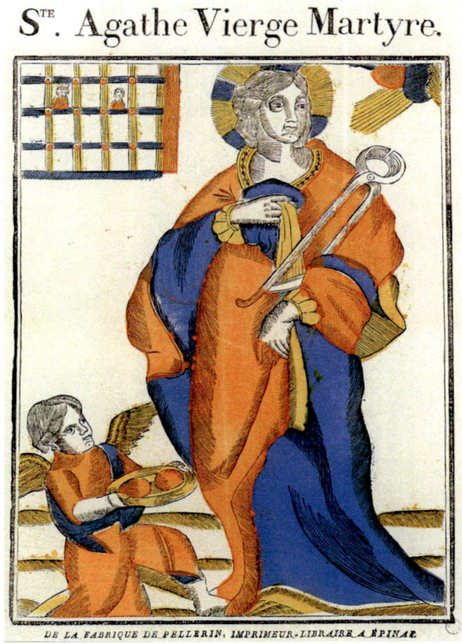

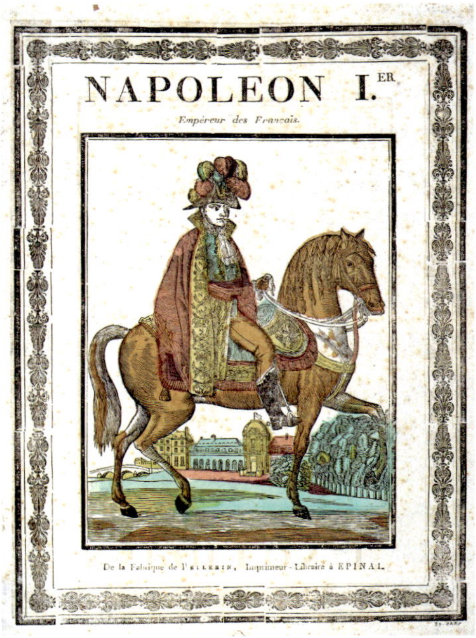

FIGURE 169

St. Agatha, Virgin Martyr [*Ste Agathe, Vierge Martyre*], before 1814. Pub. Pellerin, Épinal Stencil-colored woodcut.

FIGURE 170

Napoleon I, Emperor of the French [*Napoléon Ier. Empereur des Français*], 1804–1814. Pub Pellerin, Épinal. Stencil-colored woodcut. Musée de l'Image, Épinal, dépot MDAAC.

severe than censorship of the written word.[13] Printers and publishers (often the same person) were held responsible for their productions, and so Pellerin, like Charles Philipon, whom we encountered in previous chapters, had problems with government censors. In 1811 Pellerin was prosecuted for producing *The Willing Cuckold, or the Complacent Husband*, a traditional, if slightly risqué subject.[14] There was a huge market for Napoleonic broadsheets, second only to religious imagery in popularity; in Pellerin's 1814 catalogue, Napoleonic images make up about a quarter of his list, religious imagery about half. While Napoleonic subjects included portraits of his entire family, the hagiography focused principally on Napoleon himself; he is shown in traditional ruler mode, in full imperial regalia astride a horse, in *Napoleon I, Emperor of the French* (fig. 170). After Waterloo this market collapsed when the returning Bourbon monarchy ordered inventories of such prints destroyed. Pellerin, a committed Bonapartist, did not comply and was condemned to four months in prison and a large fine; he appealed the prison sentence successfully, but from time to time he had to pay fines for skirting the legal limits of allowable imagery.[15] In the 1830s, when the more liberal constitutional monarchy of King Louis-Philippe had replaced that of the restrictive Restoration, Pellerin became the main purveyor of Napoleonic imagery.[16]

During these decades, there were many changes in the market for popular prints, and Pellerin kept abreast of trends, constantly modernizing both imagery and production. In the aftermath of repeated revolutions that disrupted the print trade in the late eighteenth and early nineteenth century, it is understandable that publishers looked for subjects that would not be confiscated with every shift of regime. In the 1820s, Pellerin began

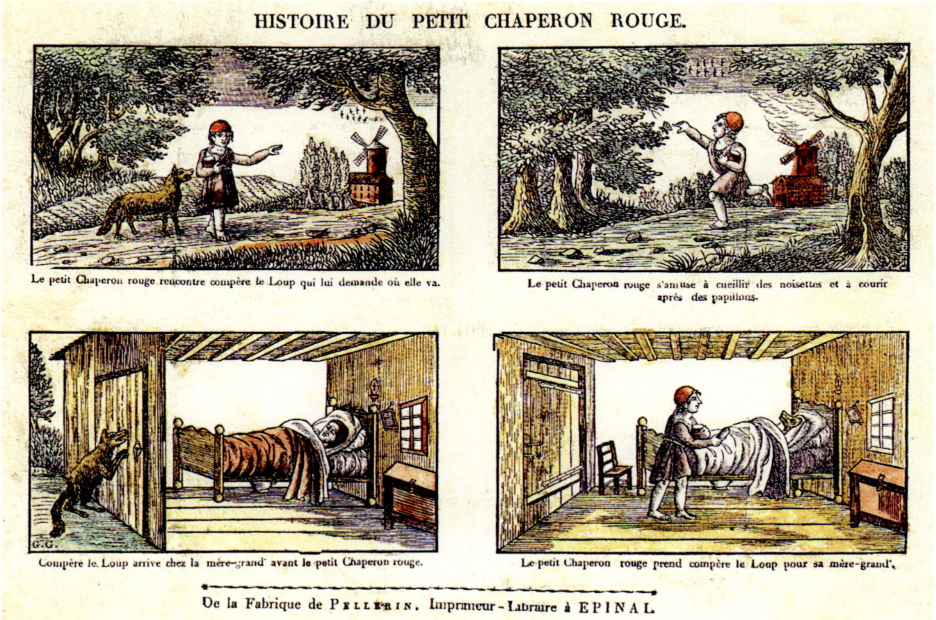

FIGURE 171
The Story of Little Red Riding Hood [*Histoire du petit chaperon rouge*], 1828. Stencil-colored woodcut. Pub. Pellerin, Épinal.

to produce broadsheets in boxed grid format, much like later comic strips. Often printed in two languages to facilitate international sales, they recounted tales and fables such as *Puss in Boots*, *Sleeping Beauty*, *Cinderella*, and *Little Red Riding Hood* (fig. 171).[17] These rapidly became central to Pellerin's inventory, and were regularly redesigned and reissued. In the course of the nineteenth century, children began to constitute a lucrative new market, and so Pellerin's production soon included songs and stories, games and cutouts.[18]

By 1846, the Maison Pellerin, now directed by Jean-Charles's son Nicolas Pellerin (1793–1868) and his son-in-law Pierre-Germain Vadet (1787–1870), was producing three and a half million prints annually, and indeed the years between 1820 and 1850 are widely considered the golden age of popular prints. After this, the weight of production shifted dramatically toward new printing techniques and new subject matter. Around 1850 the stereotypes that Pellerin had adopted in the first decade of the century were gradually replaced by lithographic stones.[19] Although lithography, discussed in chapter 1, had been invented at the turn of the century, it did not become commercially viable in France for several decades and, in periodicals, was rapidly eclipsed by wood engraving, as discussed in chapter 2. Popular print production followed the reverse course, abandoning woodcut and wood engraving for lithography. This eliminated the expense of stereotypes, since lithographic stones are capable of printing thousands of identical impressions without degradation and can then be ground down and reused. The transformation of popular print production from wood engraving via stereotypes to lithography was implemented by Jean-Charles's grandson, Charles-Nicolas Pellerin (1827–1887), who was trained in Paris by the pioneering lithographer Joseph Lemercier (1803–1887) and took over the firm

LES RÉFORMES DU MÉNAGE.

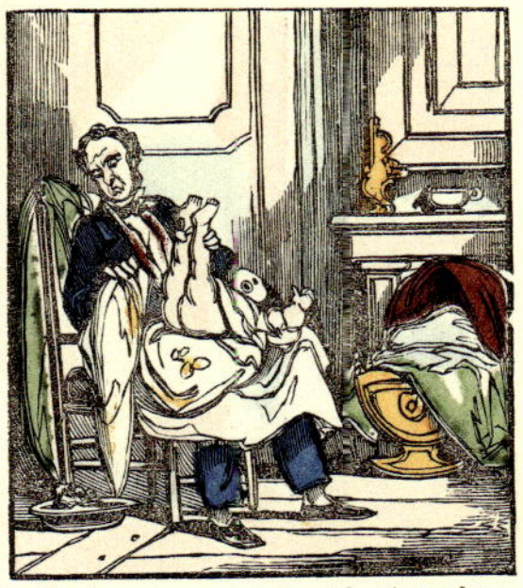

Les hommes apprendront à soigner les enfants.

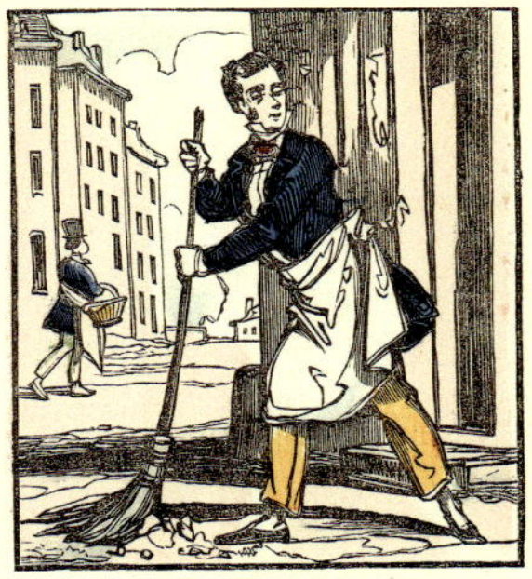

Ils balayeront et feront le ménage.

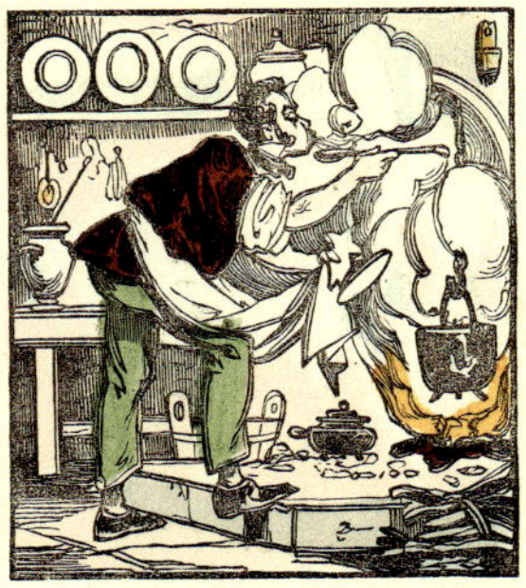

Ils feront la cuisine et éplucheront les légumes.

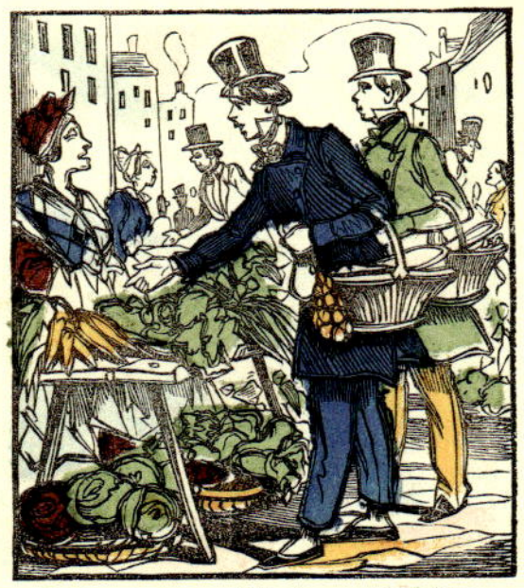

Ils iront au marché et aux provisions.

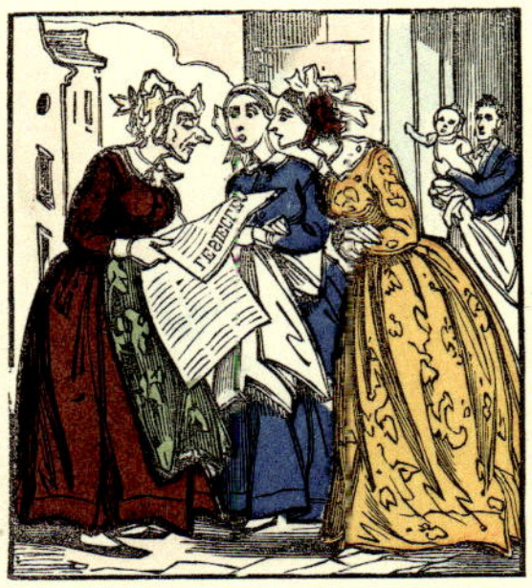

Les femmes feront les affaires et s'occuperont de politique.

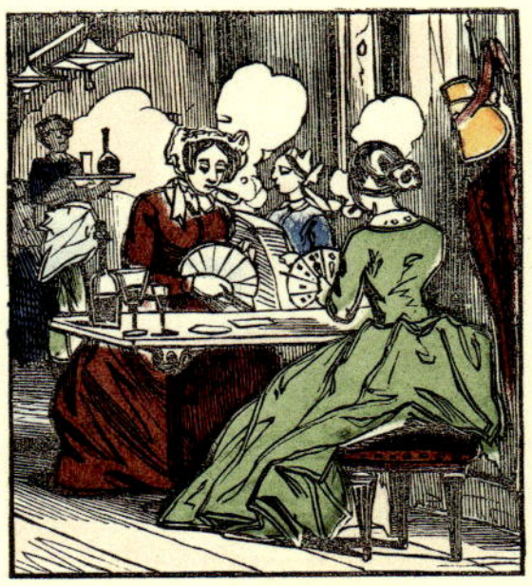

Les femmes iront à l'estaminet.

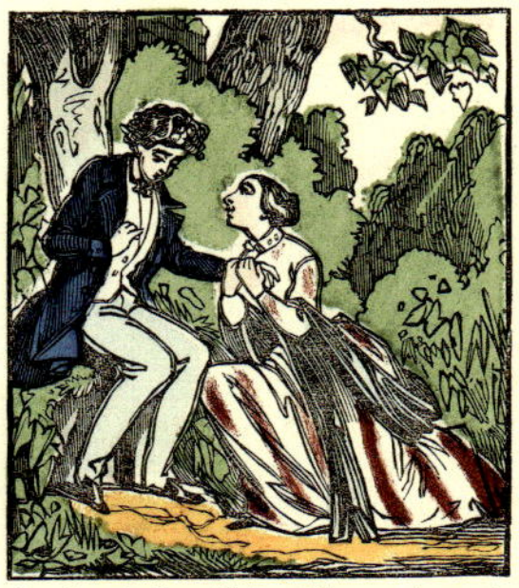

Les demoiselles séduiront les messieurs et feront des déclarations.

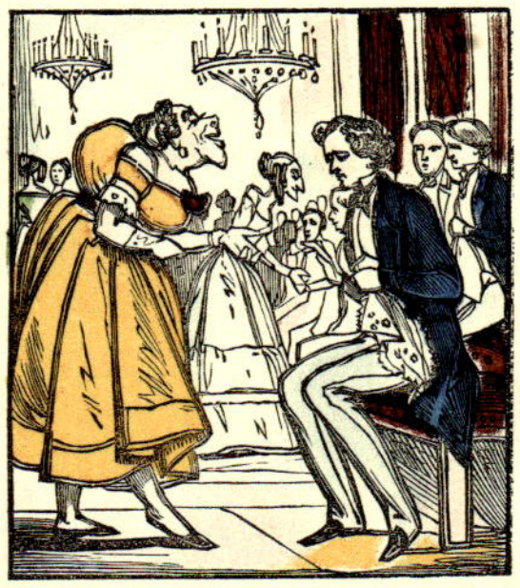

Au bal, les messieurs attendront qu'on les invite à danser.

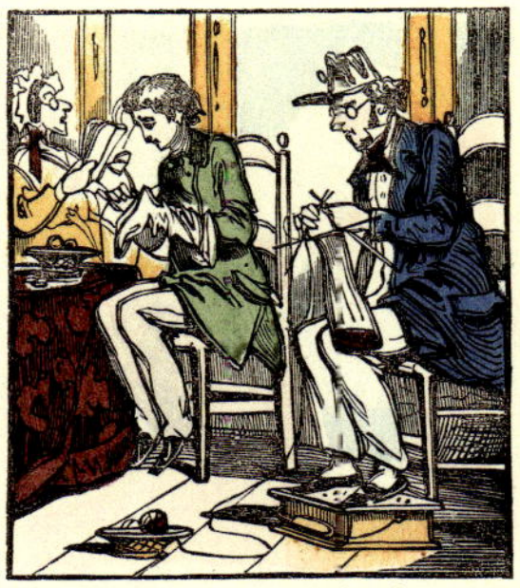

Les hommes devront savoir tricoter et raccommoder les bas.

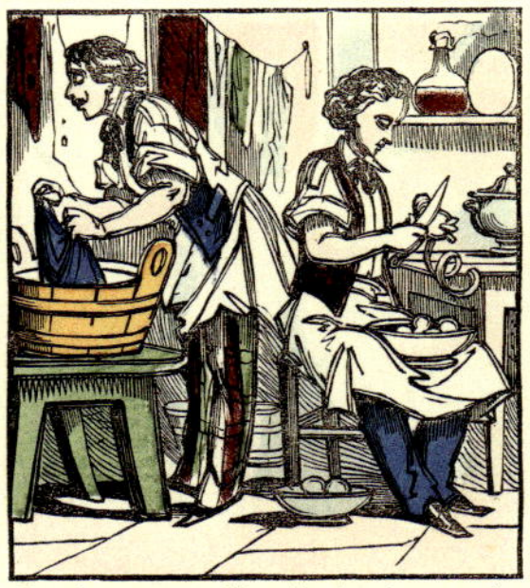

Les jeunes gens bien élevés aimeront à s'occuper de choses utiles.

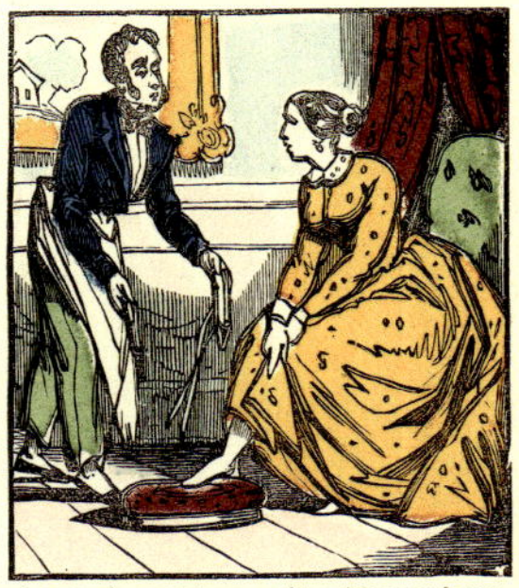

Les maris seront aux petits soins pour leurs femmes.

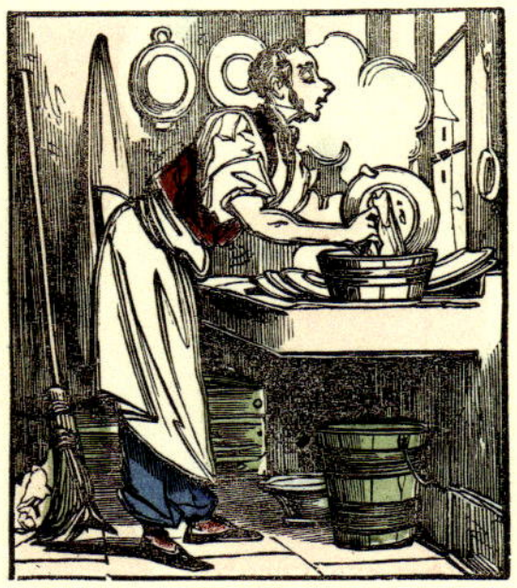

Les maris qui s'entendront bien au ménage seront très-considérés.

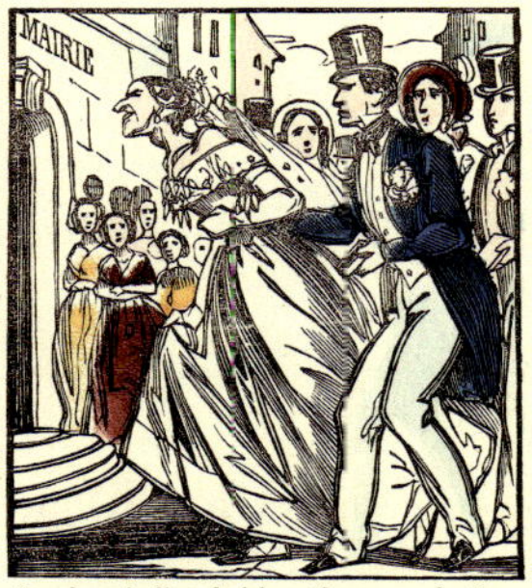

Les demoiselles choisiront leurs maris. Ces messieurs n'auront pas le droit de refuser.

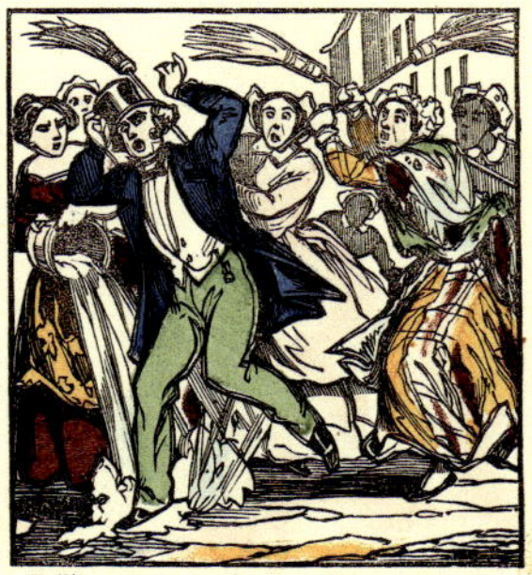

Voilà comme on traitera les garçons qui refuseraient de se marier.

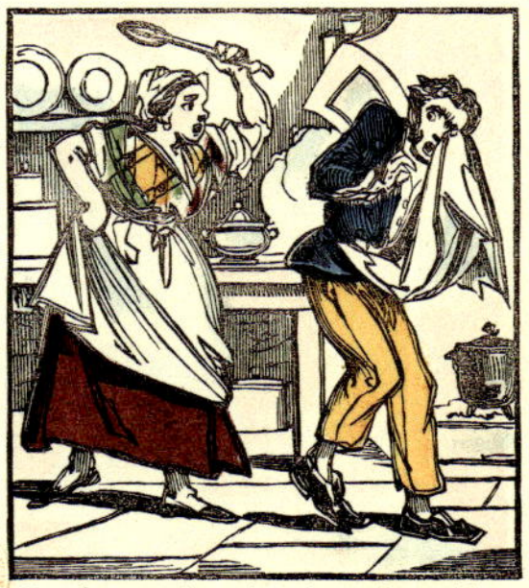

Les femmes étant les plus raisonnables, commanderont à la maison.

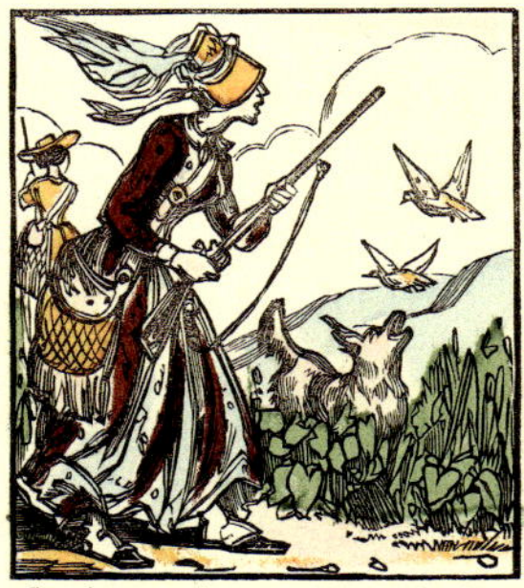

Les dames seules pourront prendre des ports d'armes.

317

Propriété de l'éditeur. (Déposé.)

Fabrique de PELLERIN, Imprimeur-Libraire, à ÉPINAL.

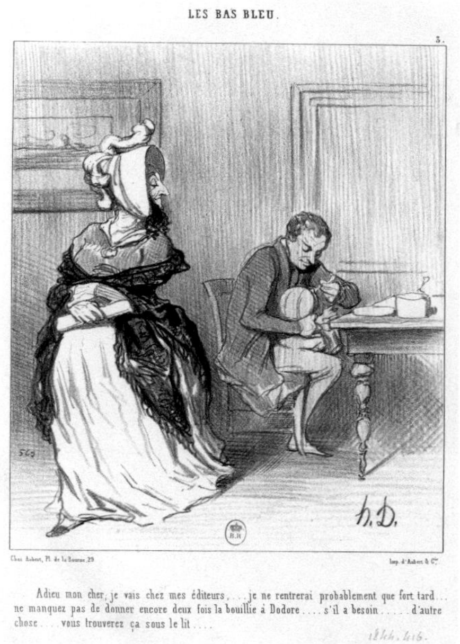

OPPOSITE **FIGURE 172**
Household Reforms [*Les réformes du ménage*], 1850. Pub. Pellerin, Épinal. Stencil-colored woodcut. Bibliothèque nationale de France.

FIGURE 173
Honoré Daumier, *Good-bye My Dear, I'm Going to My Publishers . . . I Probably Won't Be Back until Quite Late. Don't Forget to Feed Dodore Twice . . . If You Need . . . Anything Else . . . You'll Find It under the Bed . . .*, from *Bluestockings* [*Adieu, mon cher, je vais chez mes éditeurs; . . . je ne rentrerai probablement que fort tard . . . ne manquez pas de donner encore deux fois la bouillie à Dodore . . . s'il a besoin . . . d'autre chose . . . vous trouverez ça sous le lit . . . / Les bas-bleus*], no. 3, 1844. Lith. Aubert. Bibliothèque nationale de France.

in 1854.[20] Most studies of images d'Épinal conclude here, labeling the subsequent period one of degeneration because the Pellerin output underwent a transformation, not only in medium, but also in style and subject.[21]

By 1858, the Maison Pellerin had adopted the process of drawing on transfer paper with lithographic crayon, the technique that Töpffer had used for his lithographed comic albums in the 1830s.[22] By 1882, the Pellerin firm had begun using chromolithography, at least for its albums; this color printing from lithographic stones eliminated the stencil-coloring characteristic of popular prints or the laborious hand-coloring characteristic of more upscale print production.[23] Zincography, which replaced expensive limestone plates with metal ones, was introduced in 1890, quickly followed by the photomechanical processes that rapidly replaced all earlier printing techniques.

In addition to technical modernizations, the Pellerin firm made changes in subject matter, producing updated versions of traditional subjects. The 1850 *Household Reforms* perpetuated the traditional *World Upside-Down* themes (see fig. 96), but added new "upside-down" infractions: women drinking, smoking, and playing cards (fig. 172). The differences between images d'Épinal and other forms of graphic production were rapidly becoming insignificant; indeed, Daumier's own series of antifeminist caricatures, *Bluestockings*, published in *Le Charivari* in 1844, might well have served as intermediary between traditional World Upside-Down imagery and Pellerin's *Household Reforms.* Daumier continued the World Upside-Down theme of men doing housework and caring for children while women indulge in "male" occupations, but instead of the hunting shown in figure 96, a rural occupation, he substituted the more urban transgression of a woman who is an author leaving her child with her husband while she meets with her editor (fig. 173). In both cases, the modernization of this theme is a direct—and negative—response to the contemporaneous feminist movement.

Pellerin soon commissioned Parisian artists and designers to produce amusing stories retold in many vignettes, similar to early comics, but—unlike early comics—Pellerin's were in color. In addition to individual prints, the Pellerin firm published albums, reprinting and recombining its most successful imagery, a practice that Charles Philipon's Maison Aubert had initiated several decades earlier.

In the past, popular prints were drawn by workshop-trained artisans for the least educated socioeconomic classes; prints destined for the elite were produced by academically trained artists. With the weakening of the guild system in the eighteenth century, however, new government-sponsored art schools were established throughout France, resulting in a transformation in art education: the old artisanal workshop model was gradually replaced by professional training in art schools.[24] As a result, a higher level of expertise became available across the spectrum of art production, even for ephemeral illustration such as popular prints. Increasingly, all artists, whether destined to be illustrators or history painters, had the same kind of basic training, focused on anatomy, perspective, and emulation of the great art of the past. By the later nineteenth century there was no discernable difference between artists who produced images d'Épinal and other professional illustrators; many artists, in fact, did both. And so, as the Maison Pellerin moved toward modern industrial production, the artists who designed and cut these images, or who copied high-art imagery to the limits of their ability, gradually were replaced by academically trained illustrators, with the technical aspects of production left to journeymen printers.

This is not to imply that the artists of the later prints were superior to earlier ones. Among early Pellerin artists, François Georgin (1801–1863), for many years Pellerin's master designer, stands out.[25] He drew over two hundred images for Pellerin, including the celebrated series of grand-format prints of Napoleonic subjects; they were originally intended as limited editions of five thousand, but were reprinted numerous times and are still in production today.[26] Passionate and patriotic, these images contrast vividly with the sardonic tone of lithographic production seen in Charlet's *Hah! What Fun... to Be a Soldier!* (see fig. 11) and Horace Vernet's *Rotten Weather* (see fig. 38). Georgin's *Battle of Fleurus* (fig. 174) depicts a major French military victory of 1794. General Jean-Baptiste Jourdan is seen on the right surrounded by his officers and soldiers while the opposing coalition forces of Great Britain, the Habsburgs, Hanover, and the Dutch Republic, on the left, are shown in disarray. Everything on the French side ascends, even the balloon—this was the first time it was used for military reconnaissance. Everything on the coalition side collapses, even the tree. The lengthy caption beneath the image manages to assimilate this victory to the glory of Napoleon by noting that on June 16, 1815 he led the French army to a second victory at Fleurus, but was then betrayed, two days later, at Waterloo.

Georgin's figures are stiff, his colors garish, the perspective flattened and the compositions primitively symmetrical to an extreme. And yet, the charm of these

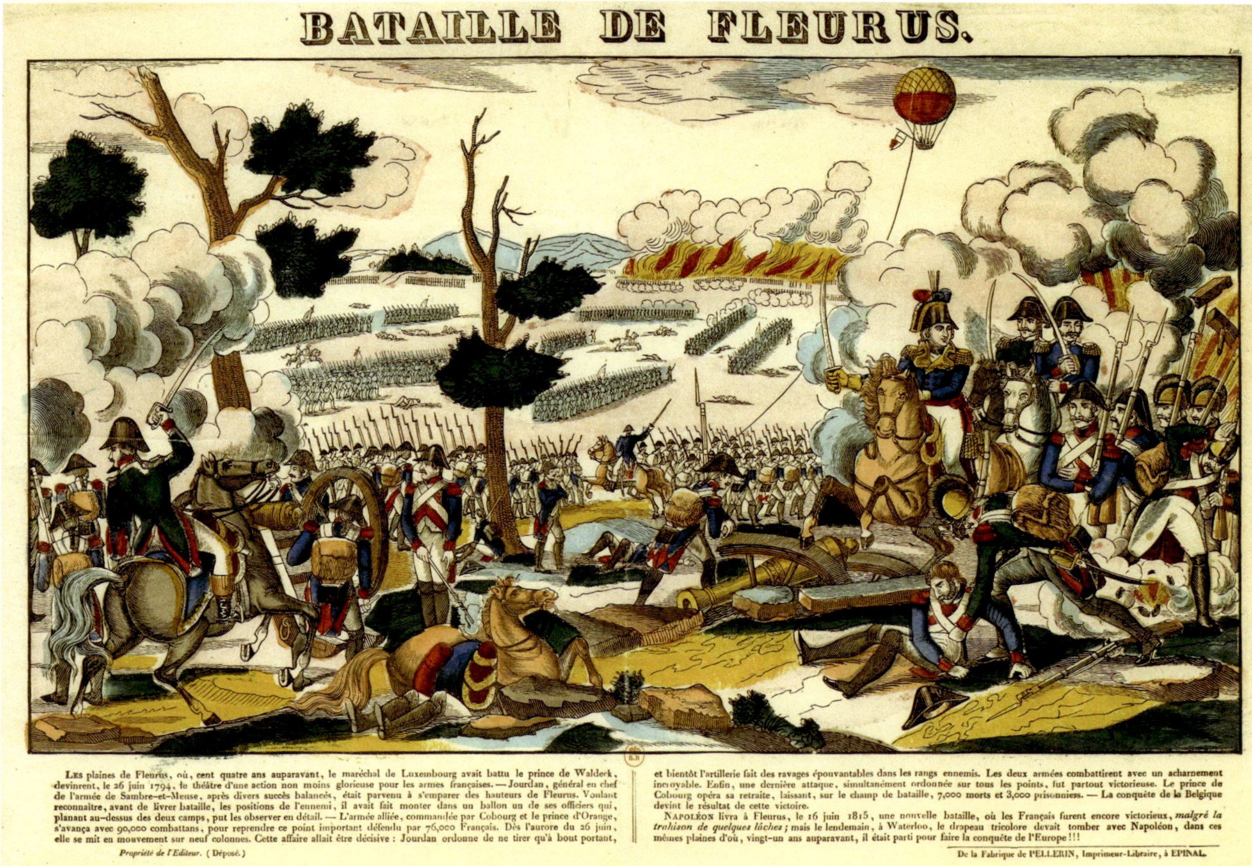

FIGURE 174

François Georgin, *The Battle of Fleurus* [*Bataille de Fleurus*], 1837. Pub. Pellerin, Épinal. Stencil-colored woodcut. Bibliothèque nationale de France.

images today—or even for a sophisticated Parisian audience in the nineteenth century—lies precisely in the nostalgia they evoke. They seem to hark back to a simpler time, that simplicity mirrored in their straightforward compositions that narrate a story without frills or sophistication.

By the last decades of the nineteenth century, many of the most prominent names in French illustration were designing for Pellerin, but, as the audience for popular prints expanded into urban settings, a curious thing happened. As academically inspired illustration became hegemonic, the traditional Épinal style of artists like Georgin assumed a "retro" significance. This earlier style seemed to incarnate simple, traditionally French virtues: sincerity, straightforwardness, and identification with what historian Herman Lebovics has called "True France," for which Épinal has always served as a potent symbol.[27] By the later nineteenth century, the crude Épinal style of Georgin that had glorified Napoleon could be referenced self-consciously to wrap politicians in its aura, much like politicians everywhere wrap themselves in their national flag. *Marshal MacMahon, Duke of Magenta* was published to celebrate the election of MacMahon, a hero of the Franco-Prussian War, as president of the French Republic (fig. 175). Its crude simplified drawing, flat bright colors, and patriotic symbols were all intended to identify his character with what were perceived as the honest and authentic virtues of this traditional French art form. He had achieved that office with the support of conservatives and monarchists, who identified their concerns with "True France"; the retro Épinal style incarnates that connection.

With the Third Republic (1870–1940), the era of democracy had permanently arrived in France, and politicians needed to court the masses of voters. In 1889, just before that year's national elections, the Parisian journal *Le Figaro* ran full-page illustrations commissioned from Pellerin in the traditional style of Épinal depicting the major parties: republicans, monarchists, Boulangists, and Bonapartists. Accompanying the first, *The Republic before the Elections*, was an editorial, "Images and Politics," that stated: "Nothing any longer escapes politics. The popular image, formerly confined to fairy tales, the naïve and old-fashioned image d'Épinal, has now, in turn, entered the fray. It has become an instrument of propaganda, so powerful in the hands of political parties that the government has ruthlessly confiscated the images of the Comte de Paris and of Prince Victor wherever they are found, as being detrimental to the security of the state."[28] The style of the accompanying images, no doubt similar to those confiscated, featured the stiff frontal poses and rigid frames of earlier nineteenth-century popular prints. The Comte de Paris and Prince Victor Napoleon are each depicted on horseback, reminiscent of Georgin's Napoleonic style (fig. 176).

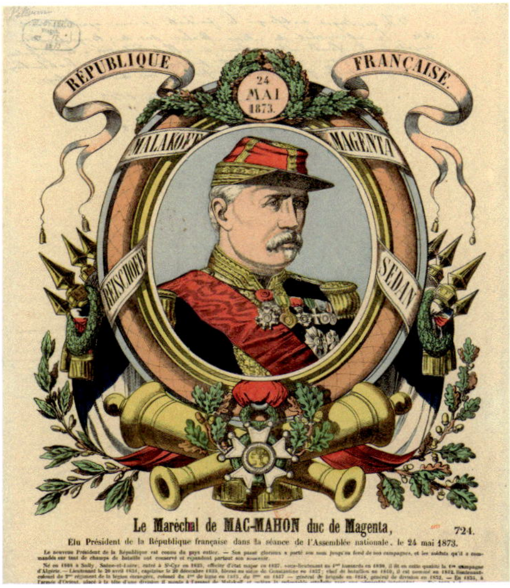

FIGURE 175

Marshal MacMahon, Duke of Magenta [*Le maréchal de Mac-Mahon, duc de Magenta*], 1873. Pub. Pellerin, Épinal. Bibliothèque nationale de France.

That style was already almost a half-century out of date, even for popular imagery, but here it was evoked intentionally. Pellerin's partner in this enterprise was Gaston Lucq, whose agency, Glucq, advertised its trade as "commercial advertising and political propaganda through popular imagery."[29] This attempt to identify prominent political figures with "True France" and its corollary, right-wing politics, through the use of images d'Épinal reached its nadir in the twentieth century, when the Vichy government adopted the style as its own. It even produced a biography of Marshal Pétain for children with crudely drawn images designed to resemble traditional popular prints.[30]

FROM POPULAR PRINTS TO COMICS

Chapters 1 and 3 discussed the transformation of caricature into the sequential narration of comics. The transformation of popular prints into comics had a parallel development. While comics shared the intention of narrating a story with two earlier art forms, history painting and book illustration, the main difference was that history painters, limited to one static image, traditionally chose a moment of high drama to represent the entire narrative.

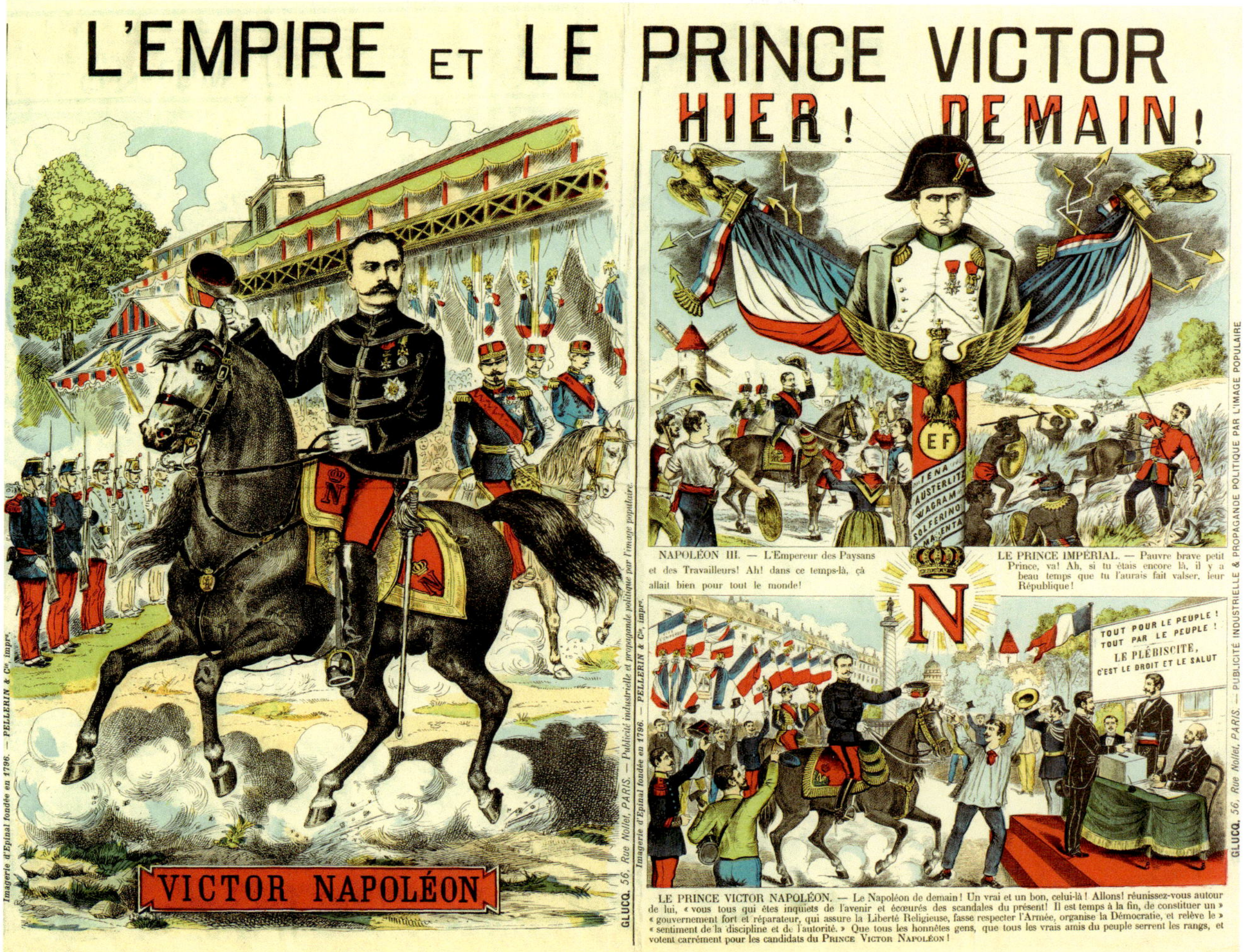

FIGURE 176

The Empire and Prince Victor, Yesterday! Tomorrow! [*L'Empire et le prince Victor, Hier! Demain!*], *Le Figaro*, March 30, 1889 (supplement). Pub. Glucq, Paris, and Pellerin, Épinal. Bibliothèque nationale de France.

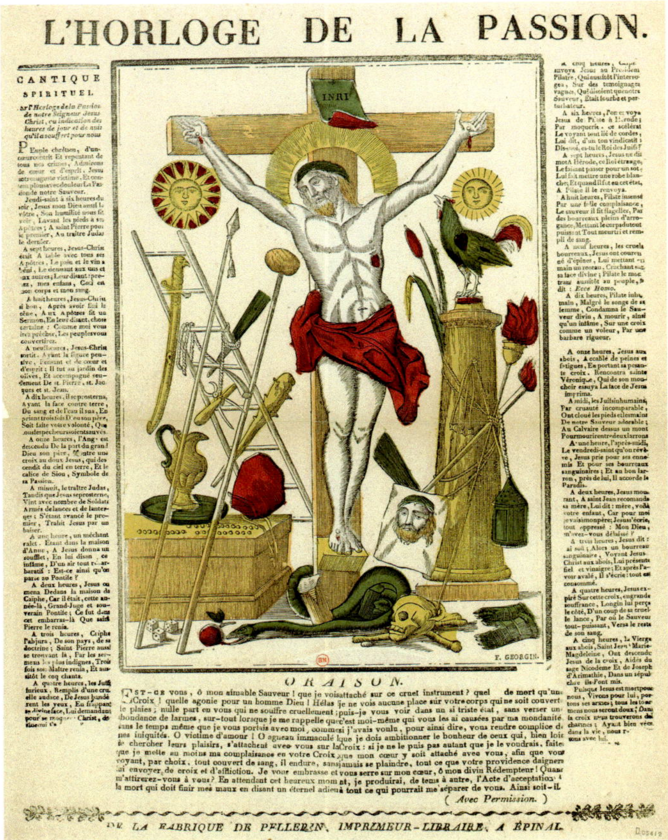

FIGURE 177
François Georgin, *The Cycle of the Passion* [*L'horloge de la passion*], 1822. Pub. Pellerin, Épinal. Stencil-colored woodcut. Bibliothèque nationale de France.

Since history painting focused on well-known events from religion, mythology, or history, it was not necessary to show every episode in the narrative, only the principal one that could symbolize the whole. Even painters who created narrative cycles chose the individual episodes carefully so that each one depicted a major event that could stand alone as a history painting. Early popular prints of religious subjects followed this same narrative strategy: in the 1822 *Cycle of the Passion* by Georgin, only the moment of Christ's crucifixion is depicted, since its audience would know well the events leading up to it; these are recounted in the extensive, hour-by-hour narrative text framing the image (fig. 177).

That popular prints such as *The Cycle of the Passion* had much in common with history painting is not surprising since printmakers at all levels often copied well-known paintings, thus extending their audience. This practice in reproductive printmaking persisted until the advent of photography, both in high-art engravings and in popular prints.[31] In this way, well-known paintings made their way into the popular repertoire. Georgin's 1831 *Crossing Mont Saint-Bernard* (fig. 178), for example, follows closely an 1806 painting by Charles Thévenin, *The French Army Crossing the St-Bernard Pass, 20 May 1800* (Musée du chateau, Versailles).[32]

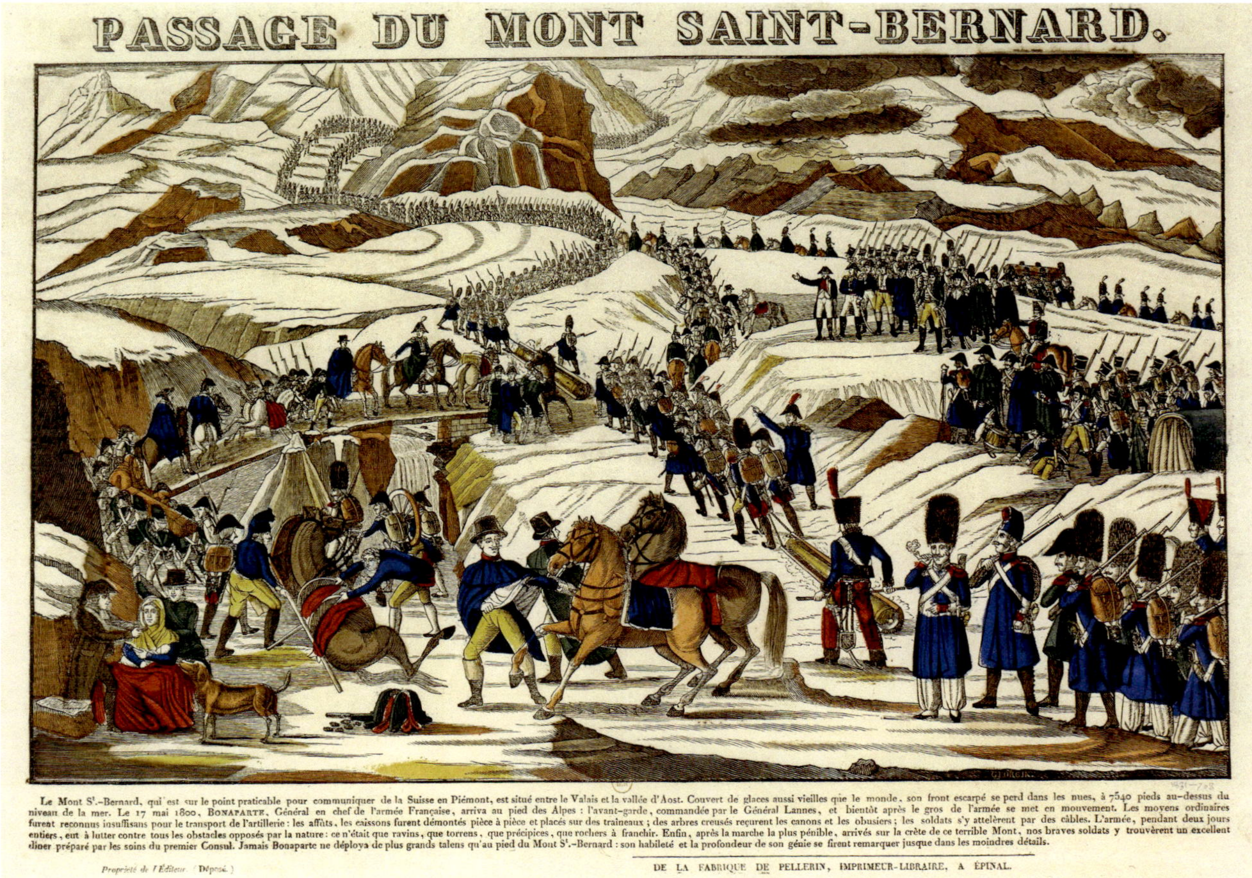

FIGURE 178
François Georgin, *Crossing Mont Saint-Bernard* [*Passage du mont Saint-Bernard*], 1831. Pub. Pellerin, Épinal. Stencil-colored woodcut. Bibliothèque nationale de France.

The transformation of single-sheet imagery into sequential narration took two routes; primary was the production of stories told through a series of individual prints, each of which depicts one episode. In chapter 3 we noted the correlation between print series such as those of Hogarth and narrative cycles of paintings. Modern comics, however, usually are composed of multiple frames arranged in several registers on a single page. This was the route more often taken by popular prints; early examples of this format predate comics but tended to have religious subjects, such as *The Creation of the World,* which dates from before 1814 (fig. 179). Here, each frame depicts a different moment: the creation of animals, the creation of Adam, the creation of Eve, the temptation and fall from grace. All are crucial moments in the story and must be read sequentially.

A variant narrative strategy, and one that led equally to comics discussed in chapter 3 and book illustration discussed in chapter 4 can be seen in Georgin's 1824 *Way of the Cross* (fig. 180). Here the exclusive focus on major events evident in *The Creation of the World,* where each episode is framed like a history painting, has been replaced by an unbroken flow of imagery that features the insignificant along with the momentous.

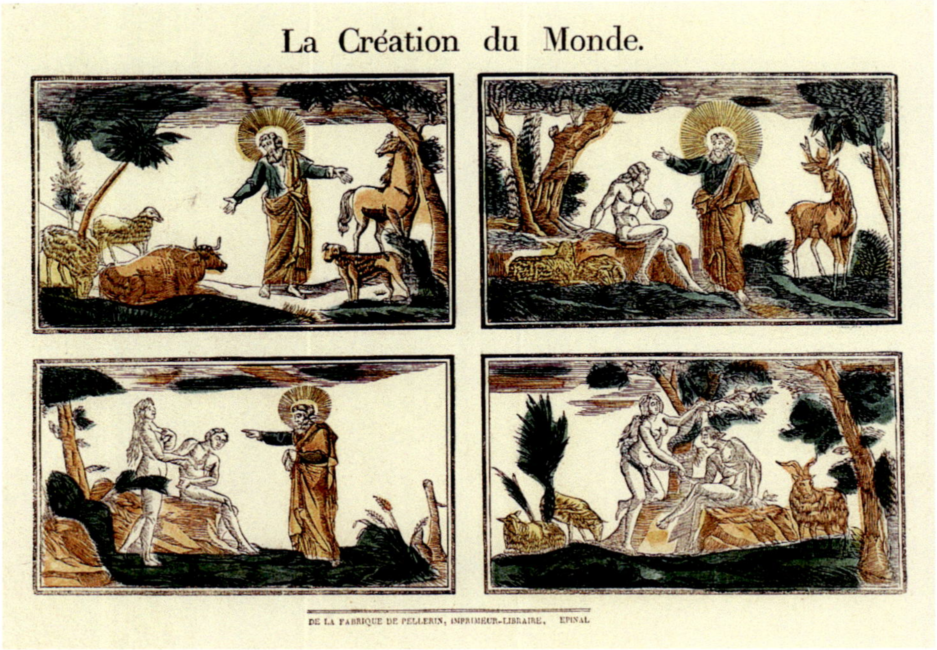

FIGURE 179
The Creation of the World [*La création du monde*], before 1814. Pub. Pellerin, Épinal. Stencil-colored woodcut.

While some of the figures here play active roles in the Passion, others chat, march along, or gawk. This shift in focus, from major events alone to the inclusion of anecdotal elements, spread rapidly in French popular prints from religious to secular imagery. What is often overlooked in this context is that, parallel to this shift in popular prints, there was a parallel shift in history painting, from the grand narratives that emphasized major events to the inclusion of minor anecdotal moments. Artists began to popularize grand-scale history painting by investing it with genre elements, small vignettes included within the depiction of the major event but subservient to it.[33] The prominent critic E.-J. Delécluze blamed David's students for creating this style, which he called *le genre anecdotique*; he claimed it had diverted the attention of the public that, until then, had been focused on more elevated painting.[34] Insofar as traditional aesthetic theory was concerned, this practice distracted from the impact and importance of the main subject, just as illustration in books diverted the reader's attention from the intellectual content of the text. Evidence of this "anecdotal genre" is prominent in the 1804 *Bonaparte Visiting the Plague-stricken in Jaffa* by Antoine-Jean Gros, one of David's most prominent students (fig. 181). Bonaparte's visit may be the subject of the painting, but a group of Arab physicians on the left and the dead and dying in the lower register of the painting form subordinate and competing centers of interest. Such vignettes enlivened and enriched the whole and made it more attractive to a larger nineteenth-century public that, truth be told, found traditional history painting, with its moral message and its emphasis on the three unities of time, place, and action, rather boring.[35] Nonetheless, the very word *vignette* had a pejorative sense throughout

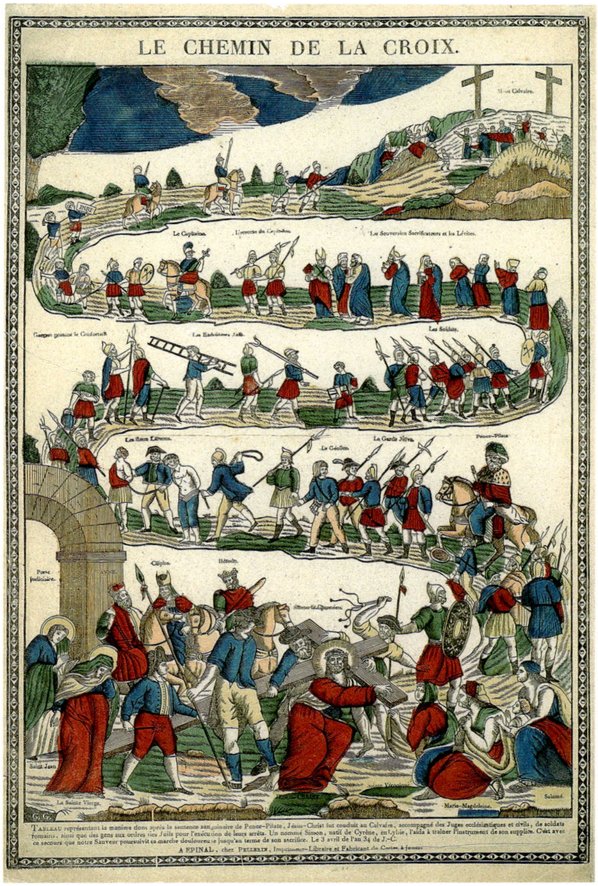

FIGURE 180
François Georgin, *The Way of the Cross* [*Le chemin de la croix*], 1824. Pub. Pellerin, Épinal. Stencil-colored woodcut. Musée de l'image, Épinal, dépot MDAAC.

the century, connoting an inconsequential decorative image, originally an ornament, but later, any illustration.[36]

The problem of the growing importance of this new hybrid form of history painting, featuring anecdotal vignettes within the larger theme, became particularly acute for Gros's younger colleague, Théodore Géricault who, as he lay dying in 1824, lamented that he had produced nothing but vignettes. Even his great painting, the 1819 *Raft of the Medusa*, was, to him, merely a vignette (fig. 182).[37] If we look at the painting in this light, we can see that it is composed of a number of smaller episodes that can be read sequentially, from the lower register where the shipwrecked lie dead, dying, or in despair, through the middle register where they are hopeful, to the upper right where salvation, in the form of a passing ship, has appeared on the horizon.

During the early decades of the nineteenth century, book illustrators were also grappling with the problem of sequential narration and the vignette. In the previous chapter, we saw how wood engraving enabled the multiplication of illustrations in novels and brought with it new illustrative strategies. Across all media, the accretion of narrative

FIGURE 181

Antoine-Jean Gros, *Bonaparte Visiting the Plague-stricken in Jaffa* [*Bonaparte visitant les pestifères à Jaffa*], 1804. Musée du Louvre, Paris.

FIGURE 182

Théodore Géricault, *The Raft of the Medusa* [*Le radeau de la Méduse*], 1819. Musée du Louvre, Paris.

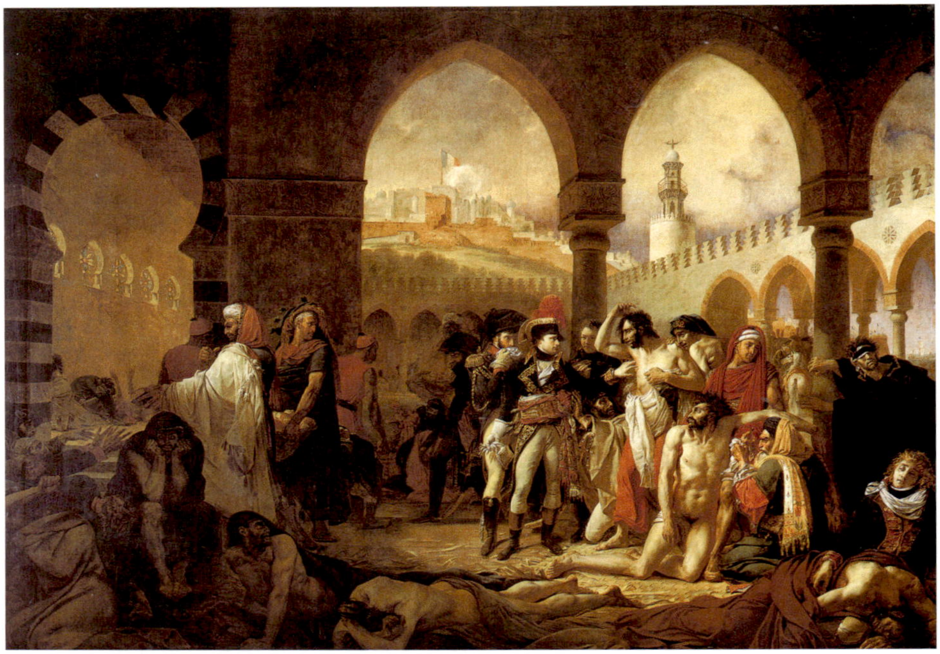

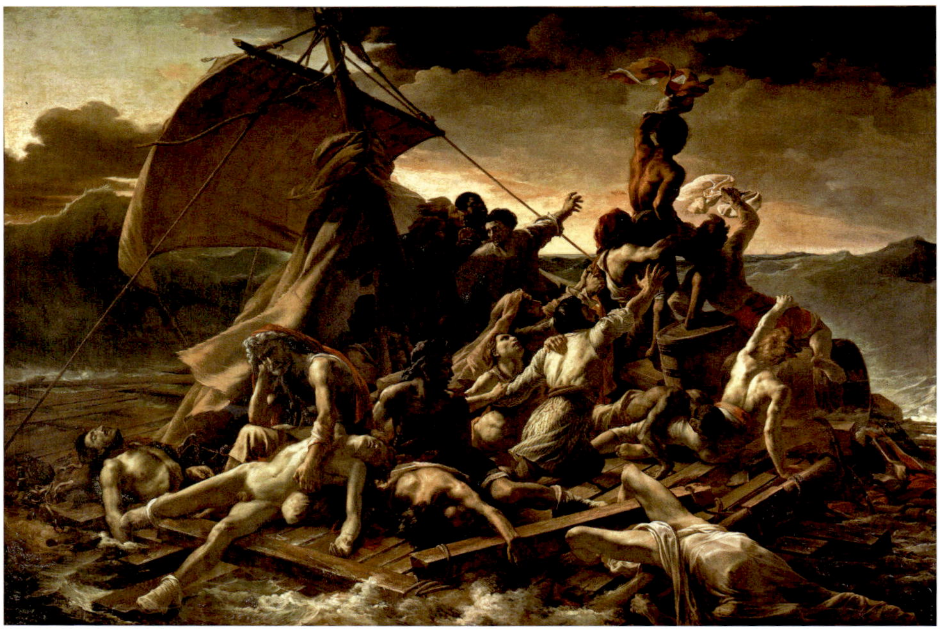

vignettes enriched the whole. These same decades saw the vignette in all its manifestations become so widespread that Pellerin adapted it to his popular prints, producing broadsheets, such as figure 171, that narrated a tale in a series of frames. In 1830, Pellerin published the well-known French medieval legend *The Story of the Four Sons of Aymon,* but now Georgin needed two prints of four frames each to complete the narrative (fig. 183). From the first frame, where with admirable symmetry the four sons of Aymon set out for the court of Charlemagne to be knighted, to the eighth where their cousins Regnault and Maugis, having liberated Jerusalem, sail home, each scene depicts an important event in the tale. In contrast to *The Cycle of the Passion* (see fig. 177), where a single image had to carry the burden of an extensive text, in this print the multiplication of images allow the text to be reduced to brief captions. From here it was a small and almost imperceptible shift to apply this same sequential narration from traditional to contemporary literature. Johann David Wyss's immensely popular 1812 novel *Der Schweitzerische Robinson, The Swiss Family Robinson,* was translated into French the following year as *Le Robinson Suisse.*[38] Pellerin's print of 1842, which recounts this story through a multiplication of frames, gives title and text in both French and German, no doubt to facilitate sales in an increasingly international print market (fig. 184).

The new imperative to sequential imagery is evident in a comparison of a print narrating a modern novel, *The Swiss Family Robinson,* to that of a traditional folktale, *The Story of Bluebeard,* published before 1811 (fig. 185). In *The Story of Bluebeard,* the central image is surrounded by a lengthy text that narrates all the various episodes of the tale, much like *The Cycle of the Passion* (see fig. 177). In *Swiss Family Robinson,* the narrative function has been taken over by a series of twelve images, each accompanied by a brief caption, exceeding in number even *The Story of the Four Sons of Aymon* (see fig. 183). Over the course of a few decades, the proportion of text to image in popular prints was gradually reversed, and the text, dethroned from its former preeminent position, became merely the caption. The question arises, of course, as to why, in all media, one image was no longer sufficient, why multiple images were now necessary to recount a story where, in the past, one iconic and well-chosen image would suffice. One explanation is the valorization of genre painting during the course of the nineteenth century. While always accepted as a category of high art, genre painting was traditionally relegated to a status lower than history painting, which, from the Renaissance, had been considered the pinnacle of art production. During the nineteenth century, however, genre paintings, by sheer force of numbers, completely overwhelmed large-scale history painting both in Salon exhibitions and in popularity.[39] These small pictures, depicting even smaller moments of everyday life, were condemned by traditionalists as pandering to a new art-viewing public that lacked the education to imagine and synthesize events and therefore insisted on being spoon-fed every aspect of the story. And yet, a different explanation could be proposed, one that, again, would cut across all media, and would attribute this newfound passion for multiple imagery to a shift in the sense of time, away from ideal moments to the pleasure derived from

HISTOIRE DES QUATRE FILS AYMON.

1.re pl.

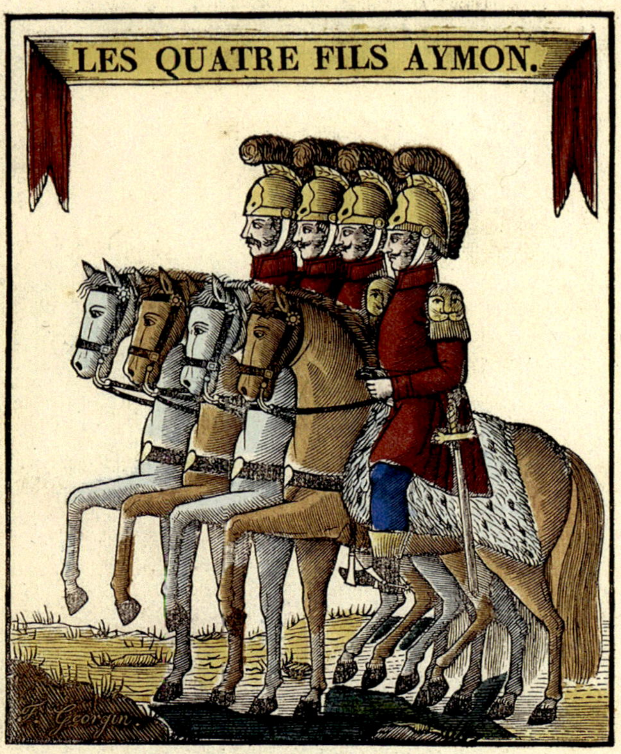

Les quatre fils Aymon se rendent à Paris, à la cour de Charlemagne, pour se faire recevoir chevaliers.

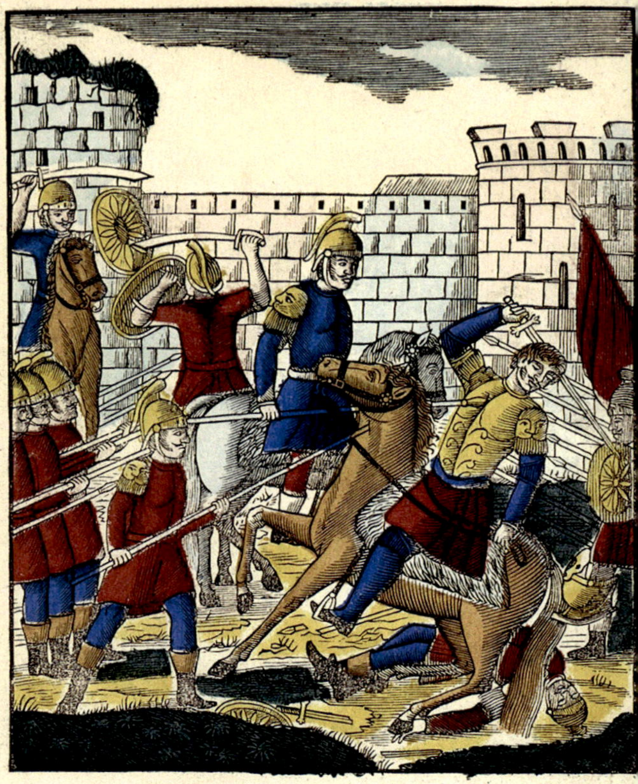

Le traître Ganélon, en embuscade dans la vallée de Soissons, tue le duc Beuves d'Aigremont, oncle des quatre fils Aymon.

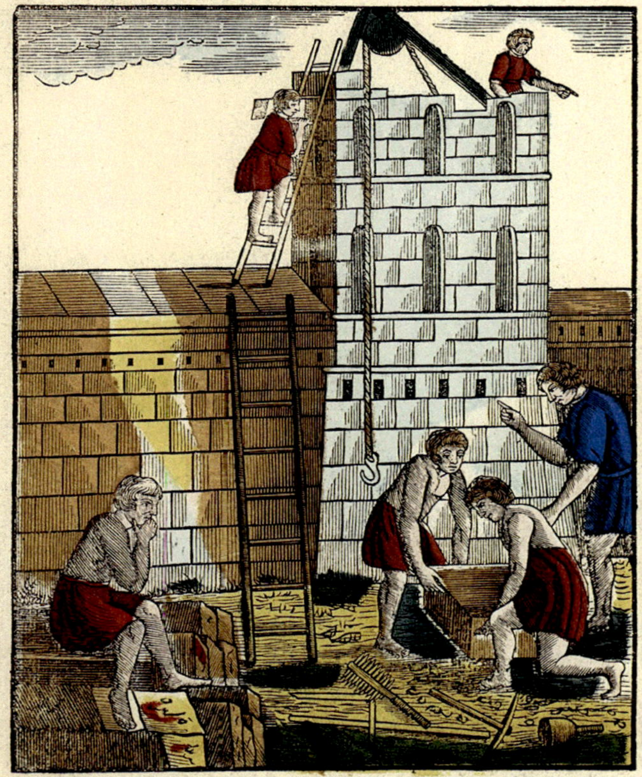

Regnault fait bâtir une forteresse qui est appelée Montaubar.

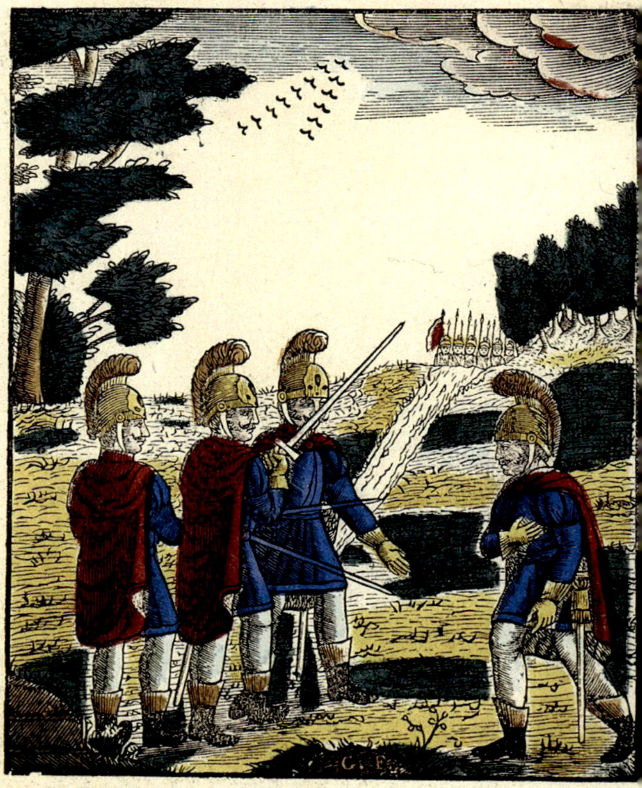

Allard, Guichard et Richard, voyant venir Foulques de Morillon avec mille combattans, croyent que Regnault les a trahis et veulent le tuer dans les plaines de Vaucouleurs.

DE LA FABRIQUE DE PELLERIN, IMPRIMEUR-LIBRAIRE, A EPINAL.

OPPOSITE **FIGURE 183**
François Georgin, *The Story of the Four Sons of Aymon* [*Histoire des quatre fils Aymon*], 1830. Pub. Pellerin, Épinal. Stencil-colored woodcut. Bibliothèque nationale de France.

FIGURE 184
The Swiss Family Robinson [*Robinson Suisse. Schweitzer-Robinson*], 1842. Pub. Pellerin, Épinal. Stencil-colored woodcut.

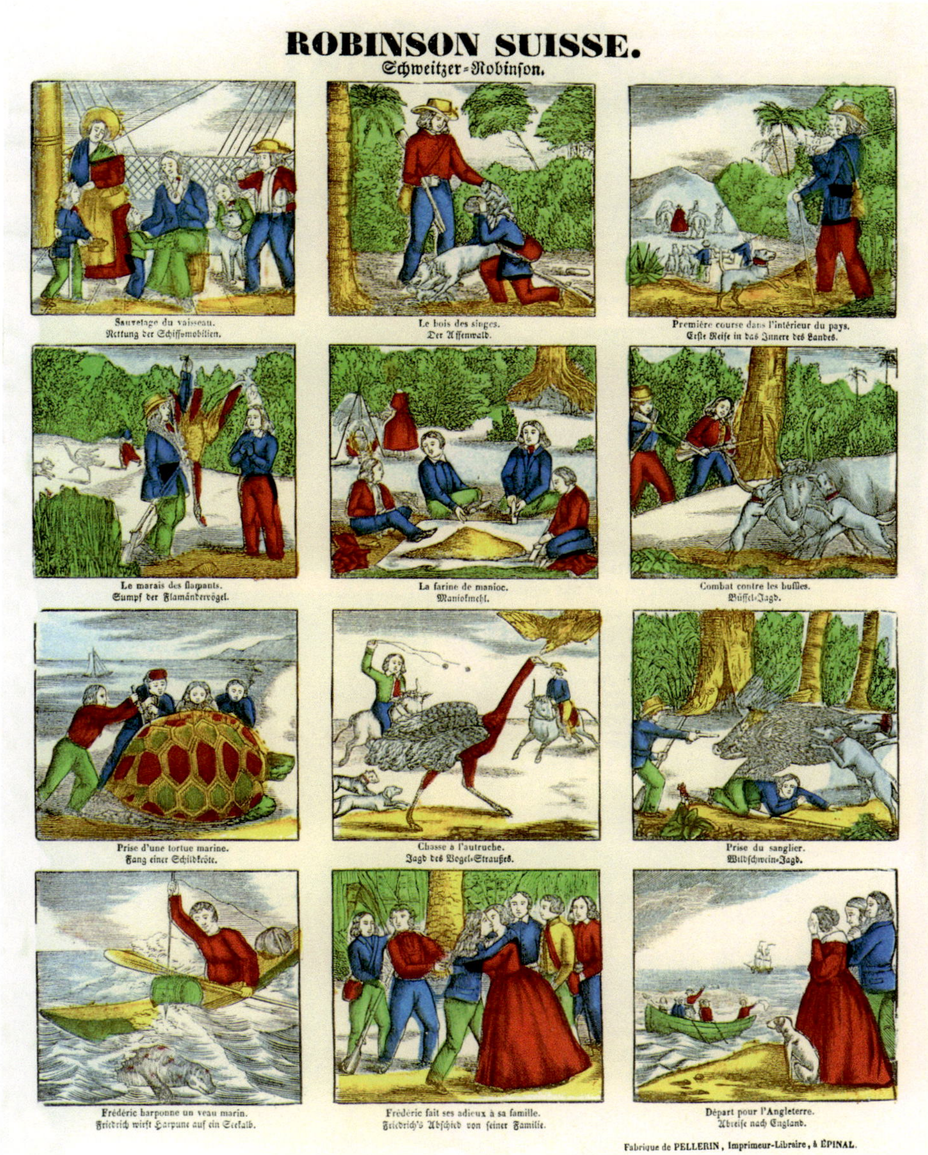

the visualization of lived experience in all its endless variety. This was the same impulse that generated the illustrated press discussed in chapter 2, new innovative journals like *Le Magasin pittoresque* (1833) and *L'Illustration* (1843) that attempted to satisfy the public's seemingly insatiable appetite for visual imagery. For in daily life we rarely experience the iconic, great dramatic moments that are the subjects of history painting, but more often the sequence of small moments characteristic of genre painting that make up the fabric of our lives. The historical novel, as well as historical genre painting, as well as comics, are all manifestations of this imperative. It was this reflection of their own lives, and even a projection of that quotidian reality onto the historical past, that the new,

HISTOIRE DE LA BARBE-BLEUE,

Avec son Portrait, tirée des CONTES DES FÉES.

UN homme fort puissant et fort riche, avait une physionomie rebutante et une barbe très-bleue; ces défauts le rendaient si laid qu'aucune fille ne pouvait se décider à l'épouser: de plus, il passait avec raison pour être très-méchant, les six premières femmes qu'il avait épousées étaient disparues sans qu'on sache ce qu'elles étaient devenues. Cela était cause que toutes les filles à qui on le proposait en mariage, refusaient de l'épouser, il avait beau faire parade de ses richesses, rien ne pouvait balancer ses défauts; la couleur de sa barbe révoltait plus que la grandeur de son château et l'entendue de ses domaines ne pouvaient éblouir: cependant, une jeune imprudente, ne consultant que sa vanite, céda à l'éclat dont elle fut frappée, et au charme des présens. On peut épouser un homme qui n'est point digne de trouver un cœur, lorsque des soins délicats, et des galanteries intéressantes, excitent la reconnaissance, ce sentiment supplée à l'amour, quand le cœur est bon et que la raison est formée; mais l'hymen n'est qu'un malheur et qu'une source de repentirs, si la magnificence qui séduit n'intéresse que la vanité; c'est le sort qu'éprouva la jeune indiscrète, et c'est bien le sujet d'une moralité. Son mari, qui joignait à la brutalité une sorte de malice, feignit d'être obligé de s'éloigner pour quelques jours, lui laissant les chefs de ses appartemens, entr'autres d'un cabinet secret où étaient enfermés les corps de ses six premières femmes; mais avec défense de n'y pas entrer: elle promit, et ne tint point parole; la curiosité la mena à la désobéissance. Le mari, à son retour, lui demanda ses clefs, qu'elle lui donna d'une main tremblante: D'où vient, lui dit-il, que la clef de mon cabinet n'est point avec les autres, il faut, dit-elle, que je l'aie laissée sur ma table. Ne manquez pas, dit la Barbe-Bleue, de me la donner. Après plusieurs remises, il fallut apporter la clef: la Barbe-Bleue l'ayant considérée, dit à sa femme, pourquoi y a-t-il du sang sur cette clef? Je n'en sais rien, répondit la pauvre femme, plus pâle que la mort. Je le sais bien, moi, reprit la

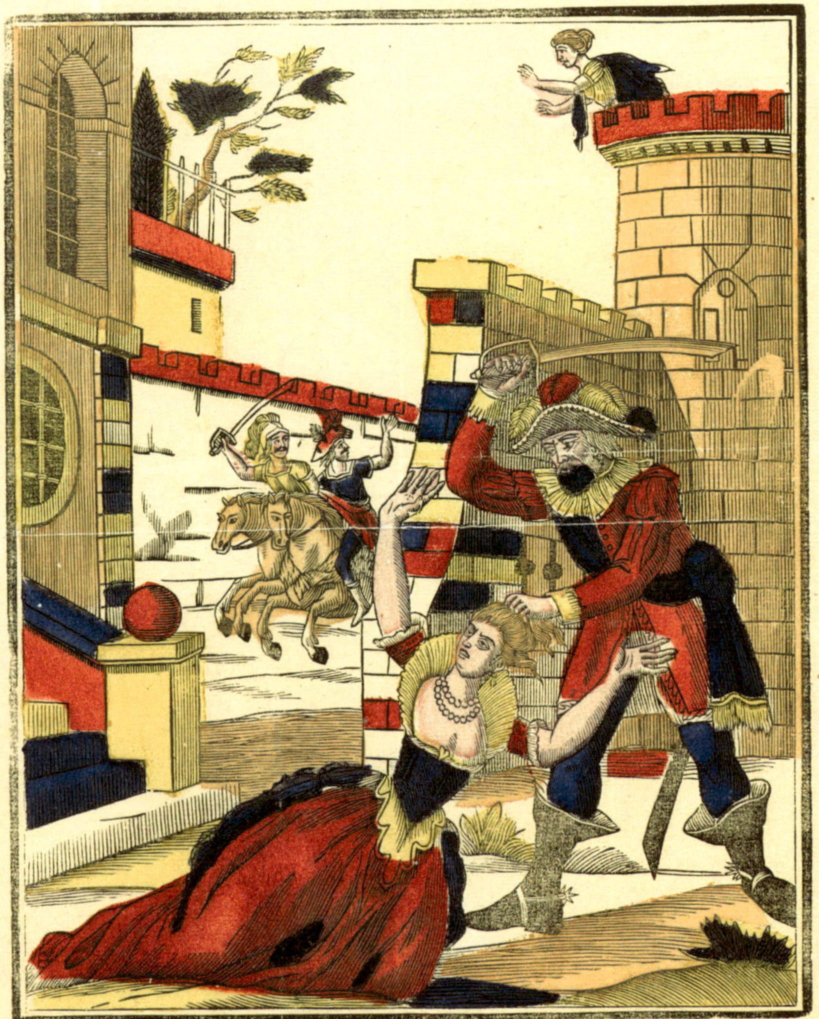

Barbe-Bleue; vous avez voulu entrer dans le cabinet; vous y entrerez et irez prendre votre place auprès des Dames que vous y avez vues. Elle se jette à genoux en demandant pardon à son mari, avec toutes les marques d'un vrai repentir, de lui avoir désobéi; elle aurait attendri un rocher: mais la Barbe-Bleue, avec son cœur de pierre, lui dit, avec une voix de tonnerre: Madame il faut mourir tout-à-l'heure. Puisqu'il faut mourir, repondit-elle en le regardant, les yeux baignés de larmes, donnez-moi un peu de tems pour prier Dieu. Je vous donne, reprit-il, un demi-quart-d'heure, mais pas un moment davantage. Lorsqu'elle fut seule, elle appella sa sœur Anne, et lui dit: monte, je te prie, sur le haut de la tour, pour voir si mes frères ne viennent point, ils m'ont promis de me venir voir aujourd'hui; si tu les vois, fais leur signe de se hâter. La sœur Anne monte sur la tour, et la pauvre affligée lui criait de tems en tems: Anne, ma sœur Anne, ne vois-tu rien venir? Je vois, lui répondit-elle, deux Cavaliers qui viennent de ce côté. Dieu soit loué! s'écria-t-elle, ce sont mes frères. Cependant la Barbe-Bleue, tenant un grand coutelas à sa main, lui criait d'une voix de tonnerre: Descenderas-tu, ou je monterai là-haut. Encore un petit moment, lui dit-elle. Un moment après, il l'appella si fort que la maison en trembla. La pauvre femme descendit et alla se jeter à ses pieds, toute éplorée et toute échevelée; cela ne sert de rien, dit la Barbe-Bleue, il faut mourir; puis la prenant d'une main par les cheveux, et de l'autre levant le coutelas en l'air, il allait lui abattre la tête; elle lui demandait encore un petit moment: non, lui dit-il, recommande-toi bien à Dieu, en levant son bras....... Dans ce moment, on heurta si fort à la porte, que la Barbe-Bleue s'arrêta tout court; on ouvrit et aussitôt on vit entrer deux Cavaliers qui, mettant l'épée à la main, coururent droit à la Barbe-Bleue, lui passèrent leurs épées au travers du corps et le laissèrent mort. La pauvre femme etait aussi morte que (*)

(*) son mari, et n'avait pas la force de se lever pour embrasser ses Frères. La Barbe-Bleue n'avait point d'héritiers, sa femme demeura maîtresse de tous ses biens, qu'elle employa à acheter des charges a ses deux frères, à marier sa sœur Anne à un jeune Gentilhomme, et à se marier elle-même à un honnête homme, qui lui fit oublier le mauvais tems qu'elle avait passé avec la Barbe-Bleue.

La curiosité, malgré tous les attraits, coûte souvent bien des regrets; on en voit tous les jours des exemples paraître. C'est, n'en déplaise au sexe, un plaisir bien léger; dès qu'on le prend il cesse de l'être, et toujours il coûte trop cher.

DE LA FABRIQUE DE PELLERIN, IMPRIMEUR-LIBRAIRE, A ÉPINAL.

FIGURE 185
The Story of Bluebeard [*Histoire de la Barbe-bleue*], before 1811. Pub. Pellerin, Épinal. Stencil-colored woodcut. Bibliothèque nationale de France.

PAGE 216 **FIGURE 186**
The Unruly Little Boys [*Les petits garçons turbulens*], 1843. Pub. Pellerin, Épinal. Stencil-colored woodcut. Musée de l'Image, Épinal.

PAGE 217 **FIGURE 187**
The Station Master's Goat [*La chèvre du chef de gare*], *Série aux armes d'Épinal*, no. 128, 1894. Pub. Pellerin, Épinal. Color lithograph.

greatly expanded, nineteenth-century public was avid to see represented, and increasingly demanded in all media—in painting, the illustrated press, popular prints, and in the new medium of comics.

As a result, there were several phases in the transformation of the popular print into the comic strip: the traditional single-image broadside (see fig. 165), then sequential narration applied first to religious imagery (see fig. 179), and then to secular subjects (see fig. 184). In the earlier phases of both religious and secular subject matter, literary narratives and legends preceded visualization, but in a final transformation—and here is where we might feel that popular imagery and comics flow together into one mighty stream—new stories were written and illustrated narrating the whimsical contemporary anecdotes that eventually became characteristic of comics.

Illustration itself was gradually becoming redefined in the course of the century. The earliest prints aimed at children were heavily didactic and moralizing, pairing naughty adventures with dire consequences. The purpose of these prints was clearly to discourage mischievous behavior, though such colorful and exuberant vignettes might well have inspired it. In an extreme example, *Unruly Little Boys*, the boy throws stones—and accidentally kills his mother; he shoots off fireworks—and burns down a barn; he teases a horse—and gets kicked in the head (fig. 186). Only gradually did the amusement value of prints overtake this earlier commitment to moral edification and stern retribution. In the last decades of the century, single-sheet comics aimed at children and adults began depicting inconsequential but lighthearted contemporaneous subjects.

In *The Station Master's Goat* of 1894, the stationmaster has been told by his doctor that he must drink a bowl of goat's milk every day (fig. 187). He buys a goat and his wife milks it for him, but one day his wife is away, so he tries milking it himself. The goat refuses to allow it, so he puts on his wife's dress and fools the goat. While he is milking, however, a train arrives and, without thinking, he runs out on the platform in his wife's dress to perform his stationmaster duties, to the great amusement of the passengers and—especially—the inspector of railways who has just descended from the train. This transformation of the subject matter of popular prints from the monumental to the quotidian, from history to genre, from didactic to entertaining is critical. Now the artist is no longer adapting themes established by previous artists and writers but has joined with comic book artists in creating a new art form celebrating the smaller moments of daily life: the comic strip as we know it. Just as genre painting had become, over the course of the nineteenth century, the predominant form of high art, so did comics become its popular art equivalent.

Alongside the new subject matter, however, the production of prints illustrating traditional literature in multiple frames persisted, although these prints were updated regularly to reflect new styles, and were often recast into contemporary dress as well. With the development of lithography and photomechanical processes in the later nineteenth century, artists were freed from the necessity of actually cutting the woodblock, or

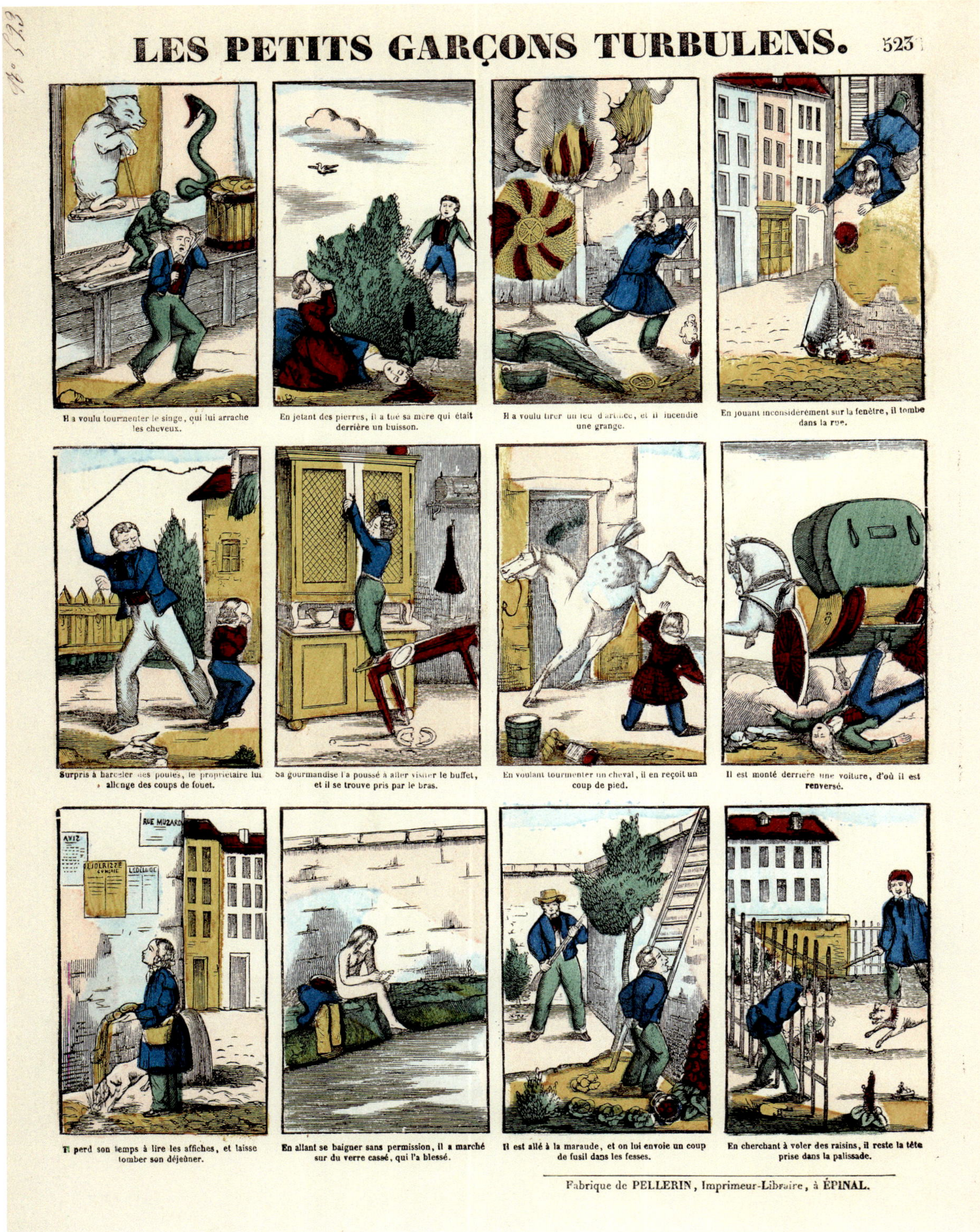
LES PETITS GARÇONS TURBULENS.
523
Il a voulu tourmenter le singe, qui lui arrache les cheveux.
En jetant des pierres, il a tué sa mère qui était derrière un buisson.
Il a voulu tirer un feu d'artifice, et il incendie une grange.
En jouant inconsidérément sur la fenêtre, il tombe dans la rue.
Surpris à harceler des poules, le propriétaire lui allonge des coups de fouet.
Sa gourmandise l'a poussé à aller visiter le buffet, et il se trouve pris par le bras.
En voulant tourmenter un cheval, il en reçoit un coup de pied.
Il est monté derrière une voiture, d'où il est renversé.
RUE MUZARD
AVIZ
Il perd son temps à lire les affiches, et laisse tomber son déjeûner.
En allant se baigner sans permission, il a marché sur du verre cassé, qui l'a blessé.
Il est allé à la maraude, et on lui envoie un coup de fusil dans les fesses.
En cherchant à voler des raisins, il reste la tête prise dans la palissade.
Fabrique de PELLERIN, Imprimeur-Libraire, à ÉPINAL.

FIGURE 186

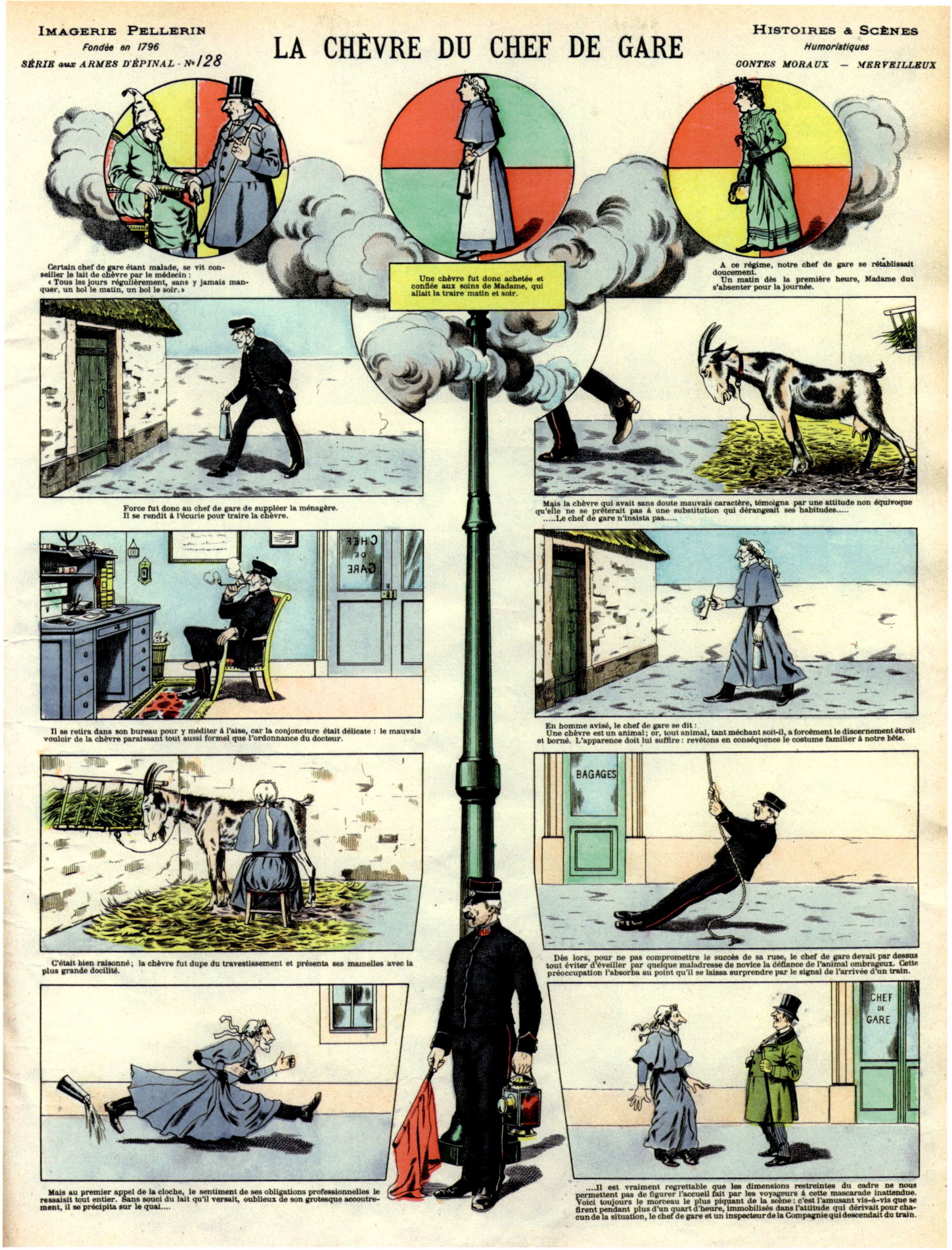
IMAGERIE PELLERIN
Fondée en 1796
SÉRIE aux ARMES D'ÉPINAL - N° 128
LA CHÈVRE DU CHEF DE GARE
HISTOIRES & SCÈNES
Humoristiques
CONTES MORAUX — MERVEILLEUX
Certain chef de gare étant malade, se vit conseiller le lait de chèvre par le médecin : « Tous les jours régulièrement, sans y jamais manquer, un bol le matin, un bol le soir. »
Une chèvre fut donc achetée et confiée aux soins de Madame, qui allait la traire matin et soir.
A ce régime, notre chef de gare se rétablissait doucement.
Un matin dès la première heure, Madame dut s'absenter pour la journée.
Force fut donc au chef de gare de suppléer la ménagère.
Il se rendit à l'écurie pour traire la chèvre.
Mais la chèvre qui avait sans doute mauvais caractère, témoigna par une attitude non équivoque qu'elle ne se prêterait pas à une substitution qui dérangeait ses habitudes.....
.....Le chef de gare n'insista pas.....
CHEF DE GARE
Il se retira dans son bureau pour y méditer à l'aise, car la conjoncture était délicate : le mauvais vouloir de la chèvre paraissant tout aussi formel que l'ordonnance du docteur.
En homme avisé, le chef de gare se dit :
Une chèvre est un animal ; or, tout animal, tant méchant soit-il, a forcément le discernement étroit et borné. L'apparence doit lui suffire : revêtons en conséquence le costume familier à notre bête.
BAGAGES
C'était bien raisonné ; la chèvre fut dupe du travestissement et présenta ses mamelles avec la plus grande docilité.
Dès lors, pour ne pas compromettre le succès de sa ruse, le chef de gare devait par dessus tout éviter d'éveiller par quelque maladresse de novice la défiance de l'animal ombrageux. Cette préoccupation l'absorba au point qu'il se laissa surprendre par le signal de l'arrivée d'un train.
CHEF DE GARE
Mais au premier appel de la cloche, le sentiment de ses obligations professionnelles le ressaisit tout entier. Sans souci du lait qu'il versait, oublieux de son grotesque accoutrement, il se précipita sur le quai....
....Il est vraiment regrettable que les dimensions restreintes du cadre ne nous permettent pas de figurer l'accueil fait par les voyageurs à cette mascarade inattendue. Voici toujours le morceau le plus piquant de la scène : c'est l'amusant vis-à-vis que se firent pendant plus d'un quart d'heure, immobilisés dans l'attitude qui dérivait pour chacun de la situation, le chef de gare et un inspecteur de la Compagnie qui descendait du train.

FIGURE 187

PAGE 220 **FIGURE 188**
The Grasshopper and the Ant [*La cigale et la fourmi*], no. 997, 1860. Pub. Pellerin, Épinal. Stencil-colored woodcut. Bibliothèque nationale de France.

PAGE 221 **FIGURE 189**
E. Phosty, *The Grasshopper and the Ant* [*La cigale et la fourmi*], *Série supérieure aux armes d'Épinal. Fables de La Fontaine*, no. 25, 1895. Pub. Pellerin, Épinal. Color lithograph. Bibliothèque nationale de France.

even designing with its limitations in mind. This further diminished the distance between artists working for Pellerin and those working for the new illustrated periodicals. A comparison of two interpretations of *The Grasshopper and the Ant* from the *Fables* of Jean de La Fontaine (1621–1695), an enduring favorite, shows the changes. The garishly colored frames of the 1860 print by an anonymous artist are arranged in the regular grid format that had been standard for decades in popular print production (fig. 188). Each frame is dominated by figures in a minimal setting. The sabot-shod ant stands and moves awkwardly, its poses clearly inspired by observation of rural life. The grasshopper is only slightly more graceful, though it is obvious that the artist intended to depict the more elegant creature who danced and sang through the summer while the industrious ant stored up food for winter. When Pellerin reissued this print over thirty years later, it was prominently signed by the artist, E. Phosty, a professional illustrator (fig. 189). Phosty references the original characters, the grasshopper and the ant, only in two narrow vignettes that frame the central upper image, eliminating the anthropomorphism and recasting the story as one of artists and philistines. The confrontation of the artist-grasshopper, now a lovely young female musician, with the philistine-ant, again depicted as a peasant with her spindle, ends tragically in Phosty's version, with the lifeless body of the improvident musician lying abandoned in the snow. This last image undercuts the stern moral judgment of La Fontaine's fable. While the sympathy of the rural viewer would be on the side of the industrious and practical ant in the earlier print, Phosty evokes sympathy for the plight of the artist, a more urban sentiment. To accomplish this, he has replaced the rigid grid with a free-form arrangement of square and rectangular frames of varying sizes, within which the figures stand gracefully, each with weight balanced on one leg, the other slightly bent. The poses are reminiscent of the contrapposto of Greek sculpture, the ideal of figure drawing taught in all art schools until modern times. Instead of the minimal schematic settings of the earlier print, Phosty gives us fully worked out landscape compositions and even adds clever trompe l'oeil touches to two of his frames, their corners curled up to emphasize that what we are seeing is an artist's vision, not reality itself. Thus one artist subtly evokes our sympathy for the plight of another: if we can appreciate the image the artist has drawn for us, how can we be unsympathetic to the poor musician? Phosty, though unknown today, drew numerous prints for Pellerin and undoubtedly had the benefit of academic art-school training. His version of *The Grasshopper and the Ant* is no longer identifiable as a popular print. It could just as easily pass as a book illustration or as a single-sheet comic, although here, since the subject is more tragic than humorous, the French term *bande dessinée*, drawn strip, would be more accurate.

The crucial moment in the transition of subject matter from popular prints to comics had come when François-Charles Pinot (1817–1874) joined the Pellerin firm in 1847.[40] Trained in the studio of the esteemed Parisian history painter Paul Delaroche, Pinot had aspired to a career as a history painter, but family necessity drew him back to Épinal and financial need resulted in his working for Pellerin. While this might have seemed an unfortunate necessity to him, and certainly historians of popular prints in France have identified this moment as the beginning of the decadence of popular prints, it can also be seen as closing the circle of popular imagery. Pinot had worked in Paris for the new weekly *L'Illustration*, established in 1843, producing drawings on contemporary themes. He introduced these subjects to the Pellerin firm where they coexisted alongside more traditional themes, but, as a result, the major scholars of French popular imagery all conclude their studies here, with the adoption of lithography and contemporary imagery.[41] Pinot's 1868 *New Year's Comedy*, produced after he left Pellerin to start his own firm in 1860, shared with popular prints the same garish coloring, but it consists of five registers of unframed vignettes that follow various classes of citizens through their social rounds of wishing friends and neighbors a Happy New Year (fig. 190). These can comfortably be called comics, although the contemporaneous term for them was *historiettes*, little stories.[42] By the 1860s, this type of imagery had become a staple of the illustrated press, but popular prints brought these single-sheet comics depicting contemporaneous subjects within reach of a more economically modest audience, one composed principally of adults.

Earlier comic albums produced by caricaturists such as Daumier, Gavarni, and Grandville, as well as by Töpffer and his French followers, had been relatively costly, but as the century progressed, technological advances made larger print runs possible. This resulted in ever cheaper books, newspapers, magazines, and prints, and Pellerin took advantage of the commercial possibilities of producing his historiettes in albums, both for children and for adults, in the same way that urban publishers churned out their albums of comics and caricatures.[43] As the Pellerin firm grew and prospered, producing ten to fifteen million prints a year in the later nineteenth century, the two audiences, urban and rural, began to merge.

Because popular prints were usually colored—stencil-colored at the beginning of the century, chromolithographed by the end—they bear a striking resemblance to twentieth-century comics. Colored comics had not yet appeared in the periodical press, however; Richard F. Outcault's *Hogan's Alley*, featuring the Yellow Kid, appeared in Joseph Pulitzer's *New York World* in 1895 and is generally acknowledged as the first newspaper comic strip in color.[44] Black-and-white serial comics had appeared earlier, as we saw in chapter 3: Nadar's *Public and Private Life of Mister Reactionary* was published in 1849 in weekly installments in the *Revue comique à l'usage des gens sérieux* (fig. 191).[45] Arranged as a multiframe banner scrolling across the bottom of each page, as many as six such pages in each issue, it bore little resemblance, other than in its sequentiality, to the garishly colored single-sheet stories already being churned out in Épinal by the same decade. *The Story of William Tell* (fig. 192), for example, seems closer to classic modern comics in appearance, despite its historical subject, than does Nadar's *Mister Reactionary*. While the genealogy of modern comics from caricature is unmistakable, as is evidenced by Nadar's comic strip, the popular print can, just as unmistakably, claim equal rights of ancestry, providing the prototype for comics in color.

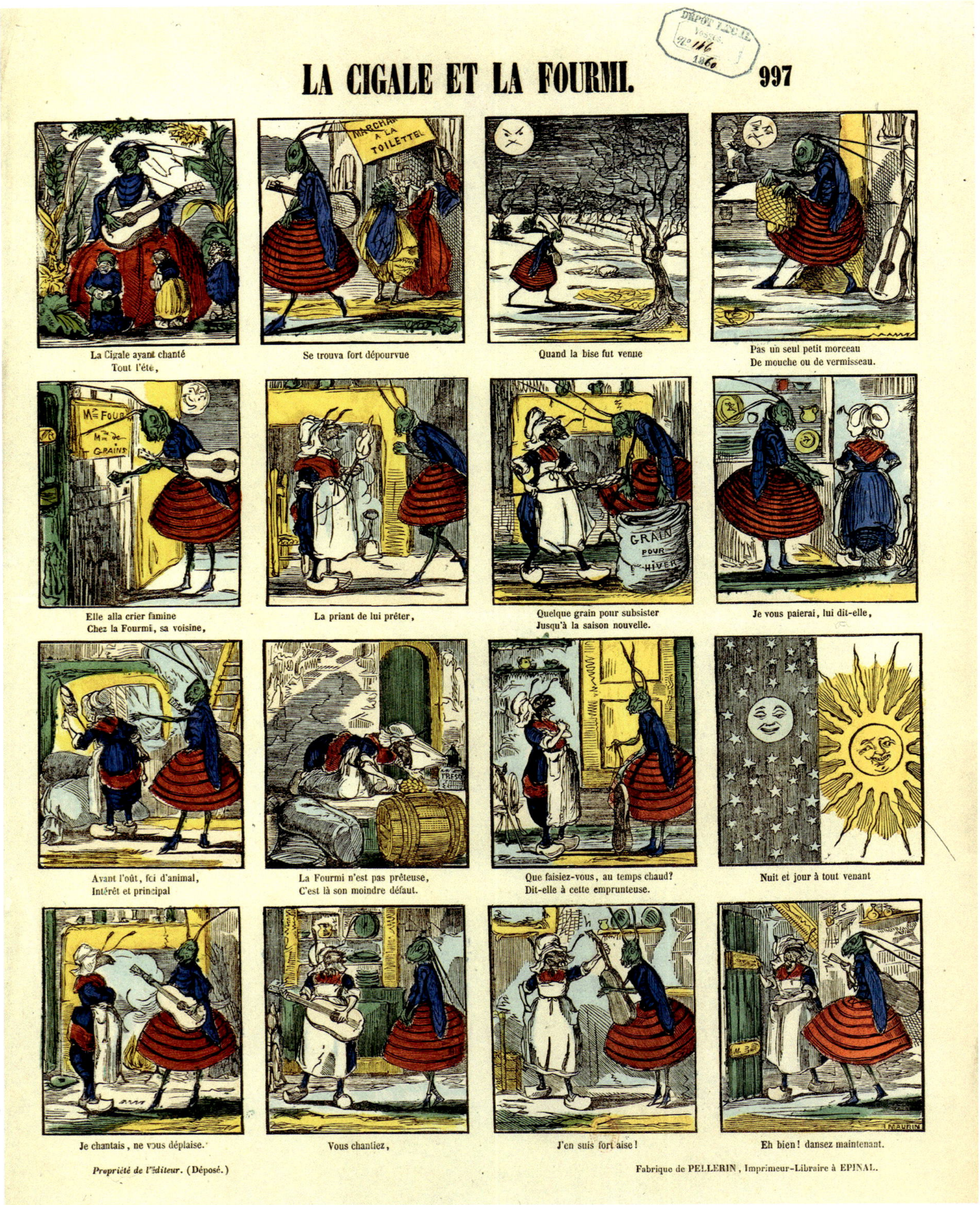
DÉPÔT LÉGAL
Vosges.
N° 116
1860
LA CIGALE ET LA FOURMI.
997
MARCHANDE À LA TOILETTE
La Cigale ayant chanté
Tout l'été,
Se trouva fort dépourvue
Quand la bise fut venue
Pas un seul petit morceau
De mouche ou de vermisseau.
Me FOURMI Mde de GRAINS
GRAIN POUR L'HIVER
Elle alla crier famine
Chez la Fourmi, sa voisine,
La priant de lui prêter,
Quelque grain pour subsister
Jusqu'à la saison nouvelle.
Je vous paierai, lui dit-elle,
Avant l'oût, foi d'animal,
Intérêt et principal
La Fourmi n'est pas prêteuse,
C'est là son moindre défaut.
Que faisiez-vous, au temps chaud?
Dit-elle à cette emprunteuse.
Nuit et jour à tout venant
Je chantais, ne vous déplaise.
Vous chantiez,
J'en suis fort aise!
Eh bien! dansez maintenant.
MAURIN
Propriété de l'Éditeur. (Déposé.)
Fabrique de PELLERIN, Imprimeur-Libraire à EPINAL.

FIGURE 188

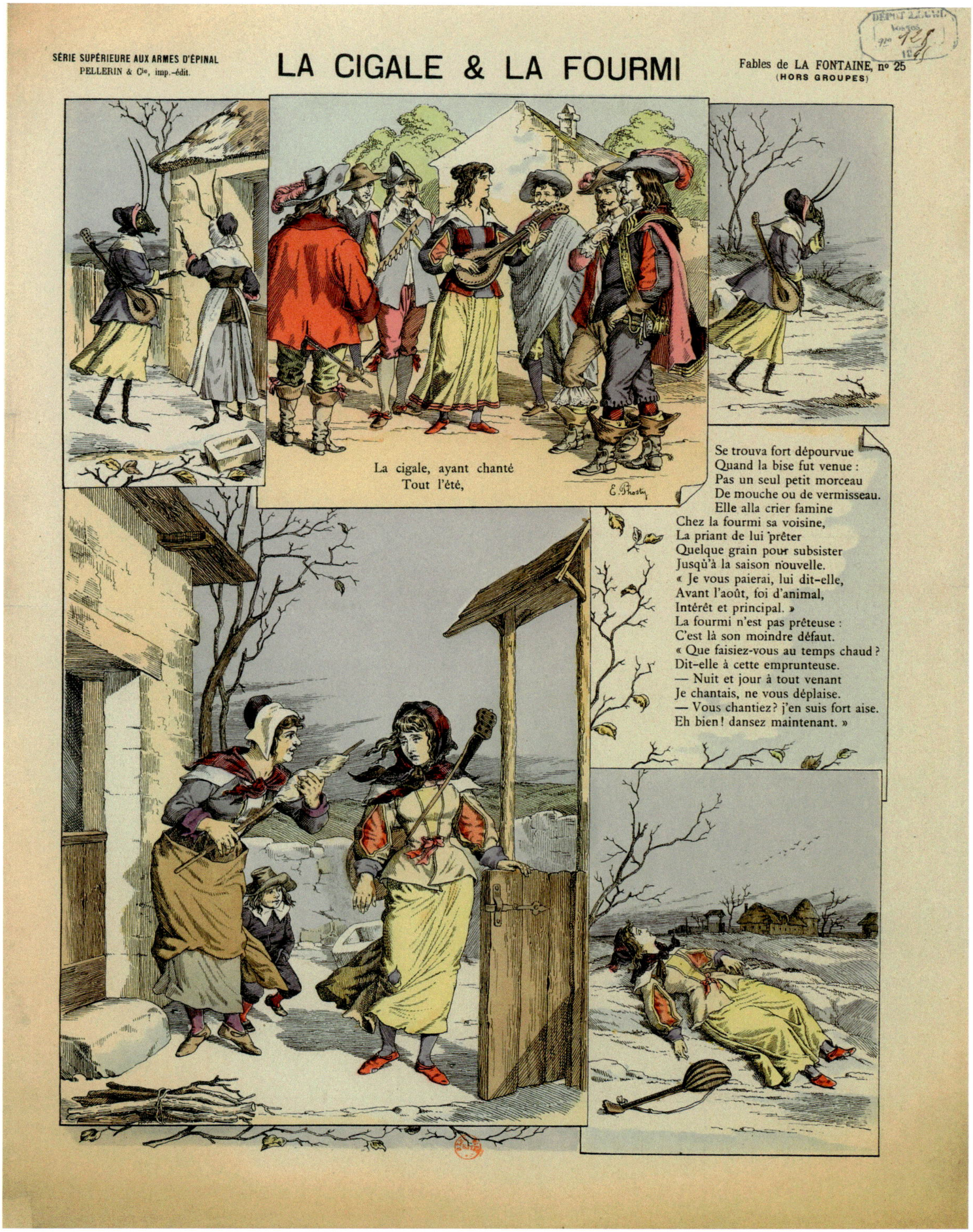
SÉRIE SUPÉRIEURE AUX ARMES D'ÉPINAL
PELLERIN & Cie, imp.-édit.
LA CIGALE & LA FOURMI
Fables de LA FONTAINE, no 25
(HORS GROUPES)
La cigale, ayant chanté
Tout l'été,
Se trouva fort dépourvue
Quand la bise fut venue :
Pas un seul petit morceau
De mouche ou de vermisseau.
Elle alla crier famine
Chez la fourmi sa voisine,
La priant de lui prêter
Quelque grain pour subsister
Jusqu'à la saison nouvelle.
« Je vous paierai, lui dit-elle,
Avant l'août, foi d'animal,
Intérêt et principal. »
La fourmi n'est pas prêteuse :
C'est là son moindre défaut.
« Que faisiez-vous au temps chaud ?
Dit-elle à cette emprunteuse.
— Nuit et jour à tout venant
Je chantais, ne vous déplaise.
— Vous chantiez? j'en suis fort aise.
Eh bien! dansez maintenant. »

FIGURE 189

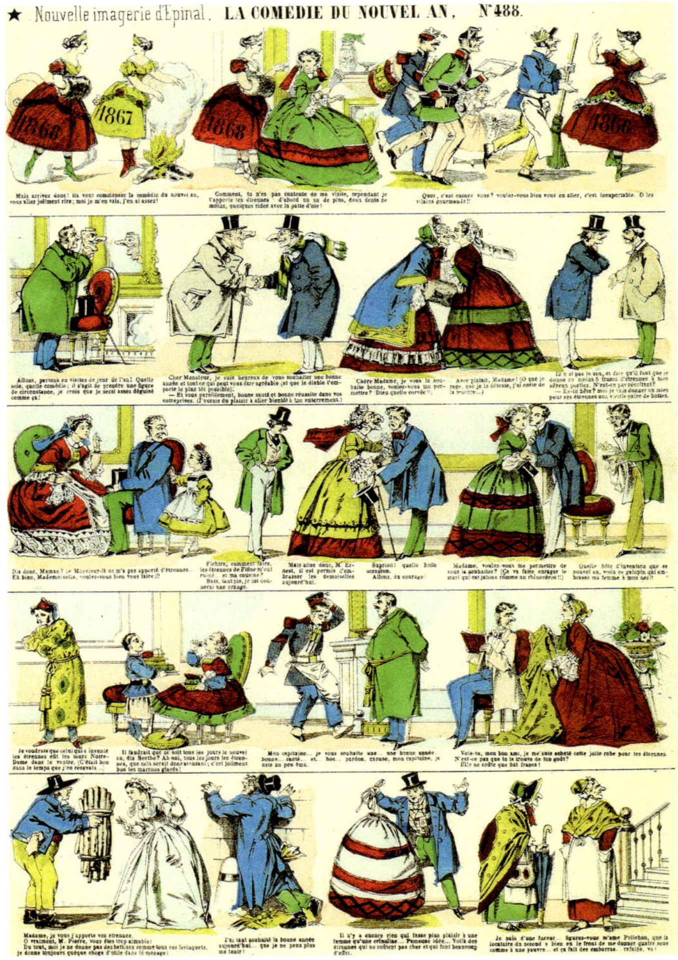

FIGURE 190
François-Charles Pinot, *New Year's Comedy* [*La comédie du nouvel an*], Nouvelle Imagerie d'Épinal, no. 488, 1868. Pub. Pinot & Sagaire, Épinal.

Pellerin's images were not published in newspapers, however, but were printed and distributed either as single sheets or in large bound albums. Of the numerous albums the firm produced, its masterpiece was undoubtedly *Aux armes d'Épinal*, a series of over five hundred numbered prints that began publication in 1889, though many of the prints had been published earlier.[46] The prints were large, more than twelve by sixteen inches, available as individual single-sheet comics or bound in albums. By 1890 the Pellerin firm was well into its fourth generation of directors from its foundation a century earlier. Its print runs had grown into the millions and were being exported all over the world, translated into many languages. In *Aux armes d'Épinal* the earlier workshop-trained artists were replaced by some of the best-known French illustrators and caricaturists, such as Draner (Jules Renard, 1823–1926), Benjamin Rabier (1864–1939), O'Galop (Marius Rossillon, 1867–1946), and Caran d'Ache (Emmanuel Poiré, 1858–1909). In style, popular prints and comics had become virtually indistinguishable. The work of Caran d'Ache provides a

FIGURE 191

Nadar, *Chapter III—Fortune Smiled on His Industrial and Philanthropic Efforts. The Public and Private Life of Mister Reactionary* [*La fortune sourit à ses efforts industriels et philantropiques. La vie publique et privée de Mossieu Réac*], *Revue comique à l'usage des gens sérieux*, March 17, 1849, 280. Wood engraving.

FIGURE 192

The Story of William Tell [*Histoire de Guillaume Tell*], 1850. Pub. Pellerin, Épinal. Stencil-colored woodcut.

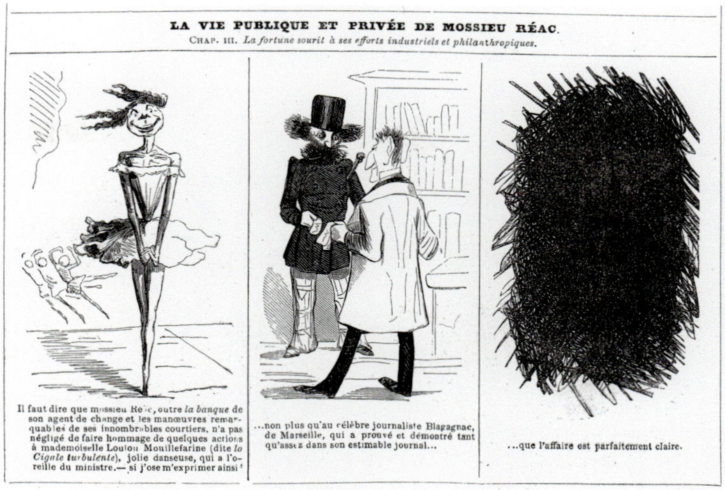

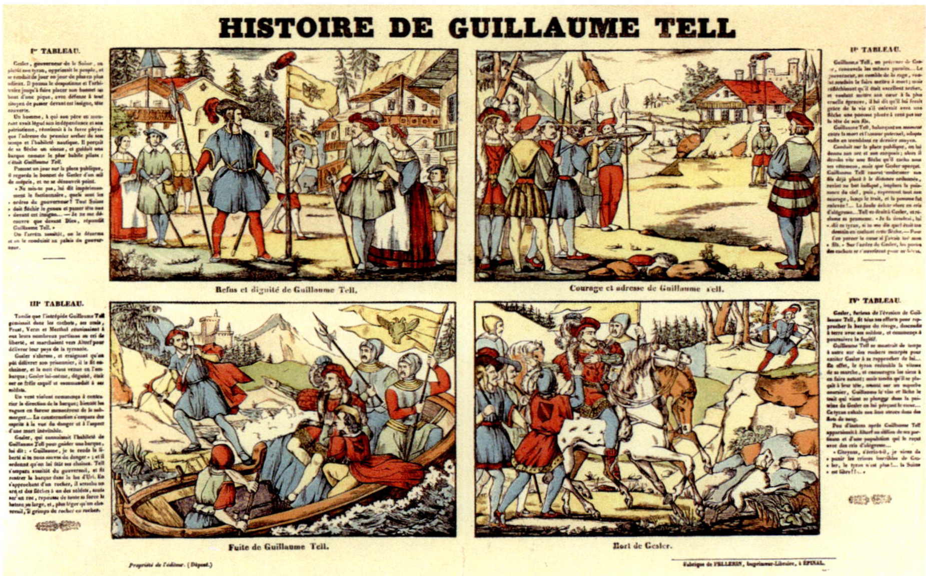

good example of this. His 1893 *The Billet* depicts the results of a military billeting whereby Captain Duracuir (Hardleather) is assigned to a very proper old marquis (fig. 193). Their dinner goes from bad to worse: the old soldier has brought along his ill-behaved dog, he has no table manners, he spills his wine, he gnaws on bones, he smokes at the dinner table. His litany of social offenses increases until, on leaving, he catches his spur on the tablecloth and pulls everything down as he departs. We can see the same concise drawing and droll humor, with a decided edge to it, in the most famous cartoon by Caran d'Ache, *A Family Dinner*, which appeared in *Le Figaro* in 1898 (fig. 194). In the first frame the assembled guests make a pact not to discuss the Dreyfus affair, an inflammatory subject; in the

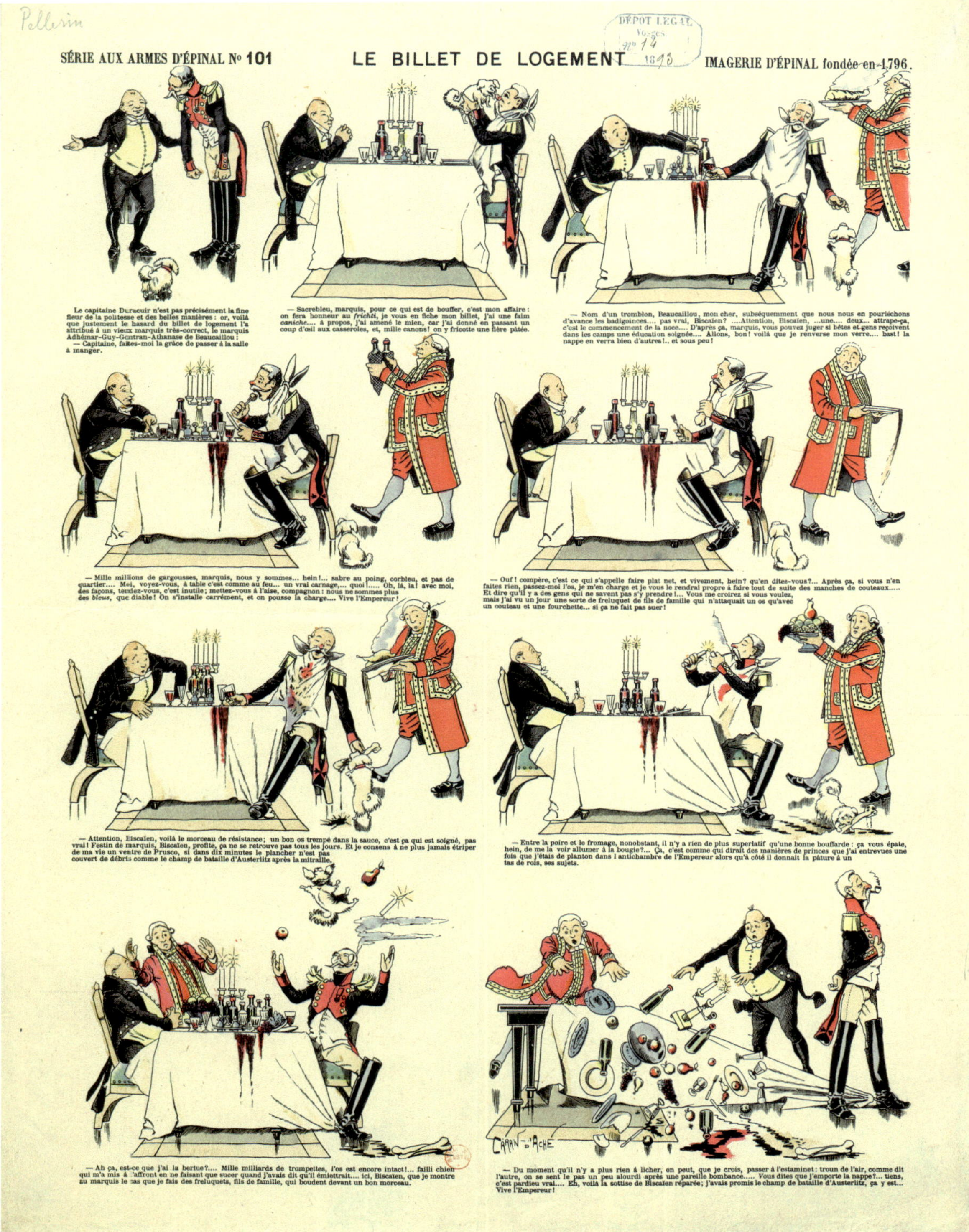
SÉRIE AUX ARMES D'ÉPINAL No 101
LE BILLET DE LOGEMENT
IMAGERIE D'ÉPINAL fondée en 1796.
DÉPOT LEGAL Vosges No 14 1893
Le capitaine Duracuir n'est pas précisément la fine fleur de la politesse et des belles manières : or, voilà que justement le hasard du billet de logement l'a attribué à un vieux marquis très-correct, le marquis Adhémar-Guy-Gontran-Athanase de Beaucaillou :
— Capitaine, faites-moi la grâce de passer à la salle à manger.
— Sacrebleu, marquis, pour ce qui est de bouffer, c'est mon affaire : on fera honneur au *frichti*, je vous en fiche mon billet, j'ai une faim *caniche*.... à propos, j'ai amené le mien, car j'ai donné en passant un coup d'œil aux casseroles, et, mille canons! on y fricotte une fière pâtée.
— Nom d'un tromblon, Beaucaillou, mon cher, subséquemment que nous nous en pourléchons d'avance les badigoinces.... pas vrai, Biscaïen? ....Attention, Biscaïen, ...une.... deux... attrape-ça, c'est le commencement de la noce.... D'après ça, marquis, vous pouvez juger si bêtes et gens reçoivent dans les camps une éducation soignée.... Allons, bon! voilà que je renverse mon verre.... bast! la nappe en verra bien d'autres!.. et sous peu!
— Mille millions de gargousses, marquis, nous y sommes... hein!... sabre au poing, corbleu, et pas de quartier.... Moi, voyez-vous, à table c'est comme au feu... un vrai carnage,... quoi!..... Oh, là, la! avec moi, des façons, tenez-vous, c'est inutile; mettez-vous à l'aise, compagnon : nous ne sommes plus des *bleus*, que diable! On s'installe carrément, et on pousse la charge.... Vive l'Empereur!
— Ouf! compère, c'est ce qui s'appelle faire plat net, et vivement, hein? qu'en dites-vous?... Après ça, si vous n'en faites rien, passez-moi l'os, je m'en charge et je vous le rendrai propre à faire tout de suite des manches de couteaux..... Et dire qu'il y a des gens qui ne savent pas s'y prendre!... Vous me croirez si vous voulez, mais j'ai vu un jour une sorte de freluquet de fils de famille qui n'attaquait un os qu'avec un couteau et une fourchette... si ça ne fait pas suer!
— Attention, Biscaïen, voilà le morceau de résistance; un bon os trempé dans la sauce, c'est ça qui est soigné, pas vrai! Festin de marquis, Biscaïen, profite, ça ne se retrouve pas tous les jours. Et je consens à ne plus jamais étriper de ma vie un ventre de Prusco, si dans dix minutes le plancher n'est pas couvert de débris comme le champ de bataille d'Austerlitz après la mitraille.
— Entre la poire et le fromage, nonobstant, il n'y a rien de plus superlatif qu'une bonne bouffarde : ça vous épate, hein, de me la voir allumer à la bougie?... Ça, c'est comme qui dirait des manières de princes que j'ai entrevues une fois que j'étais de planton dans l'antichambre de l'Empereur alors qu'à côté il donnait la pâture à un tas de rois, ses sujets.
— Ah ça, est-ce que j'ai la berlue?.... Mille milliards de trompettes, l'os est encore intact!... failli chien qui m'a mis à l'affront en ne faisant que sucer quand j'avais dit qu'il émiettrait.... ici, Biscaïen, que je montre au marquis le cas que je fais des freluquets, fils de famille, qui boudent devant un bon morceau.
CARAN D'ACHE
— Du moment qu'il n'y a plus rien à licher, on peut, que je crois, passer à l'estaminet : troun de l'air, comme dit l'autre, on se sent le pas un peu alourdi après une pareille bombance..... *Vous dites que j'emporte la nappe?*... tiens, c'est pardieu vrai.... Eh, voilà la sottise de Biscaïen réparée; j'avais promis le champ de bataille d'Austerlitz, ça y est... Vive l'Empereur!

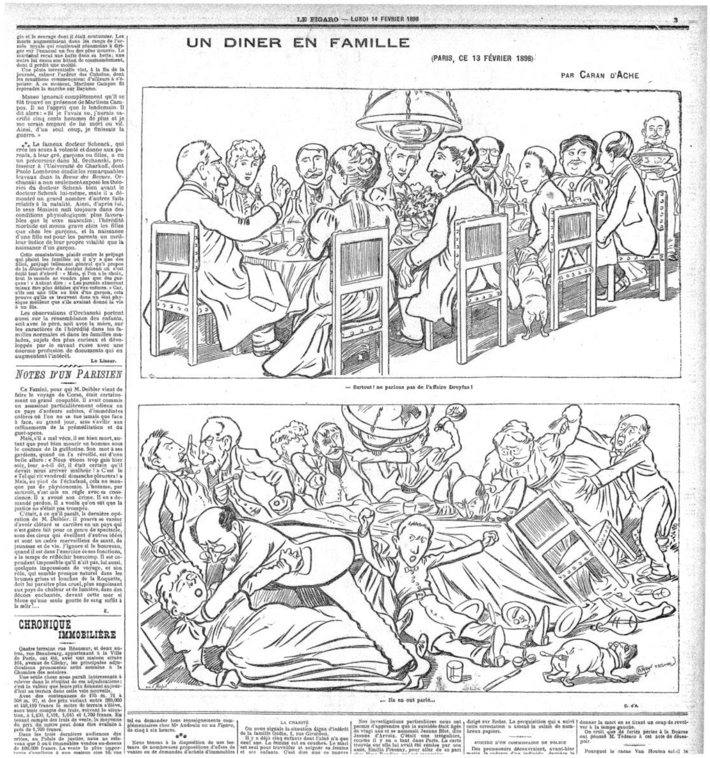

FIGURE 194
Caran d'Ache, *A Family Dinner, Paris, February 13, 1898, "—Most Important! Let's Not Talk about the Dreyfus Affair!—They Talked about It . . . "* [*Un dîner en famille, Paris, ce 13 février 1898. "—Surtout! Ne parlons pas de l'affaire Dreyfus!—Ils en ont parlé . . . "*], *Le Figaro*, February 14, 1898. Bibliothèque nationale de France.

second the dining room is in shambles, the guests are at each other's throat: they discussed it. Caran d'Ache is typical of the new generations of graphic artists who provided drawings for periodicals, for publications aimed at children, and for the new audiences of all ages that bought Pellerin's broadsheets.[47]

At the time when the seemingly separate trajectories of comics, illustration, and popular prints were merging in France, Pellerin produced, in 1888, a series of sixty plates for the Humoristic Publishing Company of Kansas City, Missouri.[48] Prints such as *The Disobedient Little Girls* (fig. 195) were simply images d'Épinal that had been popular in France decades earlier, their titles and captions now translated into English. Parallel to *Unruly Little Boys, Disobedient Little Girls* details all the childish misbehavior that results in painful, sometimes tragic, consequences: a little girl teases the dog, who bites her; she goes too near the chimney and her dress catches fire; she plays near a pond and falls in and drowns. It is ironic that in both style and content these prints were years behind what Pellerin was then producing in France. More typical of the fin-de-siècle Épinal style and content is *Baldaquin's Plume* from *Aux Armes d'Épinal*, where the compositions had long since departed from the rigid grid format characteristic of earlier prints, just as their subjects, here a whimsical and amusing tale of a soldier whose plume catches on fire, are inconsequential and devoid of any moral message (fig. 196).

PAGE 226 **FIGURE 195**
The Disobedient Little Girls, 1893–1894. Printed by Pellerin, Épinal, for the Humoristic Publishing Company, Kansas City, Missouri. Color lithograph.

PAGE 227 **FIGURE 196**
Baldaquin's Plume [*Le plumet de Baldaquin*], *Série aux armes d'Épinal*, no. 126, 1894. Pub. Pellerin, Épinal. Color lithograph.

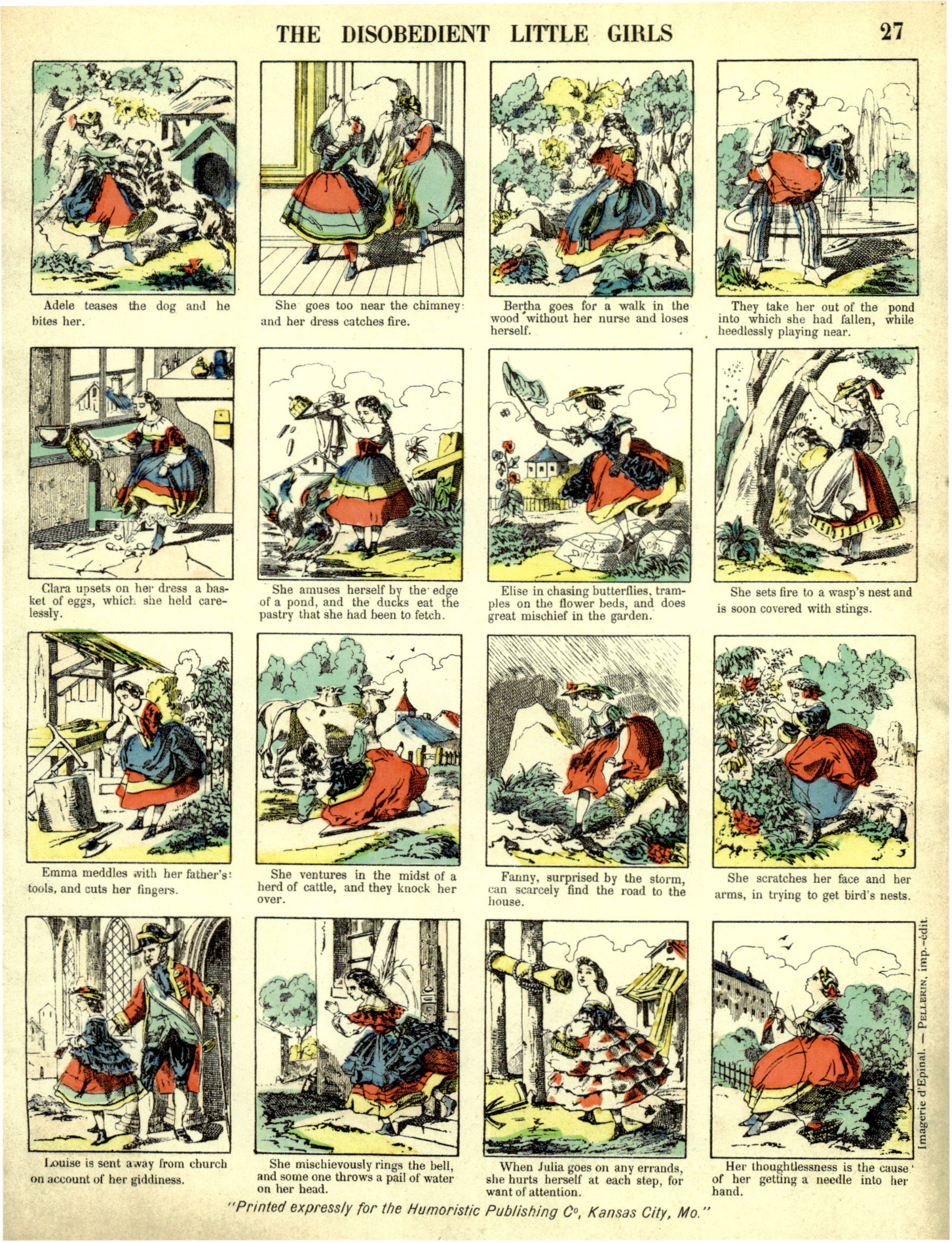
THE DISOBEDIENT LITTLE GIRLS
27
Adele teases the dog and he bites her.
She goes too near the chimney and her dress catches fire.
Bertha goes for a walk in the wood without her nurse and loses herself.
They take her out of the pond into which she had fallen, while heedlessly playing near.
Clara upsets on her dress a basket of eggs, which she held carelessly.
She amuses herself by the edge of a pond, and the ducks eat the pastry that she had been to fetch.
Elise in chasing butterflies, tramples on the flower beds, and does great mischief in the garden.
She sets fire to a wasp's nest and is soon covered with stings.
Emma meddles with her father's tools, and cuts her fingers.
She ventures in the midst of a herd of cattle, and they knock her over.
Fanny, surprised by the storm, can scarcely find the road to the house.
She scratches her face and her arms, in trying to get bird's nests.
Louise is sent away from church on account of her giddiness.
She mischievously rings the bell, and some one throws a pail of water on her head.
When Julia goes on any errands, she hurts herself at each step, for want of attention.
Her thoughtlessness is the cause of her getting a needle into her hand.
Imagerie d'Epinal. — Pellerin, imp.-édit.
"Printed expressly for the Humoristic Publishing C°, Kansas City, Mo."

FIGURE 195

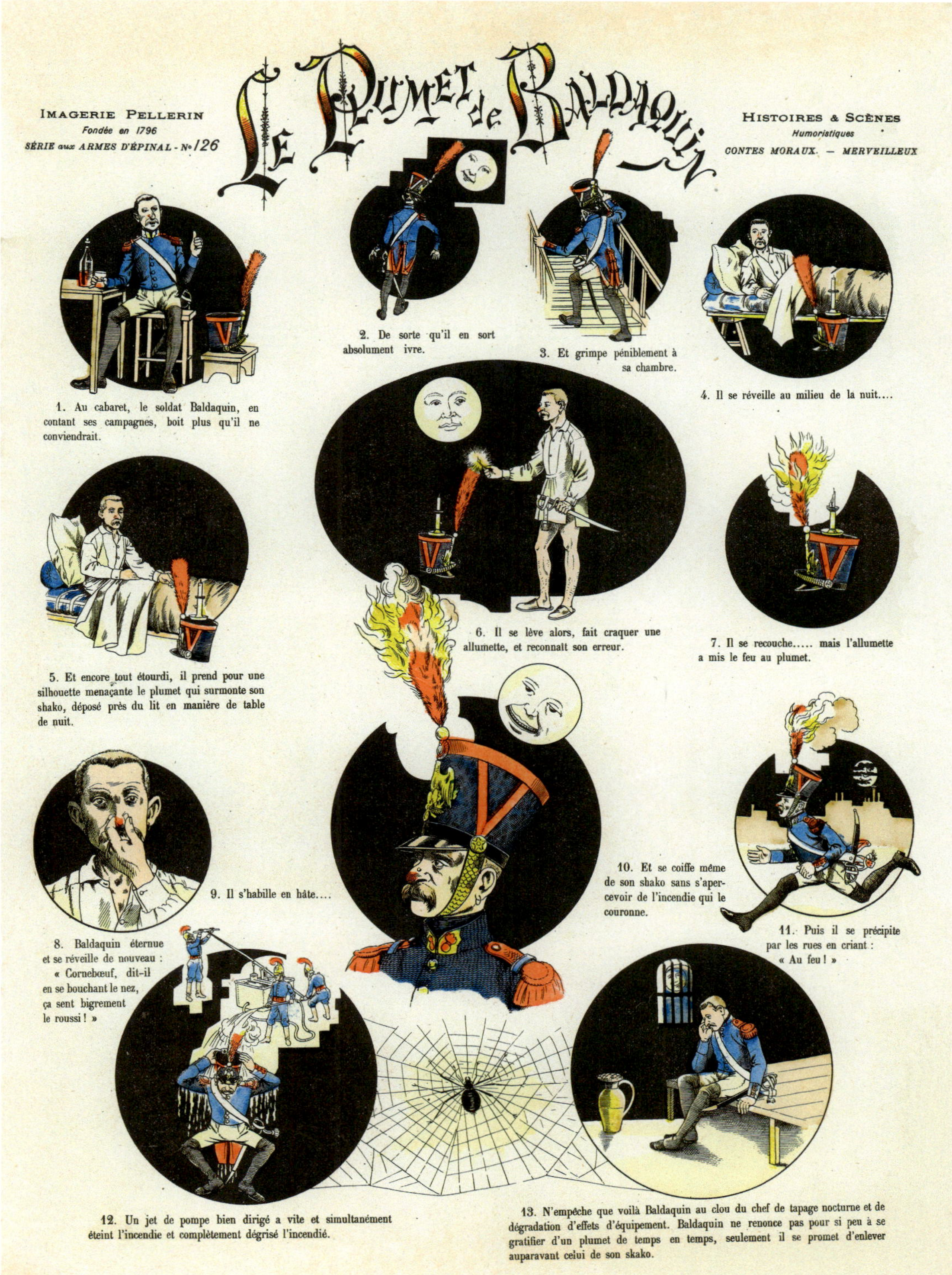
Le Plumet de Baldaquin
Imagerie Pellerin
Fondée en 1796
Série aux Armes d'Épinal - N° 126
Histoires & Scènes
Humoristiques
Contes moraux. — Merveilleux
1. Au cabaret, le soldat Baldaquin, en contant ses campagnes, boit plus qu'il ne conviendrait.
2. De sorte qu'il en sort absolument ivre.
3. Et grimpe péniblement à sa chambre.
4. Il se réveille au milieu de la nuit....
5. Et encore tout étourdi, il prend pour une silhouette menaçante le plumet qui surmonte son shako, déposé près du lit en manière de table de nuit.
6. Il se lève alors, fait craquer une allumette, et reconnaît son erreur.
7. Il se recouche..... mais l'allumette a mis le feu au plumet.
8. Baldaquin éternue et se réveille de nouveau : « Cornebœuf, dit-il en se bouchant le nez, ça sent bigrement le roussi ! »
9. Il s'habille en hâte....
10. Et se coiffe même de son shako sans s'apercevoir de l'incendie qui le couronne.
11. Puis il se précipite par les rues en criant : « Au feu ! »
12. Un jet de pompe bien dirigé a vite et simultanément éteint l'incendie et complètement dégrisé l'incendié.
13. N'empêche que voilà Baldaquin au clou du chef de tapage nocturne et de dégradation d'effets d'équipement. Baldaquin ne renonce pas pour si peu à se gratifier d'un plumet de temps en temps, seulement il se promet d'enlever auparavant celui de son skako.

FIGURE 196

Groupe V. — FEUILLE N° 45.
MÉDAILLE D'OR : MARSEILLE 1883

L'HISTOIRE DE LA VAPEUR

2e feuille : Les Bateaux à Vapeur

SÉRIE ENCYCLOPÉDIQUE GLUCQ
des Leçons de Choses Illustrées
Ouvrage adopté par la VILLE de PARIS comme Récompense dans ses Écoles.

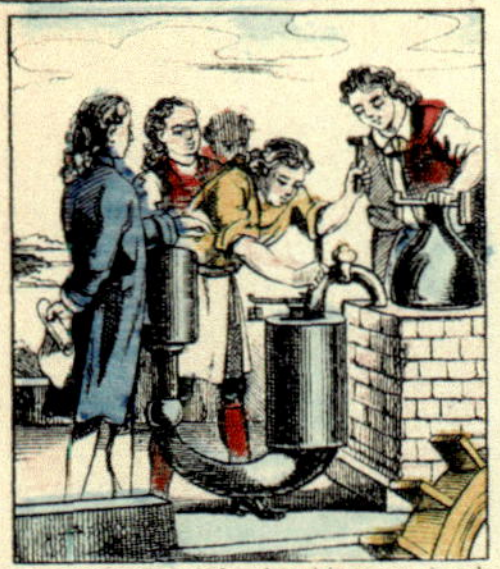

Le premier bateau dans lequel la vapeur joua le le rôle de force motrice est dû à Papin lui-même, l'inventeur de la machine à vapeur. C'est en Allemagne, à Fulda, en 1707, que ce curieux bateau fut construit et essayé. Le but de Papin était de faire marcher mécaniquement les navires contre le vent.

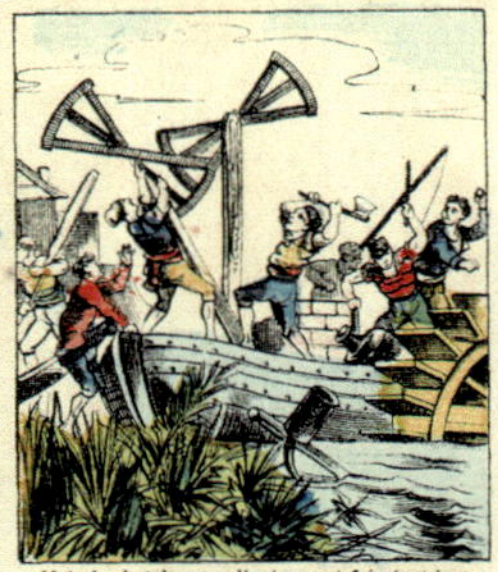

Mais les batehers ordinaires qui faisaient le service des transports de marchandises sur le fleuve du Weser, pressentant sans doute un concurrent ignoré mais redoutable dans ce navire qui marchait seul, mirent en pièces le bateau et sa machine (25 septembre 1707). Papin fut ruiné.

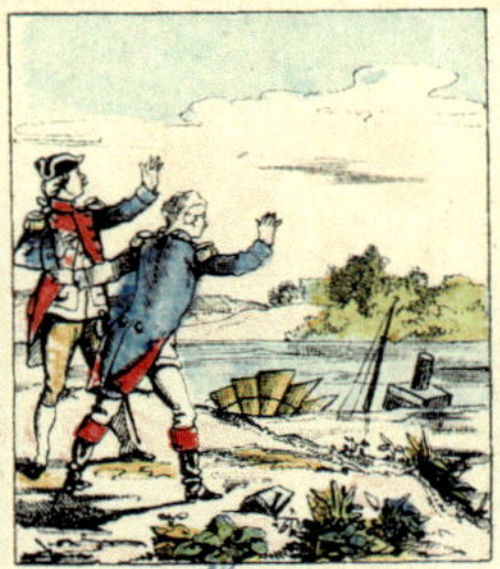

En 1776, deux officiers français, nommés d'Auxiron et de Follenai, firent établir sur la Seine un nouveau bateau à vapeur de leur invention. Mais, au mois de septembre de l'année suivante, soit par maladresse, soit par malveillance, le malheureux bateau fut trouvé coulé au fond de la Seine.

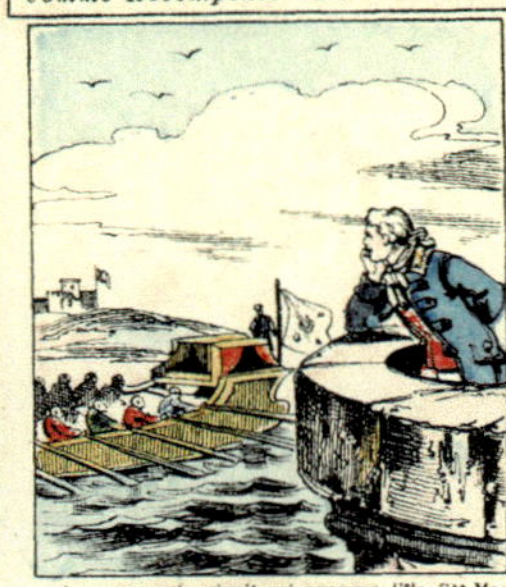

A ce moment, vivait prisonnier à l'île Ste Marguerite, à la suite d'un duel, un jeune gentilhomme franc-comtois nommé le marquis de Jouffroy. Pendant ses trop longues heures de méditation forcée, il contemplait les manœuvres des forçats qui faisaient mouvoir à la rame les Galères Royales.

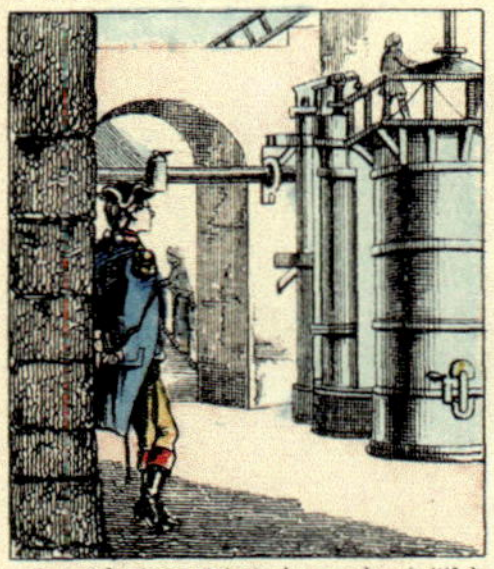

Frappé des inconvénients de ce mode primitif de propulsion des navires, et songeant à le remplacer par l'emploi de la vapeur, de Jouffroy, une fois son emprisonnement terminé, vint en 1775 à Paris et alla de suite visiter et étudier la fameuse machine à vapeur établie à Chaillot.

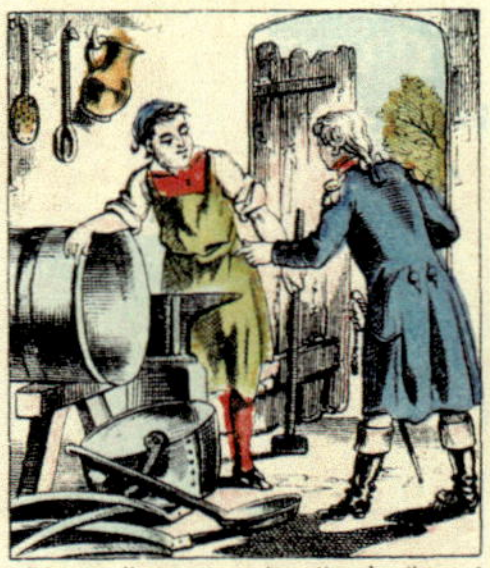

Désireux d'entourer son invention du silence et de la discrétion nécessaires, le marquis de Jouffroy se réfugia à Baume-les-Dames sur le Doubs ; et là, n'ayant pour constructeur et pour mécanicien que l'humble chaudronnier du pays, il lui fit fabriquer à la main toutes les pièces de sa nouvelle machine.

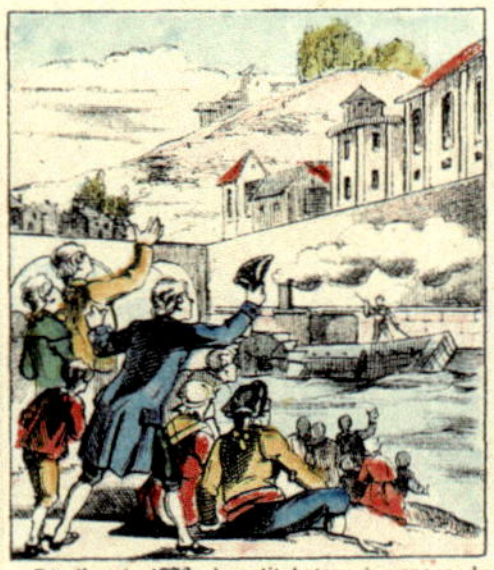

Dès l'année 1776, le petit bateau à vapeur du marquis de Jouffroy marcha sur le Doubs : et le 15 juillet 1783 il fut fait sur la Saône, à Lyon, une expérience décisive devant plus de 10,000 spectateurs enthousiasmés. La navigation à vapeur est donc bien d'origine française.

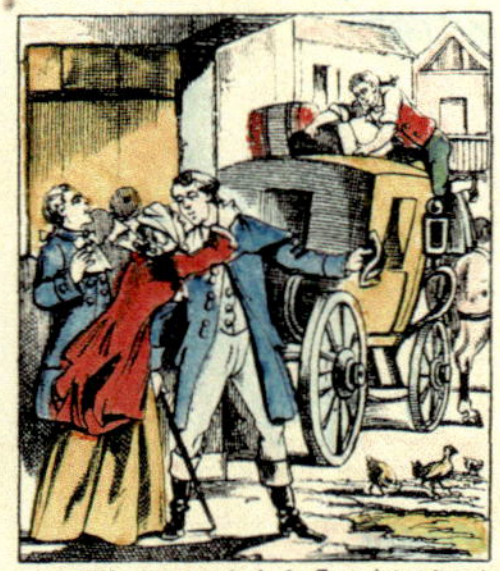

En 1790, le marquis de Jouffroy dut quitter la France : ayant voulu tirer un juste profit de sa découverte, il avait demandé au gouvernement un privilège de 30 ans pour son invention Le ministre de Calonne et l'Académie des Sciences le lui refusèrent! Tristes leçons de l'Histoire.

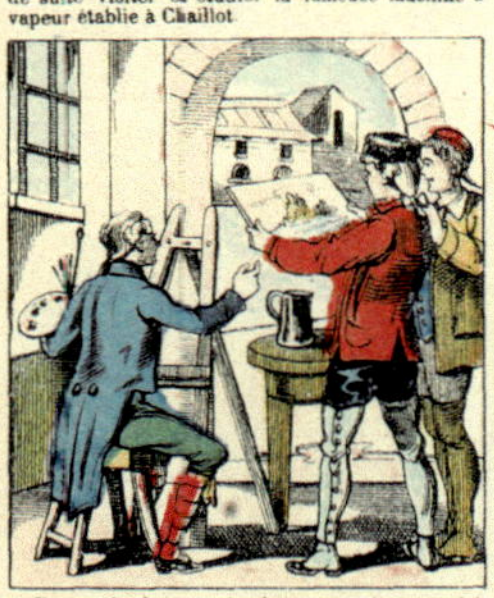

Il y avait alors en Angleterre, un jeune américain nommé Robert Fulton. Apprenti joaillier, puis dessinateur de talent, il vendait à 16 ans des tableaux pour vivre. En 1801, Fulton vint en France et y inventa les Bateaux sous-marins, le Panorama et les Torpilles.

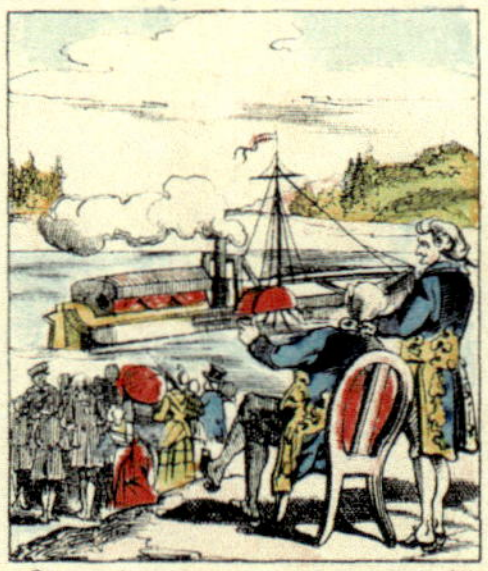

Reprenant pour son compte tous les essais déjà faits en France et en Angleterre, Fulton réussit, le 9 août 1803, à faire marcher sur la Seine, en face de toute la population et de l'Institut convoqué à cet effet, un bateau à vapeur construit par lui et qui réussit à merveille.

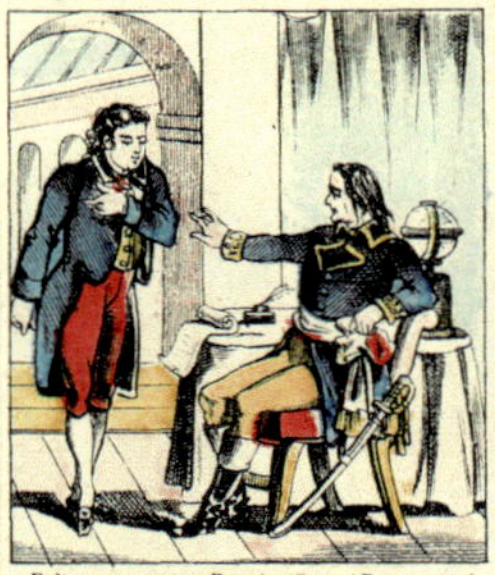

Fulton proposa au Premier Cousul Bonaparte de faire honneur de son invention à la France qui, alors en guerre avec l'Angleterre pouvait en retirer les plus importants avantages. Bonaparte ne comprit pas la valeur immense de ce projet qui, réalisé, eut peut-être changé la face du monde ! Il refusa !

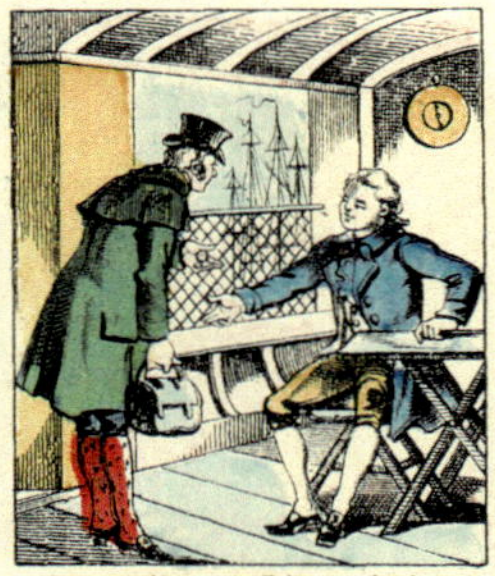

Justement découragé, Fulton résolut de porter son invention à l'Amérique, sa patrie, et partit pour New-York. Le 11 avril 1807, son navire « **LE CLERMONT** » faisait son premier voyage sur la baie d'Hudson. Un seul homme osa accompagner Fulton : ce fut un français nommé Andrieux.

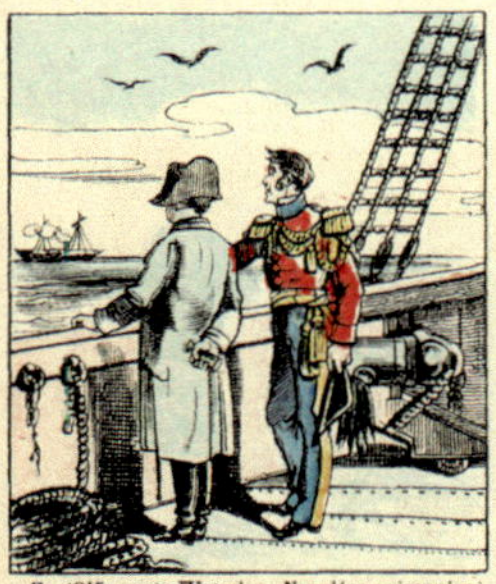

En 1815, après Waterloo, Napoléon prisonnier à bord du Vaisseau Anglais qui l'emmenait mourir à Ste Hélène, aperçut sur la mer, à l'horizon lointain, une petite fumée blanche, et demanda ce que c'était. — « Sire ! lui dit-on, c'est un navire à vapeur ! » L'Empereur se détourna. L'ombre du grand Fulton dut lui apparaître en ce moment. Il était trop tard, hélas !

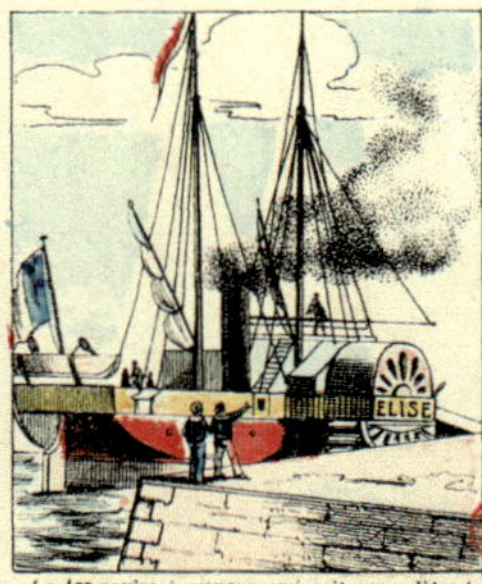

Le 1er navire à vapeur qui soit venu d'Angleterre en France à travers la Manche, est l'*Elise*, le 9 mars 1816. Le 1er navire à vapeur qui ait traversé l'Atlantique pour venir d'Amérique à Liverpool est le *Savannah* en 1819. Enfin, le 1er navire européen qui soit allé d'Irlande à New-York, est le *Sirius*, en 1838, après 17 jours de traversée.

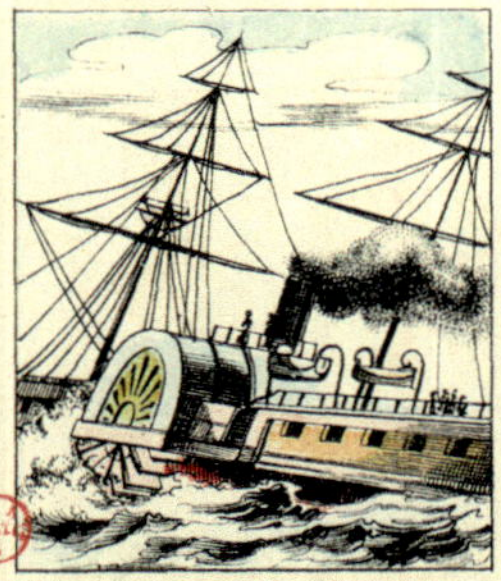

Le premier instrument de propulsion adopté pour les navires à vapeur, a été la **ROUE à AUBES**, formée de palettes de fer remplaçant les rames et pouvant tourner autour d'un axe de façon à plonger tour à tour d'une certaine profondeur dans la mer.

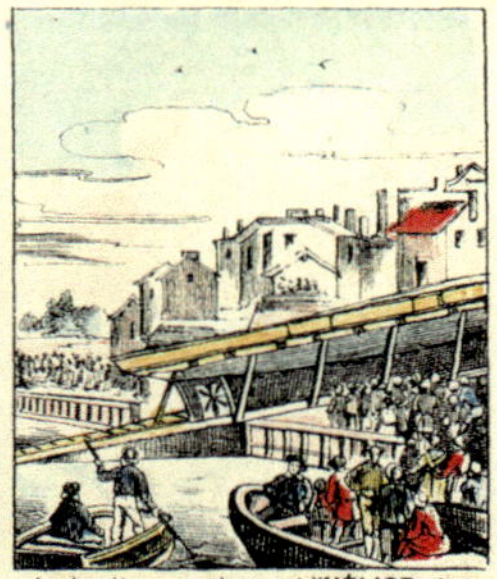

Le deuxième propulseur est l'**HÉLICE**, due au génie de notre grand et malheureux ingénieur français, Frédéric Sauvage. L'*Hélice* est un gigantesque pas de vis de bronze, qui, en tournant, se visse dans l'eau et y rencontre une résistance qui suffit à faire chasser le navire. L'*Hélice* est noyée dans l'eau et a, par conséquent, l'avantage d'être à l'abri des boulets de l'ennemi.

Typ.-Lith. de Ch. PELLERIN à Épinal. (Déposé)

GLUCQ, — 115, Boulevard Sébastopol, Paris. — Auteur-Éditeur de la série encyclopédique des Leçons de Choses Illustrées.

FIGURE 197
History of Steam No. 2. The Steam Ship, from *Glucq Encyclopedic Series of Illustrated Lessons* [*L'histoire de la vapeur, 2e feuille: Les Bateaux à vapeur / Série encyclopédique Glucq des leçons de choses illustrées*], *Group 5, no. 45*, 1884. Pub. Glucq, Paris, and Pellerin, Épinal. Color lithograph. Bibliothèque nationale de France.

PAGE 230 **FIGURE 198**
A. Chauffour, *BRILLANT DE LUXE Makes Copper Shine* [*Le BRILLANT DE LUXE rend les cuivres resplendissants*], 1895. Pub. Glucq, Paris, and Pellerin, Épinal. Color lithograph. Bibliothèque nationale de France.

PAGE 231 **FIGURE 199**
Théophile Steinlen, *Tour of Rudolphe Salis's Le Chat Noir* [*Tournée du Chat Noir de Rudolphe Salis*], 1896. Lith. Charles Verneau. Color lithograph. Zimmerli Art Museum, Rutgers University.

By the 1880s, Pellerin's production extended over the entire spectrum of popular imagery. We have already noted his collaboration with the firm of Glucq in the 1889 elections (see fig. 176), but the Glucq-Pellerin collaboration was even more pronounced in the new fields of educational publishing and advertising. Beginning in 1880 they published together *The Glucq Encyclopedic Series of Illustrated Lessons*, a series of educational broadsheets on a range of topics including industry, economics, culture, and society.[49] *The Steam Ship*, the second of two prints relating the history of steam, traces the development of steam power in navigation and is, like all the broadsheets in this series, a tour de force of illustration (fig. 197). Each of the sixteen small vignettes is fully composed, with settings varied from land to sea, interiors and exteriors, close-ups and distant views; figures move easily and naturally, working, resting, conversing. It's not surprising that the series was well received; on the upper left is written "Gold Medal, Marseilles 1883," and on the upper right, "Work Adopted by the City of Paris as an Award in Its Schools."[50]

The sincerity and integrity that were perceived as residing in traditional popular images also made them attractive as vehicles for advertising products as well as politics, and their simplified drawing and colors were ideal for a nascent technology that had not yet perfected the halftone process. The Glucq-Pellerin advertising efforts mixed the educational with the commercial in broadsheets similar to their illustrated lessons. An 1888 advertisement for Trébucien gourmet chocolate, for example, depicted its production from cocoa tree to elegantly wrapped confection in a comic-strip format virtually identical to that of the *Glucq Encyclopedic Series* broadsheets.[51] By the 1890s, however, their advertising format had developed into something more familiar to us today: exaggerated claims made for a product. Brillant de Luxe, a metal polish, was marketed in 1895 with a Pellerin image of a rural couple gesturing toward the dazzling dome of the Invalides in Paris (fig. 198). *—It's True, Isn't It, Mister Policeman, That It's Made of Gold?—Not At All, My Good Man, It's Been Polished with BRILLANT DE LUXE. It Turns Copper into Gold!* The banner beneath promises *BRILLANT DE LUXE makes copper shine.*[52] Such images became ubiquitous in the illustrated press, but never aspired to the status of the art poster movement, which, although also used for advertisement, traced its lineage more to high art than to these vernacular models.[53] A comparison of the Brillant de Luxe advertisement with a virtually contemporaneous poster by Théophile Steinlen (1859–1923) for the nightclub Le Chat Noir, shows the similarities and differences (fig. 199). Both are advertisements, but the Steinlen poster developed from a lineage of frontispieces, albums, and print portfolios dating back to the early decades of lithography, cross-fertilized with the high-art tradition. *Brillant de Luxe,* on the other hand, was the hybrid offspring of Épinal cross-fertilized with caricature and comics.

Despite this rich history of popular prints, the study of these images has congealed around a kind of golden age essentialism, with traditional imagery perceived as having been made with time-honored techniques, carved in wood, hand-colored, and sold to a simple agrarian audience. Insofar as they have been assimilated into the study of art history, it has been through their connection with Modernism.[54] By the mid-nineteenth century, they

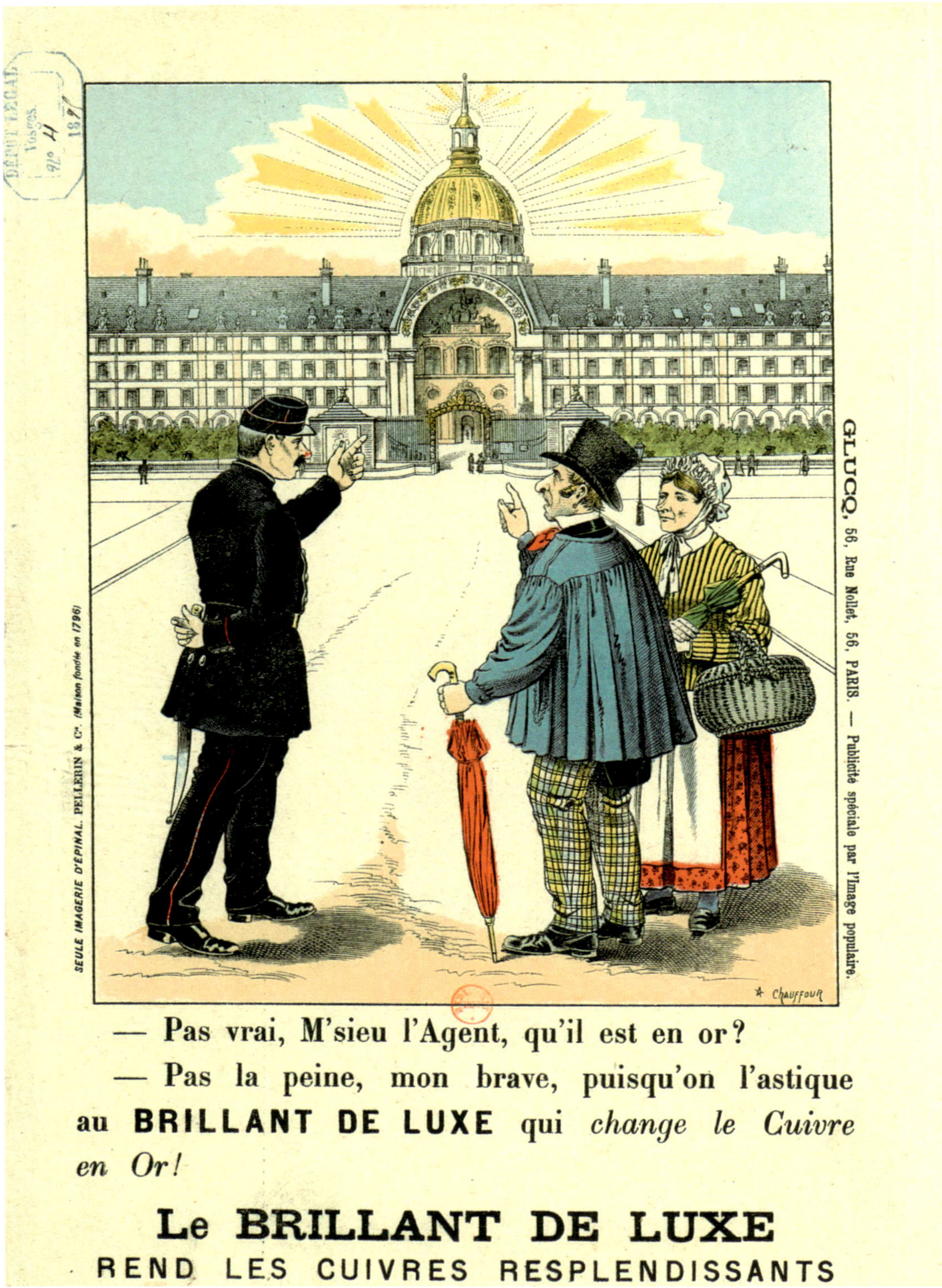
GLUCQ, 56, Rue Nollet, 56, PARIS. — Publicité spéciale par l'Image populaire.
SEULE IMAGERIE D'EPINAL. PELLERIN & Cie. (Maison fondée en 1796)
Chauffour
— Pas vrai, M'sieu l'Agent, qu'il est en or?
— Pas la peine, mon brave, puisqu'on l'astique au BRILLANT DE LUXE qui change le Cuivre en Or!
Le BRILLANT DE LUXE
REND LES CUIVRES RESPLENDISSANTS

FIGURE 198

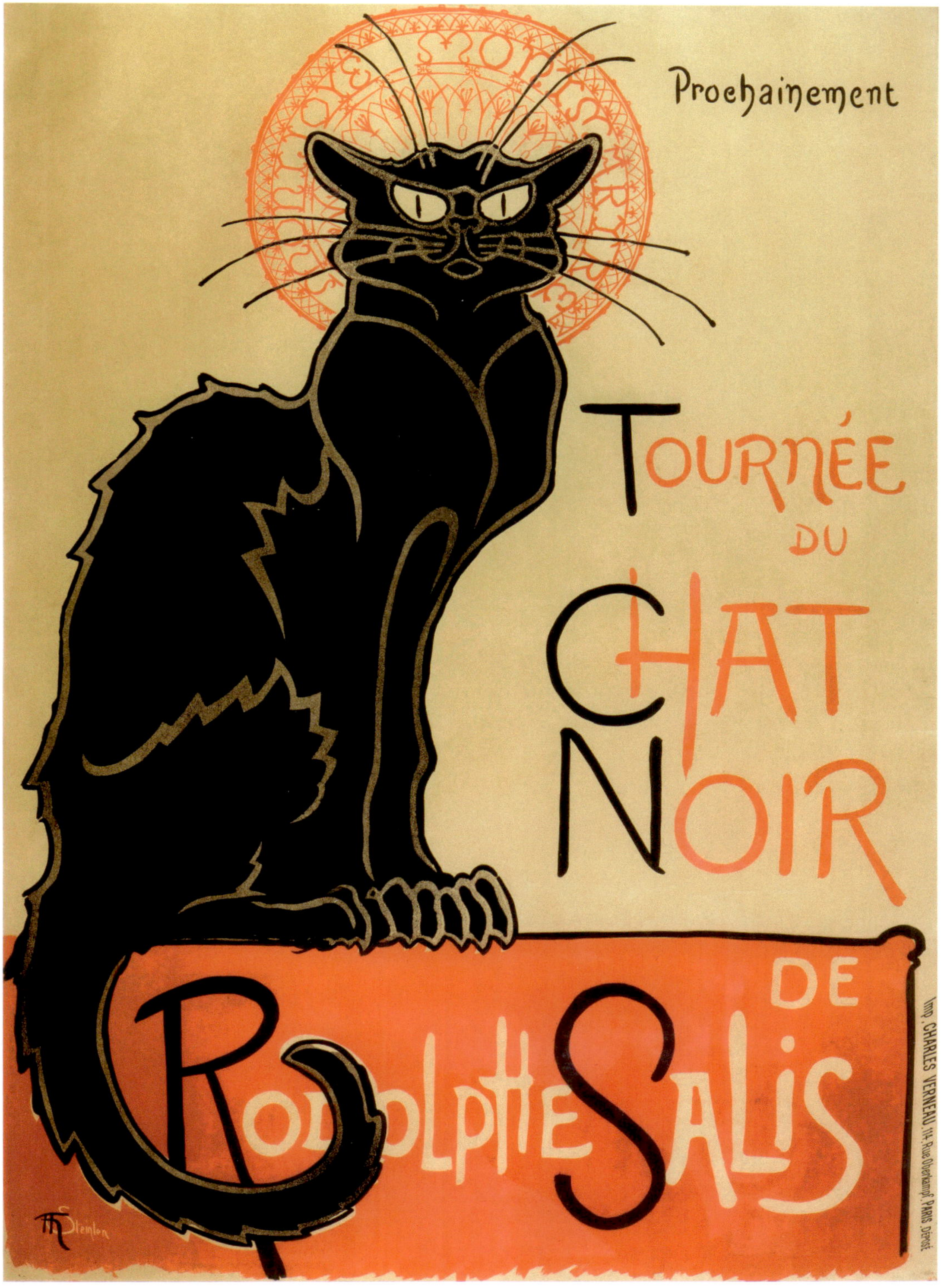
Prochainement
Tournée du
Chat Noir
de
Rodolphe Salis
Imp. CHARLES VERNEAU. 114, Rue Oberkampf, PARIS. Déposé
Steinlen

FIGURE 199

LA GUERRE 1914-1915
EN IMAGES
Faits, Combats, Episodes, Récits

Le Dieu THOR la plus barbare
d'entre les barbares divinités de la Vieille Germanie

Composition de F. CLASQUIN
PELLERIN & Cie, imp.-édit.
IMAGERIE D'ÉPINAL, N° 87

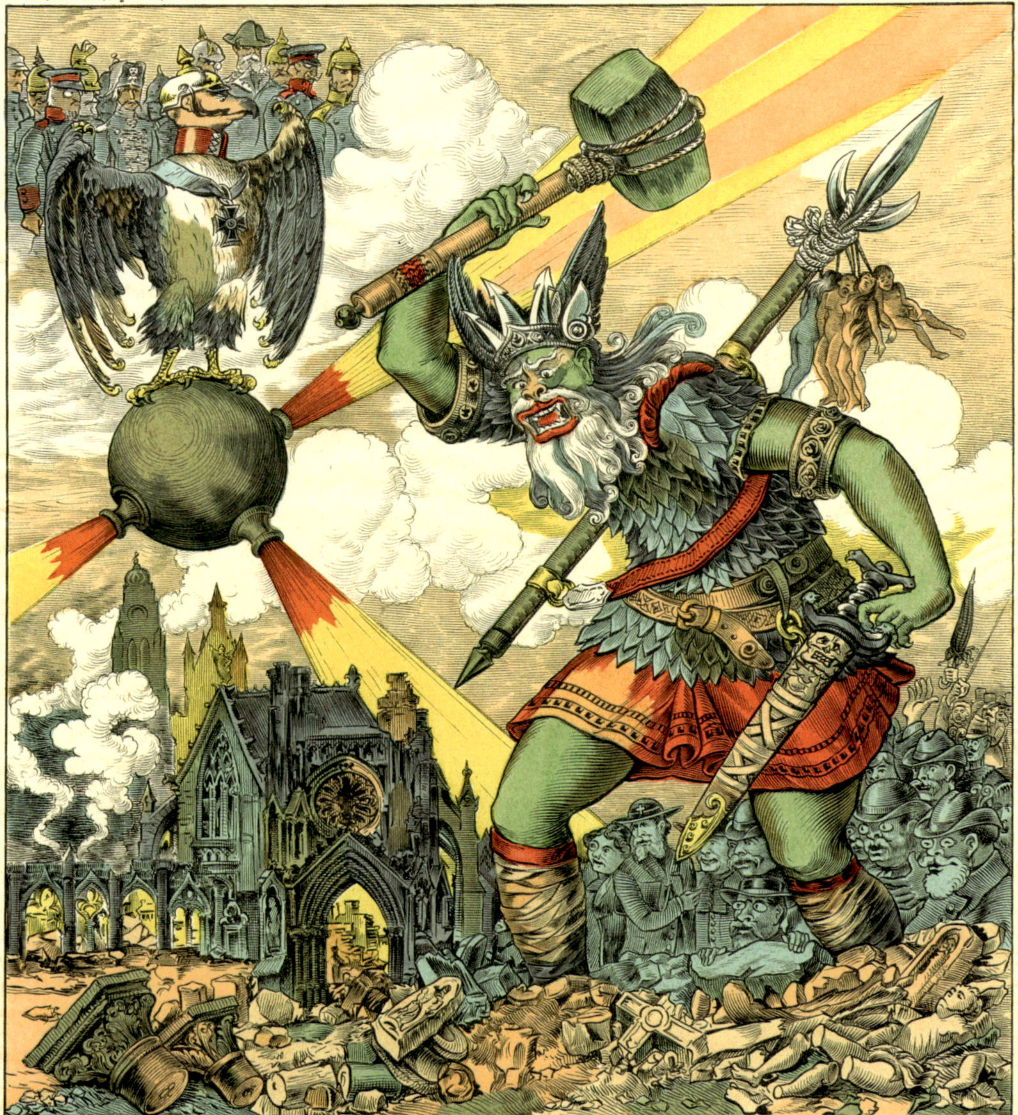

SONVIEUX

LE JOUR VIENDRA, HÉLAS! LES VIEILLES
DIVINITÉS GERMANIQUES SE LÈVERONT DE LEURS
TOMBEAUX FABULEUX ET ESSUYERONT DE LEURS
YEUX LA POUS- THOR SE DRESSERA
SIÈRE SÉCULAIRE AVEC SON MARTEAU
GIGANTESQUE ET DETRUIRA LES CATHÉDRALES
GOTHIQUES
FC 1915 HENRI HEINE 1834

BONDIEU

Car ce ne peut être que **CELUI-LA** que le **KAISER** ne cesse d'invoquer, si l'on en juge par les actes de pire vandalisme dont la Belgique et la France envahie portent le flétrissant témoignage..... Henri Heine, le poète allemand qui passa la dernière moitié de sa vie en France, avait, dès 1834, prévu, comme on voit, ce retour à la sauvagerie ancestrale..... En 1914-1915 le Kaiser commande et l'Allemagne applaudit !

FIGURE 200

François Clasquin, *The God Thor, the Most Barbarous of All the Barbarous Divinities of Old Germany* [*Le Dieu Thor, la plus barbare d'entre les barbares divinités de la vieille Germanie*], 1915. Pub. Pellerin, Épinal. Color lithograph. Bibliothèque nationale de France.

had become synonymous with anti-academic art, and from 1850 onward Champfleury (Jules Husson, 1821–1889) published studies of the subject, his *History of Popular Imagery* appearing in 1869.[55] Champfleury did not just admire the popular print as an art form, however; he used it as a battering ram against academic art, claiming that "the artistic awkwardness of these prints is closer to the work of men of genius than the second-rate works produced by art schools and false traditions."[56] Modernists like Champfleury preferred the crude to the refined, but for most contemporaries, the works of Modernist painters were as inept as images d'Épinal. Courbet's art, especially, was attacked for its resemblance to popular art, a resemblance which, in fact, was wholly intentional. Champfleury praised Courbet's painting for being like "naïve woodcuts, engraved with a clumsy chisel," concluding that "high art finds the same expression as naïve art."[57] Émile Zola defended Manet, writing: "It is said, mockingly, that the works of Édouard Manet recall Épinal prints, and there is a lot of truth in this mockery that is actually praise."[58] Picasso contributed to the Modernist tradition of using images d'Épinal to shock the art establishment; his designs for the ballet *Parade* of 1917 caused a sensation with costumes based on the crude drawing and simple abstractions of the most primitive of these prints.[59] And yet, despite this widespread conception of the populist quality of Épinal prints, we should note that Pellerin had long since abandoned the traditional style referenced by these Modernist artists and their critics. Pellerin's contemporaneous imagery, however, was much too similar to urban illustration to have any currency in the culture wars of Modernism. Even when the image d'Épinal was taken up by the Vichy government in the 1930s, the referent was always the traditional Épinal style. So it seems that the entire spectrum of culture and politics from Left to Right was in accord that the history of popular imagery had come to an abrupt halt in the mid-nineteenth century. But, of course, it did not.

The development of the image d'Épinal, and popular imagery in general, was like a great river that split into numerous tributaries fertilizing great swaths of terrain. Modernist artists from Courbet onward drew inspiration from these prints' simple forms and bright colors. The broadsheets published by Glucq and Pellerin in comic-strip format taught complex subjects to generations of school children. Insofar as popular imagery flowed into the new field of advertising, it provided both politics and commerce a "look" that, it was hoped, would inspire confidence. But it is in the field of comics that these images have had their lasting influence. In the 1830s, while urban audiences were witnessing the birth of comic books with the first publications of Rudolph Töpffer, rural audiences were already familiar with colorful prints of sequential narration. By the end of the century, the image d'Épinal and the comic strip had cross-fertilized and mutated into our modern comics.

During the First World War, Pellerin published *The God Thor, the Most Barbarous of All the Barbarous Divinities of Old Germany* by François Clasquin (1849–1917), an image whose lineage can be traced backward to the image d'Épinal and forward to our modern superheroes (fig. 200). The influence of these popular prints, should, then, be understood as seeping through all areas of visual production, a visual vernacular from Another World that helped to shape our own.

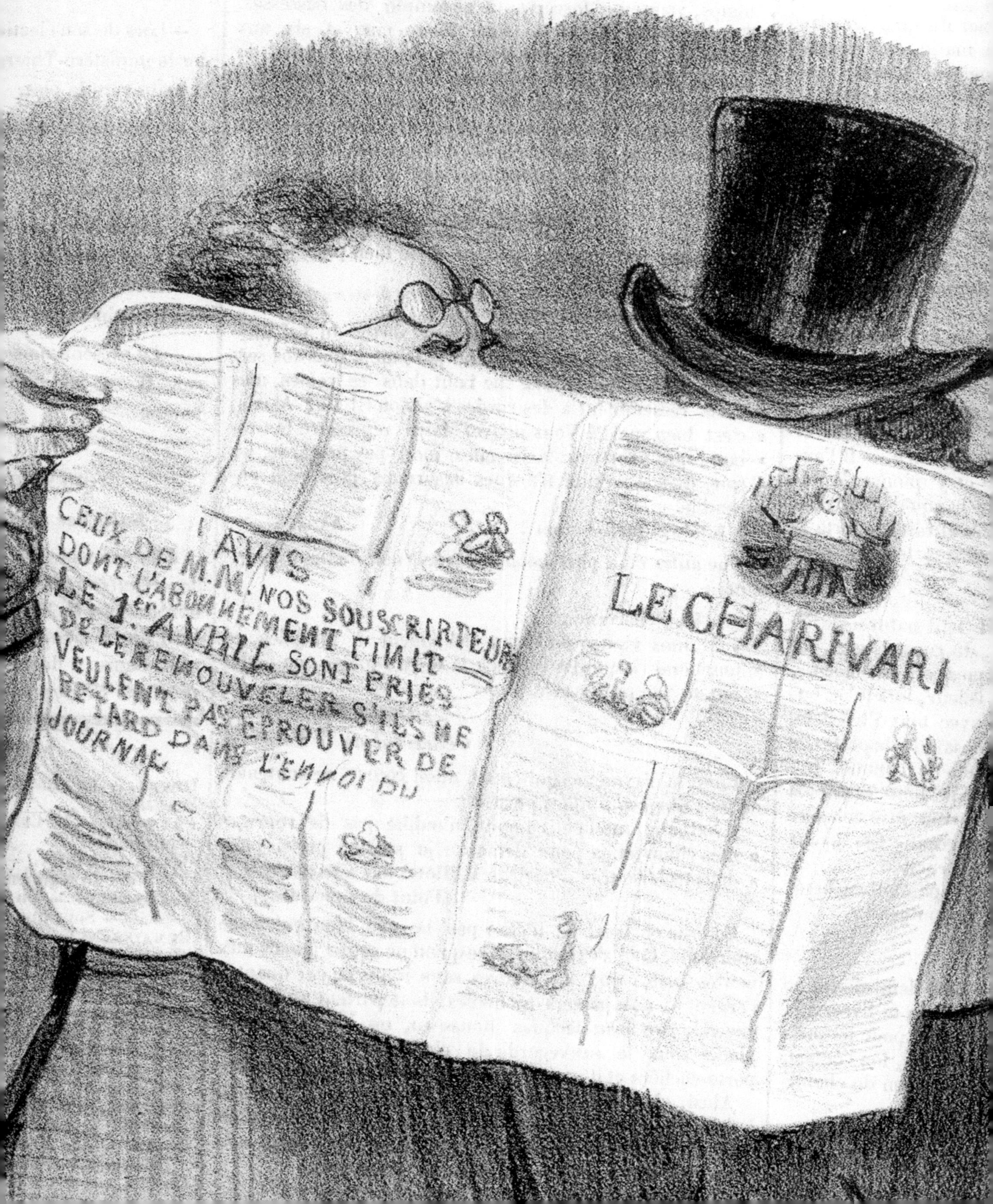
AVIS
CEUX DE M.M. NOS SOUSCRIPTEURS
DONT L'ABONNEMENT FINIT
LE 1er AVRIL SONT PRIÉS
DE LE RENOUVELER S'ILS NE
VEULENT PAS ÉPROUVER DE
RETARD DANS L'ENVOI DU
JOURNAL
LE CHARIVARI

Conclusion

A Complete Panorama

In the 1832 prospectus for *Le Charivari*, Charles Philipon described his new periodical as "A complete panorama where we continuously reproduce, with crayon and pen, every aspect of this kaleidoscopic world we live in."[1] He could well have been describing the entire spectrum of nineteenth-century illustrated print culture, the first to give us a complete visual panorama of its contemporaneous world and the last to do so exclusively with the crayon and pen and not the camera.

The sheer energy of the graphic medium, its ebullience wherein all things were possible with little risk, attracted artists of all persuasions. For some, it was a blank canvas on which to try out their ideas. For others, it was an end in itself. Motivations were as varied as the artists themselves: as a *gagne-pain,* literally a "breadwinner" that could pay the bills; as a means of getting their work known to a wider audience; as a tentative step into the cutting-edge technology of the century; and not least, as a new, largely unexplored graphic language. In this study, we have seen the enormous breadth of imagery created during these decades. Although it is customary to trace the trajectory of printed imagery as culminating in the photograph, I have proposed here an alternative history, one that foregrounds the hand-drawn image. Despite all the transformations and attractions of new media, such image-making has persisted into the twenty-first century. The hegemony of the photograph is apparent in today's periodical press, but we should also consider that it is hand-drawn imagery and not photography that has caused the most violent confrontations in the twenty-first-century world of print. Editorial cartooning in the tradition of the French press of the 1830s has lost none of its power to shock and provoke.[2]

Beginning in the later twentieth century, comics have had, if anything, a resurgence. Far from being the poor cousins of the sumptuously illustrated publications of the previous century, they have become, in the form of graphic novels, the major representatives of the illustrated book medium, along with limited-edition artists' books and children's books. With these three categories, artists' books, graphic novels, and children's literature, illustrated books have demonstrated that they can maintain their appeal over time and across the socioeconomic spectrum. While film has taken up comics' initial challenge of sequential visual narration, it has in no way obliterated its appeal.[3] Indeed, film continues to borrow heavily from the older comic-book tradition, as attested by the

DETAIL OF FIGURE 57

continuing popularity of films based on graphic novels or earlier comics superheroes. The two media continue a mutually fruitful coexistence. Popular prints, once the least favored of media, survived the challenge of urbanization and higher literacy levels by cross-fertilizing with other genres, notably comics, but they gradually infiltrated the world of advertising as well. In the late decades of the nineteenth century, the nascent enterprise of advertising bifurcated; the poster movement, influenced by high art, garnered worldwide attention and respect, while more vernacular images, informed by the clear simple design of popular prints, formed an ephemeral, although no less important, stratum of commercial imagery and graphic design.

The imagery I have discussed here cannot be encompassed under a single rubric other than that of being responsive to every shift and change in society. Graphic satire tracked every social foible, and political caricature served as gadfly to every regime. The latest fashions, the latest entertainments, unfamiliar geographical regions and customs, all became known through the work of graphic artists, lithographers, and wood engravers. Our knowledge of the visual universe of this period, and—in no small degree—of its lived social world, is due largely to their efforts.

Nor, as we have learned, is it possible to limit a discussion of illustrated print media to a single country, since media were international in scope, to a much greater extent than the high arts of painting and sculpture, which were difficult to transport and so were infrequently exhibited outside their country of origin. Lithography, the process that revolutionized the world of visual imagery, was invented by Alöys Senefelder, born in Prague, resident in Bavaria. Wood engraving, the technique that enabled the illustrated press, was largely developed by the Englishman Thomas Bewick, spread by his students; one of them, Charles Thompson, brought it to France and trained the cadre of engravers who established the hegemony of this medium in books and periodicals. The steam presses and steel plate engraving that made large editions possible came from England. Comics were invented by the Swiss Rodolphe Töpffer, but he was inspired by the Englishman Thomas Rowlandson, and the genre he initiated was developed by his followers in France. As soon as an illustrated print genre or technology was invented somewhere, it was quickly adopted everywhere. Popular illustrated books appeared almost immediately in translation in other countries, either legally, through arrangements between publishers, or illegally as plagiaries. This pattern, like a spreading inkblot, was already operative as early as 1829, when Eugène Lami and Henry Monnier's *Travels in England* was published simultaneously in Paris by Firmin-Didot and in London by Colnaghi.[4] Grandville has always been an international favorite; his illustrated books appeared throughout Europe and America and are still in print today. *Les fleurs animées* (1847) was published virtually simultaneously in New York as *The Flowers Personified*, and subsequently appeared in Barcelona as *La vida de las flores*. His *Scènes de la vie privée et publique des animaux (1842)* soon appeared in German as *Das neunzehnte Jahrhundert des Thierreichs*, and later in London and Philadelphia as *The Public and Private Life of Animals*. Traffic went both ways:

Grandville's editions of the French translations of Daniel Defoe's *Robinson Crusoe* and Jonathan Swift's *Gulliver's Travels* were soon republished in English in London and New York and in German in Stuttgart.[5]

The absence of adequate copyright protection facilitated these international exchanges, for the Berne Convention that sought to regulate them was not adopted until 1886. As a result, the first comic books, published in Geneva by Rodolphe Töpffer, were quickly plagiarized in France where they inspired the development of that genre by a younger generation of French artists.

Prints were produced through alliances between international publishers, with captions either translated or printed in both languages. Expensive and rare prints had always been sold internationally, of course, but now even modest works were widely available. Pellerin was a major figure here, selling his broadsheets internationally, even in the United States.

Technology and genres were often invented in one country, developed in another. Kenny Meadows's *Heads of the People* (1839–1840), the first of the "physiologies" that described modern "types" in pseudoscientific language and imagery, enjoyed a modest success in England. When it was plagiarized in France, however, republished as *Les Anglais peints par eux-mêmes* (1840–1841), it spawned the multivolume publication *Les Français peints par eux-mêmes* (1840–1842), reprinted throughout the century, as well as a thriving industry of similar publications written and illustrated by major writers and artists. It then completed its round-trip voyage by returning to England as Jules Janin's *Pictures of the French: A Series of Literary and Graphic Delineations of French Character* (1840). Within a few decades, these contemporary "types" had made their way into high-art painting, notably in the work of the Impressionists who borrowed subjects freely from panoramic literature and the physiologies.

As technology crossed the Channel from England to France, artists often made the reverse journey. Géricault's *English Suite* of lithographs of contemporary life resulted from his sojourn in London and was published by Charles Hullmandel, the foremost English lithographic printer; Delacroix credited a London production of *Faust* that he saw there as inspiration for his eponymous series of lithographs. Henry Monnier early in the century and Gustave Doré decades later crossed the Channel to produce some of their finest work. Doré's 1872 *London: A Pilgrimage* was an international collaboration, with text by the prominent English journalist Blanchard Jerrold (1826–1884), although Doré got prime billing. Periodical formats invented in England were quickly adopted in France: *Le Magasin pittoresque* (1833) was modeled on the *Penny Magazine* (1832), *L'Illustration* (1843) on the *Illustrated London News* (1842). And the influence went both ways; the French periodical *Le Charivari* (1832) served as model for the English *Punch* (1841), which was subtitled the *London Charivari.*

Further study of the international aspects of the nineteenth-century print trade is necessary before we can fully understand how global a phenomenon it truly was.

Charles Knight, for example, the founder and editor of the *Penny Magazine*, claimed, quite correctly, that he had produced "a revolution in popular Art throughout the world."[6] He sold stereotypes of his wood engravings to illustrated periodicals on three continents, and while it is certain that many of them appeared in French publications, research is lacking on other countries. Adolphe Goupil (1806–1893), whose trade in reproductive prints falls outside the scope of this book, established branches in all the major cities of Europe and America, and Pellerin, whose popular comics were at the low end of the print market, contracted with the Humoristic Publishing Company of Missouri to bring out English translations of his most popular broadsheets.

Industrialization is a leitmotif running through this history. In the course of the century, the material aspects of high art hardly changed at all. There were new pigments, to be sure, and new styles of art, but if anything these changes diminished the audience for high art, since the general public rarely greeted them with enthusiasm. The graphic arts, however, underwent enormous transformations that expanded their audience. All the images discussed in this study were facilitated by the new technology. They were printed on the new presses, on paper that for the first time was cheap and plentiful, and were distributed by the new railroads to an increasingly literate population. While it is often noted that huge segments of the populace visited the periodic national or regional art exhibitions, it should also be noted that an even greater percentage had regular contact with the new features of illustrated print culture.

Advances in printing enabled greater numbers of illustrations in books and periodicals, always featured as a prime selling point to a new public hungry for images. Publishers began advertising a hundred, two hundred, six hundred illustrations in a single volume; even the cheaper editions were larger and had more illustrations than most eighteenth-century books. The conceptual organization of each medium changed as a result of this increase in imagery. We have only to compare Hogarth's picture stories that had, at most, twelve images to those of Töpffer, which included ten times that many. Books customarily featured an illustrated frontispiece, but by the mid-nineteenth century could include hundreds of additional images, some full-page, some inserted within the text itself. This increase in the number of images in a book—or even on a single-sheet print—resulted in new concepts of narration. The unique, significant event, the unities of time, place, and action characteristic of both earlier book illustration and of history painting, gave way to a plethora of smaller anecdotal moments. Among the many and varied factors in this transformation we might see the twin drives of fidelity to experience and to narration, leading, on the one hand, to photography, and, on the other, to cinema. It would, however, be a mistake to read photography as the inevitable outcome, because, in fact, the multiplied images led just as inexorably to the graphic novel, a genre that has grown steadily in importance in the twenty-first century.

During the early decades of the nineteenth century, the various graphic media were in flux, as are all media at their moment of invention. Graphic satire, which in England was largely expressed through etching, in France soon settled into lithography, a medium whose ease of production—basically just drawing on stone—enabled speed of response to current events and a wider diffusion of those responses. Popular prints were traditionally woodcuts, albeit produced through stereotypes; by midcentury modern technology, again in the form of lithography, had overtaken this medium as well. Color printing developed slowly, and hand-colored or stencil-colored prints continued until the last decades of the century. Chromolithography was used at first primarily for reproductive prints, only later for posters and, even later, for art prints. The illustrated press and illustrated books had the reverse trajectory. In early examples, etchings, often hand-colored, were included separately with the text; lithographed images, when included, also had to be printed and bound in separately. Within a few decades, however, wood engraving allowed word and image to be combined on a single page, contributing to the widespread popularity of both the illustrated press and illustrated books. Sturdier steelplate engraving, replacing the more fragile copperplate, allowed larger editions, and all the new technology encouraged not just more images but larger ones: the dimensions of nineteenth-century books and periodicals are often several times larger than those of the previous century.

The different available technologies resulted in different art practices. Lithography is an immediate reflection of the artist's drawing, either directly on the stone or through the use of transfer paper; as a result, it seemed to contemporaries who witnessed its birth that it produced not replicas, but originals multiplied. But because lithographic printers did not actually make the drawing, they never attained the status enjoyed by their predecessors, the copperplate engravers, who, because they actually redrew as well as printed the artist's work, could be elected to the French Academy. Wood engravers also translated the artist's drawing into lines, incised on wood rather than copper. Nonetheless, in terms of status, wood engravers fared no better than lithographers: no lithographer or wood engraver was ever elected to membership in the Academy. An impenetrable wall built of status, rather than skill, separated technology from high art, and to the extent that technology brought art into a more popular realm, that art and its practitioners were disparaged.

This problematic relationship leads to the knotty question of the intertwining of the technological with the social and cultural. Certainly, in this early stage of capitalism, the new technology had as its goal to open and to serve new markets, as did, in their own way, the pioneers of Modernism in high art. For visionaries like the Saint-Simonians in France and the Society for the Diffusion of Useful Knowledge in England, technology would be the means to educate the populace and thereby bring about greater equality. They were convinced that widespread production and distribution of imagery through lithography or wood engraving would democratize the art of drawing, bringing it within the purview of all citizens. Édouard Charton, the founder and editor of *Le Magasin pittoresque*, wrote: *"Without drawings, it is impossible for men, whether of high or low rank, to have a complete education."*[7] Conservatives and traditionalists took the opposite point of view, equating popularization with degradation; they saw the spread of images to social and economic classes with no previous art experience as synonymous with shoddiness and the lowering of quality, "the assassination of the fine arts," as one critic complained.[8] Regardless, the success of the new media resulted from a population that had enough education to appreciate the imagery and enough disposable income to pay for it, which in turn can be attributed to the spread of Enlightenment principles. The development of markets and technology in tandem with social forces can be seen most clearly in the trajectory of popular prints, which shifted in both style and subject during the century in order to maintain and extend their market.

With advances in technology, illustrated print media became big business, at least for the nineteenth century. Print empires like those of Aubert, Pellerin, and Goupil expanded and illustrated periodicals flourished and multiplied. The economic strategy of market differentiation can be seen in the subdivision of each of the categories discussed in the preceding chapters. Books and albums were increasingly available in different editions, priced according to the number of illustrations, paper quality, and bindings. Prints could be purchased either as complete portfolios or by the sheet, either hand-colored or in black and white. Some publishers specialized in subject matter: Aubert in caricature, Pellerin in popular prints, even though they produced both single-sheet prints and albums. Others backed a variety of enterprises. Alexandre Paulin, for example, was the publisher of a wide spectrum of books, ranging from the lavishly illustrated 1835 edition of Lesage's *Gil Blas*, to periodicals such as *L'Illustration*, to—at the opposite pole— comics such as Cham's *Travel Impressions of Mr. Boniface.* Within the genres of illustrated print material there was market differentiation and segmentation as well. Comics were originally aimed at adults but, with the rise of education and income, an audience for children's literature siphoned off much of their production; Pellerin, notably, took advantage of this new market.

Artists themselves also deliberately attempted to adapt to the new market forces, often by specializing in one of these new graphic media. Gustave Doré began with caricature and comics, then abandoned them for the more lucrative and prestigious medium of book illustration. Cham, after a brilliant start in comics, abandoned that genre in favor of single-sheet caricatures that he published first in the illustrated press, then republished as albums. Only Daumier, producing cartoons regularly for *Le Charivari,* remained focused exclusively on caricature throughout his career, although he, like Doré, aspired to success in the more prestigious media of painting and sculpture. Grandville had, perhaps, the broadest career of all within graphic media. Throughout his brief life (he died at forty-four), he consistently produced cartoons and drawings for periodicals and for portfolios, as well as illustrations for his own writing and for other authors.

A notable feature of nineteenth-century illustrated print culture was its drive to veracity. While this imperative ultimately led to photography in the realm of news journalism, those media that prized expressiveness, namely graphic satire, comics, and illustration of all kinds, remained the province of the hand. Across all media, there was an insatiable appetite for visual imagery of modern life. "Types" were created that had never before been depicted, such as the office workers in Monnier's *Administrative Manners*. Religious figures and national heroes who customarily peopled the cultural *imaginaire* were replaced by fictional contemporary characters created by artists in caricature, comics, and illustrated books: Calicot, Robert Macaire, Mr. Prudhomme, and Mayeux in France; Doctor Syntax, Tom and Jerry in England; Mr. Jabot in Switzerland. But not only manners and morals were represented. The built environment that resulted from new technology, such as bridges and railways, was depicted repeatedly in lithography, popular prints, and the illustrated press as early as the 1830s. Such imagery entered the realm of high art only later, when these motifs were adopted by painters, particularly the Impressionists. The most striking example of increasing responsiveness to contemporary life is in the arena of popular prints, where the Épinal imagery that began with traditional subjects and themes ended the century as up-to-date "comics" celebrating the small pleasures and frustrations of modern life.

The nineteenth-century public, well aware of the rapidly changing world, was eager to see every aspect of that world reproduced in graphic form. Eyewitness accounts became the gold standard, from drawings that appeared in the illustrated press to the many images labeled "from the photograph" that sought to capitalize on the presumed truth of that technology. Until very late in the century such images bore only a tangential relationship to photography, notwithstanding claims to the contrary, since such images needed extensive reworking by engravers who added details such as figures and clouds that film was too slow to capture. In painting as well as illustration, genre scenes of everyday life were the most popular subjects despite being scorned by traditionally minded critics as insufficiently elevated or ideal. The illustrated press, popular prints, and comics paralleled genre painting in their focus on the manners and mores of the contemporary world, as did numerous illustrated books, including, notably, the novels of Balzac and Dickens. The illustrated press began as albums of images and rapidly became the prime instrument for disseminating knowledge of this visual world, particularly after its coverage expanded to encompass the entire panorama of contemporary life, from politics to culture, fashion to entertainment. If Stendhal could describe the novel as a mirror walking down the road, his description is even more apt for nineteenth-century illustrated print culture, which gave form and substance to the new modern world.

Of course, imagery of contemporary life could easily run afoul of the authorities, particularly where censorship was so entrenched that it was a crime even to depict the king. Publishers as well as artists were legally responsible for their imagery: Pellerin was charged with distributing Napoleonic imagery in the form of popular prints, and Philipon and Daumier both were arrested for publishing political caricatures. Censorship inspired a highly sophisticated graphic language of protest to circumvent it, the best known example of which is the pear representing King Louis-Philippe. In the periodical press, by necessity, the criticism was more oblique—an unpopular monarch could be ignored, not depicted at all or, when shown, could be represented as insignificant in scale, dwarfed by his surroundings as was done in *L'Illustration.* Despite these subterfuges, Philipon's journal *La Caricature* succumbed to the draconian 1835 Press Laws, and *Le Charivari,* its successor, had to tone down its content in order to survive; these conditions resulted in renewed focus on social rather than political subjects. Although this shift has often been interpreted as a tragic outcome, we might note that Balzac resigned from *La Caricature* before its demise because, as a novelist, his focus was always on the social more than the political. From this perspective, we can see the magnificent realist and naturalist novels of the nineteenth century as the literary parallel to the imagery and concerns introduced by illustrated print culture, a relationship that literary scholars have already begun to explore.

The development of comics should be considered in tandem with the modern novel: both developed from beginnings in picaresque narratives. The representation of narrative action in comics was initially dependent on the strategies of history painting, just as the novel itself was dependent on the form of the epic. In the course of the century, both the visual and the literary came "offstage," as it were, to refocus their narratives on individual experience. Comics as well as illustrated novels demonstrate this movement away from the paradigmatic moments of history painting to sequences of hundreds of images that tell the story from the protagonist's point of view, recounted from the inside, as it were, instead of from the vantage point of the audience or society as a whole. In graphic media, Cham and Grandville initiated the shift, with the major achievement in this mode being Grandville's *Another World,* rediscovered by the twentieth-century Surrealist movement for obvious reasons.

With an accelerating rate of technological innovation, many of the graphic media discussed here were discarded only to be reborn later into the rarified regions of high art. The etching revival restored that medium in the mid to late nineteenth century. Lithography took considerably longer: after its widespread adoption by artists in its early decades, it was plagued by its success in the mercantile realm, tainted by association with cheap chromolithographs that were largely used for reproductive, not original, imagery. Despite the poster movement and other attempts to restore its status as art, chromolithography languished under a commercial penumbra until it was revived as an art form in the mid-twentieth century only after its commercial utility had been thoroughly exhausted. The lowly woodcut, abandoned in favor of wood engraving that could produce finer lines and larger editions, was revived primarily by German Expressionist artists seeking to reconnect with their national medieval past, while in France the crude woodcut style of popular prints was adopted in painting by generations of Modernist artists from Courbet to Picasso as an attack on academic style and subject. Wood engraving never regained the luster it enjoyed before being made obsolete by the various photographic transfer methods of the late nineteenth century; it had never been a truly proletarian medium like the woodcut, and thus was regarded merely as obsolete technology. Even Max Ernst's celebrated graphic novels such as *La femme 100 têtes* (1929) and *Une semaine de bonté* (1934), composed of collages of earlier wood engravings, utilized the medium only to emphasize its retrograde look.

Most nineteenth-century painters made prints of some kind; as a result, our concept of print culture has been defined by the *peintre-graveur*, the painter-printmaker, to the exclusion of the professionals who actually dominated the arena of the graphic image, and whose work I have presented here. A wave of revisionist scholarship has included a hunt for painters' use of earlier more vernacular graphic imagery, pointing out, for example, that Courbet used popular prints, that Cézanne and Monet copied fashion illustrations, and that the Impressionists replicated in paint images from the illustrated press and from panoramic literature.[9] Looking at the printed graphic image as fodder for high art, however, reverses the direction of the influence. It was not coincidental that painters looked "below" for inspiration, or that they themselves often drew for illustrated print media. I have suggested throughout this study the primacy of the graphic, proposing that illustrated print culture formed the very foundation of imagery through which the modern world was seen and comprehended.

How should we evaluate this material? We can study it through the traditional lens of art history, pointing to formal innovations. Notable here is the language of graphic signs and symbols introduced by comic artists into the art of drawing—speed lines, close-ups, fragments, exaggerations, symbolic images, shifts in scale, and a wide variety of framing devices. This new graphic language constitutes a major achievement that has invested modern visual culture so thoroughly through both comics and cinema that its origins have gone largely unrecognized to this day. We can investigate primacy of genres, always an art historical virtue: calligrams and concrete poetry, familiar from twentieth-century avant-garde poetry, were introduced in nineteenth-century books and periodicals: *Le Charivari* was particularly fond of the calligram through which it could convey political content, like shaping text (always less subject to censorship than imagery) in the form of a pear after that image had been forbidden; its cover on February 27, 1834, was a particularly ingenious example: an image of a pear composed of the text of the most recent court verdict condemning the journal. We can also, as literary scholars have done, look more deeply into the permutations of the word-image matrix.

Art historians are uniquely situated to adopt the methodology of visual studies or social art history to try to understand how these works functioned within the culture. Albums of landscape, such as the long-running *Picturesque and Romantic Travels through Old France,* whose publication spanned the century, envisaged regions of France never before depicted. Its images, created by major artists of the period, are not just splendid examples of drawing, but also had the social function of allowing the citizenry to contemplate the possibility of being French, not just Norman or Breton or Provençal. These types of publications became especially efficacious after 1870 when, in the Third Republic, competing loyalties to one's region and one's nation constituted a social quandary. Along these same lines of nascent national identity, albums of regional landscape, customs, and costumes did the same throughout the Western world, although their role in nation building or national fragmentation has yet to be adequately recognized.[10] Imagery of the world's peoples and their cultures functioned to instill the earliest sense both of globalism and of cultural particularity. In England's *Penny Magazine* and France's *Le Magasin pittoresque*, this was done in a pedagogical spirit, to impart a basic knowledge of the diversity of peoples, geography, traditions; more upscale periodicals like the *Illustrated London News* and *L'Illustration* that focused on newsworthy contemporary events displayed more often the political attitudes of the bourgeoisie. Nonetheless, even these periodicals took firm positions against the slave trade, the burning issue of the period. In nineteenth-century illustrated print culture as a whole, we can see the early stages of global consciousness brought to a general public, a project still unfinished today. Not least, the genre images that were published in periodicals as well as in single-sheet prints, albums, comics, and physiologies replaced traditional scenes of trades and professions with up-to-date depictions of contemporaneous behavior in the modern world, both for emulation and for critique.

Many new forms of vernacular culture arose in the late nineteenth century. Advertising was developing into a new medium, photography was en route to becoming the predominant means of documenting current events, and cinema had its earliest beginnings. Popular culture was developing in numerous directions that lie outside the scope of this study, including cinema, advertising, and spectacles of various sorts, and yet, through all this, handmade images like those that I have discussed throughout this study persisted.

Even to its inhabitants, the rapidly changing universe of the nineteenth century constituted Another World. Illustrated print culture and the artists who drew it visualized that world, creating for the first time the imagery through which contemporaries and subsequent generations could comprehend it. If visualizing one's world is a way of understanding it, of negotiating it, then these graphic artists were instrumental in creating the panorama of the modern world; they gave us the eyes with which we see it.

NOTES

Articles not identified by author that appeared in *L'Illustration* (*ILL*), the *Illustrated London News* (*ILN*), *Journal des connaissances utiles* (*JCU*), *Le Magasin pittoresque* (*MP*), and *The Penny Magazine* (*PM*) are listed in the bibliography under the title of the periodical.

INTRODUCTION

1. See Grandville, *Un autre monde. Transformations, visions, incarnations, ascensions, locomotions, explorations, pérégrinations, excursions, stations, cosmogonies, fantasmagories, rêveries, folâtreries, facéties, lubies, métamorphoses, zoomorphoses, lithomorphoses, métempsychoses, apothéoses, et autres choses* (Paris: H. Fournier, 1844). Grandville was the pseudonym of Jean-Ignace-Isidore Gérard (1803–1847), who also used the pseudonym J. J. Grandville. Following the Bibliothèque nationale de France recommendation, in this text I will refer to him as Grandville.
2. Schwartz and Przyblyski, "Visual Culture's History."
3. The standard reference on the subject is Goldstein, *Censorship of Political Caricature*; this is his central thesis.
4. Nodier, Taylor, and Cailleux, *Voyages pittoresques et romantiques dans l'ancienne France*, 23 vols., 1820–78.
5. Chapter 6 in Ivins, *Prints and Visual Communication* is titled "Pictorial Statement without Syntax—The Nineteenth Century."
6. Among recent publications are Verhoogt, *Art in Reproduction*; Bann, *Distinguished Images*; and Musée Goupil, *État des lieux*, a two-volume study of the dealer Goupil, who established a lucrative international market in such prints.
7. On color lithography, in addition to Twyman, *History of Chromolithography*, see Farwell, *Lithography in Art and Commerce*, vol. 12 of her *French Popular Lithographic Imagery*. On earlier color printing, see Dackerman, *Painted Prints*, and Grasselli, *Colorful Impressions.*
8. For a comprehensive study of the poster movement, see Iskin, *Poster.*
9. For the version in *L'Illustration*, see Töpffer, *Histoire de M. Cryptogame*. The *L'Illustration* version, slightly amended, was published in Paris in book form in 1846 by J. J. Dubochet. I have followed instead the autograph versions given in *Rodolphe Töpffer* (Paris: Pierre Horay, 1975, 1996), 237–84, and *Rodolphe Töpffer: The Complete Comic Strips*, ed. and trans. David Kunzle (Jackson: University Press of Mississippi, 2007), 447 –537. A complete listing of publications of Töpffer's albums is given in *Rodolphe Töpffer, The Complete Comic Strips*, 627–50.
10. Farwell, *French Popular Lithographic Imagery*; British Museum, *Catalogue of Political and Personal Satires.*
11. Ray, *Art of the French Illustrated Book* and *The Illustrator and the Book in England*; Garnier, *L'Imagerie populaire française.*
12. Blachon, *La gravure sur bois au XIXe siècle*; Twyman, *History of Chromolithography*; Twyman, *Breaking the Mould*; Twyman, *Lithography, 1800–1850.*
13. Kaenel, *Le métier d'illustrateur*; Kunzle, *History of the Comic Strip*; Adhémar, *La France romantique*; Adhémar and Seguin, *Le livre romantique*; Le Men, "Livres illustrés et albums, 1750–1900," and "Quelques définitions romantiques de l'album," among others.
14. Kerr, *Caricature and French Political Culture*; Iskin, *Poster*; Cate, *Graphic Arts and French Society, 1871–1914*; Cate and Hitchings, *Color Revolution.*
15. Two small caveats: Sieburth's "Same Difference" reproduces numerous images without

adequate identification; art historians customarily indicate artist, title, date, medium, and publication data. Cohen's "Panoramic Literature" proposes a convincing relationship between this literature, the development of the novel, and cinema, but erroneously identifies her wood-engraved illustrations as lithographs, a medium incapable of the juxtaposition of word and image that is her focus.

16. Schwartz, *Spectacular Realities*; Schwartz and Przyblyski, *Nineteenth-Century Visual Culture Reader*; Samuels, *Spectacular Past.*
17. Kaenel, *Doré. L'imaginaire au pouvoir.* While there have been numerous Doré exhibitions, this was the first major exhibition to consider his entire oeuvre.
18. Renonciat, *La vie et l'oeuvre de J. J. Grandville.* There is also a useful catalogue raisonnée, *Grandville. Das Gesamte Werk*, ed. Gottfried Sello, 2 vols. (Munich: Rogner U. Bernhard, 1969).
19. Athanassoglou-Kallmyer, *Eugène Delacroix*; Jobert, ed., *Delacroix. Le trait romantique.*

CHAPTER 1. DRAWING'S STEPCHILD

1. The standard histories of early lithography are Twyman, *Lithography, 1800–1850*, and Twyman, *Breaking the Mould.*
2. *Specimens of Polyautography* was published by Philipp André in London in 1803; he sent his brother Peter Friedrich (later Frédéric) André to establish the lithographic business in Paris. See Twyman, *Lithography, 1800–1850*, 41–42; Twyman, *Breaking the Mould*, 120. Jacob's print, included in Senefelder, *L'art de la lithographie. Collection de plusieurs essais*, is untitled on the plate, but has always been known by this title.
3. Thiers, "De la lithographie et de ses progrès." [n'est plus resté enfoui dans des cartons, mais a été donné au public.]
4. The term "nineteenth-century media explosion" was coined by Beatrice Farwell as the title of her 1977 exhibition *The Cult of Images: Baudelaire and the Nineteenth-Century Media Explosion* at the University of California at Santa Barbara Art Museum.
5. On the issue of reproductive versus original prints, see Verhoogt, *Art in Reproduction*. See also Stein, *Artists and Amateurs*; Bann, *Distinguished Images.*
6. "Nouvelles de Paris, du 24 août," *La Quotidienne*, Aug. 25, 1819, 2; "D'une nouvelle erreur du parti des ultra," *Le Constitutionnel*, Aug. 23, 1819, 4. [Notre opinion, sur ce genre de gravure, n'a pas changé; nous le croyons toujours très-fatal à l'art du burin et nous persistons à blâmer l'abus qu'on en a fait depuis deux ans.]
7. In 1802, Peter-Friedrich (later Frédéric) André was licensed to import lithography to France; Charles Lasteyrie bought André's lithography stones in 1803, and he and Engelmann opened lithography printing businesses in Paris in 1815–1816; Twyman, *Lithography, 1800–1850*, 41–57.
8. Institut royal de France, Académie des beaux-arts, *Rapport sur la lithographie.* The report is reprinted in its entirety in Johnson, *French Lithography*, 23–40; see 36. The report was signed by Heurtier, Regnault, Guérin, Desnoyers, and Castellan. [En effet, que le peintre dessine sur la pierre, c'est lui seul qui invente, exécute avec la fougue du génie ou l'amour de la perfection; c'est son style, la manière qui lui est propre, et jusqu'à ses défauts dont il ne peut se prendre qu'à lui-même; l'on retrouve dans son ouvrage cette touche franche, prompte, spirituelle, qui est le résultat, non du tâtonnement, mais celui de l'inspiration qui conduit une main créatrice.]
9. See *gravure* and *graver*, in Rey, *Le Robert. Dictionnaire historique*, s.v., where the generic modern meaning of "print" is dated to 1829. The complete catalogue of lithography exhibited in the Salons from 1817 to 1824 was published in Johnson, *French Lithography.*
10. Miel, *Essai sur les beaux-arts*, 415–21. Partially reprinted in Johnson, *French Lithography*, 41–42. [L'impression sur le papier est le dessin même, réfléchi par la contre-épreuve, comme par un miroir; l'estampe est, pour ainsi dire, autographe; c'est la main du maître qui l'a formée; elle ne diffère en rien de l'original; elle est elle-même un original.]
11. Chaudonneret, "Le statut et la réception de la lithographie en France."
12. D. [E.-J. Delécluze], "Beaux-arts. La lithographie," 1–2.
13. On transfer paper, see Twyman, *Breaking the Mould*, 7.
14. On Vernet's *Le lancier en vedette*, see Beraldi, *Les graveurs du XIXe siècle*, 12: 215, no. 28.
15. There has been little scholarship on early lithographs by French artists. I thank John P. Lambertson for generously sharing with me his research paper "P.-N. Guérin's Classical Lithographs and Pedagogy," presented at the Consortium on the Revolutionary Era,

Charleston, S.C., Feb. 2010. On Gros's students, see Eyerman, "Im Zeichen de Grande Tradition."

16. Anne-Louis Girodet, *Coupin de la Couperie*, in Engelmann, *Recueil des essais lithographiques*. I thank Edward Vairo and Stephen Edidin for the translation.
17. Firmin-Didot, *Étude sur Jean Cousin*, 14–15, n. 3. [N'ai-je pas vu Girodet s'appliquer à dessiner sur des pierres lithographiques un grand nombre de ses compositions et suivre avec un vif intérêt les progrès de ce nouveau procédé qui le séduisait par cela même, me disait-il, qui lui permettrait de reproduire, sans le secours d'un autre interprète, et par ses propres mains, son propre ouvrage?]. This text was first reprinted by Charles Rosen and Henri Zerner in their *Romanticism and Realism*, 99. For an overview of Girodet's printmaking practice, see Rosen and Zerner, *Romanticism and Realism*, 116–35, and, more recently, Jobert, "Girodet and Printmaking."
18. On Aubry-Lecomte, see the chapter "Is Lithography an Art? Aubry-Lecomte and Lemud," in Bann, *Distinguished Images*.
19. The lithographs were listed slightly differently in the 1822 Salon catalogue, no. 1556, as "Engelmann, rue Louis le grand, no 27. Têtes tirées du tableau représentant les ombres des héros français reçues dans les palais aériens d'Ossian, dessinées par M. Aubry Lecomte, d'après M. Girodet," but the portfolio itself bore the title *Collection de têtes d'étude, d'après le tableau peint en 1801, par Mr Girodet-Trioson, membre de l'Institut & représentant les ombres des héros français reçues dans les palais aëriens d'Ossian. Lithographiées sous sa direction par Aubry-Lecomte, son élève*. See Johnson, *French Lithography* 68, no. 72, 1822/1556.
20. On Delacroix's prints, see Jobert, "Delacroix en noir et blanc," 13–19. On Delacroix's caricatures, see Athanassoglou-Kallmyer, *Eugène Delacroix*.
21. Eugène Delacroix to monsieur Ph. Burty, Mar. 1, 1862, in Delacroix, *Correspondance générale*, 4: 303–4; he wrote much the same account in his journal; see Delacroix, *Journal*, June 29, 1855, 520. [Les *Médailles* (dont j'ai retrouvé des épreuves à votre intention) ont été exposées chez les marchands, mais personne n'en a voulu.]
22. Delacroix, *Journal*, Apr. 12 [1824], 65. [Il m'a pris fantaisie de faire des lithographies d'animaux, par exemple: un tigre sur un cadavre, des vautours, etc.]
23. Delacroix to Burty, Mar. 1, 1862, in Delacroix, *Correspondance générale*, 4: 303–4. [Je me rappelle que je vis, vers 1821, les compositions de Retzsch qui me frappèrent assez: mais c'est surtout la représentation d'un drame-opéra sur *Faust*, que je vis à Londres en 1825, qui m'excita à faire quelque chose là-dessus. . . . Vous savez que Motte fut l'éditeur; il eut la malheureuse idée d'éditer ces lithographies avec un texte qui nuisit beaucoup au débit, sans parler de l'étrangeté des planches qui furent l'objet de quelques caricatures et me posèrent de plus en plus comme un des coryphées de l'*école du laid*. . . . Je ne me rappelle pas ce que j'en retirai: quelque chose comme cent francs et de plus une gravure de Lawrence, le *Portrait de Pie VII*. Toutes mes spéculations ont été dans ce goût. L'*Hamlet* mieux encore: je l'avais fait imprimer à mes frais et éditer moi-même. Le tout me coûta 5 ou 600 francs, et je ne rentrai pas dans la moitié de mes frais.]
24. Eckermann, *Conversations of Goethe with Eckermann*, 136; first cited by Trapp, *Attainment of Delacroix*, 144.
25. See, for example, Lasteyrie, *Recueil de différens genres d'impressions lithographiques* (1816); Senefelder, *Collection de plusieurs essais* (1819). Engelmann published several such collections; see, for example, Engelmann, *Essais des diverses manières de dessiner à l'encre lithographique* (1817). Johnson catalogues numerous portfolios that were shown in the Salons; see his *French Lithography*.
26. While this print from Senefelder's *Nouvelle invention lithographique* is not identified as a frontispiece, it follows the English convention of using a trompe l'œil collage print as the collection's frontispiece. See Clayton, *English Print*, 105–7.
27. *Suite d'estampes pour servir à l'histoire des moeurs et du costume des français dans le dix-huitième siècle*, 1775–83. Drawings for the first series were by Sigmund Freudeberg, the next two series by Jean-Michel Moreau le Jeune; for a brief overview of this project, see Ray, *Art of the French Illustrated Book 1700–1914*, no. 55, 97–98. The last image in the series, *True Happiness* (*Le vrai bonheur*) shows a rural family at home, as a contrast to the preceding three volumes that chronicle the lives of the elite.
28. Adolphe Thiers, "De la lithographie et de ses progrès." The complete article is given in

Johnson, *French Lithography*, 46–48. [La découverte de la lithographie a opéré une véritable révolution. La facilité qu'ont les artistes de saisir le crayon eux-mêmes, de jeter promptement sur la pierre l'idée qui les a frappés, et de livrer ainsi ces rapides esquisses à une presse plus rapide encore, les a conduits à ne rien laisser échapper de ce qui les frappe, à fixer toutes les images qui traversent leur esprit, à saisir tous les aspects fugitives que leur offre la nature. Une attitude, un air de tête, une figure originale, une scène de moeurs, une farce populaire ou militaire, ils ont tout reproduit avec une merveilleuse promptitude. Chacune de leurs impressions, rendue avec la hardiesse, la brusquerie, la vérité du premier mouvement, nous est arrivée sans l'intermédiaire d'un froid et insensible graveur; c'est le dessin même du maître qui, jeté sur la pierre, et reproduit à l'infini par la presse lithographique, nous a été transmis avec tout le génie de l'original, avec sa touche, ses négligences, son expression, son esprit.]

29. The standard text on Charlet is La Combe, *Charlet, sa vie, ses lettres et ses oeuvres*. The most recent study is the 2008 exhibition catalogue *Charlet. Aux origines de la légende napoléonienne, 1792–1845*.

30. The artist who signed as A. G. published a virtually identical drawing as *Caricatures politiques no. 51* in *Le Charivari*, July 29–30, 1833, with the same caption, "Je vous porte tous dans mon coeur," but the subject is clearly identifiable as King Louis-Philippe.

31. Delacroix published "Charlet" in *Revue des deux mondes* 37 (Jan. 1, 1862): 234–42. Baudelaire wrote about Charlet twice, in his 1857 "Quelques caricaturistes français," reprinted in *Baudelaire, Écrits sur l'art*, 2: 327–72 (see 330–34), and later in his 1863 "L'oeuvre et la vie d'Eugène Delacroix," reprinted in *Baudelaire, Écrits sur l'art*, 2: 284–326 (see 318–19).

32. On Géricault's lithography, see Bordes, "L'écurie dont je ne sortirai que cousu d'or"; Cuno, "Retour de Russie: Géricault and Early French Lithography"; Delteil, *Théodore Géricault: The Graphic Work*.

33. See Mishory, "Théodore Géricault's Grande Suite Anglaise."

34. Théodore Géricault to Dedreux-Dorcy, Feb. 12, 1821, cited in full in the 1991 Paris exhibition catalogue *Géricault*, 320. [Je suis cependant plus raisonnable que vous, puisque au moins je travaille et lithographie à force. Me voilà voué pour quelque temps à ce genre qui étant tout neuf à Londres, y a une vogue inconcevable; avec un peu plus de ténacité que je n'en ai, je suis sûr que l'on pourrait faire une fortune considérable. Je me flatte que ce ne sera pour moi que l'affiche, et que bientôt le goût des vrais amateurs qui auront ainsi à me connaître, m'emploiera à des travaux plus dignes de moi. Vous appelez cela de l'ambition; mais ma foi, il n'est rien de tel que de battre son fer tandis qu'il est chaud; et puisque je commence à être encouragé, j'envoie au diable tous les sacrés coeurs de Jésus; c'est un vrai métier de gueux à mourir de faim. J'abdique le cothurne et la sainte Écriture pour me renfermer dans l'écurie dont je ne sortirai que cousu d'or.]

35. Robert Simon has developed the theme of Géricault's interest in *faits divers* in relation to painting; see his "Géricault und die Faits Divers."

36. George Cruikshank's etching *The Second Contest between Crib and Molineux, September 28, 1811*, though unsigned, was identified by Cohn in *George Cruikshank: A Catalogue Raisonné*, 85, no. 260.

37. I am indebted to Constance McPhee and Nadine Orenstein who clarified these distinctions in their 2011 exhibition catalogue *Infinite Jest*; see especially "Shoot Folly as It Flies: Humor on Paper," 3–17.

38. Blanvillain, "Caricatures," 174. [Les caricatures sont établies partout, mais le grand bureau est chez Martinet, rue du Coq. Nous en avons périodiquement tous les quinze jours, les unes sont plaisantes, les autres morales, les autres égrillardes.]

39. For a brief survey of the history of caricature, see McPhee and Orenstein, "Shoot Folly as It Flies," *Infinite Jest*, 3–17.

40. On Hogarth's *Characters and Caricaturas*, see McPhee and Orenstein, *Infinite Jest*, 32. See also Hogarth, *Hogarth's Graphic Works*, and Paulson, *Hogarth*.

41. See "Les caricatures commandées par le comité de salut public," in *La Révolution française et l'Europe 1789–1799*, 2: 594, and Albert Boime, "Jacques-Louis David, Scatological Discourse," 67–82.

42. The "explication" below the image states: "Ce Gouvernement est représenté sous la figure d'un diable écorché tout vif, accaparant le commerce et revêtu de toutes les décorations royal, le portrait du roi se trouve au derrière du gouvernement lequel vomit sur son peuple une

multitude d'impôts avec lesquelles il le foudroye. Cette prérogative est attaché au sceptre et à la couronne."

43. On caricature of the revolutionary period, see Bindman, *Shadow of the Guillotine*, and Cuno, ed., *French Caricature and the French Revolution*, which includes Hunt, "The Political Psychology of Revolutionary Caricatures," 33–40, and Melot, "Caricature and the Revolution," 25–32. See also Taws, *Politics of the Provisional*.
44. Jouy, "Les caricatures," *Guillaume le franc-parleur* 1, no. 9 (July 9, 1814): 113–14. [La révolution inonda la France d'un déluge de caricatures, dans lesquelles chaque événement du jour, chaque séance de l'assemblée, chaque circonstance de la vie des principaux députés était tour-à-tour exposés à la risée publique. Là se fit remarquer notre infériorité dans le genre de la caricature politique où les Anglais excellent. Le défaut de goût qui a marqué si loin du but le terme qu'ils ont atteint dans les arts, loin d'être ici la matière d'un reproche, est pour eux un moyen de succès. Sans être jamais arrêtés par la crainte d'enfreindre les règles, de blesser les convenances, d'insulter au bon sens, ils donnent un plein essor à leur imagination vagabonde, et produisent des monstruosités risibles avec une fécondité sans exemple.]
45. See Melot, "Caricature and the Revolution," 25–32. For a general study of caricature, see McPhee and Orenstein, *Infinite Jest*.
46. F. [Charles-François Farcy], "Les métamorphoses.—Les contrastes," 339. [En effet, si on loue cette verve, cette force comique, cette qualité que les Anglais appellent humour, et qu'ils savent mettre dans les croquis dont il s'agit, on ne doit pas louer l'exagération de forme ou de geste qui dépasse le but, et qui, d'un dessin qui eût été spirituel, fait un dessin qui n'est plus qu'extravagant. Les Français, meilleurs dessinateurs, évitent presque toujours ce dernier excès, et comme, dieu merci, l'esprit ne leur manque pas, ils s'arrangent de manière à le laisser dominer dans le faire comme dans la pensée de leurs caricatures.]
47. There has been far too little scholarship on Boilly's prints; for a general overview see the 1988 Lille exh. cat. *Boilly 1761–1845*.
48. The main source of information on Martinet is Hautecoeur, "Une famille de graveurs et d'éditeurs parisiens," 205–340. On Mayeux, see Menon, *Complete Mayeux*. There is little information on Godissart de Cari, a caricaturist who worked for Martinet principally as creator of *Le musée grotesque*.
49. F. [Charles-François Farcy], "Les métamorphoses.—Les contrastes," 340.
50. See the 2003 exhibition catalogue from the Musée des beaux arts de Nancy *Grandville. Les Métamorphoses du jour*. An English edition of *Les Métamorphoses du jour* comprising forty-two hand-colored plates was published by Thomas McLean as *The Metamorphoses of the Day* (London: Thomas M'Lean [*sic*], 1830). I thank Lars Kokkonen for this reference.
51. On France's preeminence, see Twyman, *Breaking the Mould*, 55–57.
52. The standard study of censorship is Goldstein, *Censorship of Political Caricature*.
53. The sixth edition of the *Dictionnaire de l'Académie française* (1835) was the first to define *costume* as clothing; previously the meanings of *costume* and *coutume* were similar, encompassing both dress and behavior. See Académie française, *Dictionnaire*, s.v.
54. For the history of fashion plates, see Gaudriault, *La gravure de mode feminine*.
55. On the economic results of these social changes, see Sewell, "Empire of Fashion," 81–120.
56. Vanessa Schwartz in *Spectacular Realities* devoted a chapter to the visit to the morgue that by late century had developed into a popular form of entertainment.
57. Delpech's *Iconographie des contemporains, ou portraits des personnes dont les noms se rattachent plus particulièrement, soit par leurs actions, soit par leurs écrits, aux divers événements qui ont eu lieu en France, depuis 1789, jusqu'en 1829* was announced in the *Bibliographie de la France* on Aug. 2, 1823, with the last *livraison* announced on June 23, 1832. It was then reissued under the same title in two volumes in 1832, reprinted in 1833. [*Portraits des personnes dont les noms se rattachent plus particulièrement, soit par leurs actions, soit par leurs écrits, aux divers événements qui ont eu lieu en France, depuis 1789, jusqu'en 1829.*]
58. Nodier, Taylor, and Cailleux's *Voyages pittoresques et romantiques dans l'ancienne France* is central to Adhémar's discussion in *La France romantique*.
59. Bann discusses the significance of Normandy as the subject of the first volumes of *Voyages pittoresques* in chapter 2 of his *Distinguished Images*, "Representing Normandy."

60. See Adhémar, *La France romantique*, especially the "Bibliographie des recueils de paysages lithographiés 1817–1854," 111–31. For additional titles, see Abbey, *Scenery of Great Britain and Ireland* and Abbey, *Travel in Aquatint and Lithography*, which includes global compilations.
61. Farwell has catalogued the complete spectrum in her *French Popular Lithographic Imagery*. The standard work on social imagery is still Grand-Carteret, *Les moeurs et la caricature en France*.
62. Jouy, "Les caricatures," *Guillaume le franc-parleur* 1, no. 9 (July 9, 1814): 114. The series Jouy cited, collections of prints presenting amusing and sometimes pointed critical visions of contemporary manners and mores, had all begun in etching but continued in lithography. [Le burin ne fut pas plus libre que la plume; et l'argus de la censure, une loupe à chacun de ses yeux, ne surveilla pas avec moins d'attention les estampes que les livres. Nos dessinateurs se bornèrent à esquisser des *costumes:* dans ce genre, les collections des *Incroyables*, des *Merveilleuses*, du *Suprême bon ton*, sont recherchées comme des monuments de nos modes, d'autant plus précieux que des artistes d'un nom célèbre n'ont pas dédaigné d'y imprimer le sceau de leur talent.]
63. Balzac, "Avant-propos" in *Oeuvres complètes* (1842), 1: 14–15. [En dressant l'inventaire des vices et des vertus, en rassemblant les principaux faits des passions, en peignant les caractères, en choisissant les événements principaux de la société, en composant des types par la réunion des traits de plusieurs caractères homogènes, peut-être pouvais-je arriver à écrire l'histoire oubliée par tant d'historiens, celle des moeurs.]
64. On Martinet, see Hautecoeur, "Une famille de graveurs." On the Bergeret lithograph, see 282.
65. In 1824, Martinet retired and his daughter and son-in-law took over, renaming the firm Martinet-Hautecoeur; it continued until 1835. Hautecoeur, "Une famille de graveurs," 272, 309–10, 331.
66. Jouy, "Recherches sur l'album," *L'Hermite de la Chaussée d'Antin* 1, no. 15 (Nov. 8, 1811), 177–78. [Mais tout ce faste des magasins modernes obtient à peine quelques regards de la multitude, tandis qu'il se presse autour du modeste étalage du libraire de la rue du coq. Cette boutique a ses habitués, qui n'ont jamais mis le pied dans l'intérieur: ils se contentent d'examiner, à travers les vitres, toutes les belles choses offertes à leur curiosité; de passer en revue les caricatures nouvelles, les costumes de théâtre, les portraits d'acteurs et de musiciens, les uniformes des troupes françaises et étrangères, les mises de bon goût, les meubles de bon genre, et nous citerions telle personne de bon ton qui, de son aveu, passe plus agréablement une heure devant la boutique de Martinet, qu'à la représentation d'un des chefs-d'oeuvre de Molière.]
67. "Bulletins de Paris, jeudi 14 juillet," *Journal des arts, des sciences et de la littérature*, no. 307 (July 15, 1814): 71. The emphasis is in the original. [Martinet n'a plus seul le droit d'attirer les musards. Partout dans les rues on voit des groups nombreux réunis auteur des *expositions de caricatures*.]
68. Very little information is available on Delpech. See Hautecoeur, "Une famille de graveurs," 284; *Biographie universelle (Michaud)*, 10: "Delpech (François-Séraphin)." Delpech was licenced in 1818 and died in 1829, his widow, Marie-Marguerite-Brigitte Naudet Delpech, then took over his license; see Archives Nationales de France, F18* I -24.
69. The principal studies of Aubert are James Cuno's 1985 dissertation "Charles Philipon and La Maison Aubert," and Kerr, *Caricature and French Political Culture*.
70. On nineteenth-century color printing, see Cate, "The 1880s: The Prelude" in Cate and Hitchings, *Color Revolution*, 1–15. Michael Twyman's recent *History of Chromolithography* is an exhaustive study but focuses more on industrial and commercial rather than on art production.
71. Some examples from the *Bibliographie de la France* of series with no artist listed are *Le suprême bon-ton*, 1814–1819 (9 pl.); *Le musée grotesque*, 1817–1830 (65 pl.); *Le goût du jour*, 1813–1819 (48 pl.); see also Hautecoeur, "Une famille de graveurs," 273–86.
72. See Hautecoeur, "Une famille de graveurs," 273; and Bruson, *Théâtres romantiques*, 87–88, where he counts 2930 prints in Martinet's *Petite galerie dramatique* from 1796 to 1880.
73. The first print of *Amourettes* is listed in the *Bibliographie de la France* as by Charles Philippon [*sic*], Oct. 27, 1827, no. 961, published by Ducarme. The frontispiece, *Dix-huit Scènes d'Amourettes*, is not listed until Dec. 5, 1827, no. 1066, where the publisher is identified as Ostervald aîné. *Amourettes* nos. 19 to 26, published by Ducarme, were listed in *Bibliographie de*

la France on Feb. 16, 1828, no. 135; on Feb. 25, 1828, no. 159; there was a listing for Ch. Philippon [*sic*] *Amourettes* nos. 28–36, again published by Ducarme; no. 27 in this series was listed on Apr. 23, 1828, no. 319. On Mar. 15, 1828, *Bibliographie de la France*, no. 212, listed Philippon [*sic*] as the artist of *Souvenirs d'amourettes* nos. 7–12, published by Engelmann; there is no listing for the earlier prints in this series.

74. On Calicot, see Davis, "Entre la physiognomonie et les physiologies: le Calicot"; Keyser, *Un thème de l'imagerie parisienne sous la Restauration*; Hiner, "Monsieur Calicot."

75. *M. Mahieu* was listed in the *Bibliographie de la France*, Sept. 11, 1830, no. 998. Listings in the official record of legal deposit followed publication by weeks, or even months. On Mayeux, see Menon, *Complete Mayeux*.

76. Daumier, *Les cent-et-un Robert Macaire*. The Daumier literature is extensive; information on his prints can be can be found at the Daumier Register: http://www.daumier-register.org. See also the exhibition catalogues Daumier, *L'écriture du lithographe* (2008) and *Daumier, 1808–1879* (1999).

77. The complete caption reads: "Messieurs et dames! Les mines d'argent, les mines d'or, les mines de diamant ne sont que de la potbouille, de la ratatouille en comparaison de ma houille. . . . Mais (que vous m'allez dire) tu vends alors les actions un million? Mes actions, Messieurs, je ne les vends pas, je les donne pour 200 misérables francs, j'en donne deux pour une, je donne une aiguille, un cure-oreille, un passe-lacet et je vous donne encore ma bénédiction par dessus le marché. En avant la grosse caisse!"

78. On Monnier, see Mainardi, *Henry Monnier*; Melcher, *Life and Times of Henry Monnier*; Marie, *Henry Monnier*; and Champfleury, *Henry Monnier, sa vie, son oeuvre*.

79. For the circumstances, see Mainardi, "Dupinade." For an account of the trial, see A., "Cour d'assises. Procès du no. 35 de *La Caricature*."

80. The drawing appeared as a supplement to *La Caricature* on Nov. 24, 1831, no. 56. Redrawn by Daumier as a fundraising poster to alleviate court costs, it was published as *Les poires* in *Le Charivari* on Jan. 17, 1834.

81. The complete caption reads: "Le voyou, employé aux trognons de pommes dans les théâtres des boulevards, la croque sur les murailles pendant ses nombreux loisirs, c'est ainsi que Paris s'embellit tous les jours."

82. The London prints are in the collection of the British Museum, and carry the legend: "London Published by J. J. 1 March 1819, at 48 Strand." Since no publisher is cited, it is uncertain whether Isabey published them himself or they were plagiarized.

83. Engelmann showed Isabey's album at the 1819 Salon, no. 1479, under the full title, *Divers essais lithographiques de J. B. Isabey publiés à Paris en 1818. Cahier de la lithographie de G. Engelmann*. On Isabey's prints, see Basily-Callimaki, *J.-B. Isabey, sa vie, son temps*.

84. Carle Vernet, *Une collection de chevaux de tous les pays, montés de leur cavalier, avec leurs costumes et leurs armes*, 12 livraisons, 96 lithographs, Lith. Lasteyrie, 1817–18, and *Album lithographique par Cle Vernet*, Lith. Depech, 1821. On Carle Vernet's prints, see Dayot, *Carle Vernet*. On Horace Vernet's prints, see Brüzard, *Catalogue de l'oeuvre lithographique*, and Du Seigneur, "Appendice à la notice sur Horace Vernet."

85. On Napoleonic imagery in graphic production, see Bann, *Parallel Lines*, 14–87, and Day-Hickman, *Napoleonic Art*.

86. See, for example, Théodore Géricault, *Études de chevaux lithographiés par Géricault*, Lith. Villain, 12 pl., 1822, and *Quatre sujets divers par Géricault*, Lith. Engelmann, 4 pl., 1823. For a complete listing, see Delteil, *Théodore Géricault*.

87. Marie gives a catalogue of Monnier's lithographic production in *Henry Monnier*, 236–71. In the case of disparities, I have used the dates from the *Bibliographie de la France*. See also Mainardi, *Henry Monnier*, and Szrajber, "Henry Monnier's Letters from London."

88. Henry Monnier, *Rencontres parisiennes. Macédoine pittoresque. Croquée d'après nature, au sein des plaisirs, des modes, de l'activité, des occupations, du desoeuvrement, des travers, des vices, des misères, du luxe, des prodigalités des habitans de la capitale dans tous les rangs et dans toutes les classes de la société*, 40 pl. (Paris: Gihaut Frères, 1824).

89. On the relation of Monnier and Balzac, see Chollet, *Balzac journaliste*, 184–89.

90. See Marie, *Henry Monnier*, 236–71. [*Les Contrastes* (1824), *Modes et ridicules* (1825), *Récréations* (1826), *Esquisses parisiennes* (1827), *Moeurs administratives* (1828).]

91. *Voyage en Angleterre par Eug. Lami et H. Monnier* (Paris: Firmin Didot and Lami-Denozan; London: Colnaghi and Charles Tilt, 1829). See also Tanya Szrajber, "Henry Monnier's Letters from London."

92. "Dessin," *Le Charivari*, Nov. 25, 1833. [Le genre de grotesques, qui nous a été transmis par l'Angleterre avec le gouvernement constitutionnel, et les rosbifs aux pommes, est fort estimé des artistes.]
93. *Album lithographique* was discussed by Miel, *Essai sur les beaux-arts*. Its contributors included Carle Vernet, Horace Vernet, H. Lecomte, Gros, Demarne, Thiénon, Bourgeois, Vauzelle, Mlle Lescot, Granger, and Bessa.
94. On Gihaut, see Bouquin, *Recherches sur l'imprimerie lithographique à Paris au XIXe siècle*, 79–86. The custom of New Year's gifts, called *étrennes*, resulted in the publication of an enormous amount of light literature. See "Des albums," *Journal des artistes et des amateurs*, Mar. 23, 1828, 182–84; "Albums lithographiques," *Journal des artistes et des amateurs*, Dec. 28, 1828, 401–4; Jouy, "Des album," *L'Hermite de la Chaussée-d'Antin* 1, no. 13 (Oct. 30, 1811): 143–55; Le Men, "Quelques définitions romantiques de l'album."
95. For a brief discussion of macédoine prints, see Farwell, *Charged Image*, 74–75. Although Kerr credits Charles Philipon with introducing the macédoine print into France in 1831, such prints were already common in the 1820s; Kerr, *Caricature and French Political Culture*, 61. See also Hallett, "Medley Print."
96. [Entrez, Messieurs et Dames; c'est ici là-dedans que se fait voir la famille des/fameux Lithographantoccini, apportée du Sénégal par le célèbre capitaine/Crayonizinskhvtzp! Ces petits animaux sont parvenus, à forcede privations,/de sommeil et surtout de nourriture, à former une très belle collection d'Album,/Recueils de Croquis, Paysages, Sujets civils et militaires, Caricatures, Scènes populaires,/Idem de sociétés, Principes de dessins, Portrait au grainé doux, à l'hachure, au pointillé, etc., etc. Entrez, Messieurs et Dames, c'est l'instant de leurs exercises.]
97. Francisco de Goya, *Caricatures espagnoles. Ni plus ni moins, par Goya*, 10 pl., Lith. Motte, 1825.
98. The system of required registration and deposit was set up in 1817. See *Bulletin des lois du royaume de France*, ser. 7, vol. 5, no. 2875, Ordonnance du Roi relative aux impressions lithographiques, Oct. 8, 1817, 245. The earlier laws referred to in this law are listed in *Bulletin des lois du royaume de France*, ser. 7, vol. 2; see no. 395, Loi relative à la liberté de la presse, Oct. 21, 1814, 313–17, and no. 403, Ordonnance du Roi contenant des mesures relatives à l'impression, au dépôt et à la publication des ouvrages, etc., Oct. 24, 1814, 324–25.
99. Adhémar, *La France romantique*, 8.
100. The catalogues of the Ray collection are published as Ray, *Art of the French Illustrated Book*, and Ray, *The Illustrator and the Book in England*.
101. *Album comique de pathologie pittoresque*, 20 pl. (Paris: Ambroise Tardieu, 1823).
102. *L'Indigestion, Album comique de pathologie pittoresque*, no. 12. [L'Ambition a ordinairement les yeux plus grands que la panse.]
103. Émile Wattier, *Un an de la vie d'une jeune fille. Roman historique en XVII chapitres, écrits par son confident et lithographiés par M. Wattier*, 17 pl. (Paris: G. Engelmann and Gihaut, 1824). Other examples include Pigal's series of twelve hand-colored lithographs, *Vie d'un gamin en douze chapitres* (Paris: Gihaut frères, 1825), and Gérard-Fontallard (Henri-Gérard Fontallard), *Histoire d'une épingle en seize tableaux, racontée par elle-même*, 16 pl. (Paris: Ostervald aîné, Rittner, 1828).
104. For a more extensive discussion of Hogarth's influence and reception in France, see Mainardi, "Hogarth 'Corrected.'"
105. On the early illustrated press, see Watelet, *La presse illustrée*, Chollet, *Balzac journaliste*, and Bacot, *La Presse illustrée au XIXe siècle*.
106. The standard study of Philipon's periodicals is Kerr, *Caricature and French Political Culture*; on *La Silhouette*, see 9–13. The fullest discussion of the beginnings of *La Silhouette* is in Chollet, *Balzac journaliste*, 184–92.
107. For a discussion of Monnier's role in *La Silhouette*, see Chollet, *Balzac journaliste*, 184–92. For his relation to England, see Szrajber, "Henry Monnier's Letters from London in 1825."
108. [Chaque livraison, composée d'une feuille, papier vélin satiné, in-4, imprimée à deux colonnes, est accompagnée de deux lithographies. Treize livraisons formeront tous les trois mois un album.]
109. Its first two issues (nos. 1 and 2) have the title: *La Silhouette. Album lithographique, beaux-arts, dessins, moeurs, théâtres, caricatures.* In no. 3, this was changed to *La Silhouette. Dessins, caricatures, théâtres, moeurs, variétés, etc.*, and beginning with no. 4 its title was *La Silhouette. Journal des caricatures, beaux-arts, dessins, moeurs, théâtres, etc.*
110. *La Silhouette* was published from Dec. 24, 1829, to Jan. 2, 1831.

111. Balzac published excerpts and drafts of *Scènes de la vie privée* in *La Silhouette* 2, no. 2, 9–11, and 4, 30–32; for a complete list, see Chollet, *Balzac journaliste*, 218–19, 404.

112. The most famous image is by Bertall (Charles Albert d'Arnoux, 1820–1882): *Coupe d'une maison parisienne le 1er janvier 1845—cinq étages du monde parisien*; it appeared in *Le diable à Paris*, 2: 27, and is reproduced in Wechsler, *A Human Comedy*, 27, as *Cross-section of a Parisian House, 1 January 1845*, where its source is given as *Paris Comique, 1845*.

113. "Les Français au Missouri. Extrait d'un journal chiroquois," *La Silhouette* 1, no. 4 (1830): 27–28.

114. The caricature of Charles X was published, unsigned, in *La Silhouette* 2, no. 2 (Apr. 1, 1830): 12; after the revolution of July 1830, the image was republished as a lithographic poster in vol. 3, no. 6 (Aug. 12, 1830), as *Un Jésuite. Portrait déclaré ressemblant à Charles X par jugement du tribunal de police correctionnelle*; under the image is written "Au profit des veuves et orphelins des braves morts pour la patrie. Prix: 50c. Lith. by Ratier." It was published a third time in *La Silhouette* 4, no. 13 (Jan. 2, 1831). On the condemnation, see "Procès de la Silhouette," *La Silhouette* 3, no. 2 (July 1, 1830): 9–10. See also Kerr, *Caricature and French Political Culture*, 11–13, and Goldstein, *Censorship of Political Caricature*, 115.

115. On Philipon's beginnings, see Kerr, *Caricature and French Political Culture*, 7–8; 32–35.

116. This caption, sometimes "Ph. Inv." appeared especially in the prints of Daumier. For reproductions of proofs of several of Daumier's lithographs with glued-on handwritten captions, see McCullagh, *Dreams and Echoes*.

117. Kerr, *Caricature and French Political Culture*, 50–59.

118. *La Caricature, morale, politique et littéraire.* Prospectus et numéro-modèle, 1830; its title was changed several times to variants of this. [Aussi, depuis 1789, la caricature a été un besoin pour notre pays. Elle y est éminemment populaire; et si, jusqu'à présent, elle ne s'est pas rendue périodique, comme la pensée ou comme la plaisanterie, c'est que le prix de la gravure interdisait cette spéculation. Ce n'étaient pas les rieurs qui manquaient aux estampes, mais les estampes aux rieurs. Aujourd'hui les procédés de la lithographie ont permis de rendre presque vulgaire cette jouissance exquise que les Parisiens seul pouvaient renouveler tous les jours dans les rues, ou çà et là sur les boulevards.]

119. The announcement appeared regularly either on the first or last page. [*La Caricature* donne, par an, CENT QUATRE lithographies exécutées par les artistes les plus renommés. . . . L'administration ne met pas dans le commerce les lithographies du journal. Les marchands ne pourront les obtenir qu'en s'abonnant.]

120. Balzac published numerous drafts, excerpts, and articles in *La Caricature*, including some of his *Contes drolatiques*; for a complete list, see Chollet, *Balzac journaliste*, 454–57. See also Kerr, *Caricature and French Political Culture*, 20–24.

121. Kerr, *Caricature and French Political Culture*, 20–24.

122. The termination of *La Caricature* was announced in *Le Charivari* on Aug. 23, 1835, in "Aux abonnés du *Charivari* et de la *Caricature*." See also Kerr, *Caricature and French Political Culture* 116–20.

123. *La Caricature*, no. 105, Nov. 8, 1832. [*La Caricature*, qui ne paraît que tous les huit jours, peut et doit toujours être un journal de choix; ses dessins, faits avec soin, tirés avec précaution, et imprimés sur beau papier vélin, en font un journal de collection et de bibliothèque. Ceux de nos amis que le public a le plus encouragés seront désormais chargés *exclusivement* de l'exécution des dessins de cette oeuvre que nous tenons à honneur de maintenir au dessus de tout ce qui s'est fait et se fera dans ce genre.

Pour les établissemens publics, pour ceux d'entre les amateurs de dessins qui préfèrent la variété au mordant politique, nous créons un journal de tous les jours qui, chaque matin, publiera un croquis nouveau; et lequel, imprimé plus lestement que *la Caricature*, donnant des dessins moins terminés, pourra être livré au prix ordinaire des journaux appelés littéraires.]

124. *Le Charivari*, Dec. 21, 1832. [Journal publiant chaque jour un nouveau dessin]. *Le Charivari*, Oct. 12, 1835. [Journal publiant chaque jour un nouveau dessin (quand la censure le permet)].

125. Charles Philipon, prospectus for *Le Charivari*. [Des modes les plus nouvelles et les plus élégantes; des costumes les plus pittoresques des pièces en vogue, ainsi que des scènes les plus remarquables qu'elles auront offertes;—de sites, de monuments, de paysages, auxquels un événement quelconque viendrait attacher un intérêt de circonstance;—de scènes de moeurs;—d'esquisses des principaux tableaux des musées publics et particuliers, et de ceux que leur mérite aura fait

remarquer dans les différentes expositions;—de l'aspect des séances les plus intéressantes des deux Chambres, avec la physionomie, la pose, les tics, les gestes, en un mot toutes les habitudes parlementaires des divers orateurs; et enfin, de portraits d'acteurs, d'actrices, d'artistes en renom, de savans, de littérateurs, d'hommes publiques, de ministres, de diplomates, de pairs, de députés, de princes, de rois, de tout personnage qui, à tort ou à raison, soit en France, soit à l'étranger, éveillera un moment la curiosité publique.]

126. Charles Philipon, prospectus for *Le Charivari.* [Et puis, la satire politique, littéraire, dramatique et morale, n'envahira pas entièrement notre cadre. Après le charivari, souvent, nous donnerons la sérénade. Les sifflets pour ceux-ci, l'harmonie pour ceux-là. . . . *Le Charivari* offrira de la sorte un orchestre complet de virtuoses de tout genre.]

127. Charles Philipon, prospectus for *Le Charivari.* [Un panorama complet, où se reproduisent incessamment, par le crayon et par la plume tous les aspects de ce monde kaléidoscopique où nous vivons.]

128. "Planches. Nos. 192 et 193," *La Caricature,* no. 94 (Aug. 23, 1832). [Nous effacerons les pierres trois mois après leur premier tirage, et nous donnerons ainsi un grand prix à cette collection, qui ne tardera pas à devenir rare.]

129. *Le Charivari,* no. 4, Dec. 4, 1832. [Chez M. Aubert se trouvent toutes les lithographies qui paraissent dans le *Charivari.*]

130. *Le Charivari,* July 19, 1833. Pigal's numerous series include *Le miroir de Paris, Les moeurs populaire,* and *Les arts et métiers de Paris.* [Pour compléter la série des *Scènes familières* de M. Pigal, nous avons cru devoir offrir à nos souscripteurs le dessin formant vignette destinée à la couverture de cette charmante collection.]

131. "*Scènes familières,* par Pigal: 24 compositions de moeurs. Prix de la feuille en couleur: 75 centimes," *Le Charivari,* Nov. 21, 1833.

132. Daumier, *Album caricaturana (Les Robert Macaires)* (1838), and *Les cent-et-un Robert Macaire* (1839), were both published by Aubert. Information on all of Daumier's prints can be found on the Daumier Register: http://www.daumier-register.org.

133. *L'Artiste* cost sixty francs annually, the same as *Le Charivari.* On *L'Artiste,* see Goetz, *L' Artiste;* Damiron, "Un grand revue d'art, L'Artiste"; Damiron, "La revue 'L'Artiste.'"

134. Although single-sheet caricatures of Salon paintings had existed previously, Raimond Pelez's *Première impression du Salon de 1843,* published in *Le Charivari,* Mar. 19, 1843, was the first publication in a periodical. Credit must be given to Bertall (Charles-Albert d'Arnoux), however, for creating the concept of the caricatured Salon as an anthology of drawings; see his *Le Salon de 1843.* On the caricatured Salons, see Chabanne, *Les salons caricaturaux, and* Yang, *Les premiers salons caricaturaux au XIXe siècle.* [Un très beau cadre doré. Effet de nuit qui n'est pas clair . . . de lune, acheté subito par Mr Robertson, fabricant de cirage.]

135. S. C., "Beaux-arts. La chapelle Sixtine." [C'est aux artistes français que la lithographie doit ses progrès les plus remarquables; aussi s'est-elle popularisée chez nous avec une merveilleuse rapidité: elle a envahi tout l'art du dessin, elle a exploité tous les goûts, tous les caprices, tous les intérêts du public: beaux-arts, littérature, sciences, industrie, modes, elle a prêté à tous le secours et la facilité de son crayon. La lithographie s'est tellement vulgarisée, s'est prostituée à de tels usages, qu'elle a fini par se faire décrier et par exciter le mépris et le dédain des artistes consciencieux, de ceux qui veulent conserver toute la dignité de leur art.]

136. See Houfe, *Dictionary of British Book Illustrators and Caricaturists,* chapter 4.

137. *Le Charivari,* June 1, 1838. [Publiant chaque jour un nouveau dessin en lithographie ou gravure et des vignettes sur bois.]

CHAPTER 2. SPREADING THE NEWS

1. On the periodical press, see Kalifa et al., *La civilisation du journal,* and Bellanger et al., *Histoire générale de la presse française.* On the illustrated press, see Watelet, *La presse illustrée,* and Bacot, *La presse illustrée au XIXe siècle.*
2. See Watelet, *La presse illustrée* 1, 57; Gaudriault, *La gravure de mode feminine,* 24, 34, 193–98; Nevinson, *Origin and Early History of the Fashion Plate.*
3. Gaudriault lists 147 periodicals that included fashion plates published during this period; see Gaudriault, *La gravure de mode feminine,* 193–98.
4. Watelet, *La presse illustrée* 1, 6.
5. The standard history is by Chatto and Jackson, *Treatise on Wood Engraving.* On the revival of wood engraving, see Blachon, *La gravure sur bois au XIXe siècle.* See also Von Lintel, "Wood Engravings."
6. On Thompson, see Blachon, *La gravure sur bois au XIXe siècle,* 47–59.

7. See Barbier, "Les innovations techniques," 2: 721–29, and Feyel, "Les transformations technologiques de la presse au XIXe siècle."
8. Barbier, "Les innovations techniques," 727–28.
9. Barbier, "Les innovations techniques," 725.
10. On Girardin, see Reclus, *Émile de Girardin*; Thévenin, *Les créateurs de la grande presse*; Pellissier, *Émile de Girardin, prince de la presse*.
11. The claim that the *Penny Magazine* was an imitation of *Journal des connaissances utiles* (hereafter *JCU*) has been made by all of Girardin's biographers—though not by Girardin himself. For the claim, see Reclus, *Émile de Girardin*, 70, Thévenin, *Les créateurs de la grande presse*, 44, and Pellissier, *Émile de Girardin*, 76; for the absence of such a claim by *JCU*, see "Gravure sur bois en France et en Angleterre," *JCU*, where English precedence is acknowledged.
12. "Résumé," *JCU*, Dec. 1832, 348. *JCU* published three illustrations in April; none in May, June, or July; nine in Aug.; none in Sept.; ten in Oct.; fifteen in Nov.; Dec. contained the annual "Résumé" of previous issues, but thereafter *JCU* published illustrations regularly.
13. Circulation is given as 130,000 in "Résumé," *JCU*, Dec. 1832, 348. Reclus cites 132,000 in his *Émile de Girardin*, 69.
14. On newspaper circulation figures, see Bellanger, *Histoire générale de la presse française*, 2: 99–100. The average circulation figure was 14,700.
15. The subtitles varied as different categories of prospective readers were included or excluded. [*Journal des connaissances utiles. Politique, agricole et commercial, indiquant à tous les hommes qui savent lire. Leurs devoirs, comme: citoyen, juré, garde national, maire ou adjoint, garde champêtre ou forestier. Leurs droits, comme: contribuable, électeur communal, conseilleur municipal, électeur, éligible. Leurs intérêts, comme: père de famille, propriétaire, fermier, fabricant et commerçant, ouvrier de tous états*.]
16. Girardin, "Écoles et méthodes," *JCU*, Mar. 1833, 63–70. [Versez l'instruction sur la tête du peuple; vous lui devez ce baptême!]
17. *Le Temps* was quoted in *JCU*, Nov. 1831, 69. [Les almanachs, par la modicité de leur prix, avaient été jusqu'à ce jour la seule lecture qui fut possible à deux millions de citoyens, privés ainsi des notions les plus utiles. Cette lacune entre les journaux et les almanachs vient d'être en partie comblée par une publication mensuelle, précise et méthodique, coutant *quatre francs par année*.]
18. The French debt to the English organization was acknowledged in "Société nationale pour l'emancipation intellectuelle," *JCU*, Oct. 1831, 2.
19. "Société nationale pour l'emancipation intellectuelle," *JCU*, Oct. 1831, 2. [Ce que la Société pour l'instruction élémentaire a fait pour l'enseignement primaire des enfans, une autre association, la Société nationale pour l'émancipation intellectuelle vient de l'entreprendre pour le grand nombre d'hommes auxquels nos institutions nouvelles ont conféré des droits politiques et des attributions municipales, avant qu'ils fussent tous en état de les remplir avec discernement.]
20. The weekly *Chambers Edinburgh Journal*, founded by William Chambers, was founded on Feb. 4, 1832, a few weeks before the *Penny Magazine*, and also cost a penny, but it had no illustrations.
21. The primary source on Charles Knight is his autobiography: Knight, *Passages of a Working Life*. See also Gray, *Charles Knight: Educator, Publisher, Writer;* and Fox, *Graphic Journalism in England during the 1830s and 1840s*.
22. Knight, "Reading for All," *Penny Magazine* (hereafter *PM*), Mar. 31, 1832, 1.
23. Knight, *Passages of a Working Life*, 2: 180.
24. Knight, preface to vol. 1, *PM*.
25. The preface was published at the end of the year and bound in with that year's issues; Knight, preface to vol. 1, *PM*.
26. Knight, preface to vol. 1, *PM*.
27. Knight, preface to vol. 1, *PM*.
28. Knight, *Passages of a Working Life*, 1: viii.
29. Knight, *Passages of a Working Life*, 2: 184. On the use of reproductive wood engraving in art education, see Von Lintel, "Wood Engravings."
30. *PM*, 1832: *The Colosseum* (July 14); *The Parthenon* (Oct. 20); *The Laocoon* (Nov. 10); [Raphael] *Christ Delivering the Keys to St. Peter* (Dec. 1); *The Apollo Belvedere* (Dec. 15).
31. *PM*, 1832: [Zebra], (Mar. 31–Apr. 30 supplement); *Hippopotamus* (May 19); *Kangaroos* (May 5); *The Crocodile* (June 2); *The Giraffe* (June 30); *Giraffe Preparing to Lie Down*, Nov. 3); [elephant], (July 21); *The Boa Constrictor about to Strike a Rabbit* (Oct. 27).
32. *PM*, 1832: *Wild Bushman* (Sept. 22); *Singular Dexterity of a Goat* (June 9); *Catching Turtles on the Coast of Cuba* (Oct. 20).
33. *PM*, 1832: *Old St. Paul's Cathedral—South View* (May 12); *View of Westminster Bridge* (Sept. 1); *Falls of the Clyde* (Sept. 29); *Interior of Holyrood Chapel* (Aug. 11).

34. *PM*, 1832: *Rubens* (June 23); *Flaxman* (June 30); *Nicolas Poussin* (May 26); *Petrarch* (July 14); *Shakspeare* [*sic*] (Sept. 1); *Sir Walter Scott. From Mr. Chantrey's Bust* (Sept. 29–Oct. 31 supplement); *Portrait of Milton* (Dec. 8); *Adam Smith* (June 2); *Portrait of William Penn* (Oct. 13); *Portrait of Erasmus Reading* (Oct. 27); *Portrait of Newton* (Dec. 22).
35. *PM*, 1832: "Tea" (Apr. 28); "Tobacco" (July 14); "Statistical Notes. The Silk Trade" (Dec. 22); "The Diving Bell" (Oct. 13); "Suspension Bridges" (Apr. 30–May 31 supplement).
36. Knight, *Passages of a Working Life*, 2: 223.
37. Knight, *Passages of a Working Life*, 2: 223.
38. On the relation of Saint-Simonianism to high art, see McWilliam, *Dreams of Happiness*.
39. The principal accounts of the foundation of *Le Magasin pittoresque* (hereafter *MP*) are by Charton: "Extrait des éphémérides" and "La littérature populaire en Angleterre." See also Aurenche, *Édouard Charton*, 139–49, and Charton, *Correspondance*, 1.
40. Aurenche, *Édouard Charton*, 140.
41. Charton, "Extrait des éphémérides." [Novembre 1832.—Mes amis Jean Reynaud, Pierre Leroux, Sainte-Beuve, me conseillent d'accepter la direction d'un recueil hebdomadaire illustré, à dix centimes, que M. Lachevardière, imprimeur, se propose à publier à l'imitation du *Penny-Magazine*, récemment fondé à Londres par l'historien Charles Knight. J'hésite, je suis au barreau. Mes amis me font valoir qu'il s'agit d'un service à la rendre à la cause d'instruction et de l'éducation et que j'y suis préparé par mes travaux antérieurs, notamment comme rédacteur du *Bulletin de la Société pour l'instruction de l'éducation élémentaire* et du *Journal de la Société de la morale chrétienne*. J'accepte après l'assurance que j'aurai la collaboration de mes amis, sortis la plupart des grandes écoles. L'un d'eux, Euryale Cazeaux, depuis Inspecteur général de l'agriculture, s'associe temporairement à ma direction.]
42. Spitzer, *French Generation of 1830*.
43. Charton, "Extrait des éphémérides."
44. Charton, "À tout le monde." [C'est un vrai Magasin que nous nous sommes proposés d'ouvrir à toutes les curiosités, à toutes les bourses.]
45. This is the basic thesis of Bourdieu, *Distinction*.
46. Charton, "Extrait des éphémérides." [En Angleterre, le mot *pittoresque* a étonné: un membre de la Société Royale m'en rapproche l'emploi. Il lui semble que j'ai abusé de cette expression, et qu'elle ne doit s'entendre que de ce qui se rapporte à la peinture; mais le sens français est plus étendue, et j'ai la satisfaction de voir qu'il est rapidement adopté pour beaucoup d'ouvrages illustrés, recueils périodiques ou autres.]
47. This statement appeared as the preface to the first bound volume of *Le Magasin pittoresque*, dated Dec. 31, 1833; it is a retrospective look at the early months of publication. [Nous croyons même convenable de déclarer en tête de ce premier volume, que si nous nous sommes hasardés les premiers, sans patronage, sans prospectus, à importer en France l'idée de livrer au plus humble prix un texte varié, entremêlé de graveurs et divisé par livraisons, c'est seulement après avoir connu le succès des *Magazines* en Angleterre, et surtout celui du recueil publié à Londres, sous la haute et digne influence, par M. Charles Knight, écrivain économiste distingué, qui, par ses relations bienveillantes avec nous, a contribué à rendre moins décourageantes les premières difficultés de notre entreprise.]
48. Charton, "Extrait des éphémérides." [Décembre 1832.—Difficulté. Le *Penny Magazine* entremêle son texte de gravures de bois. En France, ce genre de gravure est depuis longtemps délaissé: on y compte à peine huit graveurs.]
49. Charton, "Jean Best." [Lorsque je fondai, au commencement de 1833, avec mon ami Pierre Cazeaux, le *Magasin pittoresque*, la plus grande difficulté que je rencontrai fut de faire exécuter chaque semaine les gravures sur bois que je considérai comme des éléments utiles à introduire dans un recueil destiné à répandre des connaissances variées. La gravure sur bois était alors un art presque oublié en France: les graveurs étaient rares: je m'adressai à trois d'entre eux, MM. Andrew, Best et Leloir, qui s'étaient associés et vivaient assez pauvrement, dans un petit réduit, près la place Saint-André-des-Arts.

    Ces messieurs, si je ne me trompe, n'avaient point d'élèves; ils m'écoutèrent avec étonnement, lorsque je leur expliquai qu'il s'agissait de livrer, chaque semaine, trois ou quatre gravures; ils se récrièrent, et me dirent nettement que la chose était impossible: passe encore si je leur avais parlé de mois au lieu de semaines! Il fallait vraiment que je n'eusse aucune idée des difficultés de leur profession.

    Ils étaient bien loin de pressentir que ma visite serait pour eux l'origine d'une grande prospérité.

J'insistai, leur promettant que, dans les premiers temps, on ne leur demanderait que des travaux faciles, et que même, pour alléger leur tâche, on entremêlerait à leurs gravures des clichés qu'on ferait venir de Londres. Je parvins à leur persuader d'essayer au moins leur zèle et leurs forces: ils appelèrent à leur aide des élèves, se mirent bravement à l'oeuvre, simplifièrent leurs procédés, et deux ans après, nous n'étions plus dans la nécessité de rien emprunter à l'Angleterre: le concours de MM. Andrew, Best et Cie pouvait désormais nous suffire amplement.]

50. Charton visited London in Dec. 1832 for this purpose; see " Extrait des éphémérides."
51. *Bulletin des lois*, ser. 9, vol. 1: no. 59, Charte constitutionelle, Aug. 14, 1830, 51–64.
52. *Bulletin des lois*, ser. 9, vol. 5: no. 236, Loi sur l'instruction primaire, June 28, 1833, 251–62. It was called the Guizot Law because François Guizot was Ministre secrétaire d'état au département de l'instruction publique. On the effects of the Guizot law, see Furet and Ozouf, *Reading and Writing: Literacy in France*, 100–48. [L'instruction primaire élémentaire comprend nécessairement l'instruction morale et religieuse, la lecture, l'écriture, les éléments de la langue française et du calcul, le système légal des poids et mesures. L'instruction primaire supérieure comprend nécessairement, en outré, les éléments de la géométrie et ses applications usuelles, spécialement le dessin linéaire et l'arpentage, des notions de sciences physiques et de l'histoire naturelle applicable aux usages de la vie; le chant, les éléments de l'histoire et de la géographie, et surtout de l'histoire et de la géographie de la France.]
53. Édouard Charton à Jean-Louis Renaudot, Mar. 3, 1838, in Charton, *Correspondance*, 1: no. 59–16, 1051–52. The letter was originally written in 1838, then certified authentic and reprinted in 1858 in Renaudot, *Les Bérangériennes*, 29–30. [Elevons l'esprit de nos concitoyens, et l'on verra commencer le véritable règne de l'Égalité. Vous par vos vers, moi par le *Magasin pittoresque*, nous aurons, je l'espère, contribué à cette grande tâche, la seule qui me paraisse préparer avec sûreté l'émancipation du people.]
54. Charton, *MP*, "De l'influence de la conversation." [La difficulté de causer partage donc en quelque sorte la société en deux classes. Ce n'est pas que l'esprit de charité n'ait toujours cherché à combler cette lacune dans les relations du riche au pauvre.... On a beau dire et beau faire, il n'en existe pas moins une ligne de séparation réelle, indépendante des préjuges politiques, et qu'on ne peut pas espérer d'effacer entièrement, même par l'enseignement élémentaire des écoles. On ne la fera disparaitre qu'à l'aide d'une certaine diffusion de connaissances variées et d'un intérêt habituel et général, qui rendra insensiblement les communications plus agréables, plus faciles, plus intimes entres toutes les classes de la société.]
55. *MP*, "Georges Cuvier. Sa vie. Ses travaux."
56. Charton, "À tout le monde." [Nous voulons qu'on y trouve des objets de toute valeur, de toux choix: choses anciennes, choses modernes, animées, inanimées, monumentales, naturelles, civilisées, sauvages, appartenant à la terre, à la mer, au ciel, à tous les temps venant de tous les pays, de l'Indostan et de la Chine, aussi bien que de l'Islande, de la Laponie, de Tombouctou, de Rome ou de Paris; nous voulons, en un mot, imiter dans nos gravures, décrire dans nos articles, tout ce qui mérite de fixer l'attention et les regards, tout ce qui offre un sujet intéressant de rêverie, de conversation, ou de l'étude.]
57. *PM*, "How to Endure Poverty"; *The Young Beggar, from Murillo* was reproduced just below the cover masthead of *PM*, Mar. 29, 1834, to illustrate the article "Murillo," 113–15.
58. The boa constrictor wood engraving from the Oct. 27, 1832, cover of the *Penny Magazine* was reused for the cover of *Le Magasin pittoresque* 1 (1833), no. 2.
59. *Mont Saint-Michel, vue prise du côté de l'est, MP* 1 (1833), no. 44, 349; *Vue de la galerie d'Orléans, au Palais-Royal, MP* 1 (1833), no. 5, 5; *Palais de Justice de Dijon, MP* 1 (1833), no. 30, 237; the *Cathédral d'Amiens, MP* 1 (1833), no. 47, 369.
60. *MP*, "Bibliothèque du roi.—Manuscrits. Portraits de Chinois célèbres"; *MP*, "Voyages. Nouvelle Zélande."
61. *MP*, "Guillaume Penn"; *MP*, "Statue de Diane à Éphèse"; *MP*, "Antiquités égyptiennes. Zodiaque circulaire de Denderah"; *MP*, "Un cimetière en Sicile."
62. *MP*, "Des animalcules microscopiques."
63. *MP*, "Alphabet manuel des sourds-muets"; *MP*, "Aérostation—Ballons.
64. *MP*, "Hygiène. Du danger des corsets trop serrés."
65. Charton, *MP*, "La littérature populaire en Angleterre."

66. *MP*, "31 décembre, 1833," ii. [donne un dégré d'utilité encore inconnu jusqu'ici à l'alliance du dessinateur et de l'écrivain.]

67. *MP*, "Des moyens d'instruction: Les livres et les images," 98–99; see 98. [Nous signalerons comme un moyen complémentaire d'instruction, presque inusité encore, les *dessins* ou les *images*. . . . Notre conviction est telle à cet égard, que nous dirions volontiers: *Sans les dessins, il est impossible d'arriver à l'éducation complète des hommes, grands et petits.* Nous attachons en effet une grande importance morale aux *images*, et nous croyons qu'elles comblent une lacune des *livres*.]

68. Édouard Charton à Émile Souvestre, Jan. 29, 1833, in Charton, *Correspondance*, 1: no. 33–1, 127–28. [un petit journal à gravures hebdomadaire.]

69. See, for example, Édouard Charton à Émile Souvestre, Aug. 14, 1833, in Charton, *Correspondance*, 1: no. 33–6, 136–37.

70. Édouard Charton à Hippolyte Fortoul, Feb. 17, 1845, in Charton, *Correspondance*, 1: no. 45–1, 476–77. [Je vais faire graver quelques vues du Midi: Toulouse, Montpellier etc.; leur musées me fourniront, j'espère, des sujets de gravures. Oserai-je vous demander le texte?]

71. Édouard Charton à Jean Reynaud, Apr. 21, 1843, in Charton, *Correspondance*, 1: no. 43–6, 425–27. [On me grave quelques scènes d'ours du Jardin des Plantes. J'aurais besoin *de 4 à 5 colonnes* sur les moeurs des ours ou sur leur histoire au Jardin des Plantes.]

72. Édouard Charton à Hippolyte Fortoul, Dec. 30, 1844, in Charton, *Correspondance*, 1: no. 44–16, 472–75. The article and illustrations appeared as "Un voyage à Pérouse," *MP* 13, no. 36 (Sept. 1845); 281–84; "Florence. De l'influence de la France sur l'art italien," *MP* 13, no. 44 (Nov. 1845): 345–47; and "Naples," *MP* 14, no. 7 (Feb. 1846): 49–51. [On grave plusieurs vues d'Italie: vous plait-il d'en faire le texte? Ce sont des vues générales de Florence, de Naples, de Pouzzoles, de Castellamare, de Sorrente et une vue de la place et de la fontaine de Pérouse. Vous savez quelle liberté de digression est permise.]

73. Édouard Charton à Hortense Charton, Aug. 29, 1845, in Charton, *Correspondance*, 1: no. 45–8, 490–92. [Lachevardière m'avait trouvé un jeune dessinateur qui devait m'accompagner et faire, pour le *Magasin*, les dessins que je lui aurais indiqués.]

74. Édouard Charton à Hippolyte Fortoul, Oct. 1846, in Charton, *Correspondance*, 1: no. 46–16, 525. [Mon cher ami, voici une des gravures sur lesquelles j'espère avoir de votre texte. C'est celle d'après Le Poussin. Malheureusement je ne puis vous envoyer les autres qui seront terminées tard, et vous savez que je compose et mets en page avant que les graveurs n'aient terminé.]

75. Édouard Charton à Hippolyte Fortoul, 1838, in Charton, *Correspondance*, 1: no. 38–22, 249. [J'ai ensuite songé qu'avec un groupe composé de figures arabes on pourrait faire une planche intéressante. Mais il faudrait un sujet d'article.]

76. Édouard Charton à Pierre-Jean David d'Angers, June 13, 1837, in Charton, *Correspondance*, 1: no. 37–13, 204–205. [Peut-être jugerez-vous à propos que, vu le peu de dimension du *Magasin*, les groupes du fronton soient dessinés séparément, après qu'une petite représentation de l'ensemble aura donné une idée de l'effet général. Ce que vous déciderez sera bien décidé.]

77. Édouard Charton à Pierre-Jean David d'Angers, July 1837, in Charton, *Correspondance*, 1: no. 37–19, 209–10. [Je crois avoir trouvé un dessinateur qui nous conviendra, et mon intention est de vous le présenter lundi après-midi.]

78. Édouard Charton à Pierre-Jean David d'Angers, July 1837, in Charton, *Correspondance*, 1: no. 37–20, 209–10. [Je vous envoie le dessin de M. Saint-Germain. Veuillez, je vous prie, m'en dire votre avis. Les graveurs n'amélioreront pas. M. Saint-Germain vient lundi à Paris. Si vous le désirez, j'irai avec lui vous voir, à une heure de l'après-midi que vous m'indiquerez. Il fera toutes les retouches que vous désirerez. J'espère mieux de ce bois que du premier.]

79. Édouard Charton à Pierre-Jean David d'Angers, July 1837, in Charton, *Correspondance*, 1: no. 37–21, 211. [Je reçois à l'instant l'épreuve de la gravure de MM. Andrew et Leloir. J'y trouve, à ma grande peine, beaucoup de petits défauts, la Patrie éborgnée, Manuel ricaneur, Voltaire un peu chargé, le bras de Bichat cassé, etc. etc. j'écris aux graveurs pour qu'ils modifient de leur mieux toutes ces choses. Je vous prie de leur adresser de votre côté vos observations sur ce qu'il serait encore possible d'améliorer. Nous n'avons plus de temps à perdre: à peine nous reste-t-il le nombre de jours nécessaires pour clicher et imprimer. Je suis désolé qu'on n'ait pas fait mieux, croyez à mon regret et à la bonne volonté que j'y ai mise, à la fois comme ami de la liberté et comme votre admirateur.]

80. The article on David d'Angers's pediment appeared in Aug. 1837, but the engraving wasn't published until Oct. See *MP*, "Le Panthéon."
81. Édouard Charton à David d'Angers, Oct. 5, 1837, in Charton, *Correspondance*, 1: no. 37–25, 214–15. [La gravure du fronton était déjà clichée et sous la presse lorsque votre avant-dernière lettre m'est arrivée: il m'a été interdit par conséquent d'y ajouter les modifications que vous désiriez.]
82. Jean-Jacques Grandville à Édouard Charton, Apr. 15, 1843, in Charton, *Correspondance*, 1: no. 42–3, 389–91. See Kaenel, "J.-J. Grandville et le *Magasin pittoresque*." Although Kaenel published the correspondence between Grandville and Édouard Charton before the Aurenche edition of Charton's collected correspondence, some of the letters appear to have been edited, and so the Aurenche edition of Charton's correspondence should be consulted.
83. Jean-Jacques Grandville à Édouard Charton, Oct. 14, 1840, in Charton, *Correspondance*, 1: no. 40–28, 342–47. [Je viens de faire envoyer à Andrew le troisième bois, que Leloir dit qu'il ne pourra graver exactement fac-similé, attendu le peu de temps que vous lui donnez; voyez donc à ce que ce motif ne soit pas une facile excuse d'échiner les trois bois.... Il faut que vous poussiez ces messieurs (sans parler de moi) à mettre ceux-ci en de bonnes mains. C'est autant votre intérêt que le mien.]
84. Jean-Jacques Grandville à Édouard Charton, Feb. 28, 1847, in Charton, *Correspondance*, 1: no. 47–4, 539–41. [Du reste, prenez, rejetez, taillez, tranchez, réunissez ceci à ce que je vous ai déjà indiqué et faites pour le mieux comme toujours. Mais pour ce bois comme pour le 1er: il est de toute nécessité que je voie le graveur qui ne pourrait jamais se tirer heureusement sans mes observations.]
85. Jean-Jacques Grandville à Édouard Charton [1841], in Charton, *Correspondance*, 1: no. 41–3, 359–61. [ogre pittoresque qui ne faites qu'une bouchée de la plus longue et rare idée du monde.... qui semblable à Saturne dévore ses artistes.]
86. Croy here uses the word *gravure* in its traditional sense of copper-plate engraving. Croy, "Les magasins pittoresques à deux sous." [comme taches d'encre noire.... Mais que deviendront les produits de la gravure, de cet art si parfait, si difficile et si digne d'encouragement? Que deviendra, même, de la lithographie, dont le perfectionnement, depuis dix ans, est si notable, les bonnes moeurs à part?... l'assassinat des beaux-arts... Laissons nos voisins d'outre-mer commercer des produits de l'imagination et faire de l'art à la mécanique; seulement, ne les imitons pas.]
87. Dupeuty, Courcy, and Alhoy, *Le Magasin pittoresque, revue en 15 livraisons* was first presented at the Théâtre des Variétés in Paris on Dec. 31, 1833.
88. The script was published as *Le Magasin pittoresque, revue en 15 livraisons, par MM.* [Charles] *Dupeuty, F.* [Frédéric] *de Courcy, Maurice Alhoy* (Paris: Marchant, 1834); see 6. The vaudeville was even mentioned in the preface to volume 3 (1835) of *Le Magasin pittoresque.* [Ne vois-je pas là vingt exemplaires de l'Encyclopédie.... C'est la mine qu'il faut exploiter.... C'est la source qu'il faut tarir... prends des ciseaux... coupe, taille, rogne... tout cela, remis à neuf, et accompagné de portraits de grands hommes et de grosses bêtes, des beautés contemporaines et de monuments gothiques formera le recueil le plus bizarre, le plus varié de notre époque à deux sous... enfin, le véritable *Magasin pittoresque!*] Ellipses are in the original.
89. *Encyclopédie, ou dictionnaire raisonné des sciences, des arts et des métiers* by Denis Diderot and Jean Le Rond d'Alembert was published between 1751 and 1780 in thirty-five volumes of text and plates.
90. Charton, *MP*, "La littérature populaire en Angleterre," 15. On newspaper circulation figures, see Bellanger et al., *Histoire générale de la presse française*, 2: 99–100, 119; see also Marchandiau, *L'Illustration, 1843–1944*, 294, and Robert, "Paysages politiques, cohérences médiatiques," in Kalifa et al., *La civilisation du journal*, 213–48.
91. Charton, "Extrait des éphémérides. [Émile de Girardin propose une association que je ne juge pas acceptable.]
92. "Feuilleton, 22 Fév. 1834. Encyclopédie pittoresque à deux sous," *Le Temps*, 1. [Je ne sais si l'âge d'or des gros livres est définitivement passé, mais il est au moins certain que le moment ne leur est pas favorable. On ne voit plus d'in-folios, plus de ces épais et vénérables volumes éclos après vingt ans d'incubation laborieuse. Tout se fait vite aujourd'hui, on pense, on écrit et l'on imprime vite.... Il ne se fait plus de livres, mais en revanche il se fait des journaux; jamais on n'en a vu pareil nombre; et parmi les journaux, ceux qui obtiennent le plus de succès

sont ceux qui s'adressent non pas aux esprits les plus élevés, mais au plus grand nombre, comme aussi aux petites bourses, qui sont aussi le plus grand nombre. Il y a le *Magasin pittoresque* qui a 60,000 abonnés, et qui se vend deux sous; il y a beaucoup de publications du même genre et du même prix qui circulent inaperçus dans la foule. Le savoir public ne gagne pas en profondeur, mais en superficie. La science, jusqu'ici hérissée, gourmée, flanquée de gros livres, retirée dans de ténébreux sanctuaires, vient en grand jour, court les rues, coudoie les passans, se fait familière et de bonne composition, se taille à la dimension des moindres cerveaux, et se fait petite pour passer partout.... Le *Magasin pittoresque* et toutes les publications qui sont venues à la suite, ont paru jusqu'à ce jour se proposer pour but l'amusement plus encore peut-être que l'instruction de leurs lecteurs, non sans doute que l'immense variété de faits dont leur répertoire se compose, et que les jolies gravures sur bois qui accompagnent le texte, en allant à l'intelligence par les yeux, ne soient destinées à répandre une foule de connaissances et de notions utiles, mais aucun ordre systématique ne coordonne les matériaux épars; c'est un enseignement, sans doute, mais un enseignement confus, anarchique, sans unité et sans lien apparent.]

93. *Illustrated London News* (hereafter *ILN*), "Our Address," May 14, 1842, 1.
94. On the foundation of *ILN*, see the articles in *Illustrated London News Jubilee Edition*, May 14, 1892, especially "The Founding of the 'Illustrated' May 14th, 1842," an account that leans heavily on the reminiscences of Vizetelly, *Glances Back through Seventy Years.* Vizetelly apparently made his manuscript available prepublication; see especially Vizetelly, 1: 221–41.
95. *ILN Jubilee*, "Founding of the 'Illustrated'"; Vizetelly, *Glances Back*, 1: 236.
96. Vizetelly, *Glances Back*, 1: 224.
97. *ILN*, "Our Address."
98. Vizetelly, *Glances Back*, 1: 226.
99. *ILN Jubilee*, "Founding of the 'Illustrated.'"
100. Jackson, "Thirty Years of Pictorial Journalism," *ILN Jubilee.*
101. *ILN*, Felicie, "The Fashions," May 14, 1842, 6.
102. *ILN*, "Abolition of the Slave Trade," Jan. 21, 1843; *ILN*, "The Disturbances in the Manufacturing Districts," Aug. 20, 1842.
103. *ILN*, "The Disturbances in the Manufacturing Districts," Aug. 20, 1842.
104. Vizetelly, *Glances Back*, 1: 232–33.
105. Vizetelly, *Glances Back*, 1: 233.
106. *ILN*, "Our Address."
107. Vizetelly, *Glances Back*, 1: 237.
108. Vizetelly, *Glances Back*, 1: 232.
109. *L'Illustration* measured 28 × 32 cm; the *Illustrated London News* was slightly larger at 29.7 × 40.8 cm.
110. Édouard Charton à Hippolyte Fortoul, May 17, 1843, in Charton, *Correspondance*, 1: 43–10, 433–36. [En souvenir d'un journal anglais que j'avais vu à Londres l'été dernier, j'avais proposé à Lachevardière de fonder un recueil pittoresque d'actualités. L'idée ne lui convint pas, mais il lui parut naturel que je fisse effort pour la réaliser avec un autre éditeur. J'avisai Dubochet et Paulin. Nous fûmes bientôt d'accord. Dubochet alla à Londres s'informer. On commença les frais, les travaux.]
111. On the founding of *L'Illustration*, see Marchandiau, *L'Illustration, 1843–1944*, 16–20; on its classification as a political journal, 52. David Kunzle sees *L'Illustration* as focused on preserving the status quo; while this was true on issues such as French colonial rule, on issues such as slavery *L'Illustration* was opposed to government policies. See Kunzle, "'L'Illustration,' journal universel (1843–1853)."
112. *L'Illustration* (hereafter *ILL*) "Notre but," 1. [Puisque le goût du siècle a relevé le mot d'*Illustration*, prenons-le! Nous nous en servirons pour caractériser un nouveau mode de la presse nouvelliste.]
113. As late as 1845, Théophile Gautier was still referring to the word *illustrateur* as a neologism; see Gautier, "Tony Johannot," 227; see also "illustrateur, " Rey, *Le Robert. Dictionnaire historique*, s.v.
114. *ILL*, "Notre but," 1. [Les livres ne parlent plus qu'à moitié, si le génie de l'artiste, s'inspirant de celui de l'écrivain, ne nous traduit leurs récits en brillantes images, et l'on dirait qu'il en est désormais de toute littérature descriptive comme de celle du théâtre, que l'on ne connait bien qu'après avoir vue représentée.... Ce que veut ardemment le public aujourd'hui, ce qu'il demande avant tout le reste, c'est d'être mis aussi clairement que possible au courant de ce qui se passe. Les journaux sont-ils en état de satisfaire ce désir avec les récits courts et incomplets auxquels ils sont naturellement obligés de s'en tenir? C'est ce qui ne paraît pas. Ils ne parviennent le plus souvent à faire entendre les choses que vaguement, tandis qu'il faudrait si bien les entendre

que chacun s'imaginât les avoir vues . . . et ayons désormais des journaux qui sachent frapper les yeux par les formes séduisantes de l'art.]

115. *ILL*, preface to vol. 1. [Un vaste annuaire où seront racontés et figurés, à leurs dates, tous les faits que l'histoire contemporaine enregistre dans ses annales: événements politiques, cérémonies publiques, grandes fêtes nationales, désastres fameux, bruits de la ville, morts illustres, biographies contemporaines, représentations théâtrales, découvertes utiles, expositions des arts et de l'industrie, publications nouvelles, faits glorieux de l'armée, modes, types, scènes populaires, etc., etc. *L'Illustration* sera, en un mot, un miroir fidèle où viendra se réfléchir, dans toute son activité merveilleuse et son agitation si variée, la vie de la société au dix-neuvième siècle.]

116. Stendhal, *Le rouge et le noir*, 414. [Un roman est un miroir qui se promène sur une grande route.]

117. Marchandiau, *L'Illustration, 1843–1944*, 27, 294.

118. See, for example *ILL*, "Revue algérienne," Mar. 18, 1843, 37–38.

119. Charton referred to ILL as "un résumé hebdomadaire de la presse avec gravures" in Édouard Charton à Hippolyte Fortoul, le 17 mai 1843, Charton, *Correspondance*, 1: no. 43–10, 433–36. Jean Reynaud à Édouard Charton, Jan. 6, 1843, in Charton, *Correspondance*, 1: no. 43–1, 415–18. [Crains les mauvaises et immorales caricatures. Quand on a le parti pris de rire, on va bien souvent jusqu'à rire sans s'en apercevoir et d'une manière plus ou moins indirect de ce qu'il faut respecter. J'ai vu quelques-unes de ces caricatures de Gavarni qui m'ont révolté, en ce qu'elles tendent à corrompre et ridiculiser ce qu'il y a de plus respectable du monde: les sentiments de famille et la gentillesse des enfants.]

120. Rodolphe Töpffer's *Histoire de M. Cryptogame* was redrawn by Cham for wood engraving, and published in eleven installments in *L'Illustration*, Jan. 25–Apr. 19, 1845.

121. *ILL*, Cham, "Arithmétique pittoresque," Apr. 13, 1844, 108. "Simplification of Arithmetical Rules" was a regular feature of *PM*, beginning on Jan. 26, 1833, and continuing weekly. *JCU* printed numerous similar articles relating particularly to finances; see, for example, *JCU*, Evon, "Instruction. Budget des familles," and "Science. Du crédit. De l'épargne." In *MP*, lessons on mathematics were given in two 1837 articles, "De la comptabilité."

122. Marchandiau, *L'Illustration, 1843–1944*, 14.

123. *ILN*, "Abolition of the Slave Trade," Jan. 21, 1843; *ILL*, "De la traite et de l'esclavage," Oct. 21, 1843. The engravings were first published in 1788 by the Plymouth chapter of the Society for the Abolition of the Slave Trade. I thank Gillian Forrester for this identification.

124. *ILL*, "Arrivée du roi au palais Bourbon," Dec. 30, 1843, 276.

125. Édouard Charton, "Jean Best," 240. The production process is further explained and illustrated in *ILL*, "Les mystères de l'*Illustration*," Mar. 2, 1844, 7–10. [Mais cette fois, quelle que fût la difficulté, M. Best ne parla plus d'impossibilité: Il s'entoura d'un groupe d'artistes exercés qui, sous son impulsion, s'engagèrent à travailler jour et nuit. Souvent, les bois dessinés de page entière furent découpés en petits morceaux et distribués entre les jeunes graveurs qui, la nuit surtout, se relayaient de deux en deux heures. C'était là, s'imposer presque un tour de force: on était épuisé de fatigue, on s'inquiétait, on voyait avec angoisse approcher la fin de la semaine; mais après tout on arrivait.]

126. Édouard Charton à Hippolyte Fortoul, May 17, 1843, in Charton, *Correspondance*, 1: 43–10, 433–36. [Les dessinateurs ne sont pas plus faciles à rallier que les écrivains. Nul ne veut sortir de son atelier. Ils dessinent sur les récits qu'on leur fait, et ils ne manquent jamais de préférer l'accessoire au principal.]

127. *ILL*, "Tremblement de terre aux Antilles," Mar. 18, 1843, 33. [Destruction de la Pointe-à-Pitre par un tremblement de terre, le 8 février 1843, à 10 heures 35 minutes du matin. —Ce dessin a été composé sur les indications de M. Lemonnier de La Croix, qui a été pendant dix années architecte voyer à la Pointe-à-Pitre, et qui n'est de retour en France que depuis deux années seulement.]

128. *ILN*, "Destruction of Pointe-à-Pitre," Apr. 29, 1843, 286. It is unclear whether the engraving was plagiarized or purchased from *L'Illustration*.

129. *ILL*, May 20, 1843, 192. [*Napoléon adoré dans un temple chinois.—Dessin fait par un témoin oculaire.*]

130. *ILL*, "Catastrophe de 8 mars à Alger," Mar. 22, 1845, 49–50. [Les deux dessins que nous publions sur cette épouvantable catastrophe, et qui nous ont été envoyés d'Alger avec le supplément de l'*Abkar* représentent l'un l'explosion, l'autre l'état des lieux après l'explosion.]

131. Édouard Charton à Hippolyte Fortoul, May 17, 1843, in Charton, *Correspondance*, 1: no. 43–10, 433–36. [Presque tout mon rôle, en ce qui est du

texte, consiste à me défendre contre l'invasion de la licence et du mensonge: on me traite de prude, de puritain.]

132. Édouard Charton à Hippolyte Fortoul, Jan. 30, 1844, in Charton, *Correspondance*, 1: no. 44–2, 452–53. [J'ai donné ma démission de rédacteur de *L'Illustration*, le 1er janvier dernier. D'une part, Lachevardière me harcelait: d'autre part le caractère et la direction du recueil me déplaisent.]
133. For a thorough overview of the development of printing techniques through the nineteenth century, see Verhoogt, *Art in Reproduction*.
134. Michiels, "Avertissement," n. p. [. . . a donné la préférence aux morceaux gravés d'après un dessin original, tracé directement sur bois par l'artiste qui l'a signé.]
135. On the introduction of photography in the illustrated press, see Gervais, "D'après photographie," 6–85; see also his "Imaging the World: *L'Illustration*," an abridged and modified version.
136. On Thibault's photographs, see the entry s.v. at http://www.musee-orsay.fr, and Gervais, "D'après photographie."
137. For statistics, see Gervais, "D'après photographie."
138. On color printing, see Cate and Hitchings, *Color Revolution*, and Twyman, *History of Chromolithography*. See also Iskin, *Poster*.
139. On these small periodicals, see Jones, *La presse satirique illustrée entre 1860 et 1890*.
140. See Gosselin, Marec, and Thomas, *Le dessin de presse à l'époque impressionniste*. For brief biographies, see Solo, *Plus de 5000 dessinateurs de presse*.
141. On Gill, see Valmy-Baysse, *André Gill, l'impertinent*.
142. On twenty-first century controversies over the hand-drawn image, see Spiegelman, "Drawing Blood."
143. Knight, *Passages of a Working Life*, 2: 328.

CHAPTER 3. THE INVENTION OF COMICS

Notes: My preliminary thoughts on early comics were first published as "The Invention of Comics" in *Nineteenth-Century Art Worldwide* (2007); much subsequent writing on the subject has followed, both by myself and others. "Mr" as an abbreviation for *Monsieur* often appears in titles of early comics, so I have left it in my translations wherever it occurs.

1. The major study of the history of comics remains that by Kunzle, *History of the Comic Strip*. For early French comics, the standard article is Mainardi, "Invention of Comics." Recent publications include Kunzle, "Caricature et bande dessinée," and Smolderen, *Origin of Comics*.
2. Rodolphe Töpffer's *Histoire de Mr Jabot* was published and dated 1833, but was not distributed until 1835; see Töpffer [R. T.], "Notice sur l'Histoire de Mr Jabot." Terminology is problematic since such works are not always comic and not always in strip format, especially when published in book format.
3. On the "World Upside-Down" theme, see Tristan, *Le monde à l'envers*.
4. See Mainardi, "Hogarth 'Corrected.'"
5. On English caricature of the period, see Donald, *Age of Caricature*. The standard reference is the British Museum *Catalogue of Political and Personal Satires*.
6. Rudolph Ackermann published "The Schoolmaster's Tour," a series of plates by Thomas Rowlandson with verses by William Combe, in his *Poetical Magazine*, May 1809 to May 1811; he then republished it as *The Tour of Doctor Syntax, in Search of the Picturesque*. For its subsequent history, see Rowlandson and Combe, *Doctor Syntax's Three Tours*, 28.
7. Rowlandson and Combe, *Second Tour of Dr. Syntax, in Search of Consolation* (1820) and *Third Tour of Dr. Syntax, in Search of a Wife* (1821).
8. The 1821 French version of *The Tour of Doctor Syntax, in Search of the Picturesque* was titled *Le Don Quichotte romantique, ou Voyage du docteur Syntaxe à la recherche du pittoresque et du romantique*, and was published with no author identified but "freely translated" (as its subtitle tells us) by Augustin Gandais with lithographs by Malapeau, probably Charles-Louis Malapeau (1795–1878).
9. The hand-colored version of Egan's *Life in London* was published in London by Sherwood, Neely, & Jones in 1821; the cheap version, published the same year in London by John Dicks, had the same images, but was poorly printed in black and white only.
10. M. S., *Diorama anglais, ou, Promenades pittoresques à Londres, renfermant les notes les plus exactes sur les caractères, les moeurs et usages de la nation anglaise*. M. S. is probably Jean-Baptiste-Balthazar Sauvan, who translated it. Pierce Egan's name does not appear on this publication, although the preface notes that the illustrations are by Cruikshank (the signatures of both George and Isaac were removed).

11. Töpffer wrote *Histoire de Mr Jabot* in 1831, had it printed in 1833 (the date that appears on the first edition), but did not allow it to be distributed until 1835, after he had been tenured as professor at the Geneva Academy. See Töpffer [R. T.], "Notice sur l'Histoire de Mr Jabot." The literature on Töpffer is extensive; Groensteen and Peeters reprint the artist's own writings and include a chronology and bibliography; see their *Töpffer: L'invention de la bande dessinée*. For the complete history of Töpffer's albums and English translations, see, *Rodolphe Töpffer: The Complete Comic Strips*, ed. David Kunzle, 627–50, and Kunzle, *Father of the Comic Strip: Rodolphe Töpffer.*

12. A complete list of Töpffer's comic books includes: *Histoire de Mr Jabot* (written in 1831), 1833; *Les amours de Mr Vieux Bois* (written in 1827), 1837; *Mr Crépin*, 1837; *Le docteur Festus* (written in 1829), 1840, *Monsieur Pencil* (written in 1831), 1840; *Histoire de monsieur Cryptogame* (written in 1830), 1845–46; *Histoire d'Albert* (written in 1844), 1845. An eighth, *Mr Trictrac*, was not published until 1937.

13. Kunzle, "Goethe and Caricature," 164–188; see 182–84. Goethe died in 1832, but *Mr. Cryptogame* was essentially complete by 1830, though not published until 1845. By intermediary of Töpffer's friend Frédéric Soret, Goethe also received copies of *Le docteur Festus*, and *Histoire de Mr Jabot*.

14. See Töpffer, "Réflexions à propos d'un programme," 150.

15. Le Men has written extensively about the transformation of albums from children's literature to adult literature; see, in particular, "De l'image au livre" and "Quelques définitions romantiques de l'album." See also Mainardi, "Des débuts de la caricature lithographique."

16. Doré, *Histoire dramatique, pittoresque et caricaturale de la sainte Russie.*

17. The pirated Paris editions were all printed under the Aubert et cie name. On the Maison Aubert, see the 1985 dissertation by James Cuno, "Charles Philipon and La Maison Aubert" and Kerr, *Caricature and French Political Culture.*

18. The catalogue *Aubert et cie, éditeurs. Place de la Bourse, 29, à Paris* (1846) lists the following titles under the heading "Livres et albums comiques": "*M. Jabot.* Aventures d'un fat, joli album oblong./*M. Crépin.* Différents systèmes d'éducation./*M. Vieux-Bois.* Tribulations amoureuses d'un vieux fat./*M. Lajaunisse.* Malheurs d'un beau garçon, par Cham (de N . . .)/*M. Lamélasse.* Histoire d'un épicier, par Cham (de N . . .)/*M. de Vertpré.* Employé retraité, par E. Forest./*M. Jobard.* Mésaventures d'un homme naïf, par Cham (de N . . .)/*Deux vieilles filles à marier.* Tribulations de famille, par Cham (de N . . .)/*Un génie incompris.* Persécutions artistiques, par Cham (de N . . .)/*Le prince Colibri et la fée Caperdulaboula, conte de fée*, par Cham (de N . . .)/*Télémaque, fils d'Ulysse*, par feu M. de Fénelon et Cham (de N . . .)./Ces onze albums, connus sous le nom de *Collection des Jabots*, figurent sur toutes les tables des salons parisiens. Prix de chaque. 6 fr." The catalogue *Étrennes de 1849, Publications d'Aubert et cie, éditeurs. Place de la Bourse, 29, à Paris* (1848), lists "*Les travaux d'Hercule.* Album dans le genre des Jabot-Crépin et autres du même format, par G. Doré. Prix 6 fr. Net 5 fr." The collection of the Bibliothèque nationale is incomplete, lacking Aubert's 1847 catalogue. Cham's *Un génie incompris* lacks a title page, and the album is sometimes known, incorrectly, by the name of its protagonist *Mr Barnabé Gogo*. There are often minor differences between the titles given in Aubert's catalogues and those printed on the publications.

19. Cham, *Histoire du prince Colibri et de la fée Caperdulaboula* (1842); Cham, *Aventures de Télémaque, fils d'Ulysse* (1842); Doré, *Les Travaux d'Hercule* (1847).

20. The advertisement for Bauger et cie in *Le Charivari*, May 30, 1840, suggests that albums are "destinés à être jetés sur les tables de salon, ou à orner la bibliothèque de l'amateur." Bauger published and sold prints, albums, and books, and advertised that customers could even make up their own albums from his stock of caricatures. On these albums, see also Mainardi, "Des débuts de la caricature lithographique."

21. On Cham, see Ribeyre, *Cham, sa vie et son oeuvre*, and Kunzle, "Cham, the 'Popular' Caricaturist."

22. *Mr. Cryptogame* was published in *L'Illustration* as *Histoire de M. Cryptogame, par l'auteur de M. Vieux-Bois, de M. Jabot, de M. Crépin, du docteur Festus, etc.* in eleven weekly installments, from Jan. 25 to Apr. 19, 1845 (some weeks were skipped). Not only were Töpffer's images reversed, but they were published three tiers per page without regard to their original placement, without their boxed formats, with some frames omitted, and with typeset legends substituted for the handwritten ones that Töpffer always

wrote out in pen lithograph.

23. [au rez de chaussée/à l'entresol/au premier/au deuxième/au troisième/au 4ème/au 5e/au 6e/au 7e/8e/9e.]
24. E. [Eugène] Forest, *Histoire de Mr de Vertpré et de sa ménagère aussi.* Eugène Forest was a caricaturist of the period who contributed to the same illustrated journals as Cham; he is sometimes misidentified as Edmund Forest, but there are no prints or publications in the Bibliothèque nationale de France by an artist of that name.
25. [Le jour du diner, Mr Anastase de Vertpré déclare a son ménagère qu'il n'a . . . . ni poisson/ni gibier/ni volailles/ni melons/!!!!]
26. The caption to plate 5 of *Un génie incompris* states: "À 10 ans M. Barnabé avait fait tant de progrès qu'il abordait le cheval au point de faire prendre ses dessins pour des Géricault."
27. The all-white pages appear in Sterne's *Life and Opinions of Tristram Shandy* in volume 7, chapter 38; volume 9, chapters 18 and 19; the all-black pages are in volume 1 between chapters 12 and 13.
28. Cham, *Impressions de voyage de monsieur Boniface, Ex-réfractaire de la 4me du 5me de la 10me.* The subtitle, *Former Insubordinate of the Fourth of the Fifth of the Tenth,* refers to the military unit in which Mr. Boniface served. By this date Cham had already begun to contribute caricatures to *L'Illustration.*
29. "Chapître XL 'Je souffre.' Un Romantique." Many of the captions contain puns, in-jokes, or word play.
30. "Chapître XIX 'O nature! Que tu es belle.'—J.-J. Rousseau."
31. The first thorough study, including an extensive bibliography, is Chadefaux, "Le salon caricatural de 1846"; see also Chabanne, *Les salons caricaturaux,* and Yang, *Les premiers salons caricaturaux au XIXe siècle.*
32. Raimond Pelez published *Première impression du salon de 1843* in *Le Charivari,* Mar. 19, 1843; Bertall published *Le salon de 1843 (Ne pas confondre avec celui de l'artiste-éditeur Challamel, éditeur-artiste). Appendice au livret. Représenté par 37 copies de Bertal* [*sic*]. *Études faites aux portes du Louvre, le 15 mars 1843,* first as *Les Omnibus, 7me livraison* (1843); it was then reissued as a separate publication, *Les Omnibus, pérégrinations burlesques à travers tous chemins* by the *Omnibus* publisher, Ildefonse Rousset, and partially reproduced in *L'Illustration,* May 13, 1843.
33. [*Une imagination poétique. Imagination plus positive. Oeuvre d'un savant coloriste. Pinceau moins fougueux et plus allemande.*]
34. [*Lever de lune sur mer (Fiat lux!). Un effet de brouillard fort bien rendu.*]
35. See Cham, *Soulouque et sa cour*; Cham, *Les aventures de monsieur Beaucoq, ex-rosier de la commune de Nanterre.*
36. In addition to my "Invention of Comics," Doré's comic books have been discussed by Renonciat, *La vie et l'oeuvre de Gustave Doré*; *9e Art: Les cahiers de la bande dessinée*; Kunzle, *History of the Comic Strip,* 2: 123–34; and Kunzle, "Caricature et bande dessinée," in Kaenel, ed., *Doré: L'imaginaire au pouvoir.*
37. [Cependant, Mr Cryptogame fait neuf fois le tour du pont sans trouver d'issue.]
38. Doré, *Trois artistes incompris et mécontents,* and Doré, *Dés-agréments d'un voyage d'agrément.*
39. The cover text reads: "*Histoire dramatique, pittoresque et caricaturale de la sainte Russie, d'après les chroniqueurs Nikon, Nestor, Sylvestre, Karamsin, Ségur, etc., etc., etc./Commentée et illustrée par 500 magnifiques gravures par Gustave Doré/Gravées sur bois par toute la nouvelle école sous la direction générale de Sotain/Graveur de l'Histoire de Russie, de batailles, de portraits, de paysages, de genre, de fleurs, d'animaux, de crustacés et de plantes rares.*" [*Dramatic and Picturesque History of Holy Russia in Caricature, from the Chronicles of Nikon, Nestor, Sylvestre, Karamsin, Ségur, etc., etc., etc./Annotated and Illustrated by 500 Magnificent Engravings by Gustave Doré/Engraved on Wood by the Entire New School under the General Direction of Sotain/Engraver of the History of Russia, of Battles, of Portraits, of Landscapes, of Genre, of Flowers, of Animals, of Crustaceans, and of Rare Plants.*]
40. The illustrated press was full of commentary about the war and several collections of caricatures were republished as albums after having appeared in journals such as *Le Charavari.* To cite only two: Delord, Carraguel, and Huart, *Messieurs les Cosaques,* with one hundred vignettes by Cham; and Cham, Daumier, and Vernier, *Les Cosaques pour rire,* with forty caricatures by the artists.
41. See À Beckett and Leech, *Comic History of England.* John Leech did two hundred illustrations for the edition, but they are dispersed throughout the two volumes, with only one or two in each chapter.

42. Doré, *Histoire de la Sainte Russie*, pl. 77. [. . . donne l'élan à l'industrie russe en encourageant d'abord la branche des forges qu'il a lui-même approfondie.]
43. The all-black vignettes appear on pl. 2 and pl. 31; pl. 7 contains five all-white vignettes.
44. Borrowings include Grandville's musical motif on pl. 179, his figurative pen and pencil on pl. 35.
45. Richard Doyle's work was often plagiarized by French illustrated journals; throughout 1849, Doré published dozens of illustrations in Doyle's style in *Le Journal pour rire.*
46. [Tiré de la collection des estampes populaires de Russie (facsimile).]
47. Some of Doré's other regional names include Personalitaslaw, Bétissgorod, Furiroslaw, Honteslaw. Although Töpffer had labeled England as Rondeterre (Roundland) instead of Angleterre (Angleland) on his map in *Le docteur Festus*, he never really exploited the comic possibilities of regional names. A recent example of this trope is "NewYorkistan" by Maira Kalman and Rick Meyerowitz, which appeared as the cover of the *New Yorker* on Dec. 10, 2001.
48. Töpffer's *Histoire de monsieur Cryptogame* (1845), redrawn by Cham for *L'Illustration*, was published as "Histoire de M. Cryptogame" in eleven weekly installments beginning on Jan. 25, 1845. The introduction and ten "chapters" of Nadar's *Vie publique et privée de mossieu Réac* began appearing in *Revue comique à l'usage des gens sérieux* on Mar. 3, 1849, and continued weekly, with interruptions.

CHAPTER 4. PATHS FORGOTTEN, CALLS UNHEARD

1. Kundera, "An Introduction to a Variation."
2. For Töpffer's discussion of "littérature en estampes," see his "Réflexions à propos d'un programme," 150.
3. An "Advertisement" signed "The Author" at the beginning of each edition of *Dr. Syntax* explained their working method. See Rowlandson and Combe, *Tour of Doctor Syntax.*
4. Gigoux, *Causeries sur les artistes*, 30–31. The edition of Gil Blas (orig. 1715–1735) with Gigoux's illustrations was published by Paulin in 1835 as Alain-René Lesage, *L'histoire de Gil Blas de Santillane* and reissued in 1838 by Dubochet. When the novel was originally published between 1715 and 1735, Lesage's title was *Les avantures de Gil Blas de Santillane*; subsequent editions modernized the spelling of *avantures* to *aventures.* [Un jour, on vint me demander cent vignettes pour une nouvelle édition de ce merveilleux livre. J'avoue que j'eux un moment d'effroi, presque. Il me semblait que je n'y trouverais jamais cent sujets de compositions. Mais, pourtant je les fis. Quelques jours après, les éditeurs m'en demandèrent trois cents de plus. Alors, moi de recommencer à lire et à croquer au fur et à mesure mes illustrations. La semaine suivante, les éditeurs s'apercevant de l'attrait que ces vignettes donnaient aux livraisons, m'en redemandèrent encore deux cents nouvelles. Bref, j'en fis six cents, et je crois que j'aurais pu continuer indéfiniment.]
5. On the vignette, see Rosen and Zerner, "The Romantic Vignette and Thomas Bewick," in their *Romanticism and Realism*, 71–97.
6. Béraldi, "Gigoux (Jean), peintre," *Les graveurs du XIXe siècle*, 7: 109–28; see 112–13. [un des cinq ou six principaux ouvrages à figures du XIXe siècle.]
7. Lagenevais, "La littérature illustrée."
8. Lagenevais, "La littérature illustrée," 653. [L'illustration est un symptôme de décadence littéraire.]
9. Lagenevais, "La littérature illustrée," 653. [Chaque art a son genre de beauté particulière.]
10. Lagenevais, "La littérature illustrée," 653, 651. [Les gravures d'ailleurs ne nuiraient pas au texte, ne rompraient pas l'unité d'impression nécessaire à toute lecture . . . (vignettes) ne font que jeter le désordre dans les pages, elles dérangent cette harmonie régulière des lignes, à laquelle l'oeil est habitué . . . le dessinateur se substitue ainsi au poète.]
11. Lagenevais, "La littérature illustrée," 648, 655, 664.
12. Lagenevais, "La littérature illustrée," 648.
13. Lagenevais, "La littérature illustrée," 651–52. Béraldi, "Johannot (Tony)," *Les graveurs du XIXe siècle*, 7: 245–77; see 271. [l'un des plus remarquables que aient jamais été publiés.]
14. Lagenevais, "La littérature illustrée," 653, 655.
15. Lagenevais, "La littérature illustrée," 664. [Comme la gravure sur bois et celle à la mécanique, comme toutes les innovations qui tendent à séduire l'acheteur par le bon marché, les *magasins pittoresques* sont nés en Angleterre, la patrie naturelle de toutes les idées commerciales.]

16. The first issue of the *Illustrated London News* was published on May 14, 1842, the first issue of *L'Illustration* on Mar. 4, 1843. Lagenevais's article appeared in the Feb. 15, 1843, issue of *Revue des deux mondes.*
17. Lagenevais, "La littérature illustrée," 648, 655.
18. On panoramic literature, see Benjamin, "II. Daguerre, or the Panoramas," *Arcades Project*, 5–6; Benjamin, "The Work of Art in the Age of Mechanical Reproduction," in Schwartz and Przyblyski, *Nineteenth-Century Visual Culture Reader*, 63–70. Benjamin, however, traces the development of this literature from lithography directly to photography, apparently unaware that it was made possible only through wood engraving. Cohen repeats Benjamin's error in her "Panoramic Literature and the Invention of Everyday Genres," where she sees it as leading to cinema. Benjamin's error is corrected in Von Lintel, "Wood Engravings."
19. There were 101 contributors to *Paris ou le livre des cent-et-un*. Judith Wechsler provides an extensive discussion of this imagery in *A Human Comedy*; see especially chapter 1, "Parisian Panorama: Codes and Classifications," 20–41. The precursor of this genre was Louis-Sebastien Mercier's 12-volume *Tableau de Paris* (1781–88), which, however, had no illustrations.
20. *Les Français peints par eux-mêmes*, 10 vols. (Paris: L. Curmer, 1840–42). The chapters were published in installments beginning in 1839. See Le Men and Abélès, *Les Français peints par eux-mêmes: Panorama social du XIXe siècle* and Cohen, "Panoramic Literature and the Invention of Everyday Genres."
21. See Dufour, "Le défenseur officieux en justice de paix," in *Les Français peints par eux-mêmes*, 2: 307–10.
22. Grandville, *Scènes de la vie privée et publique des animaux. Études de moeurs contemporains.* Lagenevais called it *Les animaux peints par eux-mêmes.*
23. Meadows, *Heads of the People* was first issued in twelve monthly installments, Nov. 1838 to Oct. 1839, then published as a book.
24. Douglas Jerrold, "The Pew-Opener," *Heads of the People*, 224–32.
25. Kenny Meadows's *Heads of the People* was plagiarized in France as *Les Anglais peints par eux-mêmes* with only the translator, Émile de La Bédollière, credited. On physiologies, see Lhéritier, *Les physiologies de 1826 à 1894*; Lhéritier, *Les physiologies 1840–1845*. There are additional articles on physiologies in *Études de Presse*, n.s. 9, no. 17 (1957); and more recently, Davis, "Entre la physiognomonie et les physiologies." For a literary approach, see Sieburth, "Same Difference."
26. Barbier, "Les innovations technologiques," 721–29.
27. Nodier, Taylor, and Cailleux, *Voyages pittoresques*; for additional such publications, see Abbey, *Scenery of Great Britain and Ireland*; Abbey, *Travel in Aquatint and Lithography*; and Adhémar, *La France romantique.*
28. See Gausseron, *Les keepsakes* and Adhémar and Seguin, *Le livre romantique*, 54–57.
29. Lagenevais, "La litterátуre illustrée," 664. [ce don d'ubiquité de M. de Balzac.]
30. On Balzac and illustration, see Kerr, *Caricature and French Political Culture*, 20–23; Chollet, *Balzac journaliste*; Contensou, *Balzac & Philipon associés.* For illustrations, see the 1842–48 Furne edition of Balzac, *Oeuvres complètes*, and the 1846 Chlendowski edition of *Petites misères de la vie conjugale.*
31. On Dickens, see Cohen, *Charles Dickens and His Original Illustrators*; and Ray, *The Illustrator and the Book in England*, which is, however, organized by artist, not author.
32. On Thackeray, see Buchanan-Brown, *Illustrations of Thackeray*; and Ray, *The Illlustrator and the Book in England*, 74–77.
33. Lagenevais, "La litteráture illustrée," 651, 655. [la gravure . . . n'a plus voulu traduire le texte, mais le dicter.]
34. Gautier, "Tony Johannot," 228. [le roi de l'illustration.] The standard biography of Johannot is Marie, *Alfred et Tony Johannot, peintres, graveurs et vignettistes.*
35. On the last pages of the novel there is a note "Au lecteur et à Tony Johannot"; the first sentence states "L'idée d'écrire ce livre nous a été suggérée par Tony Johannot" [To the reader and to Tony Johannot. The idea of writing this book was suggested to us by Tony Johannot]. Johannot, Musset, and Stahl, *Voyage où il vous plaira*, 169–70.
36. The four-page prospectus and "Avis" are bound into the copy of *Voyage où il vous plaira* held at the Bibliothèque nationale de France, Paris; see prospectus, 4. [*Voyage où il vous plaira* devant être essentiellement et par sa nature même un livre de fantaisie, on comprendra que nous laissions toute liberté aux auteurs. Leur projet

étant d'écrire et de dessiner *alternativement*, suivant qu'il y aura lieu d'écrire ou de dessiner, et que la plume ou le crayon devront être plus propres à rendre leur pensée, et le nombre des vignettes devant être considérable, relativement à l'étendue du texte, il arrivera que parmi nos livraisons, *les unes* se composeront de texte et de vignettes tout à la fois, *les autres* de vignettes seulement.]

37. Johannot, Musset, and Stahl, *Voyage où il vous plaira, livre écrit à la plume et au crayon avec vignettes, légendes, épisodes, commentaires, incidents, notes et poésie.*
38. By 1842, Töpffer had published *Histoire de Mr Jabot* (1833); *Les amours de Mr Vieux Bois* (1837); *Mr Crépin* (1837); *Le docteur Festus* (1840), and *Monsieur Pencil* (1840).
39. Johannot, Musset, and Stahl, *Voyage où il vous plaira*, 32. [Nos chevaux partirent comme le vent.]
40. Johannot, Musset, and Stahl, *Voyage où il vous plaira*, 32. [Pour sortir de la ville, il nous fallut passer devant la maison de ma fiancée! La fenêtre de sa chambre était entr'ouverte, et je vis la douce fille assise devant un clavecin que lui avait légué son oncle l'organiste. Elle chantait:]
41. Johannot, Musset, and Stahl, *Voyage où il vous plaira*, 32–33. [La fenêtre de sa chambre était entr'ouverte, et elle chantait.]
42. Johannot, Musset, and Stahl, *Voyage où il vous plaira*, 37. [Elle chantait . . . et moi je partais!]
43. Johannot, Musset, and Stahl, *Voyage où il vous plaira*, 14. Although there were numerous editions of the voyages of James Cook in both English and French, the edition that inspired Johannot was probably the much-reprinted Henri Lebrun, *Voyages et aventures du capitaine Cook*. There is no publication with the title given by Johannot, *Voyages autour du monde, par le capitaine Cook*.
44. The two full-page illustrations are unnumbered, placed between pages 14 and 15. [Je ne vois plus que navires qui se croisent sur l'immensité des mers, que chevaux et voitures qui roulent pesamment sur tous les points du globe. . . . Que courageux et hardis piétons qui traversent les sables brulants des déserts, et qui gravissent les pics escarpés des Alpes et des Cordillières!]
45. Until the mid-nineteenth century, books were often sold unbound, with the purchaser choosing the type of binding and its quality. As a result, the table of contents always included a list of illustrations with an indication of where each was to be placed, although binders' errors have resulted in different locations. In French publications this table was situated at the end of the book. See "Placement des gravures," in Johannot, Musset, and Stahl, *Voyage où il vous plaira*, 171–72.
46. Johannot, Musset, and Stahl, *Voyage où il vous plaira*, 20–21. [Brûlez, brûlez, m'écriai-je, vous qui m'avez perdu. . . . Je retombai sans force dans mon fauteuil.]
47. Sterne's *Life and Opinions of Tristram Shandy, Gentleman* was published in London in nine volumes from 1759 to 1767. Although there were numerous contemporary editions, the standard text is reprinted from an edition supervised by Sterne himself. See vols. 1 and 2 of Sterne, *Florida Edition*. A page of sheet music for "Lilliburlero" is often inserted in *Tristram Shandy*, at one of the points where Uncle Toby whistles the tune, but this was a posthumous addition and does not appear in any of the editions that Sterne supervised nor in the French editions published later.
48. *Tristram Shandy* was soon translated into several European languages: German in 1765, French in 1769, Dutch in 1777, Russian in 1790, Italian in 1805. On Sterne translations and influence in France, see Howes, *Sterne: The Critical Heritage*, 20–22, 390–95; he reprints many of the contemporary French reviews and critiques. See also Asfour, *Laurence Sterne in France*; she discusses only eighteenth-century translations, not Sterne's nineteenth-century followers.
49. Nodier, *Histoire du roi de Bohême*; the epigraph on the title page reads: "Il y avait une fois un roi de Bohême qui avait sept châteaux. Trimm." [Once upon a time there was a king of Bohemia who had seven castles. Trimm.] Corporal Trim (not Trimm) is a character in Sterne's *Tristram Shandy*.
50. For a brief appreciation of the importance of Nodier's *Histoire du roi de Bohême*, see Ray, *Art of the French Illustrated Book*, 256–58.
51. The incident of the threatened duel is recounted by Renonciat, *La vie et l'oeuvre de J. J. Grandville*, 231. Both works were published on subscription: the first installment of *Voyage où il vous plaira* was announced on Dec. 10, 1842, the last on Dec. 23, 1843; the first installment of *Un autre monde* was announced on Feb. 18, 1843, the last on Nov. 11, 1843. Dates are taken from the *Bibliographie de la France* for 1842 and 1843, an official weekly bulletin. Publication as a book usually followed the last *livraison* (Grandville) or was simultaneous with it (Johannot).

52. See the interview with Grandville reported by Gaberel, *Essai sur le caractère artistique et littéraire des oeuvres de R. Toepffer*, 4; the passage is reprinted in Kaenel, *Le métier d'illustrateur*, 284. [La plume est rebelle sous mes doigts pour former des phrases.]
53. Grandville's contract with Fournier, dated Dec. 19, 1842, stated, "Le texte sera rédigé d'après vos notes par un littérateur dont le travail restera à ma charge" [the text will be written according to your notes by an author whose work will be paid for by me]; see Renonciat, *La vie et l'oeuvre de J. J. Grandville*, 230. Renonciat's bibliography lists the contract as being in an unnamed private collection.
54. [Un autre monde: Transformations, visions, incarnations, ascensions, locomotions, explorations, pérégrinations, excursions, stations, cosmogonies, fantasmagories, rêveries, folâtreries, facéties, lubies, métamorphoses, zoomorphoses, lithomorphoses, métempsychoses, apothéoses, et autres choses.]
55. Gautier, "Grandville," 233. [son vrai chef-d'oeuvre.]
56. Grandville, *Un autre monde*, 23. [à peine au sortie de nourrice, vingt-deux mois au plus je comptais.]
57. Grandville, *Un autre monde*, 23–24. [Dans le feu d'artifice en *ré*, au moment où la fugue se termine *smorzando* par une mélodie douce et rêveuse, un ophicléide, trop chargé d'harmonie, a éclaté subitement comme une bombe lançant des noires, des blanches, des *grupetti* de notes aiguës, de croches, de doubles croches; des nuages de fumée musicale et des flammes de mélodie se sont répandus dans l'atmosphère. Plusieurs dilettanti ont eu les oreilles déchirées, d'autres ont été blessés par les éclats de la clé de *fa* et de la clé de *sol*.]
58. Mespoulet makes a convincing argument for this in her *Creators of Wonderland*, where she points out that Grandville's work was widely known in England in the decades before Tenniel drew his playing-card images for *Alice*. See also Ray, *The Illustrator and the Book in England*, 116. [Chapter 6: A vol et à vue d'oiseau. Chapter 20: Locomotions aériennes. Chapter 24: Les grands et les petits. Chapter 32: Les métamorphoses du sommeil.]
59. Grandville, *Un autre monde*, 83. [Le Louvre des marionnettes, L'ange de la peinture implorant la miséricorde divine pour le jury.]
60. Grandville, *Un autre monde*, 83–85. The title of Grandville's salon catalogue is identical in format to that of the Paris Salon, with its title only slightly altered at the end to read: *Explanation of the Works of Painting, Sculpture, Architecture, Engraving and Lithography Exhibited at the Royal Museum This Year, Including Previous Years and Future Years*. [*Explication des ouvrages de peinture, sculpture, architecture, gravure et lithographie exposés au Musée royal pour l'année présente, y compris les années précédents et celles qui suivant.*] Among the paintings listed in the catalogue are: 100—LeBlanc (Anastasius). 200—Vertchoux (Gaspard), *Une églogue de Virgile*. 410—Duflot (Neptune), *Le passage de la Mer Rouge*. 130—Shwplklmcssth (Conrad), de Munich, *Portrait de Mme P. de L., de son chien et de ses diamants*. 101—Dumortier (Nicolas), *Vagues irritées*. 9999—Baudrichon (Numa), *Une bordure de deux mille francs d'après le procédé Ruolz et Elkington*.
61. Grandville, *Un autre monde*, 85. [Rien n'est comparable à la bataille de notre grand peintre Jérôme Tulipier. Quelle mêlée! quel choc! quel tourbillon! quel ouragan! quelle trombe! Têtes furieuses, bras menaçans, sabres et épées, tout cela vit, sort de la toile et combat.]
62. Grandville, *Un autre monde*, 85. [Ce matin, en ouvrant les fenêtres pour donner de l'air à une dame qui venait de s'évanouir suffoquée par la chaleur, on a vu les oiseaux se précipiter sur le paysage de notre célèbre Thomas Gorju représentant un verger de Normandie.]
63. Grandville, *Un autre monde*, 86. [L'éclat d'un *Soleil levant* a ébloui tous les regards, jusqu'à ceux d'une taupe qui était parvenue à s'introduire dans le salon.]
64. Bertall's, *Le salon de 1843* was originally published in *Les Omnibus, 7me livraison* (1843), reissued as a separate publication as *Les Omnibus, pérégrinations burlesques à travers tous chemins* by its publisher, Ildefonse Rousset, and partially reproduced in *L'Illustration*, May 13, 1843, 173.
65. The literal translation of Grandville's rebus from image to words is "A croix moi A mi-lecteur, neuf fées Pâques homme sept," or (in good French), "Ah! Crois-moi, ami lecteur, ne fais pas comme cet [homme]." There is very little literature on rebuses, and what little exists is usually written from a literary, not art-historical, point of view. For a general history, see Delepierre, *Essai historique et bibliographique sur les rébus*.

CHAPTER 5. THE CURIOUS HISTORY OF POPULAR IMAGERY IN FRANCE

A note on dates: it is extremely difficult to date popular prints, as they were copied and reissued constantly. The dates given with illustrations are those of the *dépot légal* (if known), even though the required deposit of one copy of all prints could occur long after publication, if at all; dates before 1890 should be understood only as *terminus ante quem.*

1. The catalogue for the 1974 Hayward Gallery, London, exhibition *French Popular Imagery* has an excellent introduction to the subject by Georges-Henri Rivière; the exhibition was first shown in 1972 at the Musée national des arts et traditions populaires, Paris, as *Cinq siècles d'imagerie française.* The basic research tool for French popular prints is Garnier, *L'imagerie populaire française*; see the introduction, 1: 9–24. See also O'Connell, *Popular Print in England, 1550–1850*, and Bertarelli, *Le stampe popolari italiane.*
2. For a brief history of the Maison Pellerin, see Staub, *Histoire de l'imagerie*, 10. In 2003 a new Musée de l'image was inaugurated on the Pellerin grounds housing the departmental collection of popular prints along with changing and permanent exhibitions.
3. See Huin, "L'imagerie à Épinal," in Martin, ed. *Images d'Épinal*, 33–67; Bollème, "Une littérature perdue," in her *La bibliothèque bleue*, 7–26. See also Mandrou, *De la culture populaire aux XVIIe et XVIIIe siècles*; Muchembled, *Popular Culture and Elite Culture in France.*
4. There is some disagreement about the date when the first Pellerin arrived in Épinal, with sources giving either 1735 or 1740. Besides the general studies of French popular imagery there are several studies devoted specifically to images d'Épinal: Perrout, *Les images d'Épinal*; Mistler, Blaudez, and Jacquemin, *Épinal et l'imagerie populaire*; Musée départemental d'art ancien et contemporain, Épinal, *L'imagerie populaire française au musée d'Épinal.* Two well-illustrated studies are Bouvet, *Le grand livre des images d'Épinal*, and George, *La belle histoire des images d'Épinal.*
5. The principal source on Pellerin is Dumont, *La vie et l'oeuvre de Jean-Charles Pellerin.*
6. Dumont, *La vie et l'oeuvre de Jean-Charles Pellerin*, 41.
7. For Pellerin's sales figures, see Martin, *Images d'Épinal*, 54; Dumont, *La vie et l'oeuvre de Jean-Charles Pellerin*, 43. Dumont, *Maîtres graveurs populaires*, 54; Duchartre and Saulnier, *L'imagerie populaire*, 187–88.
8. See Musée de l'image, Épinal, *Ni tout à fait la même, ni tout à fait une autre*, an exhibition focused on the practice of popular printmakers' copies of high-art paintings, engravings, or even of each other.
9. Pellerin's 1814 catalogue is reproduced in Martin, *Images d'Épinal*, 50–51. For images and dating of *Degrés des âges*, see Garnier, *Imagerie populaire française*, 1: nos. 849, 912; 2: nos. 1170–1172.
10. Laslett, *The World We Have Lost.*
11. Dumont, *La vie et l'oeuvre de Jean-Charles Pellerin*, 41; see also Jacquemin, "Les techniques de l'imagerie populaire," in Mistler, Blaudez, and Jacquemin, *Épinal et l'imagerie populaire*, 139–74.
12. See Garnier, *L'imagerie populaire française*, 2: 9–12.
13. See Goldstein, *Censorship of Political Caricature in Nineteenth-Century France*, and his *Political Censorship of the Arts and the Press in Nineteenth-Century Europe.*
14. For the circumstances of *Le cornard volontaire, ou le mari commode* (now lost), see Dumont, *La vie et l'oeuvre de Jean-Charles Pellerin*, 46–47; Mistler, Blaudez, and Jacquemin, *Épinal et l'imagerie populaire*, 91.
15. "Une saisie d'images chez Pellerin en 1815" on the website of the Musée de l'image, Épinal: http://webmuseo.com/ws/musee-de-l-image/app/collection/expo/15. See also Dumont, *La vie et l'oeuvre de Jean-Charles Pellerin*, 57–62; Musée départemental d'art ancien et contemporain, Épinal, *L'imagerie populaire française au musée d'Épinal*, 37.
16. For an extensive discussion of Napoleonic imagery, see Day-Hickman, *Napoleonic Art.*
17. For illustrations, see Garnier, *L'imagerie populaire française*, 2: 232–53, nos. 918–1021. [Le chat botté. La belle au bois dormant. La petite cendrillon. Le petite chaperon rouge.]
18. Duchartre and Saulnier, *L'imagerie populaire*, 188–89; Dumont, *Maîtres graveurs populaires*, 64.
19. Dumont, *Maîtres graveurs populaires*, 29; Duchartre and Saulnier, *Imagerie populaire*, 189.
20. Hayward Gallery, *French Popular Imagery*, 130.
21. The reading of French popular prints that identified, even romanticized, them as woodblock prints created by semiskilled artisans for a rural audience was established in the first major study, by Champfleury (Jules Husson), *Histoire de l'imagerie populaire* (1869); it was further entrenched by the second major study, by Pierre-Louis

Duchartre and René Saulnier, *L'imagerie populaire: Les images de toutes les provinces françaises du XVe siècle au second empire* (1925). All the major scholars of French popular imagery (Champfleury, Duchartre and Saulnier, Garnier) conclude their studies with the adoption of lithography and contemporary imagery in the mid-nineteenth century. Duchartre and Saulnier explicitly identified the later decades as a period of degeneration; see Duchartre and Saulnier, *L'imagerie populaire*, 188.

22. Jacquemin, "Les techniques de l'imagerie populaire," 139–74.
23. Bouvet, *Le grand livre des images d'Épinal*, 13.
24. Pevsner lists twenty-nine government-sponsored art schools in France with their date of establishment; see his *Academies of Art*, 142. The earliest regional art school was established in Rouen in 1741, the last in Toulon and Orléans in 1786; the École des beaux-arts in Paris was founded in 1648 by Mazarin as a national institution and has remained the most prestigious of the schools.
25. On Georgin, see Descaves, *L'humble Georgin.*
26. On Napoleonic prints, see also Perrout, *Les images d'Épinal*, 108; Staub, *Histoire de l'imagerie*, 5–6; and Day-Hickman, *Napoleonic Art*. Many of Georgin's prints are still published by l'Imagerie d'Épinal.
27. Lebovics, *True France.*
28. "Les images et la politique," *Le Figaro, Supplément*, Mar. 30, 1889. The series was never completed, including only "La république devant les éléctions," "La monarchie et le comte de Paris," and "L'empire et le prince Victor." Prince Victor and the Comte de Paris were both pretenders to the throne, Bonapartist and Orleanist, respectively; in the fragile Third Republic, both were considered enemies. [Rien n'échappant plus à la politique, l'image populaire jadis confinée dans les contes de fées, la naïve et vieille image d'Épinal, est entrée à son tour dans la carrière. Elle a devenue un instrument de propagande, si puissant entre les mains des partis que le gouvernement a fait saisir impitoyablement partout où il les a rencontrées les images du comte de Paris et du prince Victor, comme attentatoires à la sécurité de l'État.]
29. Glucq's slogan, "publicité industrielle & propaganda politique par l'image populaire," appeared on all his publications. On Glucq and Pellerin's foray into advertising, see the exhibition catalogue from Imagerie d'Épinal, *L'imagerie publicitaire*; Marle, "Glucq et l'imagerie Pellerin d'Epinal"; and Staub, *Histoire de l'imagerie*, 10.
30. On the Pétain images, see Golan, *Modernity and Nostalgia*, 155; Lebovics, *True France*, 173. G. Ambroselli's *La vie du maréchal* (1944) was a children's coloring book with the plates drawn in the early nineteenth-century Épinal style.
31. For a thorough discussion of reproductive printmaking, see Verhoogt, *Art in Reproduction*; see also Bann, *Distinguished Images*, and Von Lintel, "Wood Engravings."
32. For an extensive discussion of this practice in popular art, see Musée de l'image, Épinal, *Ni tout à fait la même*; for Charles Thévenin's *Passage du grand Saint-Bernard par l'armée française le 20 mai 1800*, see 142–43.
33. See, for example, O'Brien, *After the Revolution*, 237–38. On history painting, see Haskell, *History and Its Images.* Literary scholars have followed a parallel course. Samuels sees the writing of history increasingly resembling the spectacles of popular culture; see his *Spectacular Past.*
34. See Delécluze, *Louis David*, 244; Delécluze had been a student of David's but gave up painting for art criticism.
35. Traditionalist critics such as Delécluze constantly bemoaned the public's lack of interest in history painting; see Delécluze, *Louis David*, 244–45. For an extensive discussion on the fate of history painting, see Mainardi, *End of the Salon*, 9–35.
36. See Rey, *Le Robert, dictionnaire historique*, s.v. "vignette."
37. When Géricault's friends who were present denied his lament, citing *The Raft of the Medusa*, Géricault said, "Bah! une vignette!" See Lachèvre, *Détails intimes sur Géricault*, in *Géricault raconté*, 267. I thank Philippe Bordes for calling my attention to Géricault's use of the word *vignette.* On Géricault, see Bordes, "'L'écurie dont je ne sortirai que cousu d'or,'" and Simon, "Géricault und die Faits Divers."
38. *Der Schweitzerische Robinson* was translated into French by Isabelle de Montolieu as *Le Robinson suisse, ou Journal d'un père de famille naufragé avec ses enfants*. The French translation was published first in Lausanne in 1813, then in Paris by Artus Bertrand in 1814 with numerous reprint editions.
39. For an excellent survey of the theory and practice of genre painting in France, see Vottero, *La peinture de genre en France, après 1850*. The popularity of genre painting relative to history painting is discussed throughout Mainardi, *Art*

and Politics of the Second Empire and Mainardi, *End of the Salon.*

40. On François-Charles Pinot, see Perrout, *Les images d'Épinal*, 120–33; Hayward Gallery, *French Popular Imagery*, 130; Staub, *Histoire de l'imagerie*, 10; Duchartre and Saulnier, *L'imagerie populaire*, 189. In 1860 Pinot left Pellerin to establish Pinot & Sagaire, his own printing firm that, since it also was located in Épinal, competed with Pellerin for the distinction of producing images d'Épinal; he was bought out by Pellerin in 1888. There has yet to be a monograph on Pinot.
41. Duchartre and Saulnier, *L'imagerie populaire*, 188. Perrout, *Les images d'Épinal*, 131–33; Staub, *Histoire de l'imagerie*, 10. See n. 21 above.
42. See Martin, "Émergence d'un nouveau marché," in his *Images d'Épinal*, 137–42.
43. On the emerging market for children, see Martin, "Le petit musée des enfants," in Martin, *Images d'Épinal*, 146–200.
44. See Carlin, "Masters of American Comics," 27; Groensteen, "Le patrimoine européen du neuvième art," 13.
45. Rodolphe Töpffer's *Histoire de M. Cryptogame* (1845) was redrawn by Cham for *L'Illustration*, where it was published in eleven weekly installments beginning on Jan. 25, 1845. The eleven installments of Nadar's *Vie publique et privée de Mossieu Réac* began appearing in *Revue comique à l'usage des gens sérieux* on Mar. 3, 1849.
46. There is little scholarship on Pellerin's later production; Martin, *Images d'Épinal* is one of the few publications even to consider it.
47. On Caran d'Ache, see Groensteen, *Les années Caran d'Ache.*
48. There is virtually no scholarship on Pellerin's Humoristic Publishing Company project, though it is mentioned briefly in Martin, *Images d'Épinal*, 157. The complete series is in the Special and Area Studies Collection of the University of Florida; for the list of sixty titles in the Humoristic series, see http://ufdc.ufl.edu/UF00078633/00001/citation.
49. See the Musée de l'image, Épinal, 1997 exhibition catalogue, *Série encyclopédique Glucq des leçons de choses illustrées*; Marle, "Glucq et l'imagerie Pellerin d'Épinal"; and the 3 volume reprint of the series by Imagerie d'Épinal, *Série encyclopédique Glucq des leçons de choses illustrées.*
50. [Medaille d'or: Marseilles 1883. Ouvrage adopté par la ville de Paris comme recompenses dans les écoles.]
51. See *Le chocolat des gourmets, Trébucien, cours de Vincennes, Paris*, Pub. Glucq, Paris, and Pellerin, Épinal, 1888, http://gallica.bnf.fr.
52. [—Pas vrai, M'sieu l'Agent qu'il est en or?—Pas la peine, mon brave, puisqu'on l'astique au BRILLANT DE LUXE qui change le Cuivre en Or!— Le BRILLANT DE LUXE rend les cuivres resplendissants.]
53. For a thorough study of the poster movement, see Iskin, *Poster.*
54. The classic article on the relationship between these images and Modernism is Schapiro, "Courbet and Popular Imagery." For the historiography of popular prints in France, see Mainardi, "Popular Prints for Children."
55. Champfleury, *Histoire de l'imagerie populaire*; the chapters were first published in a variety of journals, beginning with *Le National* in 1850.
56. Champfleury, *Histoire de l'imagerie populaire*, xi–xii. [moins barbares que l'art médiocre de nos expositions, où une habileté de main universelle fait que deux mille tableaux semblent sortis d'un même moule. Telle maladresse artistique est plus rapprochés de l'oeuvre des hommes de génie que ces compositions entre-deux, produits des écoles et des fausses traditions.]
57. Champfleury, *Grandes figures d'hier et d'aujourd'hui*, 244. [ces naïves images sur bois, taillées par un couteau maladroit. . . . L'art savant trouve le même accent que l'art naïf.]
58. Zola, *Édouard Manet, Étude biographique et critique*; Zola, *Mes haines. Causeries littéraires et artistiques*, 344–45. [On a dit, par moquerie, que les toiles d'Édouard Manet rappelaient les gravures d'Épinal et il y a beaucoup de vrai dans cette moquerie qui est un éloge.]
59. See Silver, *Esprit de corps*, 124–25.

CONCLUSION

1. Philipon, prospectus for *Le Charivari.* [un panorama complet, où se reproduisent incessamment, par la crayon et par la plume tous les aspects de ce monde kaléidoscopique où nous vivons.]
2. Art Spiegelman's "Drawing Blood" was published in *Harper's Magazine* in 2006, well before the Jan. 7, 2015, massacre at the Parisian satirical magazine *Charlie Hebdo*, but it is still the most insightful commentary on political cartooning.
3. For a recent insightful study of the relationship between film and comics, see Smolderen, *Origins of Comics.*

4. See *Voyage en Angleterre par Eug. Lami et H. Monnier* (Paris: Firmin Didot and Lami-Denozan; London: Colnaghi and Charles Tilt, 1829).
5. Grandville, *Les fleurs animées* (Paris: Gabriel de Gonet, 1847); *The Flowers Personified*, 2 vols., trans. N. Cleaveland (New York: R. Martin, 1847–49); *La vida de las flores* (Barcelona: Celestina Verdaguer, 1878). Grandville, *Scènes de la vie privée et publique des animaux. Études de moeurs contemporains*, ed. P.-J. Stahl, 2 vols. (Paris: J. Hetzel et Paulin, 1842); *Das neunzehnte Jahrhundert des Thierreichs.* 2 vols. (Leipzig: Friedrich Volckmar [1845]); *The Public and Private Life of Animals*, trans. J. Thomson (London: Sampson Low; Philadelphia: J. B. Lippincott, 1877). Grandville illustrations for Daniel Defoe, *Les aventures de Robinson Crusoe* (Paris: H. Fournier aîné, 1840); republished as Daniel Defoe, *The Life and Adventures of Robinson Crusoe* (London: Routledge, 1853); and *The Pictorial Robinson Crusoe* (New York: D. Appleton, 1841). Grandville, illustrations for Jonathan Swift, *Voyages de Gulliver dans les contrées lointaines*, 2 vols. (Paris: Furne, H. Fournier, 1838); republished as Jonathan Swift, *Reisen in unbekannte Länder* (Stuttgart: Adolph Krabbe, 1839); and Jonathan Swift, *Travel into Several Remote Nations of the World… by Lemuel Gulliver* (London: Hayward, 1840); and Johathan Swift, *Reisen in unbekannte Länder* (Stuttgart: Adolph Krabbe, 1839).
6. Knight, *Passages of a Working Life*, 2: 223.
7. Charton, in *MP*, "Des moyens d'instruction. Les livres et les images"; see 98. [*Sans les dessins, il est impossible d'arriver à l'éducation complète des hommes, grands et petits.*] Emphasis in the original.
8. Croy, "Les magasins pittoresques à deux sous." [l'assassinat des beaux-arts.]
9. On Courbet, see Schapiro, "Courbet and Popular Imagery," and Nochlin, "Gustave Courbet's Meeting." On Cézanne, see Dombrowski, "Emperor's Last Clothes." On Impressionism, see Roskill, "Early Impressionism and the Fashion Print," and Isaacson, "Impressionism and Journalistic Illustration."
10. For example, the classic study of this process, Weber's *Peasants into Frenchmen*, discusses virtually every factor contributing to the consciousness of modern citizenship except art.

BIBLIOGRAPHY

Articles not identified by author that appeared in *L'Illustration* (*ILL*), *The Illustrated London News* (*ILN*), *Journal des connaissances utiles* (*JCU*), *Le Magasin pittoresque* (*MP*), and *The Penny Magazine* (*PM*) are listed under the title of the periodical.

A. "Cour d'assises. Procès du no. 35 de *La Caricature*, audience du 14 novembre 1831." *La Caricature*, Nov. 17, 1831.

À Beckett, Gilbert Abbott, and John Leech. *The Comic History of England: From Julius Caesar to George II*. 2 vols. London: Punch, 1847–48.

Abbey, John Roland. *Travel in Aquatint and Lithography, 1770–1860*. 2 vols. London: Curwen, 1956–57.

——. *Scenery of Great Britain and Ireland in Aquatint and Lithography, 1770–1860*. London: Curwen, 1952.

Académie française. *Dictionnaire de l'Académie francaise*. 2 vols. Paris: Firmin-Didot, 1835.

Adam, Victor. *Un an de la vie d'un jeune homme. Histoire veritable en 17 chapitres, écrit par lui-même*. Lith. Langlumé. Paris: Sazerac and Duval, 1824.

Adhémar, Jean. *La France romantique. Les lithographies de paysage au XIXe siècle*. Paris: Somogy, 1997.

——. "Les Imprimeurs lithographes au XIXe siècle," *Nouvelles de l'estampe* 24 (Nov.–Dec. 1975): 8–9, 12.

——. "Introduction." *French Popular Imagery*. London: Hayward Gallery, 1974. Exh. cat. Also published as *Cinq siècles d'imagerie française*. Paris: Musées nationaux, 1972.

Adhémar, Jean, and Jean-Pierre Seguin. *Le livre romantique*. Paris: Éditions du Chêne, 1968.

Album comique de pathologie pittoresque. 20 pl. Lith. Langlumé. Paris: Ambroise Tardieu, 1823.

Album du Magasin pittoresque. Cent gravures choisies dans la collection. Paris: Magasin pittoresque, 1862.

"Albums lithographiques," *Journal des artistes et des amateurs*, Dec. 28, 1828, 401–4.

Ambroselli, Gérard. *La vie du maréchal, petit album à colorier par les enfants de France*. Limoges: Imagerie du maréchal, 1944.

Les Anglais peints par eux-mêmes. See [Meadows, Kenny].

Archives nationales de France F18* I -24: Enregistrement des brevets puis déclarations des imprimeurs en lithographie et en taille douce (exécution de l'ordinance royale du 8 octobre 1817). 28 février 1818–14 octobre 1879.

Asfour, Lana. *Laurence Sterne in France*. London: Continuum, 2008.

Athanassoglou-Kallmyer, Nina. *Eugène Delacroix: Prints, Politics and Satire, 1814–1822*. New Haven: Yale University Press, 1991.

Aubert et cie, eds. *Étrennes de 1849. Publications d'Aubert et cie, éditeurs. Place de la Bourse, 29, à Paris* [1848].

Aubert et cie, éditeurs. Place de la Bourse, 29, à Paris [1846]. [Catalogue.]

Auberty, Jacqueline. *Charlet, sa vie, son oeuvre*. Paris: [Auberty], 1945.

Aubry-Lecomte, Hyacinthe. *Collection de têtes d'étude, d'après le tableau peint en 1801, par Mr Girodet-Trioson, Membre de l'Institut & représentant les ombres des héros français reçues dans les palais aëriens d'Ossian. Lithographiées sous sa direction par Aubry-Lecomte, son élève*. 16 pl. Paris: G. Engelmann, 1821.

Aurenche, Marie-Laure. *Édouard Charton et l'invention du* Magasin pittoresque *(1833–1870)*. Paris: Honoré Champion, 2002.

Bacot, Jean-Pierre. *La presse illustrée au XIXe siècle. Une histoire oubliée*. Limoges: PULIM, 2005.

Balzac, Honoré de. *Oeuvres complètes de M. de Balzac*. 17 vols. Paris: L. Furne, 1842–48.

——. *Petites misères de la vie conjugale, Illustrées par Bertall*. Paris: Chlendowski, 1846

——. "Avant-propos." In Balzac, *Oeuvres complètes de M. de Balzac*, 1: 7–32. Paris: L. Furne, 1842.

——. "Scènes de la vie privée." *La Silhouette* 2, no. 2 (1830): 9–11; no. 4 (1830): 30–32.

Bann, Stephen. *Distinguished Images: Prints in the Visual Economy of Nineteenth-Century France.* New Haven: Yale University Press, 2013.

——. *Parallel Lines: Printmakers, Painters and Photographers in Nineteenth-Century France.* New Haven: Yale University Press, 2001.

Barbier, Frédéric. "Les innovations technologiques." In Roger Chartier and Henri-Jean Martin, eds., *Histoire de l'édition française 2: Le Livre triumphant*, 721–29. Paris: Fayard/Cercle de la Librairie, 1990.

Basily-Callimaki, Mme de. *J.-B. Isabey, sa vie, son temps, 1767–1855, suivi du catalogue de l'oeuvre gravée par et d'après Isabey.* Paris: Frazier-Soye, 1909.

Baudelaire, Charles. *Baudelaire. Écrits sur l'art.* Ed. Yves Florenne. 2 vols. Paris: Gallimard, 1971.

——. "L'oeuvre et la vie d'Eugène Delacroix. Au rédacteur de *L'Opinion nationale*" [1863]. In Baudelaire, *Baudelaire. Écrits sur l'art* 2: 284–326.

——. "Quelques caricaturistes français" [1857]. In Baudelaire, *Baudelaire. Écrits sur l'art* 2: 327–72.

Bellangé. Hippolyte. *Croquis lithographiques par H. Bellangé.* 15 pl. Lith. Villain. Paris: Gihaut, 1824.

Bellanger, Claude, Jacques Godechot, Pierre Guiral, and Fernand Terrou. *Histoire générale de la presse française.* 5 vols. Paris: PUF, 1969–76.

Bellenger, Sylvain, ed. *Girodet, 1767–1824.* Paris: Gallimard, Musée du Louvre, 2005. Exh. cat.

Benjamin, Walter. *The Arcades Project.* Trans. Howard Eiland and Kevin McLaughlin. Cambridge: Belknap Press, 1999.

——. "The Work of Art in the Age of Mechanical Reproduction (1936)." In Vanessa R. Schwartz and Jeannene M. Przyblyski, eds. *The Nineteenth-Century Visual Culture Reader*, 63–70. New York: Routledge, 2004.

Béraldi, Henri. *Les graveurs du XIXe siècle. Guide de l'amateur d'estampes modernes.* 10 vols. Paris: L. Conquet, 1885–92.

Bertall [Charles-Albert d'Arnoux]. "Le salon de 1843." *L'Illustration*, May 13, 1843, 173.

——. *Le salon de 1843 (ne pas confondre avec celui de l'artiste-éditeur Challamel, éditeur-artiste). Appendice au livret, représenté par 37 copies de Bertal [sic] Études faites aux portes du Louvre, le 15 mars 1843. Les Omnibus, 7me livraison;* and *Les Omnibus, pérégrinations burlesques à travers tous chemins.* Paris: Rousset, 1843.

Bertarelli, Achille. *Le stampe popolari italiane.* Milan: Biblioteca Universale Rizzoli, 1974.

Bibliothèque nationale de France. *Bibliographie de la France. Journal général de l'imprimerie et de la librairie.* Paris: Pillet, 1814–1971. Online image listings search at http://artfl-project.uchicago.edu/content/image-france.

——. *Inventaire du fonds français après 1800.* 14 vols., ed. Jean Laran. Paris, 1930–67.

Bindman, David. *The Shadow of the Guillotine: Britain and the French Revolution.* London: British Museum, 1989. Exh. cat.

Biographie universelle (Michaud), ancienne et moderne. Paris: Mme C. Desplaces, 1855.

Blachon, Rémi. *La gravure sur bois au XIXe siècle. L'âge du bois debout.* Paris: Éditions de l'amateur, 2001.

Blanvillain, J.-F.-C. "Caricatures." In *Le Pariséum, ou Tableau de Paris en l'an XII* (1804). Paris: Henrichs, n.d.

Boilly, Louis-Leopold. *Recueil de grimaces.* 96 pl. Lith. Delpech. Paris: Delpech, 1823–28.

Boilly 1761–1845. Un grand peintre français de la Révolution à la Restauration. Lille: Musée des beaux-arts, 1988. Exh. cat.

Boime, Albert. "Jacques-Louis David, Scatological Discourse in the French Revolution, and the Art of Caricature." In *French Caricature and the French Revolution, 1789–1799*, ed. James Cuno, 67–82. Los Angeles: University of California, Grunwald Center for the Graphic Arts, 1988. Exh. cat.

Bollème, Geneviève. "Une littérature perdue." In *La bibliothèque bleue: La littérature populaire en France du XVIIe au XIXe siècle*, 7–26. Paris: Gallimard/Julliard, 1971.

Bordes, Philippe. "'L' écurie dont je ne sortirai que cousu d'or': Painters and Printmaking from David to Géricault." In *Théodore Géricault: The Alien Body: Tradition in Chaos*, ed. Serge Guilbaut, Maureen Ryan, and Scott Watson, 145–59. Vancouver: Morris and Helen Belkin Art Gallery, University of British Columbia, 1997. Exh. cat.

Bouquin, Corinne. *Recherches sur l'imprimerie lithographique à Paris au XIXe siècle: l'imprimerie Lemercier, 1803–1901.* Thèse d'État, Université de Paris 1. Paris: C. Bouquin, 1993.

Bourdieu, Pierre. *Distinction: A Social Critique of the Judgement of Taste.* London: Routledge and Kegen Paul, 1996.

Bouvet, Mireille-Bénédicte. *Le grand livre des images d'Épinal.* Paris: Solar, 1996.

Boys, Thomas Shotter. *Picturesque Architecture in Paris, Ghent, Antwerp, Rouen, etc. Drawn from Nature on Stone by Thomas Shotter Boys.* Lith. Hullmandel. 25 pl. London, 1839.

British Museum, London. *Catalogue of Political and Personal Satires Preserved in the Department of*

Prints and Drawings in the British Museum. Ed. F. G. [Frederick George] Stephens and M. D. [Mary Dorothy] George. 10 vols. London: British Museum, 1870–1954.

Bruson, Jean-Marie. *Théâtres romantiques à Paris. Collections du musée Carnavalet.* Paris: Paris-Musées, 2012.

Brüzard, L. M. *Catalogue de l'oeuvre lithographique de Mr J.- E.-Horace Vernet.* Paris: J. Gratiot, 1826.

Buchanan-Brown, John. *The Illustrations of Thackeray.* North Pomfret, Vt.: David and Charles, 1979.

Bulletin des lois du royaume de France, ser. 9, vol. 5. Paris: Imprimerie royale, 1833. No. 236, Loi sur l'instruction primaire, June 28, 1833, 251–62.

——. Ser. 9, vol. 1. Paris: Imprimerie royale, 1831. No. 59, Charte constitutionelle, Aug. 14, 1830, 51–64.

——. Ser. 7, vol. 5. Paris: Imprimerie royale, 1818. No. 2875, Ordonnance du Roi relative aux impressions lithographiques, Oct. 8, 1817, 245.

——. Ser. 5, vol. 2. Paris: Imprimerie royale, 1815. No. 395, Loi relative à la liberté de la presse, Oct. 21, 1814, 313–17; and No. 403, Ordonnance du Roi contenant des measures relatives à l'impression, au dépôt et à la publication des ouvrages, etc., Oct. 24, 1814, 324–27.

"Bulletins de Paris, jeudi 14 juillet." *Journal des arts, des sciences et de la littérature*, no. 307, July 15, 1814, 71.

La Caricature, morale, politique et littéraire. Paris, 1830–35.

"Les caricatures commandées par le Comité de salut public." In *La Révolution française et l'Europe, 1789–1799.* 3 vols. 2: *L'Evénement Révolutionnaire*, 594. Paris: Galeries nationales du Grand Palais, Paris and RMN, 1989. Exh. cat.

Carlin, John. "Masters of American Comics: An Art History of Twentieth-Century American Comic Strips and Books." In *Masters of American Comics*, ed. Paul Karasik and Brian Walker. New Haven: Yale University Press, 2005. Exh. cat.

Carroll, Lewis. *Alice's Adventures in Wonderland.* London: Macmillan, 1866.

Cate, Phillip Dennis, ed. *The Graphic Arts and French Society 1871–1914.* New Brunswick, N.J.: Rutgers University Press, 1988. Exh. cat.

Cate, Phillip Dennis, and Sinclair Hamilton Hitchings. *The Color Revolution. Color Lithography in France, 1890–1900.* Santa Barbara, Calif.: Peregrine Smith, and New Brunswick, N.J.: Rutgers University Art Gallery, 1978. Exh. cat.

Chabanne, Thierry, ed. *Les salons caricaturaux.* Les dossiers du musée d'Orsay 41. Paris: Réunion des musées nationaux, 1990. Exh. cat.

Chadefaux, Marie-Claude. "Le salon caricatural de 1846 et les autres salons caricaturaux." *Gazette des beaux-arts*, Mar. 1868, 161–76.

Cham [Amédée de Noé]. *Les aventures de monsieur Beaucoq, ex-rosier de la commune de Nanterre.* Paris: Arnauld de Vresse, 1856.

——. *Messieurs les Cosaques. Relation charivarique, comique et surtout véridique des hauts faits des Russes en Orient. Par MM. Taxile Delord, Clément Carraguel et Louis Huart. 100 vignettes par Cham.* Paris: Plon, 1854.

——. *Soulouque et sa cour.* Paris: Le Charivari, 1850.

——. "Arithmétique pittoresque." *L'Illustration*, Apr. 13, 1844, 108.

——. *Impressions de voyage de monsieur Boniface, Ex-réfractaire de la 4me du 5me de la 10me.* Paris: Paulin, 1844.

——. *Histoire du prince Colibri et de la fée Caperdulaboula.* Paris: Aubert, 1842.

——. *Aventures de Télémaque, fils d'Ulysse, par Fénelon et Cham.* Paris: Aubert, 1842.

——. *Miroir du calicot.* Paris: Arnauld de Vresse, 1841.

——. *Un génie incompris [Mr Barnabé Gogo].* Paris: Aubert, 1841

——. *Histoire de Mr Jobard.* Paris: Aubert, 1840.

——. *Deux vieilles filles vaccinées à marier.* Paris. Aubert, 1839.

——. *Histoire de Mr Lajaunisse.* Paris: Aubert, 1839.

——. *Mr Lamélasse.* Paris: Aubert, 1839.

Cham, Honoré Daumier, and Charles Vernier. *Les Cosaques pour rire. Album de quarante caricatures par MM. Cham, Daumier & Charles Vernier.* 40 pl. Lith. Destouches. Paris: Le Charivari, and Martinet, 1854.

Champfleury [Jules Husson]. *Henry Monnier, sa vie, son oeuvre, avec un catalogue complet de l'oeuvre et 100 gravures fac-similés.* Paris: Dentu, 1879.

——. *Histoire de la caricature sous la Republique, l'Empire et la Restauration.* Paris: Dentu, 1874.

——. *Histoire de l'imagerie populaire.* Paris: Dentu, 1869.

——. *Histoire de la caricature moderne.* Paris: Dentu, 1865.

——. *Grandes figures d'hier et d'aujourd'hui.* Paris: Poulet-Malassis et de Broise, 1861.

Le Charivari. Paris, 1832–1937.

Charlet, Nicolas-Toussaint. *Croquis lithographiques par Charlet.* 16 pl. Lith. Villain. Paris: Gihaut, 1824.

Charlet. Aux origines de la légende napoléonienne, 1792–1845. Boulogne-Billancourt: Bibliothèque Paul-Marmottan; La Roche-sur-Yon: Musée municipal, 2008; Paris: Bernard Giovanagneli. Exh. cat.

Chartier, Roger, and Henri-Jean Martin, ed. *Histoire de l'édition francaise.* 3 vols. 2: *Le livre triumphant.* 3: *Le temps des éditeurs: Du romantisme à la belle époque.* Paris: Fayard/Cercle de la Librairie, 1990.

Charton, Édouard. *Édouard Charton. Correspondance générale.* Ed. Marie-Laure Aurenche. 2 vols. 1: *1824–1859.* 2: *1860–1890.* Paris: Honoré Champion, 2008.
——. "Extrait des éphémérides d'une histoire du *Magasin pittoresque.*" *Magasin pittoresque,* added to the Jan. 1888 issue, n. p.
——. "Jean Best." *L'Illustration.* Oct. 11, 1879, 240.
——. "La littérature populaire en Angleterre." *Le Magasin pittoresque* 28 (1860), no. 2 (Jan. 1860): 14–15.
——. "Des moyens d'instruction. Les livres et les images." *Le Magasin pittoresque* 1 (1833), no. 13, 98–99.
——. "De l'influence de la conversation." *Le Magasin pittoresque* 1 (1833), no. 1, 2–3.
——. "À tout le monde." *Le Magasin pittoresque* 1 (1833), no. 1, 1.
Chatto, William Andrew, and John Jackson. *A Treatise on Wood Engraving, Historical and Practical.* London: Charles Knight, 1839; rev. 1841, 1861, 1881.
Chaudonneret, Marie-Claude. "Le statut et la réception de la lithographie en France dans les années 1820–1840." In *L'estampe, un art multiple à la portée de tous?* ed. Sophie Raux et al., 323–31. Villeneuve-d'Ascq: Presses universitaires du Septentrion, 2008.
Chollet, Roland. *Balzac journaliste. Le tournant de 1830.* Paris: Klincksieck, 1983.
Clayton, Timothy. *The English Print, 1688–1802.* New Haven: Yale University Press, 1997.
Cohen, Jane Rabb. *Charles Dickens and His Original Illustrators.* Columbus: Ohio State University Press, 1980.
Cohen, Margaret. "Panoramic Literature and the Invention of Everyday Genres." In Leo Charney and Vanessa Schwartz, eds., *Cinema and the Invention of Modern Life,* 227–52. Berkeley: University of California Press, 1995.
Cohn, Albert M. *George Cruikshank: A Catalogue Raisonné of the Work Executed during the Years 1806–1877.* London: Bookman's Journal, 1924.
Contensou, Martine, ed. *Balzac & Philipon associés. Grands fabricants de caricatures en tous genres.* Paris: Paris-Musées: 2001. Exh. cat.
Croy, Raoul de. "Les magasins pittoresques à deux sous." *Journal des artistes et des amateurs* 7/2, no. 18 (Nov. 3, 1833): 281–83.
Cuno, James. "Retour de Russie: Géricault and Early French Lithography." In *Théodore Géricault: The Alien Body: Tradition in Chaos,* ed. Serge Guilbaut, Maureen Ryan, and Scott Watson, 116–35. Vancouver: Morris and Helen Belkin Art Gallery, University of British Columbia, 1997. Exh. cat.
——. "Charles Philipon and La Maison Aubert: The Business, Politics, and Public of Caricature in Paris, 1820–1840." Ph.D. diss., Harvard University, 1985. Ann Arbor, Mich.: University Microfilms International, 1985.
——. "The Business and Politics of Caricature: Charles Philipon and La Maison Aubert." *Gazette des beaux-arts* 106 (Oct. 1985): 95–112.
——. "Charles Philipon, La Maison Aubert, and the Business of Caricature in Paris, 1829–1841." *Art Journal* 43 (Winter 1983): 347–54.
Cuno, James, ed. *French Caricature and the French Revolution, 1789–1799.* Los Angeles: University of California, Grunwald Center for the Graphic Arts, 1988. Exh. cat.

D. See Delécluze, Étienne-Jean.
Dackerman, Susan. *Painted Prints: The Revelation of Color.* University Park: Penn State University Press for the Baltimore Museum of Art, 2002. Exh. cat.
Damiron, Suzanne. "La revue 'L'Artiste.' Histoire administrative. Présentation technique. Gravures romantiques hors-texte." *Bulletin de la société de l'histoire de l'art français,* 1952, 131–42.
——. "Un grand revue d'art, *L'Artiste*: Son rôle dans le movement artistique au XIXe siècle; ses illustrations hors texte." Thèse complèmentaire, Sorbonne, 1946.
Daumier, Honoré. *Les cent-et-un Robert Macaire, composés et dessinés par M. H. Daumier, sur les idées et les légendes de M. Ch. Philipon.* 2 vols. Paris: Aubert, 1839.
Daumier Register: http://www.daumier-register.org.
Daumier, l'écriture du lithographe. Paris: Bibliothèque nationale de France, 2008. Exh. cat.
Daumier, 1808–1879. Paris: Réunion des musées nationaux, 1999. Exh cat.
Davis, Peggy. "Entre la physiognomonie et les physiologies: le Calicot, figure du panorama parisien sous la Restauration." *Études françaises* 49, no. 3 (2013): 63–85.
Day-Hickman, Barbara Ann. *Napoleonic Art: Nationalism and the Spirit of Rebellion in France (1815–1848).* Newark: University of Delaware Press, 1999.
Dayot, Armand. *Carle Vernet. Étude sur l'artiste suivie d'un catalogue de l'oeuvre gravé et lithographié et du catalogue de l'exposition retrospective de 1925.* Paris: Le Goupy, 1925.
Delacroix, Eugène. *Journal, 1822–1863,* ed. André Joubin. Paris: Plon, 1982.
——. *Correspondance générale,* ed. André Joubin. 5 vols. Paris: Plon, 1936–38.

——. "Charlet." In *Revue des deux mondes* 37 (Jan. 1, 1862): 234–42.
Delécluze, Etienne-Jean [D.]. "Beaux-arts. La lithographie." *Journal des débats*, Jan. 31, 1825, 1–2.
Delécluze, Étienne-Jean. *Louis David, son école et son temps. Souvenirs.* Paris: Macula, 1983 [orig. 1855].
Delepierre, Octave. *Essai historique et bibliographique sur les rébus.* London: Wertheimer, Lea, 1870.
Delord, Taxile, Clément Carraguel, and Louis Huart. *Messieurs les Cosaques. Relation charivarique, comique et surtout véridique des hauts faits des Russes en Orient. Par MM. Taxile Delord, Clément Carraguel et Louis Huart. 100 vignettes par Cham.* 2 vols. Paris: V. Lecou, 1854–55.
Delpech, François-Séraphin. *Iconographie des contemporains, ou portraits des personnes dont les noms se rattachent plus particulièrement, soit par leurs actions, soit par leurs écrits, aux divers événements qui ont eu lieu en France, depuis 1789, jusqu'en 1829.* 2 vols. Paris: Delpech, 1832.
Delteil, Loys. *Théodore Géricault: The Graphic Work: A Catalogue Raisonné*, ed. Alan Hyman. San Francisco: Alan Wofsy Fine Arts, 2010.
"Des albums." *Journal des artistes et des amateurs*, Mar. 23, 1828, 182–84.
Descaves, Lucien. *L'humble Georgin, imagier d'Épinal.* Paris: Librairie de Paris, Firmin-Didot, 1932.
"Dessin." *Le Charivari*, Nov. 25, 1833.
Devéria, Achille. *Le goût nouveau.* 24 pl. Lith. Lemercier. Paris: Tessari, Aumont; London: Charles Tilt, 1831.
Le diable à Paris: Paris et les Parisiens. 2 vols. Paris: J. Hetzel, 1845–46.
Dickens, Charles. *The Posthumous Papers of the Pickwick Club.* London: Chapman and Hall, 1837.
Diderot, Denis, and Jean Le Rond d'Alembert. *Encyclopédie, ou dictionnaire raisonné des sciences, des arts et des métiers.* 35 vols. Paris: Briasson, 1751–80.
Dombrowski, André. "The Emperor's Last Clothes: Cézanne, Fashion and *l'année terrible.*" *Burlington Magazine* 148, no. 1242 (Sept. 2006): 586–94.
Le Don Quichotte romantique, ou Voyage du docteur Syntaxe à la recherche du pittoresque et du romantique. See Rowlandson, Thomas.
Donald, Diana. *The Age of Caricature: Satirical Prints in the Reign of George III.* New Haven: Yale University Press, 1996.
Doré, Gustave. *L'oeuvre graphique de Gustave Doré.* Ed. Gabriele Forberg. 2 vols. Paris: Arthur Hubschmid, 1976.
——. *Histoire dramatique, pittoresque et caricaturale de la sainte Russie.* Paris: J. Bry ainé, 1854.
——. *Dés-agréments d'un voyage d'agrément.* Paris: Aubert, 1851.
——. *Trois artistes incompris et mécontents. Leur voyage en province… et ailleurs!! Leur faim dévorante et leur déplorable fin.* Paris: Aubert, 1850.
——. *Les travaux d'Hercule.* Paris: Aubert, 1847.
Doré, Gustave, and Blanchard Jerrold. *London: A Pilgrimage.* London: Grant, 1872.
Duchartre, Pierre-Louis, and René Saulnier. *L'imagerie populaire. Les images de toutes les provinces françaises du XVe siècle au second empire.* Paris: Gründ, 1925.
Dufour, Émile. "Le defenseur officieux en justice de paix." In *Les Français peints par eux-mêmes.* 10 vols. 2: 307–10. Paris: L. Curmer, 1840–42.
Dumont, Jean-Marie. *Les maîtres graveurs populaires, 1800–1850.* Épinal: Pellerin, 1965.
——. *La vie et l'oeuvre de Jean-Charles Pellerin, 1756–1836.* Épinal: Pellerin, 1956.
"D'une nouvelle erreur du parti des ultra." *Le Constitutionnel*, Aug. 23, 1819, 4.
Dupeuty, Charles, Frédéric de Courcy, and Maurice Alhoy. *Le Magasin pittoresque, revue en 15 livraisons, par MM.* [Charles] *Dupeuty, F.* [Frédéric] *de Courcy, Maurice Alhoy.* Paris: Marchant, 1834.
Du Seigneur, Jehan. "Appendice à la notice sur Horace Vernet. Ses tableaux et ses lithographies." *Revue universelle des arts* 17 (1863): 343–57.

Eckermann, Johann Peter. *Conversations of Goethe with Eckermann.* Ed. J. K. Moorhead. Trans. John Oxenford. London: Dent; New York: Dutton, 1930.
Egan, Pierce. *Life in London; or, The Day and Night Scenes of Jerry Hawthorn, Esq., and His Elegant Friend Corinthian Tom, Accompanied by Bob Logic, the Oxonian, in Their Rambles and Sprees through the Metropolis.* London: Sherwood, Neely, & Jones, 1821.
——. *Boxiana; or, Sketches of Ancient and Modern Pugilism. From the Days of the Renowned Broughton and Slack to the Heroes of the Present Milling Era.* 6 vols. London: G. Smeeton, 1812–29.
[Egan, Pierce]. *Diorama anglais, ou Promenades pittoresques à Londres, renfermant les notes les plus exactes sur les caractères, les moeurs et usages de la nation anglaise, par M. S.* Trans. Jean-Baptiste-Balthazar Sauvan. Paris: Jules Didot l'aîné, Baudouin frères, 1823.
Engelmann, Godefroy. *Album chromolithographique, ou Recueil d'essais du nouveau procédé d'impression lithographique en couleurs.* Mulhouse: Engelmann; Paris: J. Risler fils; Leipzig: Del Vecchio, 1837.
——. *Essais des diverses manières de dessiner à l'encre lithographique.* Paris: Engelmann, 1817.

——. *Recueil des essais lithographiques dans les différents genres de dessin tels que manière de crayon, de la plume, du pinceau et de lavis exécutés par le procédé de G. Engelmann, directeur de la Société lithographique de Mulhouse.* Paris: chez l'auteur [1816].

[Engelmann, Godefroy]. Obituary of Godefroy Engelmann, in *Le Lithographe. Journal des artistes et des imprimeurs* [Paris] 2 (1839): 223.

Études de presse, n.s. 9, no. 17 (1957) [issue on physiologies].

Eyerman, Charlotte. "Im Zeichen de Grande Tradition: Lithographen im Atelier des Baron Gros, 1816–1835." In Stefan Germer and Michael F. Zimmermann, eds., *Bilder der Macht, Macht der Bilder: Zeitgeschichte in Darstellungen des 19. Jahrhunderts*, 176–91. Munich: Klinkhardt and Biermann, 1997.

F. [Charles-François Farcy]. "Les métamorphoses.—Les contrastes." *Journal des artistes et des amateurs*, Nov. 30, 1828, 339–42.

Farwell, Beatrice. *Lithography in Art and Commerce.* Vol. 12 of Farwell, *French Popular Lithographic Imagery, 1815–1870*, 1997.

——. *French Popular Lithographic Imagery, 1815–1870.* 12 vols. Chicago: University of Chicago Press, 1981–97.

——. *The Charged Image: French Lithographic Caricature, 1816–1848.* Santa Barbara, Calif.: Santa Barbara Museum of Art, 1989. Exh. cat.

——. *The Cult of Images: Baudelaire and the Nineteenth-Century Media Explosion.* Santa Barbara: University of California at Santa Barbara Art Museum, 1977. Exh. cat.

"Feuilleton, 22 Fév. 1834. Encyclopédie pittoresque à deux sous." *Le Temps. Journal des progrès politiques, scientifiques, littéraires et industriels*, Feb. 22, 1834, 1.

Feyel, Gilles. "Les transformations technologiques de la presse au XIXe siècle." In Dominique Kalifa, Philippe Régnier, Marie-Ève Thérenty, and Alain Vaillant, eds., *La civilisation du journal. Histoire culturelle et littéraire de la presse française au XIXe siècle*, 97–139. Paris: Nouveau Monde, 2011.

Firmin-Didot, Ambroise. *Étude sur Jean Cousin suivie de notices sur Jean Leclerc et Pierre Woëiriot.* Paris: Firmin-Didot, 1872.

Fontaine, Laurence. *Histoire du colportage en Europe, XVe–XIXe siècle.* Paris: Albin Michel, 1993.

Forest, E. [Eugène]. *Histoire de Mr de Vertpré et de sa ménagère aussi.* Paris: Aubert, 1840.

Fox, Celina. *Graphic Journalism in England during the 1830s and 1840s.* New York: Garland, 1988.

"Les Français au Missouri. Extrait d'un journal chiroquois." *La Silhouette* 1, no. 4 (1830): 27–28.

Les Français peints par eux-mêmes. Encyclopédie morale du dix-neuvième siècle. 10 vols. Paris: L. Curmer, 1840–42.

French Popular Imagery. London: Hayward Gallery, 1974. First published as *Cinq siècles d'imagerie française.* Paris: Musées nationaux, 1972. Exh. cat.

Furet, François, and Jacques Ozouf. *Reading and Writing: Literacy in France from Calvin to Jules Ferry.* Cambridge: Cambridge University Press; Paris: La Maison des sciences de l'homme, 1977.

Gaberel, Jean. *Essai sur le caractère artistique et littéraire des oeuvres de R. Toepffer.* Geneva: Gruaz, 1846.

Garnier, Nicole, ed. *L'imagerie populaire française.* Musée national des arts et traditions populaire and Bibliothèque nationale de France. 2 vols. Paris: RMN, 1990.

Gaudriault, Raymond. *La gravure de mode feminine en France.* Paris: Éditions de l'amateur, 1983.

Gausseron, Bernard-Henri. *Les keepsakes et les annuaires illustrés de l'époque romantique. Essai de bibliographie.* Paris: E. Rondeau, 1896.

Gauthier, Maximilien. *Achille et Eugène Devéria.* Paris: H. Floury, 1925.

Gautier, Théophile. "Grandville." In his *Portraits contemporains. Littérateurs, peintres, sculpteurs, artistes dramatiques*, 231–33. Paris: Charpentier, 1874 [orig. *La Presse*, Mar. 24, 1847].

——. "Tony Johannot." In his *Portraits contemporains. Littérateurs, peintres, sculpteurs, artistes dramatiques*, 226–30. Paris: Charpentier, 1874 [orig. *La Presse*, June 16, 1845].

George, Henri. *La belle histoire des images d'Épinal.* Paris: Cherche Midi, 2005.

Gérard-Fontallard [Henri-Gérard Fontallard]. *Histoire d'une épingle en seize tableaux, racontée par elle-même.* 16 pl. Paris: Ostervald aîné, Rittner, 1828.

Géricault, Théodore. *Géricault raconté par lui-même et par ses amis.* Ed. Pierre Courthion. Geneva: Pierre Cailler, 1947.

——. *Quatre sujets divers par Géricault.* 4 pl. Lith. Engelmann. Paris: Mme Hulin. 1823.

——. *Études de chevaux lithographiés par Géricault.* 12 pl. Lith. Villain. Paris: Gihaut, 1822.

——. *Various Subjects Drawn from Life and on Stone by J. Géricault* [*The English Suite*]. 13 pl. Lith. Charles Hullmandel. London: Rodwell and Martin, 1821.

Géricault. Paris: Galeries nationales du Grand Palais and RMN, 1991. Exh. cat.

Gervais, Thierry. *L'illustration photographique. Naissance du spectacle de l'information, 1843–1914.* 2 vols. Thèse de doctorat, École des hautes études en sciences sociales, 2007. Paris: T. Gervais, 2007.

———. "Imaging the World: *L'Illustration:* The Birth of the French Illustrated Press and the Introduction of Photojournalism in the Mid-19th Century." *Medicographia* 27, no. 1 (2005): 97–106.

———. "D'après photographie. Premiers usages de la photographie dans le journal *L'Illustration* (1843–1859)." *Études photographiques* 13 (2003): 56–85.

Gigoux, Jean. *Causeries sur les artistes de mon temps.* Paris: C. Lévy, 1885.

Girardin, Émile de. "Écoles et méthodes. Instruction nationale et professionnelle." *Journal des connaissances utiles*, Mar. 1833, 63–70.

Girodet, Anne-Louis. *Collection de têtes d'étude.* See Aubry-Lecomte, Hyacinthe.

Glucq. *Série encyclopédique Glucq des leçons de choses illustrées.* 3 vols. Épinal: Imagerie d'Épinal, 1985–92 [reprint edition].

Godissart de Cari. *Le musée grotesque*. 64 pl. Paris: Martinet, 1817–30.

Goetz, Adrien. *L'Artiste, une revue de combat des années romantiques (1831–1948).* 2 vols. Thèse, Université de Paris 4. Paris: A. Goetz, 1998.

Golan, Romy. *Modernity and Nostalgia: Art and Politics in France Between the Wars.* New Haven: Yale University Press, 1995.

Goldstein, Robert Justin. *Censorship of Political Caricature in Nineteenth-Century France.* Kent, Ohio: Kent State University Press, 1989.

———. *Political Censorship of the Arts and the Press in Nineteenth-Century Europe.* Basingstroke, Hampshire, UK: Macmillan, 1989.

Gosselin, Gérard, Yannick Marec, and Martine Thomas. *Le dessin de presse à l'époque impressionniste 1863–1908, de Daumier à Toulouse-Lautrec.* Paris: Éditions Jean di Sciullo/Democratic Books, 2010.

Le goût du jour. 48 pl. Paris: Martinet, 1813–20.

Goya, Francisco de. *Caricatures espagnoles. Ni plus ni moins, par Goya.* 10 pl. Paris: Motte, 1825.

Grand-Carteret, John. *Les moeurs et la caricature en France.* Paris: La librairie illustrée, 1888.

Grandville [Jean-Ignace-Isidore Gérard]. *Grandville. Das Gesamte Werk*. Ed. Gottfried Sello. 2 vols. Munich: Rogner U. Bernhard, 1969.

———. *Les fleurs animées.* Text by Taxile Delord. Paris: Gabriel de Gonet, 1847.

———. *Un autre monde. Transformations, visions, incarnations, ascensions, locomotions, explorations, pérégrinations, excursions, stations, cosmogonies, fantasmagories, rêveries, folâtreries, facéties, lubies, métamorphoses, zoomorphoses, lithomorphoses, métempsychoses, apothéoses, et autres choses.* Paris: H. Fournier, 1844.

———. *Scènes de la vie privée et publique des animaux. Études de moeurs contemporains.* Ed. P.-J. Stahl. 2 vols. Paris: J. Hetzel et Paulin, 1842.

———. *The Metamorphoses of the Day*. London: Thomas M'Lean [McLean], 1830.

———. Les *métamorphoses du jour*. 71 pl. Lith. Langlumé. Paris: Bulla and Martinet, 1828–29.

Grandville. Les métamorphoses du jour. Nancy: Musée des beaux-arts de Nancy, 2003. Exh. cat.

Graselli, Margaret Morgan. *The Printmaking Revolution in Eighteenth-Century France.* Washington, D.C.: National Gallery of Art, 2003. Exh. cat.

Gray, Valerie. *Charles Knight: Educator, Publisher, Writer.* Aldershot, UK: Ashgate, 2006.

Gretton, Tom. "The Pragmatics of Page Design in Nineteenth-Century General-Interest Weekly Illustrated News Magazines in London and Paris." *Art History* 33, no. 4 (Sept. 2010): 680–709.

Grivel, Marianne. *Le commerce de l'estampe à Paris au XVIIIe siècle.* Geneva: Droz, 1986.

Groensteen, Thierry. "Le Patrimoine européen du neuvième art." In *Maîtres de la bande dessinée européenne*, ed. Thierry Groensteen. Paris: Bibliothèque nationale de France/Seuil, 2000. Exh. cat.

———. *Les années Caran d'Ache.* Angoulême: Musée de la bande dessinée, 1988. Exh. cat.

Groensteen, Thierry, and Benôit Peeters, eds. *Töpffer. L'invention de la bande dessinée.* Paris: Hermann, 1994.

Gusman, Pierre. *La gravure sur bois en France aux XIXe siècle.* Paris: A. Morancé, 1929.

Hallett, Mark. "The Medley Print in Early Eighteenth-Century London." *Art History* 20, no. 2 (June 1997): 214–37.

Hannoosh, Michele. *Baudelaire and Caricature: From the Comic to an Art of Modernity.* University Park: Penn State University Press, 1992.

Haskell, Francis. *History and Its Images: Art and the Interpretation of the Past.* New Haven: Yale University Press, 1993.

Hautecoeur, Louis. "Une famille de graveurs et d'éditeurs parisiens: Les Martinet et les Hautecoeur (XVIIIe et XIXe siècles)." In *Paris et Île-de-France. Mémoires*, vol. 18–19 (1967–68). Paris: Fédération des sociétés historiques et archéologiques de Paris et de l'Île-de-France, 1970.

Henker, Michael, Karlheinz Scherr, Elmar Stolpe, eds. *Von Senefelder zu Daumier: Die Anfänge der lithographischen Kunst.* Munich: Haus der Bayerischen geschichte, 1988. Also published as *De Senefelder à Daumier, Les débuts de l'art lithographique.* Exh. cat.

Hiner, Susan. "Monsieur Calicot: French Masculinity between Commerce and Honor." *West 86th: A Journal of Decorative Arts, Design History, and Material Culture* 19, no. 1 (2012): 32–60.

Hogarth, William. *Hogarth's Graphic Works.* Ed. Ronald Paulson. 3rd rev. ed. London: Print Room, 1989.

Houfe, Simon. *The Dictionary of British Book Illustrators and Caricaturists, 1800–1914.* Rev. ed. Woodbridge: Antique Collectors Club, 1981.

Howes, Alan B., ed. *Sterne: The Critical Heritage.* London: Routledge and Kegan Paul, 1974.

Huin, Bernard. "L'imagerie à Épinal." In Martin, *Images d'Épinal*, 33–67. Paris: RMN, 1995. Exh. cat.

Hults, Linda C. *The Print in the Western World: An Introductory History.* Madison: University of Wisconsin Press, 1996.

Hunt, Lynn. "The Political Psychology of Revolutionary Caricatures." In *French Caricature and the French Revolution 1789–1799*, ed. James Cuno, 33–40. Los Angeles: University of California, Grunwald Center for the Graphic Arts, 1988. Exh. cat.

The Illustrated London News. London, 1842–2003.

——. "Destruction of Pointe-à-Pitre." Apr. 29, 1843, 286.

——. "Abolition of the Slave Trade and the Right of Search." Jan. 21, 1843.

——. "The Disturbances in the Manufacturing Districts." Aug. 20, 1842, 232–33, 236–37.

——. Felicie, "The Fashions. May 14, 1842, 6.

——. "Our Address." May 14, 1842, 1.

The Illustrated London News Jubilee Edition. "The Founding of the 'Illustrated' May 14th 1842: A Chapter in the History of Journalism," May 14, 1892, n. p.

L'Illustration. Journal universel. Paris, 1843–1944.

——. "Catastrophe de 8 mars à Alger." Vol. 5, Mar. 22, 1845, 49–50.

——. "Les mystères de l'*Illustration*." Vol. 3, Mar. 2, 1844, 7–10.

——. "Arrivée du roi au Palais Bourbon. Vol. 2, Dec. 30, 1843, 276.

——. "De la traite et de l'esclavage." Vol. 2, Oct. 21, 1843.

——. "Revue algérienne." Vol. 1, Mar. 18, 1843, 37–38.

——. "Tremblement de terre aux Antilles." Vol. 1, Mar. 18, 1843, 33.

——. Preface to vol. 1. Sept. 1, 1843, n. p.

——. "Notre but." Vol. 1, Mar. 4, 1843, No. 1, 1.

"Les Images et la politique." *Le Figaro*, Mar. 30, 1889 supplement.

Imagerie d'Épinal. *Série encyclopédique Glucq des leçons de choses illustrées.* 3 vols. Épinal: Imagerie d'Épinal, 1985–92. [reprint edition.]

——. *L'imagerie publicitaire. De la propagande religieuse à l'imagerie publicitaire.* Épinal: Imagerie d'Épinal, 1987. Exh. cat.

Imagerie Pellerin. *Histoire de l'imagerie... une entreprise bicentenaire.* Épinal: Imagerie Pellerin, 1991.

Institut royal de France, Académie des beaux-arts. *Rapport sur la lithographie, et particulièrement sur un recueil de dessins lithographiés par M. Engelmann.* Paris: Firmin-Didot, 1816. Repr. in Johnson, *French Lithography*, 23–40.

Isaacson, Joel. "Impressionism and Journalistic Illustration." *Arts Magazine* 56, no. 10 (June 1982): 95–115.

Isabey, Jean-Baptiste. *Caricatures de J. J. à Paris.* 12 pl. Lith. Charles Motte. Paris: Alphonse Giroux, 1818.

——. *Divers essais lithographiques de J. B. Isabey publiés à Paris en 1818.* Paris: G. Engelmann, 1818.

Iskin, Ruth E. *The Poster: Art, Advertising, Design, and Collecting, 1860s–1890s.* Hanover, N.H.: Dartmouth College Press, 2014.

Ivins, William. *Prints and Visual Communication.* Cambridge: Harvard University Press, 1953.

Jackson, Mason. "Thirty Years of Pictorial Journalism." In *Illustrated London News Jubilee Edition*, May 14, 1892.

Jacquemin, André. "Les techniques de l'imagerie populaire." In Mistler, Blaudez, and Jacquemin, *Épinal et l'imagerie populaire*, 139–74.

Janin, Jules, et al. *Pictures of the French: A Series of Literary and Graphic Delineations of French Character.* London: Wm. S. Orr, 1840.

Jerrold, Douglas. "The Pew-Opener." In Kenny Meadows, *Heads of the People, or Portraits of the English, Drawn by Kenny Meadows*, 224–32. London: Robert Tyas, 1840.

Jobert, Barthélémy. "Girodet and Printmaking." In Sylvain Bellenger, ed. *Girodet, 1767–1824*, 149–77. Paris: Gallimard, Musée du Louvre, 2005. Exh. cat.

——. "Delacroix en noir et blanc." In Jobert, ed., *Delacroix. Le trait romantique*, 13–19. Paris: Bibliothèque nationale de France, 1998. Exh. cat.

Jobert, Barthélémy, ed. *Delacroix. Le trait romantique.* Paris: Bibliothèque nationale de France, 1998. Exh. cat.

Johannot, Tony, Alfred de Musset, and P.-J. Stahl. *Voyage où il vous plaira, livre écrit à la plume et au crayon avec vignettes, légendes, épisodes, commentaires, incidents, notes et poésie.* Paris: J. Hetzel, 1843.

Johnson, W. McAllister. *French Lithography: The Restoration Salons, 1817–1824.* Kingston, Ontario: Agnes Etherington Art Centre, 1977.

Jones, Philippe. *La presse satirique illustrée entre 1860 et 1890.* Paris: Institut français de presse, 1966.

Journal des connaissances utiles. Politique, agricole et commercial, indiquant à tous les hommes qui savent lire. Leurs devoirs, comme citoyen, juré, garde national, maire ou adjoint, garde champêtre ou forestier. Leurs droits, comme contribuable, électeur communal, conseilleur municipal, électeur, éligible. Leurs intérêts, comme père de famille, propriétaire, fermier, fabricant et commerçant, ouvrier de tous états. [subtitle varied] Paris, 1831–48.

——. "Science. Du crédit. De l'épargne—et de l'amortissement." Feb. 1, 1834, 31–33.

——. "Gravure sur bois en France et en Angleterre." Oct. 1833, 276–79.

——. M. N. Evon. "Instruction. Budget des familles. Économie pour le présent; richesse pour l'avenir." Apr. 1, 1832, 96–100.

——. "Resumé." Dec. 1832, 348.

——. [Review of press from *Le Temps*]. Nov. 1831, 69.

——. "Société nationale pour l'emancipation intellectuelle." Oct. 1831, 2.

Journal des dames et des modes. Frankfurt, 1798–1848.

Jouy, Étienne de. "No. IX—9 juillet 1814. Les caricatures." *Guillaume le franc-parleur, ou Observations sur les moeurs françaises au commencement du XIXe siècle.* 5 vols. 1: 107–17. Paris: Pillet, 1816–17.

——. "No. XIII–30 octobre 1811. Des album"; "No. XV–8 novembre 1811. Recherches sur l'album et sur le chiffonier sentimental." *L'Hermite de la Chaussée d'Antin, ou Observations sur les moeurs et les usages parisiens au commencement du XIXe siècle.* 5 vols. 1: 143–55; 1: 167–78. Paris: Pillet, 1815.

Kaenel, Philippe. *Le Métier d'illustrateur 1830–1880. Rodolphe Töpffer, J.-J. Grandville, Gustave Doré.* 2nd rev. ed. [orig. 1996]. Geneva: Librairie Droz, 2005.

——. "J.-J. Grandville et le *Magasin pittoresque*." *Gazette des beaux-arts,* Feb. 1990, 77–98.

Kaenel, Philippe, ed. *Doré. L'imaginaire au pouvoir.* Paris: Musée d'Orsay/Flammarion and Ottawa: Musée des beaux-arts du Canada, 2014. Exh. cat.

Kalifa, Dominique, Philippe Régnier, Marie-Ève Thérenty, and Alain Vaillant. *La civilisation du journal. Histoire culturelle et littéraire de la presse française au XIXe siècle.* Paris: Nouveau Monde, 2011.

Kerr, David S. *Caricature and French Political Culture, 1830–1848: Charles Philipon and the Illustrated Press.* Oxford: Oxford University Press, 2000.

Keyser, Édouard de. *Un thème de l'imagerie parisienne sous la Restauration: Le combat des montagnes et la guerre des calicots.* Paris: Le Vieux Papier, 1964.

Knight, Charles. *Passages of a Working Life.* 3 vols. London: Bradbury and Evans, 1864–65.

——. 1. Preface to vol. 1 (Dec. 18, 1832), *Penny Magazine,* n.p.

Kundera, Milan. "An Introduction to a Variation." *The New York Times,* Jan. 6, 1985.

Kunzle, David. "Caricature et bande dessinée autour du *Journal pour rire,* 1848–1855." In Philippe Kaenel, ed., *Doré. L'Imaginaire au pouvoir,* 39–63. Paris: Musée d'Orsay/ Flammarion, 2014. Exh. cat.

——. *Father of the Comic Strip: Rodolphe Töpffer.* Jackson: University Press of Mississippi, 2007.

——. *History of the Comic Strip.* Vol. 1: *The Early Comic Strip.* Vol. 2: *The Nineteenth Century.* Berkeley: University of California Press, 1973–90.

——. "Goethe and Caricature: From Hogarth to Töpffer." *Journal of the Warburg and Courtauld Institutes* 48 (1985): 164–88.

——. "Cham, the 'Popular' Caricaturist." *Gazette des beaux-arts,* Dec. 1980, 213–24.

——. "'L'Illustration,' journal universel (1843–1853). Le premier magazine illustré en France. Affirmation du pouvoir de la bourgeoisie." *Nouvelles de l'estampe,* no. 43 (1979): 8–16.

La Combe, Joseph-Félix Leblanc de. *Charlet, sa vie, ses lettres et ses oeuvres.* Paris: Paulin and Le Chevalier, 1856 [orig. *Revue contemporaine,* Jan. 31, Feb. 15, 1854].

Lachèvre, Henry. *Détails intimes sur Géricault.* Rouen, 1870. Reprinted in *Géricault raconté par lui-même et par ses amis.* Ed. Pierre Courthion. Vésenas-Geneva: Pierre Cailler, 1947.

Lagenevais, F. de [Henri Blaze de Bury]. "La littérature illustrée." *Revue des deux mondes,* n.s. 1, no. 3 (Mar. 1843): 645–71.

Lambertson, John P. "P.-N. Guérin's Classical Lithographs and Pedagogy." Unpublished paper presented at Consortium on the Revolutionary Era, Charleston, S.C., Feb. 2010.

Lami, Eugène, and Henry Monnier. *Voyage en Angleterre.* Paris: Firmin-Didot/Lami-Denozan and London: Colnaghi, 1829–30.

La Lanterne magique. Journal des choses curieuses et amusantes. Paris, 1833–1836.

Laslett, Peter. *The World We Have Lost.* New York: Scribner, 1965.

Lasteyrie, Charles de. *Recueil de différens genres d'impressions lithographiques qui peuvent avoir une application utile aux sciences et aux arts méchaniques et libéraux.* Paris: Lasteyrie, 1816.

Leblanc, Henri. *Catalogue de l'oeuvre complet de Gustave Doré.* Paris: Charles Bosse, 1931.

Lebovics, Herman. *True France: The Wars over Cultural Identity, 1900–1945.* Ithaca: Cornell University Press, 1992.

Lebrun, Henri. *Voyages et aventures du capitaine Cook.* Paris: Lavigne, 1837.

Le Men, Ségolène. "Livres illustrés et albums, 1750–1900." In *Livres d'enfance, livres de France,* ed. Annie Renonciat and Viviane Ezratty, 65–76. Paris: Hachette, 1998.

——. "Le romantisme et l'invention de l'album pour enfants." *Revue française d'histoire du livre.* Nos. 82–83 (1994): 145–75.

——. "Quelques définitions romantiques de l'album." *Art et métiers du livre.* Special issue on albums of prints. No. 143 (Feb. 1987): 40–47.

——. "De l'image au livre. L'éditeur Aubert et l'abécédaire en estampes." *Nouvelles de l'estampe.* Special issue on alphabet books. No. 90 (Dec. 1986): 17–30.

——. "Du livre d'église au livre d'étrennes." *Art et métiers du livre.* Special issue on P-.J. Hetzel. No. 139 (May 1986): 17–24. Paris: Technorama, 1986.

Le Men, Ségolène, and Luce Abélès. *Les Français peints par eux-mêmes: Panorama social du XIXe siècle.* Paris: Réunion des musées nationaux, 1993. Exh. cat.

Lesage, Alain-René. *Histoire de Gil Blas de Santillane* [1715–35]. Paris: Paulin, 1835; repr. Paris: Dubochet, 1838.

Lhéritier, Andrée. *Les physiologies 1840–1845. Bibliographie descriptive.* 1958; Paris: Service international de microfilms, 1966.

——. *Les physiologies de 1826 à 1894. Contribution à l'étude du livre illustré au 19ème siècle.* Paris, 1955.

M. S. *Diorama anglais, ou Promenades pittoresques à Londres.* See Egan, Pierce.

[*Le Magasin pittoresque*]. *Album du Magasin pittoresque. Cent gravures choisies dans la collection.* Paris: Magasin pittoresque, 1862.

Le Magasin pittoresque. Paris, 1833–1938.

——. 5 (1837). "Le Panthéon." No. 32 (Aug. 1837): 249–51; no. 40 (Oct. 1837): 319–20.

——. 5 (1837). "De la comptabilité." No. 7 (Feb. 1837): 53–54; no. 12 (Mar. 1837): 89–93.

——. 1 (1833). "Antiquités egyptiennes. Zodiaque circulaire de Denderah." No. 40, 313–15.

——. 1 (1833). "Un cimetière en Sicile." No. 40, 315–17.

——. 1 (1833). "Bibliothèque du roi.—Manuscrits. Portraits de Chinois célèbres." No. 39, 306–9; no. 42, 333–35.

——. 1 (1833). "Alphabet manuel des sourds-muets." No. 38, 300.

——. 1 (1833). "Guillaume Penn." No. 26, 208.

——. 1 (1833). "Statue de Diane à Éphèse." No. 26, 208.

——. 1 (1833). "Voyages. Nouvelle Zélande." No. 24, 191–92; no. 28, 219–21.

——. 1 (1833). "Aérostation–Ballons." No. 21, 163.

——. 1 (1833). "Des animalcules microscopiques." No. 19, 145.

——. 1 (1833). "Hygiène. Du danger des corsets trop serrés." No. 13, 99.

——. 1 (1833). "Georges Cuvier. Sa vie. Ses travaux. Histoire de l'homme fossile." No. 1, 5–6.

——. 1 (1833). Preface to vol. 1. Dec. 31, 1833, i–ii.

Le Magasin pittoresque, revue en 15 livraisons, par MM. [Charles] *Dupeuty, F.* [Frédéric] *de Courcy, Maurice Alhoy.* Paris: Marchant, 1834.

Le Magasin universel. Paris, 1833–40.

Maidment, Brian. *Comedy, Caricature and the Social Order, 1820–1850.* Manchester: Manchester University Press, 2013.

Mainardi, Patricia. "Hogarth 'Corrected': Modern Moral Subjects in France." In Cynthia Ellen Roman, ed., *Hogarth's Legacy,* 143–65. Farmington, Conn.: Lewis Walpole Library in association with Yale University Press, 2016.

——. "Dupinade." In Vanessa R. Schwartz and Jason E. Hill, eds., *Getting the Picture: The Visual Culture of the New,* 16–18. London: Bloomsbury, 2015.

——. "From Popular Prints to Comics." *Nineteenth-Century Art Worldwide* 10, no. 1 (Spring 2011).

——. "Paths Forgotten, Calls Unheard: Nineteenth-Century Book Illustration." *Van Gogh Studies* 3 (2010), 163–97.

——. "Popular Prints for Children . . . and Everyone Else." *Princeton University Library Chronicle* 71, no. 3 (Spring 2010): 357–91.

——. "The Invention of Comics." *Nineteenth-Century Art Worldwide* 6, no. 1 (Spring 2007).

——. "Des débuts de la caricature lithographique à la Restauration." In Jean-Yves Mollier, Martine Reid, and Jean-Claude Yon, eds., *Repenser la Restauration,* 211–22. Paris: Nouveau Monde, 2005.

——. *Henry Monnier: The Comedy of Modern Life.* New York: CUNY Graduate Center Gallery, 2005. Exh. cat.

———. *Husbands, Wives, and Lovers: Marriage and Its Discontents in Nineteenth-Century France.* New Haven: Yale University Press, 2003.
———. *The End of the Salon: Art and the State in the Early Third Republic.* Cambridge: Cambridge University Press, 1993.
———. *Art and Politics of the Second Empire: The Universal Expositions of 1855 and 1867.* New Haven: Yale University Press, 1987.
Mandrou, Robert. *De la culture populaire aux XVIIe et XVIIIe siècles. La bibliothèque bleue de Troyes* [1964]. Rev. ed. Paris: Stock, 1975.
Marchandiau, Jean-Nöel. *L'Illustration, 1843–1944. Vie et mort d'un journal.* Toulouse: Bibliothèque historique Privat, 1987.
Marie, Aristide. *Henry Monnier, 1799–1877.* Paris: H. Floury, 1931.
———. *Alfred et Tony Johannot, peintres, graveurs et vignettistes.* Paris: H. Floury, 1925.
Marle, Anne. "Glucq et l'imagerie Pellerin d'Épinal: de l'image comme outil pédagogique à l'image publicitaire: histoire d'un échec," In Jean-René Gaborit, ed. *Le pouvoir de l'image*, 123–38. Paris: CTHS, 2012.
Martin, Denis. "Emergence d'un nouveau marché. De l'histoire en tableaux aux historiettes." In Denis Martin, ed., *Images d'Épinal*, 137–42. Québec: Musée de Québec, and Paris: RMN, 1995. Exh. cat.
———. "Le petit musée des enfants," in Denis Martin, ed., *Images d'Épinal*, 146–200. Québec: Musée de Québec, and Paris: RMN, 1995. Exh. cat.
Martin, Denis, ed. *Images d'Épinal.* Québec: Musée de Québec, and Paris: RMN, 1995. Exh. cat.
McCullagh, Suzanne Folds, ed. *Dreams and Echoes: Drawings and Sculpture in the David and Celia Hilliard Collection.* New Haven: Yale University Press, and Chicago: Art Institute of Chicago, 2013. Exh. cat.
McIntosh, DeCourcy E. "New York's Favorite Pictures: Intersections in Publishing and Art Dealing in the 1870s." In Ján Bakos, ed., *Artwork through the Market: The Past and the Present*, 139–53. Bratislava: VEDA, 2004.
McPhee, Constance C., and Nadine M. Orenstein. *Infinite Jest: Caricature and Satire from Leonardo to Levine.* New York: Metropolitan Museum of Art, and New Haven: Yale University Press, 2011. Exh. cat.
McWilliam, Neil. *Dreams of Happiness: Social Art and the French Left, 1830–1850.* Princeton: Princeton University Press, 1993.
Meadows, Kenny. *Heads of the People, or Portraits of the English, Drawn by Kenny Meadows.* London: Robert Tyas, 1840.
[Meadows, Kenny]. *Les Anglais peints par eux-mêmes.* Trans. Émile de La Bédollière. 2 vols. Paris: L. Curmer, 1840–41.
Melcher, Edith. *The Life and Times of Henry Monnier, 1799–1877.* Cambridge: Harvard University Press, 1950.
Melot, Michel. "Caricature and the Revolution." In *French Caricature and the French Revolution 1789–1799*, ed. James Cuno, 25–32. Los Angeles: University of California, Grunwald Center for the Graphic Arts, 1988. Exh. cat.
———. *L'Illustration.* Geneva: Skira, 1984.
Menon, Elizabeth K. *The Complete Mayeux: Use and Abuse of a French Icon.* Bern: P. Lang, 1998.
Mercier, Louis-Sebastien. *Tableau de Paris.* 12 vols. Amsterdam, 1783–88.
Mespoulet, Marguerite. *Creators of Wonderland.* New York: Arrow, 1934.
Michiels, Alfred. "Avertissement." *Album du Magasin pittoresque. Cent gravures choisies dans la collection.* Paris: Magasin pittoresque, 1862.
Miel, François. *Essai sur les beaux-arts, et particulièrement sur le salon de 1817, ou Examen critique des principaux ouvrages d'art exposés dans le cours de cette année, avec trente-huit gravures au trait, par E. F. A. M. Miel.* Paris: Pélicier, 1817–18.
Mishory, Alex. "Théodore Géricault's Grande Suite Anglaise: Sources, Iconography, and Significance." Ph.D. diss., City University of New York, 1997.
Mistler, Jean, François Blaudez, and André Jacquemin. *Épinal et l'imagerie populaire.* Paris: Hachette, 1961.
Mollier, Jean-Yves, Martine Reid, and Jean-Claude Yon, eds. *Repenser la Restauration.* Paris: Éditions du nouveau monde, 2005.
Monnier, Henry. *Moeurs administratives.* 18 pl. Lith. Delpech. Paris: Delpech, 1828.
———. *Rencontres parisiennes. Macédoine pittoresque. Croquée d'après nature, au sein des plaisirs, des modes, de l'activité, des occupations, du desoeuvrement, des travers, des vices, des misères, du luxe, des prodigalités des habitans de la capitale dans tous les rangs et dans toutes les classes de la société.* 40 pl. Paris: Gihaut frères, 1824.
Monnier, Henry, and Eugène Lami. *Voyage en Angleterre par Eug. Lami et H. Monnier.* Paris: Firmin Didot and Lami-Denozan; London: Colnaghi and Charles Tilt, 1829.
Morienval, Jean. *Les créateurs de la grande presse en France.* See Thévenin, Henri.
La Mosaïque. Livre de tous les temps et de tous les pays. Paris, 1833–36.

Muchembled, Robert. *Popular Culture and Elite Culture in France, 1400–1750.* Trans. Lydia Cochrane. Baton Rouge: Louisiana State University, 1985.

Musée de l'image, Épinal. "Une saisie d'images chez Pellerin en 1815." http://webmuseo.com/ws/musee-de-l-image/app/collection/expo/15.

——, 2009. *Ni tout à fait la même, ni tout à fait une autre, ou des chefs-d'oeuvre commes modèles.* Exh. cat.

——, 1997. *Série encyclopédique Glucq des leçons de choses illustrées.* Exh. cat.

Musée départemental d'art ancien et contemporain, Épinal. *L'Imagerie populaire française au musée d'Épinal,* ed. Bernard Huin. Épinal: Musée départemental des Vosges, 1989.

Musée des familles. Lectures du soir. Paris, 1833–1900.

Musée Goupil, Bordeaux. *État des lieux.* 2 vols. Bordeaux: Musée Goupil, 1994–2000.

Nadar [Gaspard-Félix Tournachon]. "La vie publique et privée de mossieu Réac." *Revue comique à l'usage des gens sérieux,* Mar. 3, 1849ff.

9e Art: Les cahiers de la bande dessinée 3 (Jan. 1998) [issue on Gustave Doré].

Nevinson, John L. "Origin and Early History of the Fashion Plate." *United States National Museum Bulletin* 250. Washington, D.C.: Smithsonian Press, 1967.

Nochlin, Linda. "Gustave Courbet's Meeting: A Portrait of the Artist as a Wandering Jew." *Art Bulletin* 1949, no. 3 (Sept. 1967): 209–22.

Nodier, Charles. *Histoire du roi de Bohême et de ses sept châteaux.* Paris: Delangle frères, 1830.

Nodier, Charles, Justin Taylor, and Alphonse de Cailleux, eds. *Voyages pittoresques et romantiques dans l'ancienne France.* 23 vols. Paris: Gide fils, A.-F. Lemaître, Thierry frères, 1820–78.

Noon, Patrick J. ed. *Crossing the Channel: British and French Painting in the Age of Romanticism.* Also published as *Constable to Delacroix: British Art and the French Romantics.* London: Tate, 2003. Exh. cat.

"Nouvelles de Paris, du 24 août." *La Quotidienne,* Aug. 25, 1819, 2.

O'Brien, David. *After the Revolution: Antoine-Jean Gros, Painting and Propaganda under Napoleon.* University Park: Pennsylvania State University Press, 2006.

O'Connell, Sheila. *The Popular Print in England, 1550–1850.* London: British Museum Press, 1999.

Paris ou le livre des cent-et-un. 15 vols. Paris: Ladvocat, 1831–34.

Paulson, Ronald. *Hogarth.* 3 vols. New Brunswick: Rutgers University Press, 1991–93.

Pelez, Raimond [R. P.]. "Première impression du salon de 1843." *Le Charivari,* Mar. 19, 1843.

"Pellerin/Humoristic Publishing Company." Special and Area Studies Collection of the University of Florida. http://ufdc.ufl.edu/UF00078633/00001/citation.

Pellissier, Pierre. *Émile de Girardin, prince de la presse.* Paris: Denöel, 1985.

The Penny Magazine. London, 1832–1945.

——. 3. "Murillo." No. 127 (Mar. 29, 1834): 113–15.

——. 2. "Simplifications of Arithmetical Rules." No. 52 (Jan. 26, 1833): 26–27.

——. 1. Preface to vol. 1 (Dec. 18, 1832).

——. 1. "Decimal Fractions." No. 40 (Nov. 17, 1832): 322–24.

——. 1. "How to Endure Poverty." No. 5 (Apr. 28, 1832): 38.

——. 1. "Reading for All." No. 1 (Mar. 31, 1832): 1.

Perrout, René. *Les images d'Épinal.* Nancy: Édition de la Revue Lorraine illustrée, Rev. ed. 1914. [orig. 1912].

Petite galerie dramatique. 2,930 pl. Paris: Martinet, 1796–1880.

Pevsner, Nikolas. *Academies of Art, Past and Present.* Cambridge: Cambridge University Press, 1940; repr. New York: Da Capo, 1973.

Philipon, Charles. *Amourettes.* 36 pl. Paris: Ducarme, Ostervald aîné, 1827–28.

——. *Souvenirs d'amourettes.* 12 pl. Paris: Engelmann, 1828.

——. Prospectus for *Le Charivari.* Paris: Aubert, 1832.

Pigal, Edmé-Jean. *Vie d'un gamin en douze chapitres.* 12 pl. Paris: Gihaut frères, 1825

R. P. See Pelez, Raimond.

Ray, Gordon N. *The Illustrator and the Book in England from 1790 to 1914.* New York: Pierpont Morgan Library and Oxford University Press, 1976; repr. Mineola, N.Y.: Dover, 1991.

——. *The Art of the French Illustrated Book, 1700 to 1914.* 2 vols. New York: Pierpont Morgan Library; Ithaca: Cornell University Press, 1982; repr. Mineola, N.Y.: Dover, 1986.

Reclus, Maurice. *Émile de Girardin. Le créateur de la presse moderne.* Paris: Hachette, 1934.

Renaudot, J.-L. *Les Bérangériennes. Chansons de J.-L. Renaudot.* Avallon: Renaudot, 1860.

Renié, Pierre-Lin. *Une image sur un mur. Images et décoration intérieure au 19e siècle.* Bordeaux: Musée Goupil, 2005. Exh. cat. Partially reprinted as "The Image on the Wall: Prints as Decoration in Nineteenth-Century Interiors." *Nineteenth-Century Art Worldwide*, Autumn 2006. http://www.19thc-artworldwide.org.

Renonciat, Annie. *La vie et l'oeuvre de J. J. Grandville.* Courbevoie: ACR and Paris: Vilo, 1985.

——. *La vie et l'oeuvre de Gustave Doré.* Paris: ACR, 1983.

La Révolution française et l'Europe, 1789–1799. 3 vols. Paris: Galeries nationales du Grand Palais/RMN, 1989. Exh. cat.

Rey, Alain, ed. *Le Robert. Dictionnaire historique de la langue française.* 2 vols. Paris: France loisirs, 1994.

Ribeyre, Félix. *Cham, sa vie et son oeuvre.* Paris: E. Plon, Nourrit, 1884.

Rivière. Georges-Henri. Preface to *French Popular Imagery.* London: Hayward Gallery, 1974. Also published as *Cinq siècles d'imagerie française.* Paris: Musées nationaux, 1972. Exh. cat.

Robert, Vincent. "Paysages politiques, cohérences médiatiques." In Dominique Kalifa, Philippe Régnier, Marie-Ève Thérenty, and Alain Vaillant, *La civilisation du journal. Histoire culturelle et littéraire de la presse française au XIXe siècle,* 213–48. Paris: Nouveau Monde, 2011.

Roman, Cynthia Ellen, ed. *Hogarth's Legacy.* Farmington, Conn.: Lewis Walpole Library in association with Yale University Press, 2016.

Rosen, Charles, and Henri Zerner. *Romanticism and Realism: The Mythology of Nineteenth-Century Art.* New York: Viking, 1984.

Roskill, Mark. "Early Impressionism and the Fashion Print." *Burlington Magazine* 112 (June 1970): 390–93, 395.

Rowlandson, Thomas, and William Combe. *Doctor Syntax's Three Tours: In Search of the Picturesque, Consolation and a Wife. The Original Edition, Complete and Unabridged, with the Life and Adventures of the Author, Now First Written, by John Camden Hotten.* London: J. C. Hotten [1868]).

——. *The Third Tour of Doctor Syntax, in Search of a Wife.* London: R. Ackermann, 1821.

——. *The Second Tour of Doctor Syntax, in Search of Consolation.* London: R. Ackermann, 1820.

——. *The Tour of Doctor Syntax, in Search of the Picturesque.* London: R. Ackermann, 1812.

[Rowlandson, Thomas, and William Combe]. *Le Don Quichotte romantique, ou Voyage du docteur Syntaxe à la recherche du pittoresque et du romantique: Poëme en XX chants traduit librement de l'anglais par M.* [Auguste] *Gandais et orné de 26 gravures.* Paris: l'auteur, Pélicier, 1821.

S. C. "Beaux-Arts. La chapelle Sixtine. Lithographie, par M. Sudre, d'après M. Ingres." *L'Artiste,* 1st ser., vol. 7 (1834): 1.

Samuels, Maurice. *The Spectacular Past: Popular History and the Novel in Nineteenth-Century France.* Ithaca: Cornell University Press, 2004.

Schapiro, Meyer. "Courbet and Popular Imagery." In Meyer Schapiro, *Modern Art: 19th and 20th Centuries,* 47–85. New York: G. Braziller, 1978.

Schwartz, Vanessa. *Spectacular Realities: Early Mass Culture in Fin-de-Siècle Paris.* Berkeley: University of California Press, 1998.

Schwartz, Vanessa R., and Jeannene M. Przyblyski. "Visual Culture's History: Twenty-First-Century Interdisciplinarity and Its Nineteenth-Century Objects." In Schwartz and Przyblyski, eds., *The Nineteenth-Century Visual Culture Reader,* 3–14. New York: Routledge, 2004.

Schwartz, Vanessa R., and Jeannene M. Przyblyski, eds. *The Nineteenth-Century Visual Culture Reader.* New York: Routledge, 2004.

Sello, Gottfried. *Grandville, Das Gesamte Werk.* 2 vols. Munich: Rogner und Bernhard, 1969.

Senefelder, Alöys. *Nouvelle invention lithographique. Porte-feuille lithographique, recueil de sujets de divers genres, dessinés et imprimés sur planches lithographiques. Nouvellement inventées pour la multiplication de tous dessins, écritures, cartes et plans topographiques, oeuvres de musique, etc. etc. etc.* Paris: A. Senefelder, 1823.

——. *L'art de la lithographie, ou instruction pratique contenant la description claire et succincte des différens procédés à suivre pour dessiner, graver et imprimer sur pierre; précédée d'une histoire de la lithographie et de ses divers progrès.* Paris: Treuttel et Würtz, 1819.

——. *L'art de la lithographie. Collection de plusieurs essais en dessins et gravures pour server de supplément à l'instruction pratique de la lithographie par Alloys Senefelder.* Paris: A. Senefelder, 1819.

Sewell, William H., Jr. "The Empire of Fashion and the Rise of Capitalism in 18th-Century France." *Past & Present.* No. 206 (Feb. 2010): 81–120.

Sieburth, Richard. "Same Difference: The French Physiologies, 1840–1842." In Norman F. Cantor, ed., *Notebooks in Cultural Analysis: An Annual Review,* 163–200. Durham: Duke University Press, 1984.

La Silhouette. Paris, 1829–31. Subtitle varied: Nos. 1–2: *Album lithographique, beaux-arts, dessins, moeurs, theâtres, caricatures;* no. 3: *Dessins, caricatures, theâtres, moeurs, variétés, etc.*"; no. 4 onward: *Journal des caricatures, beaux-arts, dessins, moeurs, théâtres, etc.*

——. "Procès de la Silhouette." Vol. 3, no. 2 (July 1, 1830): 9–10.

Silver, Kenneth E. *Esprit de Corps: The Art of the Parisian Avant-Garde and the First World War, 1914–1925.* Princeton: Princeton University Press, 1989.

Simon, Robert. "Géricault und die Faits Divers." In Stefan Germer and Michael Zimmermann, eds., *Bilder der Macht, Macht der Bilder: Zeitgeschichte in Darstellungen des 19. Jahrhunderts,* 192–207. Munich: Klinkhardt and Biermann, 1997.

Smolderen, Thierry. *The Origin of Comics from William Hogarth to Winsor McCay.* Trans. Bart Beaty and Nick Nguyen. Jackson: University Press of Mississippi, 2014.

Solo. *Plus de 5000 dessinateurs de presse et 600 supports en France de Daumier à l'an 2000.* Vichy: Aedis, 2004.

Specimens of Polyautography. 2 vols., 12 pl. London: P. André and J. Heath, 1803.

Spiegelman, Art. "Drawing Blood: Outrageous Cartoons and Cartoons of Outrage." *Harper's Magazine,* June 2006, 43–52.

Spitzer, Alan B. *The French Generation of 1830.* Princeton: Princeton University Press, 1987.

Staub, Eric. Preface to Imagerie Pellerin, *Histoire de l'imagerie... une entreprise bicentenaire.* Épinal: Imagerie Pellerin, 1991.

Stein, Perrin. *Artists and Amateurs: Etching in 18th-Century France.* New York: Metropolitan Museum of Art, and New Haven: Yale University Press, 2013. Exh. cat.

Stendhal [Marie-Henri Beyle]. *Le rouge et le noir. Chronique du XIXe siècle* [1830]. Ed. Béatrice Didier. Paris: Gallimard, 1972.

Sterne, Laurence. *The Life and Opinions of Tristram Shandy, Gentleman* [1759–67]. Vols. 1 and 2 of Melvyn New and Joan New, eds., *The Florida Edition of the Works of Laurence Sterne.* 8 vols. Gainesville: University Press of Florida, 1978–2008.

Suite d'estampes pour servir à l'histoire des moeurs et du costume des français dans le dix-huitième siècle. 3 suites. Paris: Prault, 1775–83.

Le suprême bon ton. 9 pl. Paris: Martinet, 1814–19.

Szrajber, Tanya. "Henry Monnier's Letters from London in 1825." *Print Quarterly* 31, no. 3 (Sept. 2014): 299–308.

Taws, Richard. *The Politics of the Provisional: Art and Ephemera in Revolutionary France.* University Park: Pennsylvania State University Press, 2013.

Thackeray, William Makepeace. *Vanity Fair. A Novel without a Hero.* London: Bradbury and Evans, 1848.

Thévenin, Henri [writing as Jean Morienval]. *Les créateurs de la grande presse en France: Émile de Girardin, H. de Villemessant, Moïse Millhaud.* Paris: Spes, 1934.

Thiers, Adolphe. "De la lithographie et de ses progrès." *La Pandore,* no. 259 (Mar. 30, 1824); no. 263 (Apr. 3, 1824). Repr. in Johnson, *French Lithography,* 46–48.

Töpffer, Rodolphe. *Rodolphe Töpffer: The Complete Comic Strips.* Ed. and trans. David Kunzle. Jackson: University Press of Mississippi, 2007.

——. *Rodolphe Töpffer.* Paris: Pierre Horay, 1975, 1996. [Facsimiles of all the published albums.]

——. *Histoire de M. Cryptogame.* Paris: J. J. Dubochet, 1846.

——. *Histoire d'Albert par Simon de Nantua.* Geneva, 1845.

——. "Histoire de M. Cryptogame, par l'auteur de M. Vieux-Bois, de M. Jabot, de M. Crépin, du Docteur Festus, etc." *L'Illustration. Journal Universel,* Jan. 25–Apr. 19, 1845 [redrawn by Cham].

——. *Essai de physiognonomie.* Geneva: Schmidt, 1845.

——. *Le docteur Festus* [1829]. Paris: Abraham Cherbuliez, 1840.

——. *Monsieur Pencil.* Paris: Abraham Cherbuliez, 1840.

——. [R. T.] "Notice sur l'histoire de Mr Jabot." Repr. in Groensteen and Peeters, *Töpffer: L'invention de la bande dessinée,* 161–63 [orig. *Bibliothèque universelle de Genève,* no. 18 (June 1837): 334–37].

——. *Les amours de Mr Vieux Bois.* Geneva: Frutiger, 1837.

——. *Mr Crépin.* Geneva: Frutiger, 1837.

——. "Réflexions à propos d'un programme" (1836). Repr. in Groensteen and Peeters, *Töpffer. L'invention de la bande dessinée,* 144–60. Paris: Hermann, 1994.

——. *Histoire de Mr Jabot.* Geneva: J. Freydig, 1833.

Trapp, Frank Anderson. *The Attainment of Delacroix.* Baltimore: Johns Hopkins University Press, 1970.

Trésor de la langue française: Dictionnaire de la langue du XIXe et du XXe (1789–1960), ed. Paul Imbs. 16 vols. Paris: CNRS, 1977.

Tristan, Frédérick. *Le monde à l'envers.* Paris: Atelier Hachette/Massin, 1980.

Twyman, Michael. *A History of Chromolithography: Printed Colour for All.* London: British Library/Oak Knoll, 2013.

——. *Breaking the Mould: The First Hundred Years of Lithography.* London: British Library, 2001.

——. *Lithography, 1800–1850.* London: Oxford University Press, 1970.

Valmy-Baysse, Jean. *André Gill, l'impertinent*. Paris: Felin, 1991.

Verhoogt, Robert. *Art in Reproduction: Nineteenth-Century Prints after Lawrence Alma-Tadema, Jozef Israëls, and Ary Scheffer.* Trans. Michelle Hendriks. Amsterdam: Amsterdam University Press, 2007.

Vernet, Carle. *Album lithographique par Cle Vernet.* Paris: Depech, 1821.

——. *Une collection de chevaux de tous les pays, montés de leur cavalier, avec leurs costumes et leurs armes.* 12 installments, 96 pl. Lith. Lasteyrie. Paris: Gastel, 1817–18.

Vernet, Horace. *Croquis lithographiques par H. Vernet.* 50 pl. Paris: Delpech, 1818.

Vizetelly, Henry. *Glances Back through Seventy Years: Autobiographical and Other Reminiscences.* 2 vols. London: Kegan Paul, Trench, Trübner, 1893.

Von Lintel, Amy M. "Wood Engravings, 'The Marvellous Spread of Illustrated Publications,' and the History of Art." *Modernism/modernity* 19, no. 3 (Sept. 2012): 515–42.

Vottero, Michaël. *La peinture de genre en France, après 1850.* Rennes: Presses universitaires de Rennes, 2012.

Watelet, Jean. *La presse illustrée en France, 1814–1914.* 2 vols. Villeneuve d'Ascq: Presses universitaires du Septentrion, 2002.

Wattier, Émile. *Un an de la vie d'une jeune fille. Roman historique en XVII chapîtres, écrits par son confident et lithographiés par M. Wattier*. Paris: G. Engelmann and Gihaut, 1824.

Weber, Eugen. *Peasants into Frenchmen: The Modernization of Rural France, 1870–1914.* Stanford: Stanford University Press, 1976.

Wechsler, Judith. *A Human Comedy: Physiognomy and Caricature in 19th-Century Paris.* Chicago: University of Chicago Press, 1982.

Wolff, Albert. "Édouard Manet." *Le Figaro,* May 1, 1883, 1.

Yang, Yin-Hsuan. *Les premiers salons caricaturaux au XIXe siècle*. Paris: Presses universitaires de Paris Ouest, 2011.

Zola, Émile. *Édouard Manet. Étude biographique et critique.* Paris: Dentu, 1867. Repr. in Émile. Zola, *Mes haines. Causeries littéraires et artistiques*, 287–372. Rev. ed. Paris: G. Charpentier, 1879.

ILLUSTRATION CREDITS

The photographers and the sources of visual material other than the owners indicated in the captions are as follows. Every effort has been made to supply complete and correct credits; if there are errors or omissions, please contact Yale University Press so that corrections can be made in any subsequent edition.

Photo courtesy of the author: Figs. 3, 5, 9, 12, 22, 25, 30, 31, 34, 35, 42, 44, 46, 51, 52, 58, 60, 61, 62, 63, 64, 74, 75, 77, 79, 81, 82, 83, 85, 86, 88, 89, 90, 91, 92, 93, 99, 100, 103, 104, 105, 106, 107, 113, 114, 115, 119, 121, 122, 123, 124, 125, 126, 127, 128, 129, 130, 131, 133, 136, 148, 151, 152, 153, 154, 155, 156, 160, 164, 166, 168, 169, 171, 179, 184, 187, 190, 191, 192, 195, 196.

Photo: Imaging Department © President and Fellows of Harvard College: Figs. 6, 26, 55.

Photo: Wikimedia: Figs. 24, 181, 182, 199.

© Musée de l'Image/cliché H. Rouyer: Figs. 170, 180.
© Musée de l'Image/cliché E. Erfani: Fig. 186.

INDEX

Page numbers in **bold** refer to illustrations.